Quantum Dots, Nanoparticles and Nanowires

MATERIALS RESEARCH SOCIETY
SYMPOSIUM PROCEEDINGS VOLUME 789

Quantum Dots, Nanoparticles and Nanowires

Symposium held December 1–5, 2003, Boston, Massachusetts, U.S.A.

EDITORS:

P. Guyot-Sionnest
University of Chicago
Chicago, Illinois, U.S.A.

H. Mattoussi
Naval Research Laboratory
Washington, D.C., U.S.A.

U. Woggon
Universität Dortmund
Dortmund, Germany

Z.-L. Wang
Georgia Institute of Technology
Atlanta, Georgia, U.S.A.

Materials Research Society
Warrendale, Pennsylvania

This work was supported in part by the Office of Naval Research under Grant Number N00014-04-1-0067. The United States Government has a royalty-free license throughout the world in all copyrightable material contained herein.

Single article reprints from this publication are available through University Microfilms Inc., 300 North Zeeb Road, Ann Arbor, Michigan 48106

CODEN: MRSPDH

Published by:

Materials Research Society
506 Keystone Drive
Warrendale, PA 15086
Telephone (724) 779-3003
Fax (724) 779-8313
Web site: http://www.mrs.org/

Manufactured in the United States of America

CONTENTS

SYNTHESIS, SHAPE CONTROL AND CHARACTERIZATION

QD-BIOCONJUGATES: DESIGN AND USE IN BIOTECHNOLOGICAL APPLICATIONS

POSTER SESSION II: QDs AND NANOPARTICLES FOR BIOLOGICAL APPLICATIONS

QUANTUM DOTS AND WIRES: STRUCTURE, SPECTROSCOPY, AND TRANSPORT

*Invited Paper

SELF-ASSEMBLED QDs, NANOPARTICLES AND NANOWIRES

METALLIC, MAGNETIC AND SEMICONDUCTOR NANOPARTICLES: GROWTH AND CHARACTERIZATION

*Invited Paper

POSTER SESSION III

ELECTRICAL, ELECTRONIC PROPERTIES AND DEVICES I

ELECTRICAL, ELECTRONIC PROPERTIES AND DEVICES II

POSTER SESSION IV

CARBON NANOTUBES (CNT) AND RELATED PROPERTIES

POSTER SESSION V: CARBON NANOTUBES

PREFACE

Nanostructures of semiconductors and metals show novel optical and transport properties. Semiconductor quantum dots, for example, show striking size-dependent optical and electronic properties when their dimensions become comparable to or smaller than the Bohr exciton radius, due to quantum confinement of the charge carriers. These nanostructures offer the perspective of designing materials properties with unprecedented flexibility and control. This has motivated much research in the synthesis and characterization of new materials, along with deriving novel properties and devices. There has also been a growing interest in the conjugation of colloidal QDs and metallic nanoparticles with biomolecules, and their use in a variety of bio-oriented applications.

Symposium N, "Quantum Dots, Nanoparticles and Nanowires," held December 1–5 at the 2003 MRS Fall Meeting in Boston, Massachusetts, was an excellent opportunity for scientists with various expertise in the growth, characterization and applications of inorganic nanostructures, such as quantum dots, nanowires and nanorods, to gather and share the latest developments in the field. The weeklong symposium was extremely well attended with over 110 oral presentations and 140 poster presentations. With over 300 submissions, Symposium N was the largest of the overall Meeting.

Reports published in this proceedings were peer-reviewed. They include reports on techniques to prepare and characterize novel materials, investigations of novel optical and electronic properties, and novel applications, such as those that are biologically inspired. Topics covered in this proceedings include: Synthesis and characterization of semiconductor quantum dots, nanoparticles and nanowires using wet chemistry and molecular beam approaches; Synthesis, characterization and novel properties of metallic nanostructures; Optical properties of neutral and charged excitons and exciton complexes in self-assembled QDs; Nanoscale devices and sensors based on nanostructures and their properties; Design and characterization of QD-bioconjugates and their use in assay developments. Reports are organized by a combination of subjects and sessions.

The organizers thank all contributors for their effort and hard work to make this symposium an exciting experience and the published proceedings a valuable reference. The organizers also acknowledge the financial support from Evident Technology and the Office of Naval Research (ONR).

P. Guyot-Sionnest
H. Mattoussi
U. Woggon
Z.-L. Wang

March 2004

MATERIALS RESEARCH SOCIETY SYMPOSIUM PROCEEDINGS

Volume 762— Amorphous and Nanocrystalline Silicon-Based Films—2003, J.R. Abelson, G. Ganguly, H. Matsumura, J. Robertson, E. Schiff, 2003, ISBN: 1-55899-699-0
Volume 763— Compound Semiconductor Photovoltaics, R. Noufi, D. Cahen, W. Shafarman, L. Stolt, 2003, ISBN: 1-55899-700-8
Volume 764— New Applications for Wide-Bandgap Semiconductors, S.J. Pearton, J. Han, A.G. Baca, J-I. Chyi, W.H. Chang, 2003, ISBN: 1-55899-701-6
Volume 765— CMOS Front-End Materials and Process Technology, T-J. King, B. Yu, R.J.P. Lander, S. Saito, 2003, ISBN: 1-55899-702-4
Volume 766— Materials, Technology and Reliability for Advanced Interconnects and Low-k Dielectrics—2003, A. McKerrow, J. Leu, O. Kraft, T. Kikkawa, 2003, ISBN: 1-55899-703-2
Volume 767— Chemical-Mechanical Planarization, M. Oliver, D. Boning, D. Stein, K. Devriendt, 2003, ISBN: 1-55899-704-0
Volume 768— Integration of Heterogeneous Thin-Film Materials and Devices, H.A. Atwater, M. Levy, M.I. Current, T. Sands, 2003, ISBN: 1-55899-705-9
Volume 769— Flexible Electronics—Materials and Device Technology, B.R. Chalamala, B.E. Gnade, N. Fruehauf, J. Jang, 2003, ISBN: 1-55899-706-7
Volume 770— Optoelectronics of Group-IV-Based Materials, T. Gregorkiewicz, R.G. Elliman, P.M. Fauchet, J.A. Hutchby, 2003, ISBN: 1-55899-707-5
Volume 771— Organic and Polymeric Materials and Devices, P.W.M. Blom, N.C. Greenham, C.D. Dimitrakopoulos, C.D. Frisbie, 2003, ISBN: 1-55899-708-3
Volume 772— Nanotube-Based Devices, P. Bernier, S. Roth, D. Carroll, G-T. Kim, 2003, ISBN: 1-55899-709-1
Volume 773— Biomicroelectromechanical Systems (BioMEMS), C. Ozkan, J. Santini, H. Gao, G. Bao, 2003, ISBN: 1-55899-710-5
Volume 774— Materials Inspired by Biology, J.L. Thomas, L. Gower, K.L. Kiick, 2003, ISBN: 1-55899-711-3
Volume 775— Self-Assembled Nanostructured Materials, C.J. Brinker, Y. Lu, M. Antonietti, C. Bai, 2003, ISBN: 1-55899-712-1
Volume 776— Unconventional Approaches to Nanostructures with Applications in Electronics, Photonics, Information Storage and Sensing, O.D. Velev, T.J. Bunning, Y. Xia, P. Yang, 2003, ISBN: 1-55899-713-X
Volume 777— Nanostructuring Materials with Energetic Beams, S. Roorda, H. Bernas, A. Meldrum, 2003, ISBN: 1-55899-714-8
Volume 778— Mechanical Properties Derived from Nanostructuring Materials, H. Kung, D.F. Bahr, N.R. Moody, K.J. Wahl, 2003, ISBN: 1-55899-715-6
Volume 779— Multiscale Phenomena in Materials—Experiments and Modeling Related to Mechanical Behavior, K.J. Hemker, D.H. Lassila, L.E. Levine, H.M. Zbib, 2003, ISBN: 1-55899-716-4
Volume 780— Advanced Optical Processing of Materials, I.W. Boyd, M. Dinescu, A.V. Rode, D.B. Chrisey, 2003, ISBN: 1-55899-717-2
Volume 781E—Mechanisms in Electrochemical Deposition and Corrosion, J.C. Barbour, R.M. Penner, P.C. Searson, 2003, ISBN: 1-55899-718-0
Volume 782— Micro- and Nanosystems, D. LaVan, M. McNie, A. Ayon, M. Madou, S. Prasad, 2004, ISBN: 1-55899-720-2
Volume 783— Materials, Integration and Packaging Issues for High-Frequency Devices, P. Muralt, Y.S. Cho, J-P. Maria, M. Klee, C. Hoffmann, C.A. Randall, 2004, ISBN: 1-55899-721-0
Volume 784— Ferroelectric Thin Films XII, S. Hoffmann-Eifert, H. Funakubo, A.I. Kingon, I.P. Koutsaroff, V. Joshi, 2004, ISBN: 1-55899-722-9

MATERIALS RESEARCH SOCIETY SYMPOSIUM PROCEEDINGS

Volume 785— Materials and Devices for Smart Systems, Y. Furuya, E. Quandt, Q. Zhang, K. Inoue, M. Shahinpoor, 2004, ISBN: 1-55899-723-7

Volume 786— Fundamentals of Novel Oxide/Semiconductor Interfaces, C.R. Abernathy, E. Gusev, D.G. Schlom, S. Stemmer, 2004, ISBN: 1-55899-724-5

Volume 787— Molecularly Imprinted Materials—2003, P. Kofinas, M.J. Roberts, B. Sellergren, 2004, ISBN: 1-55899-725-3

Volume 788— Continuous Nanophase and Nanostructured Materials, S. Komarneni, J.C. Parker, J. Watkins, 2004, ISBN: 1-55899-726-1

Volume 789— Quantum Dots, Nanoparticles and Nanowires, P. Guyot-Sionnest, N.J. Halas, H. Mattoussi, Z.L. Wang, U. Woggon, 2004, ISBN: 1-55899-727-X

Volume 790— Dynamics in Small Confining Systems—2003, J.T. Fourkas, P. Levitz, M. Urbakh, K.J. Wahl, 2004, ISBN: 1-55899-728-8

Volume 791— Mechanical Properties of Nanostructured Materials and Nanocomposites, R. Krishnamoorti, E. Lavernia, I. Ovid'ko, C.S. Pande, G. Skandan, 2004, ISBN: 1-55899-729-6

Volume 792— Radiation Effects and Ion-Beam Processing of Materials, L. Wang, R. Fromknecht, L.L. Snead, D.F. Downey, H. Takahashi, 2004, ISBN: 1-55899-730-X

Volume 793— Thermoelectric Materials 2003—Research and Applications, G.S. Nolas, J. Yang, T.P. Hogan, D.C. Johnson, 2004, ISBN: 1-55899-731-8

Volume 794— Self-Organized Processes in Semiconductor Heteroepitaxy, R.S. Goldman, R. Noetzel, A.G. Norman, G.B. Stringfellow, 2004, ISBN: 1-55899-732-6

Volume 795— Thin Films—Stresses and Mechanical Properties X, S.G. Corcoran, Y-C. Joo, N.R. Moody, Z. Suo, 2004, ISBN: 1-55899-733-4

Volume 796— Critical Interfacial Issues in Thin-Film Optoelectronic and Energy Conversion Devices, D.S. Ginley, S.A. Carter, M. Grätzel, R.W. Birkmire, 2004, ISBN: 1-55899-734-2

Volume 797— Engineered Porosity for Microphotonics and Plasmonics, R. Wehrspohn, F. Garcial-Vidal, M. Notomi, A. Scherer, 2004, ISBN: 1-55899-735-0

Volume 798— GaN and Related Alloys—2003, H.M. Ng, M. Wraback, K. Hiramatsu, N. Grandjean, 2004, ISBN: 1-55899-736-9

Volume 799— Progress in Compound Semiconductor Materials III—Electronic and Optoelectronic Applications, D. Friedman, M.O. Manasreh, I. Buyanova, F.D. Auret, A. Munkholm, 2004, ISBN: 1-55899-737-7

Volume 800— Synthesis, Characterization and Properties of Energetic/Reactive Nanomaterials, R.W. Armstrong, N.N. Thadhani, W.H. Wilson, J.J. Gilman, Z. Munir, R.L. Simpson, 2004, ISBN: 1-55899-738-5

Volume 801— Hydrogen Storage Materials, M. Nazri, G-A. Nazri, R.C. Young, C. Ping, 2004, ISBN: 1-55899-739-3

Volume 802— Actinides—Basic Science, Applications and Technology, L. Soderholm, J. Joyce, M.F. Nicol, D. Shuh, J.G. Tobin, 2004, ISBN: 1-55899-740-7

Volume 803— Advanced Data Storage Materials and Characterization Techniques, J. Ahner, L. Hesselink, J. Levy, 2004, ISBN: 1-55899-741-5

Volume 804— Combinatorial and Artificial Intelligence Methods in Materials Science II, R.A. Potyrailo, A. Karim, Q. Wang, T. Chikyow, 2004, ISBN: 1-55899-742-3

Volume 805— Quasicrystals 2003—Preparation, Properties and Applications, E. Belin-Ferré, M. Feuerbacher, Y. Ishii, D. Sordelet, 2004, ISBN: 1-55899-743-1

Volume 806— Amorphous and Nanocrystalline Metals, R. Busch, T. Hufnagel, J. Eckert, A. Inoue, W. Johnson, A.R. Yavari, 2004, ISBN: 1-55899-744-X

Volume 807— Scientific Basis for Nuclear Waste Management XXVII, V.M. Oversby, L.O. Werme, 2004, ISBN: 1-55899-752-0

Prior Materials Research Society Symposium Proceedings available by contacting Materials Research Society

Optical and Electronic Properties of Nanoparticles I

Mat. Res. Soc. Symp. Proc. Vol. 789 © 2004 Materials Research Society N1.8

Structural and optical properties of CdSe, CdTe and CdSeTe nanoparticles dispersed in SiO_2 films

P. Babu Dayal *[1], N. V. Rama Rao[1], B. R. Mehta [1], S. M. Shivaprasad [2] and P. D. Paulson [3]
[1] Thin Film Laboratory, Department of Physics, Indian Institute of Technology Delhi, New Delhi-16, India.
[2] Surface Physics Group, National Physical Laboratory, New Delhi-42, India.
[3] Institute of Energy Conversion, University of Delaware, Newark, USA.

Abstract

$CdSe_xTe_{1-x}$ nanoparticles (with different stoichiometry ratio x) dispersed in silicon dioxide films have been grown by magnetron sputtering technique followed by thermal annealing. Effect of thermal annealing conditions on the structural, compositional, optical and electronic properties of nanoparticles has been studied using GAXRD, XPS, TEM, and spectroscopic ellipsometry techniques. A structural transformation in the nanoparticle core mediated purely by surface layer effects in the case of CdTe and a spontaneous self-organization of nanoparticles into nanorods in the case of CdSe via fractal growth has been observed. Preliminary observations from the ellipsometry measurements carried out on some of these nanoparticle films shows a blue shift of absorption edge.

Introduction:

Electronic and optical properties of semiconductor nanoparticles can be controlled by varying size, shape and surface layers [1-4]. In recent years, chacogenide semiconductor nanoparticles have attracted considerable attention due to remarkable electronic response and ultra-fast optical gain [5-9]. In the case of semiconductor nanoparticles dispersed in glass matrix, the transparent glass matrix provides an effective way of maintaining the individual nanoparticle characteristics and preserving the surface structure [10-13]. This is advantageous in comparison to chemically capped or nanoparticles dispersed in polymeric materials [14]. Although size-dependent properties of semiconductor nanoparticles of I-VII, II-VI, III-V and IV groups in glass matrix have been widely reported [15-16], the synthesis parameters used for controlling size also effect the composition of binary or ternary phases. This is especially important as different elements (Cd, Se and Te) have widely varying vapor pressure and diffusion co-efficient in the glass matrix. Thus it is difficult to control nanoparticle size, without varying the composition or surface structure. To some extent, this problem has been circumvented by using elemental Cd, Se and Te along with SiO_2 targets for growing binary and ternary $CdSe_xTe_{1-x}$ semiconductor nanoparticles in glass films by keeping the area of elemental targets in proportion to the sputtering yield. This method of making semiconductor nanoparticles using elemental targets seems to be superior in comparison to conventional melting and quenching techniques [17]. The central objective of this study is to give an overview of the results of the structural, compositional, optical, and electronic characterization of $CdSe_xTe_{1-x}$ nanoparticles dispersed in SiO_2films prepared by the above method.

Experimental:

Magnetron sputtering technique has been used to grow semiconductor nanoparticles in SiO_2 films. High purity elemental Cd, Se and Te pieces fixed onto a 2-inch diameter SiO_2 slab has been used as the sputtering target. Substrate rotation at 8 rpm has been employed to achieve deposition uniformity. Argon partial pressure of 3.5×10^{-2} Torr at a flow rate of 10 sccm is maintained during sputtering. A power of 120 watt from a 13.56 MHz RF generator has been used for sputtering. In the following discussion, samples having $CdSe_xTe_{1-x}$ nanoparticles dispersed in SiO_2 films will be referred to as $CdSe_xTe_{1-x}$:SiO_2 samples. Post deposition heat treatment has been done in vacuum at 5×10^{-6} Torr, in air and in N_2 ambients at atmospheric pressure at temperatures ranging from 200 to 600 ^{0}C. The structural characterization of CdSexTe1$_{-x}$:SiO2 films has been performed using a x-ray diffractometer (Rigaku RU-200 B, Japan) with a glancing angle attachment (GAXRD). The composition of various phases has been studied with a Perkin-Elmer-1257 x-ray photoelectron spectrometer (XPS) using Mg K_α radiation having energy at 1253.6 eV. Optical absorption (OA) studies have been done using Perkin-Elmer Lembda 900 spectrophotometer. Spectroscopic Ellipsometry (SE) measurements were carried out using variable angle spectroscopic ellipsometer, VASE® (J.A.Woollam Co., Inc) at 57.5, 62.5 and 67.5^0 incident angles in the 0.725-4.6 eV photon energy range. Electron microscopy studies were carried out using JEOL JEM 200 CX TEM operated at 200 KeV (with a resolution of 0.35 nm in point TEM mode) by depositing approximately a 100 Å thin films directly onto the amorphous carbon coated grids (200 mesh) followed by thermal annealing.

Results and discussion:

i) CdTe nanoparticles:

As shown in Fig. 1 (i), GAXRD spectrum of T0 (as-deposited) sample has amorphous nature and is possibly an admixture of Cd, Te and SiO_2 phases. The spectrum of the sample T6 (vacuum annealed at 600^{0}C for 2 hours) shows predominant peaks corresponding to the cubic (fcc) phase of CdTe. In the case of sample T7 (annealed in air ambient at 600^{0}C for 2 hours) peaks corresponding to the hexagonal CdTe and monoclinic cadmium tellurium oxide ($CdTeO_3$) phases are observed [18]. XPS spectra of T0, T6 and T7 samples taken after sputter cleaning for about 10-20 minutes are shown in Fig. 1 (ii). The splitting of Te $3d_{5/2}$ peak inT6 and T7 samples [curves (b) and (c) respectively] indicates two types of bonding configurations for Te. In case of T6 and T7 samples, the peak 1 at 573.5 eV corresponds to the presence of CdTe (which is probably shifted from the bulk value 572.7 eV) and peak 2 at 576.8 eV is due to $CdTeO_3$, matches well with the standard values [19]. In the Te 3d $_{5/2}$ spectra, the higher intensity of peak 2 and peak 1 is in T7 sample indicates the formation of a thicker $CdTeO_3$ capping layer in comparison to amorphous interfacial layer in T6 sample. In O1s core levels, spectra of T0, T6 and T7 samples [curves (d), (e) and (f) respectively] the position of peak 3 (531 eV) and peak 4 (533 eV) indicates that oxygen is present in the form of $CdTeO_3$ and SiO_2 phases respectively [39]. XPS analysis shows that the CdTe:SiO_2 concentration ratio in samples T6 is 12:88. Estimated nanoparticle density in these samples is approximately $8 \times 10^{11}/cm^3$. Electron micrograph of sample T6 given in Fig. 3 (i) clearly shows the nanoparticle character of CdTe in

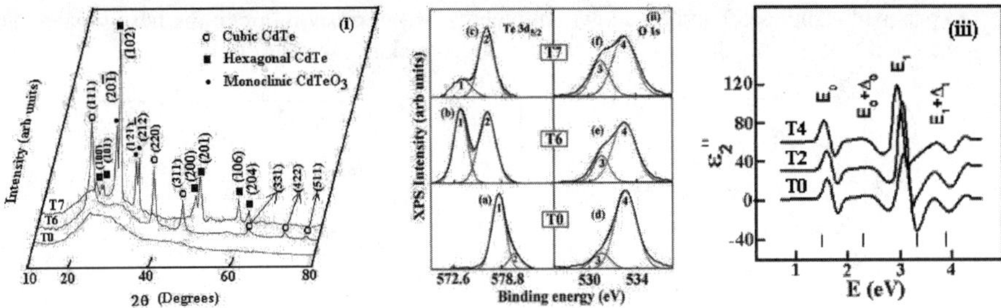

Fig. 1: (i) X-ray diffractogram of T0, T6 and T7 $CdTe:SiO_2$ samples, (ii) XPS core level spectra of Te and O_2 of T0, T6 and T7 $CdTe:SiO_2$ nanoparticle films and (iii) ε_2'' spectra of T0, T2 and T4 samples showing E_0, $E_0+\Delta_0$, E_1 and $E_1+\Delta_1$ transitions. Vertical lines are the corresponding bulk values. (For clarity, spectra of samples T2 and T4 have been shifted by 30 and 60 units along the positive y-axis, respectively)

$CdTe:SiO_2$ samples. GAXRD and XPS studies shows that, diffusion of Cd and Te in SiO_2 matrix results in coalescence leading to the growth of CdTe nanoparticles having cubic structure with an amorphous $CdTeO_3$ interfacial layer. On annealing the same T0 sample in air at 600^{0}C, CdTe nanoparticles having hexagonal structure covered by a crystalline $CdTeO_3$ capping layer are observed. The presence of large concentration of stacking faults in hexagonal CdTe nanoparticles leads to a cubic to hexagonal structural transformation due purely to surface effects. The kinetics of formation of amorphous and crystalline surface layers along with the existence of stacking faults has been well demonstrated elsewhere [20].

The effect of nanoparticle size on optical properties of $CdTe:SiO_2$ films has been studied using VASE on T0, T2 and T4 samples (T2 and T4 being vacuum annealed samples at 200 and 400^{0}C respectively). The method of extracting the optical functions of individual CdTe nanoparticles in SiO_2 films has been described elsewhere [21]. The dielectric response [second derivative of dielectric constant ε_2 spectra (ε_2'')] of the CdTe nanoparticles given in Fig. 1 (iii) show four critical points (corresponding to the zero of ε_2'') E_0, $E_0+\Delta_0$, E_1 and $E_1+\Delta_1$ respectively corresponding to the allowed optical transitions occurring between Γ_8^v-Γ_6^c, Γ_7^v-Γ_6^c, $L_{4,5}^v$-L_6^c and L_6^v-L_6^c energy levels in the electronic band structure of CdTe. E_0 is the optical absorption gap of CdTe at Γ point and E_1 is the absorption edge at the L point corresponding to the center and edge of the first Brillouin zone, respectively. E_0 energy is blue shifted by a very small amount and approaches the bulk value with the increase in size of nanoparticles.

ii) CdSe nanoparticles:

GAXRD spectrum of $CdSe:SiO_2$ samples shown in the Fig. 2 (i). The XRD spectrum of sample S0 (as-deposited) indicates the amorphous behavior analogous to that observed in $CdTe:SiO_2$ nanoparticle samples. The GAXRD spectrum of samples S5 (annealed in vacuum at 500^{0}C for 2 hours) and S7 (air annealed at 600^{0}C for 2 hours) shows predominant peaks corresponding to the hexagonal and cubic phases of CdSe respectively. The hkl values assigned to peaks of samples S5 and S6 shown in the Fig. 2 (i) corresponding to the hexagonal and cubic phases of CdSe nanoparticles in agreement with standard literature [24].

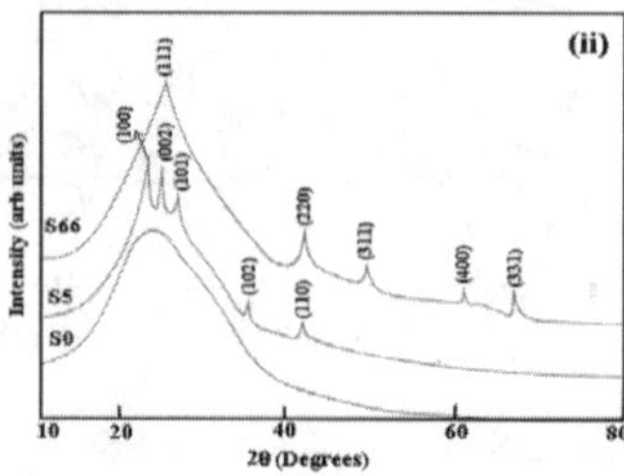

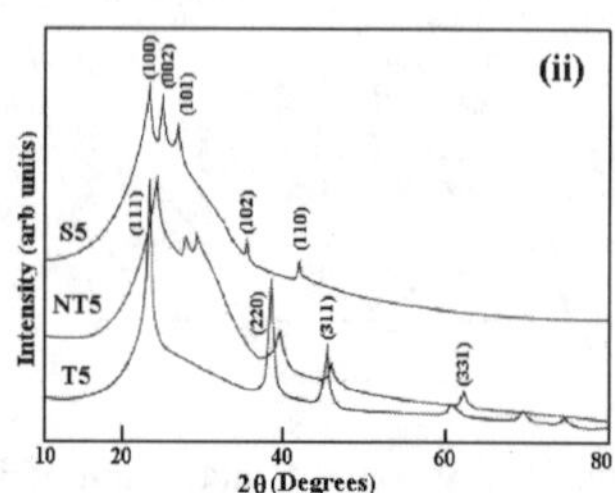

Fig. 2: (i) GAXRD spectra of S0, S5 and S6 CdSe:SiO_2 nanoparticle films and (ii) GAXRD spectrum of $CdSe_xTe_{1-x}$ (x=0.5, surface area) nanoparticle of sample ST5 annealed in nitrogen ambient in comparison to CdTe and CdSe nanoparticle samples T5 and S5 prepared under similar conditions.

Electron micrographs of CdSe:SiO_2 sample S4 (annealed to 400^0C for 2 hours in vacuum) shown in Fig. 3 (ii) presents a very interesting observation. Nanorod structures having an average particle size 13-19 nm and an aspect ratio of 1-26 have been observed in sample S4. This is in contrast to the nanoparticle structure observed in case of CdTe:SiO_2 samples in Fig. 3 (i). Strong and long-range dipole-dipole interactions among the individual CdSe nanoparticles present in the S0 samples might be driving these CdSe nanoparticles to aggregate and align themselves to form fractals [25,26]. The decrease in viscosity of glass matrix with higher annealing temperature make the diffusion process easier allowing the spontaneous self-organization of CdSe nanoparticles into quasi 1-D nanorods. These nanorod structures have been persistently observed up to 600^0C annealing in vacuum in the case of sample S6.

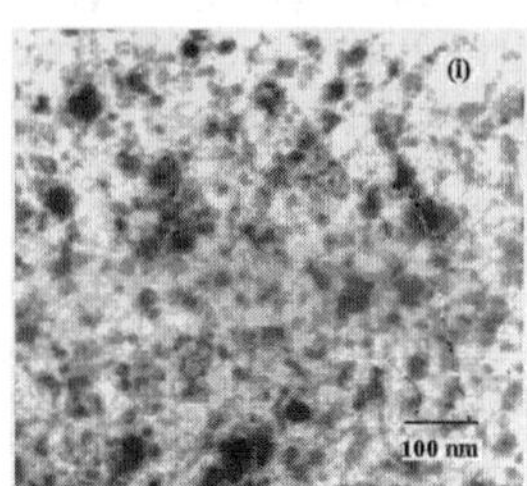

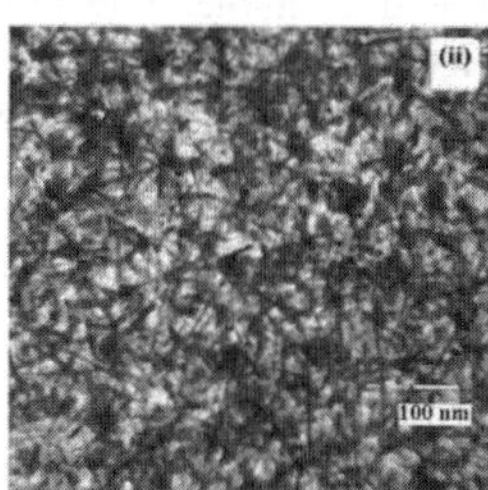

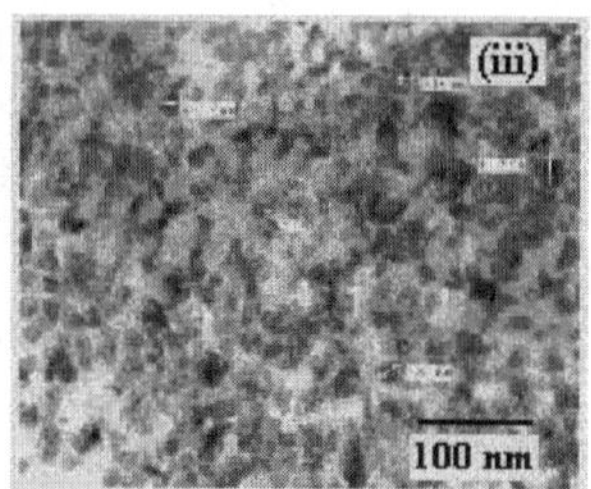

Fig. 3: Electron micrographs of (i) CdTe:SiO_2 nanoparticles in sample T4, (ii) CdSe:SiO_2 nanorods in sample S4 and (iii) $CdSe_xTe_{1-x}$ nanoparticles in ST5 sample annealed in nitrogen ambient in similar conditions.

iii) $CdSe_xTe_{1-x}$ nanoparticles:

As-deposited and annealed $CdSe_xTe_{1-x}$: SiO2 nanoparticle films have been prepared having targets surface area x= 0.25, 0.5 and 0.75. For x=0.5, GAXRD spectrum given in the Fig. 3 (ii) shows the formation of CdSeTe nanoparticles evidenced by the shift in (111) peak [27]. Electron micrograph studies were also carried out to directly observe the CdSeTe nanoparticles. Fig. 3 (iii) shows an electron micrograph of as deposited $CdSe_xTe_{1-x}$ films annealed at 400^0C in nitrogen ambient (sample ST4) showing the nanoparticle characteristics. Due to the higher vapor pressure of Se in comparison to Te for a given temperature [28], a change in the stoichiometry was observed in CdTeSe sample prepared by annealing in vacuum. Annealing the CdSeTe samples in N2 ambient was observed to reduce this problem.

Conclusions:

$CdSe_xTe_{1-x}$ ($0 \leq x \leq 1$) nanoparticles dispersed in SiO_2 films have been synthesized by sputter deposition followed by thermal annealing. In the case of CdTe nanoparticles, an amorphous interfacial layer formed during vacuum annealing transforms the crystal structure from cubic to hexagonal surrounded by crystalline $CdTeO_3$ capping layer formed during air annealing. An accurate determination of the optical properties using spectroscopic ellipsometry has resulted in a first time and unambiguous study of the size dependent shift of E_0 and $E_0+\Delta_0$ energies at Γ point in CdTe nanoparticles. CdSe nanorods having an average diameter 13-19 nm and aspect ratio of 1-26 have also made in glass matrix .In all the three, CdTe, CdSe and CdSeTe nanoparticles systems, the optical absorption edge has been blue shifted to higher energies due to quantum confinement effect. This study shows that the present method can be used for growing semiconductor nanoparticle ternary phase.

Acknowledgements:

Authors (PBD and BRM) are grateful to BRNS, Department of Atomic Energy (DAE) BRNS, Mumbai, India for providing financial assistance to carry out this work. PBD specifically acknowledge the Counsel of Scientific and Industrial Research (CSIR), India for awarding Senior Research Fellowship. PBD also acknowledge CSIR and Indian National Science Academy (INSA) providing funds for attending MRS fall 2003 meeting held in Boston.

References:

1. S. I. Ijima, Nature (London) **354**, 56 (1991).
2. Z. Tang, N. A. Kotov, M. Giergig, Science **297**, 237 (2002).
3. X. Gao, W. C. W. Chan and S. Nie, J. Bio-med. Optics **7**, 532 (2002).
4. W. H. Huynh, J. J. Ditlmer and A. P. Alivisatos, Science **295**, 2425 (2002)
5. M. Nirmal, D. Norris, M. Kuno, M.G. Bawendi, Al. l. Efros and M. Rosen, Phys. Rev. Lett. **75**, 3728 (1995).
6. M. B. Mohamed, C. Burda and M. A. El. Sayeed, Nano. Lett. **1**, 589 (2001).
7. S. Link and M. A. El. Sayeed, J. Appl. Phys. **92**, 6799 (2002).
8. L. S. Li, J. Hu, N. Yang and A. P. Alivisatos, Nano. Lett. **1**, 349 (2001).
9. X-Y. Wang, J-Y. Zhang, A. Wazzal, M. Darragh and M. Xino, Appl. Phys. Lett. **81** (2002) 4829.
10. J. H. Simmons, J. Non-cryst. Solids **239**, 1 (1998) [Morey Award Paper]

11. R. Resifeld, J. Alloys. Comp. **341**, 56 (2002)
12. Al. Meldrum, R. F. Haglund Jr, L. A. Boatner and C. W. White, Adv. Mater. **13**, 1431 (2001).
13. I. V. Bondar et al, Semiconductors **36**, 298 (2002)
14. W. Caseri, Macromol. Rapid. Commun. **21**, 705 (2000)
15. Al. L. Efros and M. Rosen, Annu. Rev. Mater. Sci. **30**, 475 (2000)
16. U. Woggon in *Optical properties of semiconductor quantum dots* (Springer, Heidelberg, p-16, 1997)
17. K. Tsunetomo, H. Nasu, H. Katiyama, A Kawabuchi, Y. Osaka, Jpn. J. Appl. Phys. **28**, 1928 (1989)
18. Joint Committee on Powder Diffraction Standards data cards, Cubic CdTe-150770, Hexagonal CdTe-800088 and Monoclinic $CdTeO_3$-491757.
19. http://hrdata.nist.gov/xps/elm_comp_res.asp
20. P. Babu Dayal, B. R. Mehta, S. M. Shivaprasad and Y. Aparna, Appl. Phys. Lett. **81**, 4254 (2002)
21. P. Babu Dayal, B. R. Mehta and P. D. Paulson. (Communicated to Appl. Phys. Lett)
22. J. H. Simmons, J. Non-cryst. Solids **239**, 1 (1998) [Morey Award Paper]
23. L. M. Roth and B. Lax, Phys, Rev. Lett. **3**, 217 (1959)
24. Joint Committee on Powder Diffraction Standards data cards, Cubic CdSe:16-191, Hexagonal CdSe:77-2307 and $CdSeO_3$-491757.
25. Z. Tang, N. A. Kotov, M. Giergig, Science 297, 237 (2002).
26. Q. Peng, Y. Dong, Z. Deng and Y. Li, In-org. Chem. **41**, 5249 (2002)
27. M. M. EL-Nahass, A.A. M. Farag and H.E.A. El-Sayed, Appl. Phys. A **77**, 819 (2003)
28. David Rlide in *Handbook of Chemistry and Physics* (CRC Press, 82 nd edition, 2001-2002)

Optical and Electronic Properties of Nanoparticles II

Mat. Res. Soc. Symp. Proc. Vol. 789 © 2004 Materials Research Society N2.5

Lasing from CdSe/ZnS Quantum Rods in a Cylindrical Microcavity

Miri Kazes, David Y. Lewis, Yuval Ebenstein, Taleb Mokari and Uri Banin
Institute of Chemistry and the Center for Nanoscience and Nanotechnology,
The Hebrew University of Jerusalem,
Jerusalem 91904, Isreal

ABSTRACT

Lasing from CdSe/ZnS quantum rods and quantum dots both in solution and in a film is studied by utilizing a high Q cylindrical microcavity, showing Whispering Gallery Mode (WGM) lasing. CdSe/ZnS quantum rods, in comparison to quantum dots, exhibited remarkably reduced lasing thresholds. In addition, polarization measurements revealed that quantum rods have a linear polarized lasing, in contrast to quantum dots that show no preferable lasing polarization. Furthermore, an efficient and reproducible method is employed for preparation of nanocrystal films inside capillaries by laser irradiation for achieving robust lasing. Further irradiation of the film resulted in a room temperature stable lasing over hundreds of pump pulses, lasing thresholds as low as 0.02mJ and lasing intensities that are three orders of magnitude larger than the saturated fluorescence intensity. This was successfully applied to CdSe/ZnS quantum rod samples of varied dimensions and was also demonstrated for quantum dot samples.

INTRODUCTION

Semiconductor nanocrystals show potential for applications ranging from photonics and resonator on-a-chip devices to biosensing. An obvious advantage of such nano particles is the remarkable spectral coverage for luminescence that is afforded via the quantum confinement effect by merely controlling the size and composition of the nanocrystals using well-developed colloidal synthesis [1,2]. Recent synthesis efforts led to the shape control of such colloidal prepared CdSe nanocrystals by a modification of the colloidal synthesis to obtain rod shaped particles – quantum rods [3,4,5]. In contrast to the spherical shaped quantum dots, quantum rods have linearly polarized emission as demonstrated by single crystal measurements and theoretical calculations [6,7]. Nanocrystal's low dimensionality is expected to provide low lasing thresholds and lasing that is insensitive to temperature thus making the nanocrystals interesting new candidates as optical gain materials. However, there are also very fast loss mechanisms and mainly the fast Auger recombination process that prevents efficient lasing [8,9]. In order to overcome the losses we utilize glass capillary tubes that serve as both convenient micro-containers for the samples i.e. the gain medium, as well as the microcavity for Whispering Gallery Mode (WGM) lasing [10,11]. The short round-trip times in the microcavities and the very high quality factors make it possible to compete with rapid loss mechanisms that counteract population inversion necessary for lasing [12].

EXPERIMENTAL DETAILS

The CdSe quantum dots were grown using the methods of colloidal nanocrystal synthesis utilizing high temperature pyrolysis of organometallic precursors in coordinating solvents and are overcoated by hexadecylamine (HDA) and trioctylphosphine oxide (TOPO) [3]. The growth of a few monolayers of ZnS to form a core/shell structure enhances the fluorescence quantum yield from ~5 % to 60 %. Quantum rods are prepared as published elsewhere, where the addition of phosphonic acids results in rod growth through a surfactant controlled mechanism [3,4]. Growing a thin ZnS shell provides rod samples with quantum yield of 20 to 40% depending on rod-dimensions [5].

For experiments preformed on a nanocrystal solution, an optical fiber (125 micron in diameter) that was inserted into a glass capillary (200 micron in diameter) was used as the cylindrical microcavity. 3mg of nanocrystals were dissolved in 100μL of anhydrous Hexanes to form a highly concentrated solution that was loaded into the capillary by simple capillary action. The capillaries were prepared within a dry box and sealed by epoxy glue.

In the case of laser prepared films, 200 micron and 153 micron capillaries that served as the cylindrical microcavity were loaded with a concentrated nanocrystal solution in Toluene. The laser prepared films were produced by illuminating the capillary with a sequence of pump pulses with intensity of ~3 mJ resulting in the formation of a dense solid film on the edges of the irradiated spot. After the preparation process, the threshold for lasing from the pre-prepared film was characterized.

For emission and lasing studies, the capillary tubes were pumped from the side using the second harmonic of a Nd:YAG laser at 532 nm (beam radius w~0.3 mm), with the emission collected at 90 degrees and detected using a spectrograph/CCD setup. All experiments were carried out in ambient conditions.

RESULTS AND DISCUSSION

The characteristics of samples used in this study are presented in figure 1 showing transmission electron microscope (TEM) images of CdSe/ZnS quantum rods of different size and of a quantum dot sample. The good size distribution and shape control that is observed is essential for achieving lasing by reducing the inhomogeneous broadening of the gain profile.

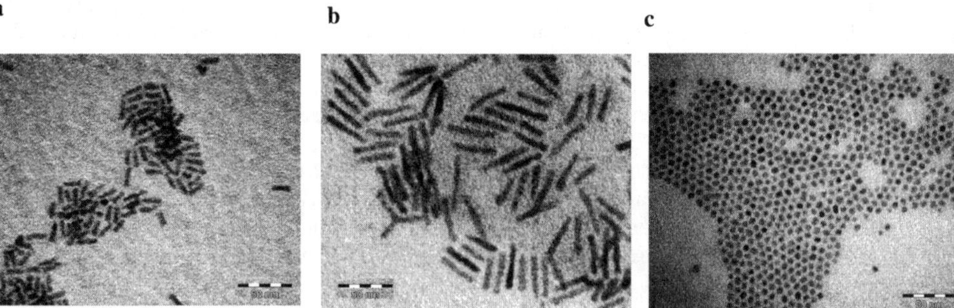

Figure 1: TEM images of quantum rod sample of (a) 4nm x 14nm (diameter x length) (b) 4nm x 24nm and (c) quantum dot sample of 4nm in diameter.

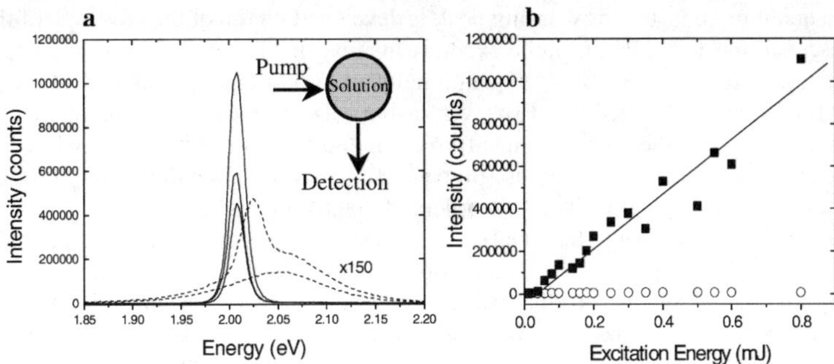

Figure 2: a) Lasing in quantum rods, 4nm in diameter and 24nm in length, at different pump powers. The pump intensities from low to high are: 0.01mJ, 0.02mJ, 0.4mJ, 0.55mJ, 0.8mJ. The laser peak at 2eV appears above a threshold of 0.02mJ. The right inset is a schematic top view of the cylindrical microcavity experimental setup as described in the text. **b)** The intensity of the lasing peak (filled squares) and the fluorescence (empty circles) versus the pump power. An abrupt change in intensity is apparent above the threshold (The line is a guide for the eye).

Figure 2a presents results of lasing for a pre-prepared film of CdSe/ZnS quantum rods, 4nm in diameter and 24nm in length, within the capillary tube. Emission spectra collected from a laser pre-prepared film are shown for several laser excitation intensities where each spectrum corresponds to a single laser shot. At low excitation intensities (e.g. 0.01 mJ), the spectrum resembles that measured for the emission of quantum rods in solution (Fig. 2-left inset).At higher intensities, starting around 0.02 mJ, a narrow peak to the red of the fluorescence maximum starts to emerge. At even higher intensities, the lasing peak completely dominates the emission. Figure 2b shows the dependence of the intensity of the lasing peak and fluorescence peak on pump power. The lasing shows clear threshold behavior manifested as an abrupt change of slope at the onset of laser action and lasing intensities that are nearly three orders of magnitude larger then the saturated fluorescence intensity.

Such robust lasing films within the capillary tube were reproducibly obtained by utilizing the pump laser [13]. By extensive high power irradiation of the quantum rods solution in the capillary, a build up of the emission is starting to occur while forming a solid film on the cylindrical microcavity inner walls. Such a pre-prepared area then yields robust lasing and shows the typical threshold behavior as demonstrated in Figure 2. Lasing was demonstrated using this preparation procedure also for another rod sample with dimensions of 4x14 nm and for quantum dot sample of 3.5nm in diameter, proving the applicability of our approach to different nanocrystal samples. However, the achieved threshold values vary depending on the quality of the film/capillary boundary formed that could be improved by further optimization of this procedure.

Lasing was also observed for CdSe/ZnS nanocrystals in solution. In this case, where an optical fiber surrounded by the nanocrystals solution and enclosed by the capillary served as the cylindrical microcavity, the lasing was observed on the fiber outer surface. Figure 3 presents results for lasing from quantum dots of 4nm in diameter in a Hexane solution. While at low excitation powers the typical fluorescence spectrum is observed, at higher excitation powers, as for the quantum rods, a narrow lasing peak is developed on top of the envelope of the fluorescence spectrum. Pump-probe experiments revealed induced absorption features in the potential gain region for CdSe/ZnS quantum dots in solution suggested to limit the possibility of achieving lasing in this medium. However, lasing was achieved here by pumping close to the band gap and due to the high Q value afforded by the cylindrical microcavity [12,14]. Both in this configuration and in the laser pre-prepared film configuration the lasing threshold for quantum dots was typically about two orders of magnitude higher than the lasing threshold for quantum rods (Figure 3b). The remarkably reduced lasing thresholds exhibited for CdSe/ZnS quantum rods, in comparison to quantum dots, is essentially assigned to increased oscillator strength, increased Stokes shift that reduces losses due to reabsorption, and slower Auger rates which are dependent on size. The difference between quantum rods and quantum dots is related to the transition from zero-dimensions to quasi-one dimensions, which directly impacts their optical and electronic properties [6].

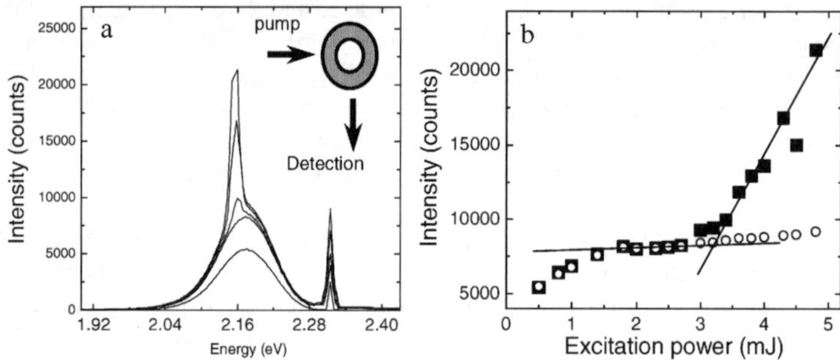

Figure 3: a) Lasing from quantum dots solution of 4nm in diameter. The spectra taken for different pump powers from low to high: 0.5mJ, 2.5mJ, 3.4mJ, 4.8mJ. Above a threshold of ~3mJ a narrow lasing peak emerges at 2.16eV (The peak at 2.33eV is the scatter from the pump laser). The right inset describes schematically the experimental configuration where an optical fiber that is inserted into the capillary serves as the cylindrical microcavity. The gray area represents the quantum dot solution that serves as the gain medium. b) The intensity of the lasing peak (filled squares) and the fluorescence (empty circles) versus the pump power. The change in the slopes of the lines clearly indicates the threshold behavior for lasing.

An additional marked difference between lasing in dots versus rods in shown in Figure 4 presenting results of polarization measurements from the cylindrical microcavity loaded with a quantum dot solution (Figure 4a) versus a quantum rod solution (Figure 4b). Lasing from the quantum rod solution is linearly polarized (VV, denoted by vertical arrow) in parallel to the polarization of the excitation beam that was parallel to the capillary long axis. In contrast, the quantum dots solution did not show any preferable polarization. The polarization in rod lasing is assigned to preferable excitation of rods aligned parallel to the pump laser polarization.

The WGM form of the observed lasing, in contrast to radial mode lasing, was verified by the spacing detected for the lasing modes. High resolution spectra taken for the 200 micron capillary, for a 153 micron capillary and for a 125 micron fiber inserted within the 200 micron capillaries showed average mode spacing of 0.32nm, 0.5nm, and 0.62nm, respectively. All three spectra show good agreement with the theoretical peak structure, corresponding to WGMs that are best resolved for the cavity with the smallest diameter and hence largest spacing. See equation 1.

$$\Delta\lambda \sim \lambda_n^2 / (m_2 2\pi r) \qquad (1)$$

Where λ_n the detected mode wavelength, r the cavity radius and m_2 the refractive index at the lasing interface.

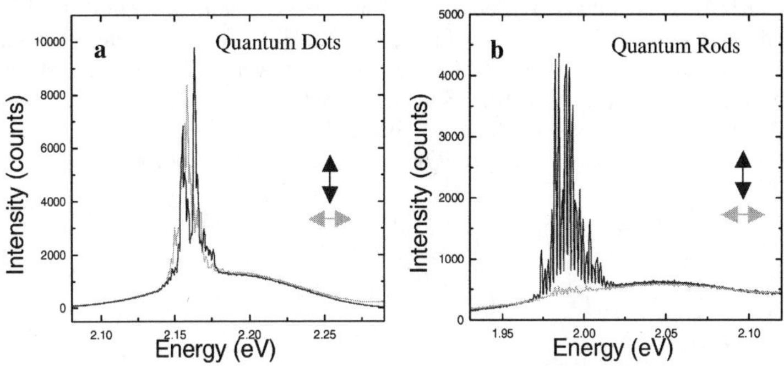

Figure 4: Polarization measurement of lasing from quantum dots (a) and quantum rods (b). The excitation laser is polarized parallel to the capillary long axis. Detection polarization parallel to the pump laser is denoted VV (Vertical arrow) and detection polarization perpendicular to the pump laser is denoted VH (Horizontal arrow). Lasing from quantum dots show no polarization dependence while for quantum rods the lasing is clearly linearly polarized in parallel to the pump laser.

CONCLUSIONS

WGM Lasing was detected for CdSe/ZnS quantum rods and quantum dots, both in solution and in a film utilizing the high Q factor of the cylindrical microcavity. A reproducible method of preparation of films of quantum rods in a cylindrical microcavity, that exhibit robust lasing is reported. Quantum rods show lasing thresholds that are significantly lower than thresholds detected for quantum dots. Furthermore, Polarization measurements show that lasing from quantum rods is linearly polarized. These properties may prove advantageous for the utility of quantum rods as novel optical gain medium.

REFERENCES

1. Y. W. Cao, U. Banin, *J. Am. Chem. Soc.* **122**, 9692 (2000).
2. L. Qu, Z. A. Peng, X. Peng, Nano Letters, Vol. 1, 6, 333 (2001).
3. X. G. Peng, L. Manna, W. D. Yang, J. Wickham, E. Scher, A. Kadavanich, A. P. Alivisatos, *Nature*, 404, 59 (2000).
4. L. Manna, E. C. Scher, A. P. Alivisatos, *J. Am. Chem. Soc.*122, 12700 (2000).
5. T. Mokari, U. Banin, Chem. Matt.15, 3955 (2003).
6. J. Hu, L. S. Li, W. Yang, L. Manna, L. W. Wang, A. P. Alivisatos, *Science*, 292, 2060 (2001).

7. E. Rothenberg, Y. Ebenstein, M. Kazes, U. Banin – to be published.
8. V. I. Klimov, A. A. Mikhailovsky, D. W. McBranch, C. A. Leatherdale, M. G. Bawendi, *Science* **287**, 1011 (2000).
9. H. Giessen, U. Woggon, B. Fluegel, G. Mohs, Y.Z. Hu, S.W. Koch, and N. Peyghambarian, *Optics Letters* **21**, 1043 (1996).
10. H. J. T. Moon, Y. Chough, K. An, *Phys. Rev. Lett.* 85, 3161 (2000).
11. M. Kazes, D. Y. Lewis, Y. Ebenstein, T. Mokari, U. Banin, *Adv. Mat.* 4, 317 (2002).
12. M. Mazumder, D. Q. Chowdhury, S. C. Hill, R. K. Chang, *Optical Processes i* *Microcavities*, edited by R. K. Chang and A. J. Campillo, (Advanced Series in Applie Physics, Vol. 3, World Scientific, Singapore, 1996) pp. 209-221.
13. M. Kazes, D. Y. Lewis, U. Banin – to be published.
14. S. Link, M. A. El-Sayed, *J. App. Phys. Vol.* 92, 11, 6799 (2002).

Mat. Res. Soc. Symp. Proc. Vol. 789 © 2004 Materials Research Society N2.10

Energy exchange between optically excited Silicon Nanocrystals and Molecular Oxygen

E. Gross, D. Kovalev, N. Künzner, J. Diener, and F. Koch
Technische Universität München, Physik Department E16,
85747 Garching, Germany
V.Yu. Timoshenko
Faculty of Physics, Moscow State M.V. Lomonosov University,
119992 Moscow, Russia
Minoru Fujii,
Department of Electrical and Electronics Engineering,
Faculty of Engineering, Kobe University,
Rokkodai, Nada, Kobe 657-8501, Japan

ABSTRACT

We report on the photosensitizing properties of optically excited Silicon (Si) nanocrystal assemblies that are employed for an efficient generation of singlet oxygen. Spin triplet state excitons confined in Si nanocrystals transfer their energy to molecular oxygen (MO) adsorbed on the nanocrystal surface. This process results in a strong suppression of the photoluminescence (PL) from the Si nanocrystal assembly and in the excitation of MO from the triplet ground state to singlet excited states. The high efficiency of the energy transfer if favored by a broad energy spectrum of photoexcited excitons, a long triplet exciton lifetime and a highly developed surface area of the nanocrystal assembly. Due to the specifics of the coupled system Si nanocrystal – oxygen molecule all relevant physical parameters describing the photosensitization process are accessible experimentally. This includes the role of resonant and phonon-assisted energy transfer, the dynamics of energy transfer, and its mechanism.

INTRODUCTION

The transfer of electronic excitation energy between electronically coupled systems is one of the fundamental problems in modern physics of condensed matter [1,2]. It describes the intermediate sequences between the primary event of electronic excitation and the final processes that utilize the energy of electrons. Despite a general knowledge acquired in the field of excitation transfer processes, the high complexity of the systems considered inhibits a detailed understanding of the microscopic processes occurring during the transfer of energy. For this reason more simple configurations are demanded to uncover the mechanism of electronic excitation transfer.

Semiconductor nanostructures represent a promising approach to overcome the difficulties in determining the basic parameters of electronic energy transfer processes. The main advantage of these systems is the huge flexibility and degree of freedom in their fabrication. The sizes, shapes and the geometrical arrangements of the nanostructures as well as their chemical composition and the quality of the interfaces are well controllable. These artificial structures combine molecular and solid state properties and allow one a description of the physical phenomena by taking advantage of separately developed concepts. To study energy transfer processes in detail, we used microporous silicon (PSi) as a semiconductor nanostructure. PSi consists of a Si nanocrystal skeleton and an interconnected pore network with an average structure size of 2-5 nm [3]. Due to quantum confinement effects the effective bandgap energies of the nanocrystal assembly range from 1.12 eV up to 2.5 eV [3]. The intrinsic electronic structure of photogenerated excitons confined in Si nanocrystals is characterized by two possible spin configurations: a spin singlet state and a spin triplet state, that is lowered in energy. At cryogenic temperatures only the triplet exciton state is occupied and a radiative exciton lifetime in the millisecond range is observed [3].

The important role of MO in various scientific fields is directly related to its particular electronic structure (inset of figure 1). The $^3\Sigma$ ground state (superscript denotes the spin multiplicity) of MO has spin triplet multiplicity and is chemically inert [4]. The lowest excited states, $^1\Delta$ and $^1\Sigma$, have spin singlet nature (singlet oxygen) and are chemically highly reactive [4]. Thus, singlet oxygen is successfully employed in technical and medical processes, such as water disinfection or photodynamic cancer therapy. However, the direct photoexcitation of singlet oxygen is forbidden by spin selection rules. The electronic excitation of MO requires photosensitizers as intermediate light-absorbing substances that subsequently transfer the energy to oxygen molecules.

EXPERIMENTAL DETAILS

PSi layers have been prepared by electrochemical etching of (100) oriented, Boron-doped bulk Si wafers with a resistivity of 10 Ω∃cm in a 1:1 by volume mixture of hydrofluoric acid (50 wt% in water) and ethanol. The current density and etching time ranges from 25 mA/cm^2 to 100 mA/cm^2 and from 1 to 10 minutes, respectively. The thickness of the porous layer varies from 6 μm to 30 μm with a mean nanocrystal size of 2-10 nm. The experiments have been performed either in vacuum or with a well-defined amount of oxygen in the sample chamber. Optical excitation of the Si nanocrystals is performed using an Ar$^+$-laser (E_{exc} = 2.41 eV) operating in cw mode. The spectrally resolved intensity of the emitted light is measured with a monochromator equipped with a charge-coupled device or a nitrogen-cooled Ge-detector. All spectra are corrected on the spectral sensitivity of the optical system. Magnetooptical experiments have been performed in a cryostat equipped with a superconducting solenoid that provides magnetic fields up to 11 T. For excitation of the sample and collection of the emitted light fiber optics has been used.

RESULTS AND DISCUSSION

The strong interaction of photoexcited Si nanocrystals with oxygen molecules is demonstrated by figure 1. The low temperature PL spectrum of PSi measured in vacuum (dashed lines) is characterized by a broad, featureless emission band located in the visible region. It reflects the wide band gap distribution of the Si nanocrystal assembly that can be controlled by varying the etching parameters during the anodization of bulk Si. A larger current density during electrochemical etching results in smaller average sizes of the nanocrystals. This causes a blue shift of the PL emission band due to stronger exciton confinement. The PL emission bands

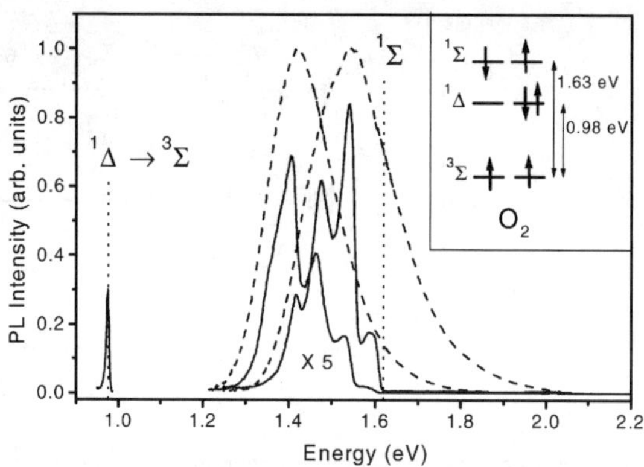

Figure 1. Emission spectrum of PSi in vacuum (dashed line) and in the presence of adsorbed oxygen molecules (solid line). The two PL emission bands correspond to two different PSi samples. Inset: Electronic spin configurations and spectroscopic labelling of molecular oxygen states. T = 5 K, E_{exc} = 2.41 eV. Partial pressure of oxygen at 293 K was 10^{-2} mbar.

shown in figure 1 correspond to two distinct PSi layers that have been prepared using different etching current densities (left PL spectrum: 30 mA/cm^2, right PL spectrum: 40 mA/cm^2). The adsorption of oxygen molecules on the nanocrystal surface drastically modifies the PL emission properties of both samples (figure 1, solid lines). The PL intensity is strongly suppressed and a spectroscopic fine structure is resolved. Above the energy that coincides with the excitation energy of the $^3\Sigma$ - $^1\Sigma$ transition in MO complete PL quenching (at least four orders of magnitude) occurs. The modifications of the PL from PSi are accompanied by an infrared emission at 0.98 eV. It results from the radiative relaxation of the $^1\Delta$ state to the $^3\Sigma$ ground state of oxygen molecules. These observations are clear evidence for the energy transfer from excitons confined in Si nanocrystals to oxygen molecules. Obviously, the structural properties of the Si nanocrystal assembly, i.e. the specific size distribution of the nanocrystals does not influence the energy transfer process. The exact shape of the quenched spectra is governed by the initial PL emission band of the nanocrystal assembly measured in vacuum. It turns out that the spectral positions of the signatures coincide for samples having different PL bands. The PL fine structure observed for the quenched spectra provides important information on the electronic energy transfer process. Quenching is strongest for nanocrystals having bandgap energies that coincide exactly with the $^3\Sigma$ - $^1\Sigma$ transition of MO. From figure 1 it is evident that nanocrystals whose band gaps do not match resonantly the excitation energies of the $^1\Delta$ and $^1\Sigma$ excited states of MO participate in the energy transfer as well. The excess of the exciton energy with respect to the energies of the $^1\Delta$ and $^1\Sigma$ states is released via the emission of phonons.

In figure 2 the mechanism proposed for the energy transfer from excitons to MO is sketched. Below the nanocrystal bandgap real electronic states are absent and energy dissipation should be governed by multiphonon emission rather than a phonon cascade [5]. This process is most probable for phonons having the highest density of states. In bulk Si these are transverse optical phonons near the center of the Brillouin zone with an energy of ≈ 63 meV [6]. If the

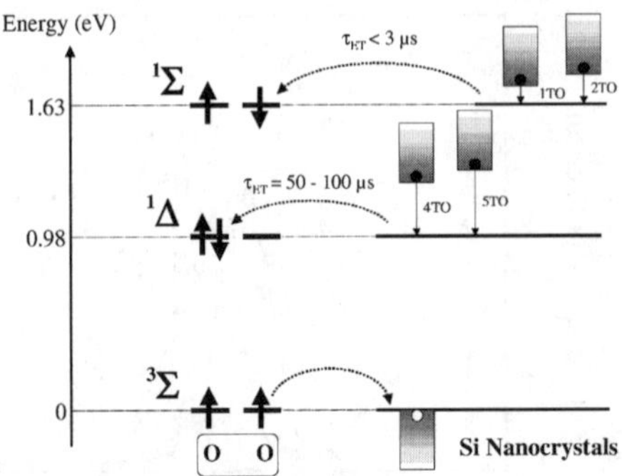

Figure 2. Energy-level diagram of molecular oxygen and of the Si nanocrystal assembly. Principle steps of the energy transfer process are sketched. Direct electron exchange (dotted arrows) is accompanied by multiphonon emission (solid arrows). Electrons photoexcited in Si nanocrystals are exchanged with electrons belonging to MO. This process results in the formation of singlet MO states and compensation of the holes confined in Si nanocrystals. The measured time of energy transfer τ_{ET} is indicated.

bandgap energy of Si nanocrystals does not coincide with the excitation energy of a MO singlet state plus an integer number of the energy of those phonons, an additional emission of acoustic phonons is required to conserve the energy. This process has a smaller probability and the efficiency of energy exchange is reduced. Consequently, equidistant maxima and minima in the quenched spectrum appear and evidence the phonon-assisted energy transfer [7]. The coupling strength of excitons to the $^1\Sigma$ state is much stronger than for the $^1\Delta$ state, since the $^3\Sigma \rightarrow {}^1\Sigma$ excitation of MO conserves the total orbital momentum. It results in an almost complete quenching of the PL emission above the $^1\Sigma$ state energy and inhibits the observation of a phonon-related fine structure.

The dynamics of the energy transfer is defined from the PL decay time of the Si nanocrystal assembly. Without oxygen molecules being present on the nanocrystal surface, the PL decay time from PSi is in the millisecond range [3]. If MO is adsorbed, energy transfer represent a nonradiative decay channel for excitons. This results in quenching of the PL emission and a significant shortening of the PL decay time occurs. By measuring the transients of the PL from PSi in vacuum and in oxygen ambient we determined the energy transfer time τ_{ET} [8]. Depending on the level of PL suppression τ_{ET} for the $^1\Delta$ state is in the range of 50-100 μs, while for the $^1\Sigma$ state τ_{ET} is shorter than 3 μs.

During energy transfer photoexcited electrons in Si nanocrystals are exchanged with electrons belonging to MO. This process results in the formation of singlet MO states and in the compensation of the holes confined in Si nanocrystals. To confirm the exchange mechanism of electrons the quenched PL spectrum has been monitored at different strength of magnetic field (figure 3). For energy transfer to occur the exchanged electrons must have the proper spin orientation [9] since the excitation of singlet oxygen states ($^1\Delta$, $^1\Sigma$) from the triplet ground state ($^3\Sigma$) requires “spin-up” states. A magnetic field introduces a common quantization axis for the

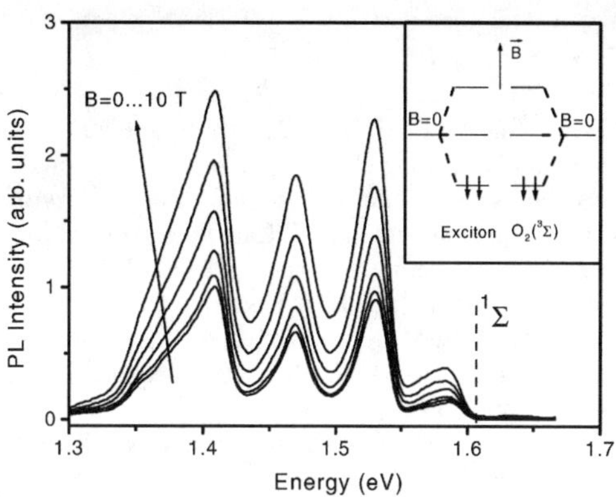

Figure 3. Quenched PL spectrum of PSi at various strength of magnetic field. The magnetic field increment is equal to 2 T. T = 5 K. Inset: Zeeman splitting of the triplet exciton state and the $^3\Sigma$ ground state of molecular oxygen. The spin orientations for electrons in the lowest lying levels are indicated by arrows. Energy transfer from exciton to ground state MO via electron exchange among these states is prohibited by conservation of the total magnetic quantum number.

spins and the triplet exciton states and the $^3\Sigma$ state of MO are Zeeman splitted (inset figure 3). At low temperatures "spin-down" states are preferentially occupied and the number of possible states participating in the electron exchange is reduced [10]. This results in a decreased energy transfer rate and weaker PL quenching is observed.

CONCLUSIONS

To conclude, we presented spectroscopic evidence of the efficient energy transfer from excitons confined in Si nanocrystals to MO. The experiments show that PSi layers have remarkable photosensitizing properties due to a broad energy spectrum of photoexcited excitons expanding over the entire visible spectral range, a long exciton lifetime and a large internal surface area. We believe that the result of this research offers a novel area of further application of the most commonly used semiconductor.

ACKNOWLEDGMENTS

This work is supported by Industrial Technology Research Grant Program in '02 from New Energy and Industrial Technology Development Organization (NEDO), Japan and Deutsche Forschungsgemeinschaft (KO 1966/5-1). V.Yu. Timoshenko and M. Fujii are grateful to the AvH Foundation.

REFERENCES

1. V. M. Agranovich, and M. D. Galadin, *Electronic excitation energy transfer in condensed matter* (North-Holland Publishing Company, Amsterdam, New York, Oxford, 1982).
2. D. L. Andrews, and A. A. Demidov, *Resonance Energy Transfer* (Wiley Publishers, 1999).
3. A. G. Cullis, L. T. Canham, and P. D. J. Calcott, J. Appl. Phys. **82**, 909 (1997).
4. P. H. Krupenie, J. Phys. Chem. Ref. Data **1**, 423 (1972).
5. V. N. Abakumov, V. I. Perel, and I. N. Yassievich, in *Nonradiative Recombination in Semiconductors*, edited by V. M. Agranovich and A. A. Maradudin, Modern Problems in Condensed Matter Science Vol. 33 (North-Holland, Amsterdam, 1991).
6. W. Weber, Phys. Rev. B. **15**, 4789 (1977).
7. D. Kovalev, E. Gross, N. Künzner, and F. Koch, V. Yu. Timoshenko, M. Fujii, Phys. Rev. Lett. **89**, 137401 (2002).
8. E. Gross, D. Kovalev, N. Künzner, J. Diener, F. Koch, V. Yu. Timoshenko, and M. Fujii, Phys. Rev. B **68**, 115405 (2003).
9. D. L. Dexter, J. Chem. Phys. **21**, 836 (1953).
10. D. R. Kearns, and A. J. Stone, J. Chem. Phys. **55**, 3383 (1971).

Poster Session I

Mat. Res. Soc. Symp. Proc. Vol. 789 © 2004 Materials Research Society N3.2

Preparation of Tungsten Bronze Nanowires

Hang Qi, Cuiying Wang, Jie Liu
Department of Chemistry, Duke University, Durham, NC

Abstract

Two simple techniques to prepare tungsten bronze nanowires are reported in this paper. Tungsten bronze nanowires with a diameter of less than 100nm and a length of more than 10 μm have been successfully prepared by employing these techniques. A Vapor-Liquid-Solid (VLS) mechanism has been proposed to explain the growth of these tungsten bronze nanowires.

Introduction

One-dimensional (1D) nanostructure materials, such as nanotubes and nanowires have attracted much attention due to their unique properties that originate from their high surface areas and low dimensionality [1]. Many efforts have been taken on the synthesis of various 1D nanoscale materials, especially those of semiconductor materials for the potential applications as building blocks for nanoelectronic devices [2,3]. Homogenous semiconductor nanowires have already been used as devices such as field effect transistor (FET). For more sophisticated devices, heterogeneous semiconductor nanowires (or heterogeneous one-dimensional structures) are required [4,5]. The key to prepare these sophisticated devices is the achievement of heterojunction, the junction of two kinds of materials [6,7].

Heterojunction obtained by two crossed nanowires has been reported [7-10]. In these reports, crossed nanowire junctions were made by using a layer-by-layer fluidic alignment strategy. However, junctions obtained by this strategy were not very solid for the two nanowires were merely put together crosswise. The goal of our work is to prepare heterojunction in an individual nanowire, which consists of two different kinds of material. Our strategy is modifying the chemical composition of a section of a given nanowire.

To change the chemical composition of a section of a given nanowire, chemical reaction must be localized in a nanometer-sized region. We achieve this by using Atomic Force Microscopy (AFM) Based Electrochemically "Dip-Pen" Nanolithography (E DPN) technique, which is developed from "Dip-Pen" Nanolithography (DPN) technique [11,12].

However, nanowires of some kinds of material may be broken into a series of nanoparticles after modification by E-DPN, for the crystal structure of these material changes a lot during the modification. Thus it is critical to choose a proper system. The system we have chosen for such a nanowire is tungsten oxide and tungsten bronze.

Tungsten oxide and tungsten bronze have been studied intensively for more than one century for electrochromic and other interesting properties [13,14]. The word "bronze" has been applied to nonstoichiometric crystalline of transitional metal oxide, especially those ternary systems. The crystalline structure of tungsten trioxide is based on corner-shared W — O octahedra [15]. Within the structure, the tunnels of various sizes and shapes are generated by different arrangements of the octahedra. Tungsten bronze can be achieved by partially reducing

W(VI) to W(V) in tungsten oxide crystalline; alkali metal, alkaline earth metal or other monovalent ions are inserted into the tunnels during such a reducing process [16].

$$xM^{+} + xe^{-} + WO_3 = M_xWO_3 \qquad (0 < x < 1)$$

Optical and electrical properties of tungsten bronze are different from those of tungsten oxide and vary with the concentration of inserted ion [17,18]. Tungsten oxide is insulator while tungsten bronze is semiconductor. Thus a heterojunction can be achieved in a nanowire consisting of two sections made of tungsten bronze and tungsten oxide as well as two sections with different concentrations of inserted ion.

Tungsten oxide and tungsten bronze system has an advantage over others for the AFM chemical modification by E-DPN technique of nanowires. The AFM tip modification only changes the concentration of inserted ion of this system, and does not change the tungsten oxide framework. Thus the wire-like shape will not be destroyed by AFM tip modification and the nanowire can still be employed as a device after modification. Moreover, AFM modification of tungsten oxide film has already been reported [19,20].

However, most of the reports about this system are about the bulk material till now. No method has been reported about preparing tungsten bronze nanowires and only a few methods have been reported about tungsten oxide. Therefore, we report here the first step toward our goal: the preparation of tungsten bronze nanowires. We have chosen tungsten bronze nanowires instead of tungsten oxide because it is difficult to prepare tungsten oxide nanowires long enough for AFM modification. Two techniques to prepare tungsten bronze nanowires will be discussed next.

Experimental

1. Tungsten plate as tungsten source

The preparation was carried in a traditional furnace with a horizontal one-inch quartz tube. Tungsten plate used was bought from Aldrich (99.9+%). The preparation includes three heating stage. The first is at 700°C in 460sccm hydrogen 10 min; the second is at 700°C in 330sccm argon for 2 hr, and the last is at 600°C under 330sccm argon flow for 3hr, in this stage the tungsten plate is covered by KI powder.

2. Tungsten carbonyl as tungsten source

Put 0.4 g tungsten carbonyl (Acros 99%) in a silicon-wafer-capped alumina crucible as tungsten source. The silicon wafer is coated with SiO_2 layer and serves as the substrates of nanowires growth. Prior to the preparation, put two drops of KI acetone solution on the silicon wafer and then heat the silicon wafers at 400°C immediately to evaporate the solvent. The preparation was carried at 625°C under 25sccm Ar flow for 3 hr.

3. Characterization of the sample

The as-grown product was analyzed by following instruments without removing nanowires from the tungsten (silicon) plate: Scanning Electron Microscopy (SEM, FEI Company XL-FEG/SFEG) at 30kV for images; X-ray Photoelectron Spectroscopy (XPS, Riber LAS-3000) and Energy Dispersive X-ray spectrometer (EDX) attached to an SEM (Hitachi S-3200N) for chemical composition analysis; Raman spectra is measured by Dilor Triple Spectrograph XY

Raman Spectrometer and the excitation wavelength is 514.5nm; I-V curve was measured by 4145A Semiconductor Parameter Analyzer (Hewlett Packard).

Results and Discussion

1. Tungsten plate as the tungsten source

After being heated with KI, the tungsten plates showed a blue color and tungsten plates heated without KI showed gold or black at same condition. The blue material can be removed by sonication for 2min, while the gold or black material cannot be removed even after 2hr sonicating. Fig.1 is the SEM images of them. The blue material is nanowires with diameters of less than 100 nanometers and lengths of several micrometers (Fig. 1) while the gold (black) one is very short nanowires or nanorods (Fig. 1 insertion) made of tungsten oxide as reported by Gu et al [21].

The EDX result showed that the blue material consisted of K, O and W. The absence of iodine in the sample suggests that the potassium cannot come from KI residue. Quantitative analysis showed the ratio of these elements was K : W : O = 0.1 : 1 : 1.5. The original XPS spectrum showed two peaks that were not well separated and with a shoulder at the low bonding energy side of the larger peak (Fig.2). This XPS spectrum can be deconvoluted into two groups with two peaks each. The shaded group of peaks at 34.2eV and 36.6eV in Fig.2 suggest that there is W(V) in the sample and the sample is tungsten bronze [13,16,22]. Raman spectrum confirms this suggestion (Fig. 3). The spectrum of the blue sample shows a peak around 960 cm^{-1}, while the spectrum of tungsten oxide shows nothing in the same region. 960

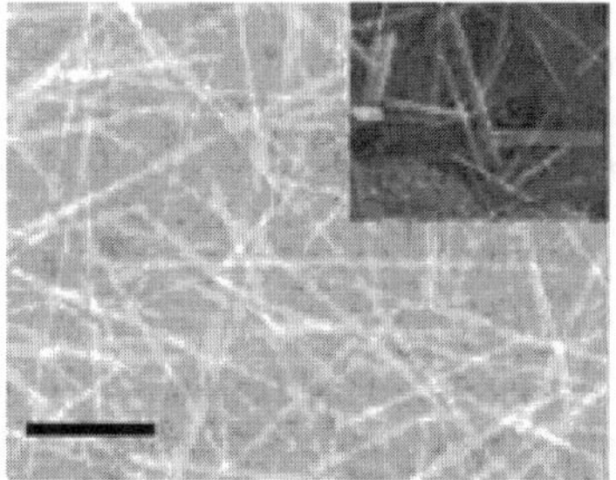

Fig. 1 SEM image of the "blue material", the insertion is the SEM image of "gold or black material". The bar is 1μm

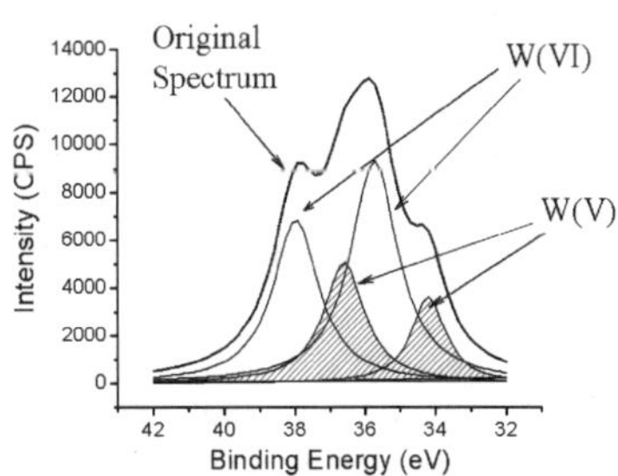

Fig.2 XPS spectrum of the blue material

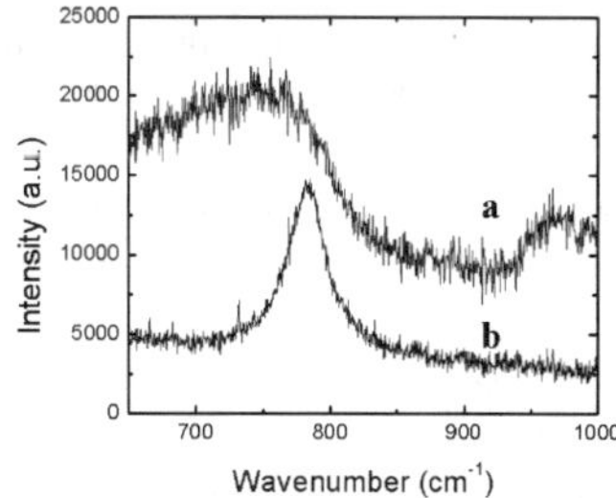

Fig.3 Raman spectra of a) blue material and b) tungsten oxide

cm^{-1} peak is due to W(VI)=O bonds which are absent in crystal tungsten oxide. After some W(VI) was reduced to W(V), the tungsten oxide lattice was distorted and W(VI)=O bonds formed [13,23,24]. However, ratio between tungsten and oxygen is 1/3 and the quantitative analysis suggests excessive tungsten. The excess of tungsten is due to that the detection depth of EDX is several micrometers and is larger than that of our tungsten bronze sample. So the EDX result includes the signals from both tungsten bronze nanowires and tungsten metal substrate. Thus the excessive amount of tungsten was shown in the result.

Tungsten bronze is semiconductive or metallic because of the potassium ion insertion while tungsten oxide is nonconductive. The conductivity measurement shows that the resistance of a piece of tungsten bronze nanowire with a diameter of 100nm and length of 13.6μm is 926Ω. This result indicates that tungsten bronze nanowire gotten by our method is metallic, for the conductivity of this piece of tungsten bronze nanowire is $6.81\times10^{-7}\Omega\cdot m$ and conductivities of metals are in the range of 10^{-8}~$10^{-6}\Omega\cdot m$.

However, the nanowires prepared by this method cannot meet the requirement for nanodevice fabrication. Because nanowires grow on conductive tungsten plate in this method, we must remove the nanowires from the wafer before the tungsten bronze nanowire can be employed as part of a circuit. Moreover, tungsten bronze nanowire prepared from this method is not pure long thin wire. There is always some short thick one, which makes it difficult to purify the nanowires.

2. Growth mechanism of nanowires

The VLS crystal growth mechanism was first proposed by Wagner and Ellis in 1964 for silicon whisker growth and has been widely used to guide the preparation of various kinds of one-dimensional nanostructures [25,26]. We use this mechanism to explain the growth tungsten bronze nanowires, which is important to improve our method.

At a growth temperature of 625°C, tiny KI particles melt and form droplets, although its melting point is higher than this (681°C). These tiny droplets dissolve tungsten oxide and tungsten, as well as absorb oxygen from the atmosphere. Follow reactions occur within these droplets:

$$2W + 3O_2 = 2WO_3$$

$$WO_3 + x\,KI = K_xWO_3 + x/2\,I_2$$ [27,28]

When tungsten bronze reaches saturation in this system, it will participate from the droplet in the form of nanowires. Continuous feeding of tungsten and oxygen elements into the liquid droplet can sustain the growth of the nanowire. In this proposed mechanism, the size of the nanowires is directly related to the size of the initial droplet of liquid. Therefore, the experimental conditions must be carefully controlled to avoid the formation of large liquid droplets.

Several experiments provide support for this proposed mechanism. First, the influence of the temperature on the formation of nanowires was studied. It was noticed that no nanowires were formed when the reaction temperature was higher than the melting point of the potassium salts (melting point of KI is 681°C). Under such conditions, the KI on the tungsten surface was completely melted into liquid, and no small droplets were available to act as the template for the formation of nanowires. Second, nanowires obtained by this method can be removed easily by sonicating the sample for two minutes. Nanowires were attached to the wafer by KI catalyst particles, which dissolve in water. So tungsten bronze nanowires can be removed from the wafer by dissolving those KI particles.

3. Tungsten carbonyl as the tungsten source

KI plays a very important role in the preparation of tungsten bronze nanowires based on the discussion of mechanism, and must be taken into account when we improve the method. In our improved method, KI particles were obtained by heating KI solution at a high temperature. Because the boiling point of solvent used is low, solvent can be removed at a high temperature as soon as possible and KI particles can be obtained as small as possible. Tungsten carbonyl was employed as the tungsten source instead of the tungsten plate. Si wafer coated SiO_2, which was a insulating substrate was employed for nanowires growth.

The sample of this method shows some small islands at the positions of the KI particles among a fine yellow particle sea. The fine yellow particles are made of tungsten metal or tungsten oxide. Tungsten bronze nanowires with a diameter of about 70nm and a length of 15 μm can be found around these islands (Fig 4). These nanowires grew from a pack of particles and sometimes connected two packs of particles like bridges. They are isolated from each other. It is not difficult to find a single piece of "clean" nanowire with more than 10 μm long, and there is no other material around it. All these features make tungsten bronze nanowires obtained by the improved method more suitable for AFM modification than ever.

Fig.4 SEM image of the tungsten carbonyl soured method product

Conclusion

In conclusion, we reported two techniques to prepare metallic tungsten bronze nanowires in this paper, and a VLS mechanism was proposed to explain the growth of these nanowires. KI had an important role in these methods. KI particles served as a catalyst, reductant and source of potassium, and their size controlled the diameter of the nanowires. Preparing KI particles with a very small diameter and depositing them in a controlled fashion is the key to prepare nanodevice-usable tungsten bronze nanowires.

Acknowledgements

This work is in part supported by a startup fund from Duke University and Grant #49620-02-1-0188 from AFOSR. The authors thank Professor Gleb Finkelstein at duke for help in the SEM characterization and Eric Harley at UNC Chapel Hill for the help with Raman spectra.

References

[1] T. Ruecks, K. Kim, E. Joselevich, G. Y. Tseng, C. Cheung, C. M. Lieber, *Science* **2000**, *289*, 94.

[2] Z. Pan, Z. Dai, Z. Wang, *Science* **2001**, *291*, 1947.

[3] J. H. Song, B. Messer, Y. Wu, *J. Am. Chem. Soc.* **2001**, *123*, 9714.

[4] Y. Huang, X. Duan, Y. Cui, C. M. Liber, *Nano. Lett.* **2002**, *2*, 101

[5] X. Duan, Y. Huang, R. Agarwal, C. M. Liber, *Nature* **2003**, *421*, 241

[6] C. M. Liber, *Nano. Lett.* **2002**, *2*, 81

[7] X. Duan, Y. Huang, Y. Cui, J. Wang, C. M. Liber, *Nature* **2001**, *409*, 66

[8] Y. Huang, X. Duan, Q. Wei, C. M. Liber, *Science* **2001**, *291*, 630

[9] Y. Huang, X. Duan, Y. Cui, L. J. Lauhon, K. Kim, C. M. Liber, *Science* **2001**, *294*, 1313

[10] Y. Cui, C. M. Liber, *Science* **2001**, *291*, 851

[11] R. D. Piner, J. Zhu, F. Xu, S. H. Hong, C. A. Mirkin, *Science* **1999**, *283*, 661

[12] S. H. Hong, C. A. Mirkin, *Science* **2000**, *288*, 1808

[13] C. G. Granquist, *Hankbook of Inorganic Electronicchromic Materials,* Elsevier, Amsterdam **1995**

[14] S. K. Deb, Phil. Mag. **1973**, *22*, 801

[15] E. M. MaCarron, *J. Chem. Soc. Chem. Commun.* **1986**, 336

[16] J. I. Jeong, J. H. Hong, J. H. Moon, J. S. Kang, Y. Fukuda, *J. Appl. Phys.* **1999**, *79*, 9343

[17] S.-H. Lee, H. M. Cheong, E. C. Tracy, A. Mascarenhas, A. W. Czanderna, S. K. Deb, *Appl. Phys. Lett.* **1999**, *75*, 1541

[18] S.-H. Lee, H. M. Cheong, J.-G. Zhang, A. Mascarenhas, D. K. Benson, S. K. Deb, *Appl. Phys. Lett.* **1999**, *74*, 242

[19] H. Qiu, Y. F. Lu, Z. H. Mai, *J. Appl. Phys.* **2002**, *91*, 440

[20] I. Turyan, U. O. Krasovec, B. Orel, *Adv. Mater.* **2000**, *12*, 330

[21] G. Gu, B. Zheng, W. Q. Han, S. Roth, J. Liu, *Nano Lett.* **2002**, *2*, 849.

[22] K. Bange, T. Gambke, *Adv. Mater.* **1990**, *2*, 10

[23] S.-H. Lee, H. M. Cheong, E. C. Tracy, A. Mascarenhas, A. W. Czanderna, S. K. Deb, *Appl. Phys. Lett.* **1999**, *75*, 1541

[24] E. Salje, *Acta Cryst.* **1975,** *A 31*, 360

[25] R. S. Wagner, W. C. Ellis, *Appl. Phys. Lett.* 1964, *4*, 89

[26] Z. Pan, Z. Dai, C. Ma, and Z. L. Wang, *J. Am. Chem. Soc.* 2002, *124*, 1817.

[27] B. Vasudeva, J. Gopalakrishnan, *J. Chem. Soc., Chem. Commun.,* 1986, 1644

[28] L. E. Conroy, G. Podolsky, *Inorg. Chem.* 1968, *7*, 614

Mat. Res. Soc. Symp. Proc. Vol. 789 © 2004 Materials Research Society N3.5

New Zinc and Cadmium Chalcogenide Structured Nanoparticles

S. M. Daniels,[a] P. O'Brien[a]* N. L. Pickett[a] and J. M. Smith[b]

[a]Department of Chemistry and The Manchester Materials Science Centre, University of Manchester, UK. *E-mail:* paul.obrien@man.ac.uk

[b]School of Engineering and Physical Sciences, Heriot-Watt University, Edinburgh, UK.

Abstract

The growth of 2D quantum dot quantum well (QDQW) nanocrystals in which a shell of CdSe is grown onto cores of ZnS and capped with a further shell of ZnS is reported. The red shift in the interband absorption and photoluminescence spectrum of the quantum dots (QDs) indicates relocalization of carriers from confinement in the ZnS core to the CdSe shell. The change in interband absorption energy utilizing the effective mass approximation with spherical symmetry was modeled, enabling an estimate of the CdSe thicknesses grown. 1.8nm and 2.5nm ZnS cores were selected as the base on which to grow the CdSe shells. Despite the 12% lattice mismatch between ZnS and CdSe, our results indicate that we have successfully grown CdSe shells approximately three monolayers thick onto 2.5nm ZnS core. Anything beyond a single monolayer of CdSe could not be grown onto the 1.8nm core, although some success was observed by incorporating a CdS graded layer in-between the ZnS core and CdSe shell. The effect of ZnS shell thickness on photoluminescence efficiency has also been studied with optimum shell thicknesses showing quantum yields as high as 52%. Growth of these nanocrystals represents a significant step in the development of strained nanocrystalline heterostructures.

Introduction

Compound semiconductor nanoparticles have been the subject of intense research over the last ten years, generated by their novel optical and electronic properties. The trend in nanomaterial synthesis, amongst others, has mainly been confined to II-VI semiconductor materials known as quantum dots where quantum-confinement effects are large and in the visible part of the spectrum. They are fundamentally important because they possess properties found in-between that of molecular and bulk states of matter. Greater understanding of the physical properties of these materials can be linked to developments in solvothermal preparative routes that allow monodisperse quantum dots to be easily prepared with high optical quality and a high degree of crystallinity.[1]

Zero-dimensional (0D) and one-dimensional (1D) structures have been grown and studied extensively. Colloidal 2D systems, however, are rare (thin films on the other hand have been relatively well studied). Most of the properties of 2D semiconductors have yet to be studied. The first and hence most studied system were of CdS/HgS/CdS by Eychmüller *et al*,[2-4] grown by the substitution of Cd for Hg on the core surface to deposit one monolayer of HgS. The CdS/HgS/CdS emitted a red band-edge emission originating from the HgS layer. The poor particle quality prevented detailed study of the expected 2D semiconductor properties, such as thickness dependent absorption and emission, from being studied. More recently, Little *et al* have grown ZnS/CdS/ZnS QDQW's using a similar growth technique to that of Eychmüller. The electronic structure of this system has been investigated in much more detail.[5-7] The large lattice mismatch between ZnS and CdSe is large enough that in CdSe/ZnS systems, only a few monolayers of ZnS can be grown before a reduction in quantum yield (QY) is observed, indicative of a breakdown in the ZnS lattice structure. [8] The synthetic methodology for growth of CdSe/ZnS is such that very high quantum yield of 20-50% are routine,[9] with recent reports of 85% quantum yield indicate nearly a complete absence of nonradiative decay mechanisms.[10] While CdSe/ZnS core-shells have some of the highest quantum yields, the effects of carrier escape onto surface states is still apparent from the spectral drift and intermittency effects present in photoluminescence studies on single nanocrystals.[11-14] Such random phenomena represents a severe hindrance to the development of application based on single nanocrystal emitters.

Here we report the growth of QDQWs with ZnS as the core material of two different sizes, 1.8nm and 2.5nm, CdSe as the well material, and ZnS as the capping material. We model the ground state luminescence and the electron/hole wavefunctions using a simple effective mass model. Layer thicknesses are estimated by comparison of the peak absorption wavelengths with the ground state exciton energy.

Experimental details

QDQW's are synthesized by the thermolysis of precursors of the type $Li_4[M_{10}E_4(SPh)_{16}]$,[15] where M=Cd/Zn and E=Se/S, in the coordinating solvent oleic acid. The precursors were added at temperatures ranging from 180 °C to 200 °C. The dots were isolated at each individual step with a chloroform/methanol mix (ratio 1:5) and collected as dry powders.

Synthesis of ZnS cores

Two sizes of ZnS cores were grown, 1.8nm and 2.5nm. To grow 1.8nm cores, 75ml oleic acid was degassed at 100°C for 30mins before being put under nitrogen and then heating to 200°C. 0.3175g (0.124mmol) $Li_4[Zn_{10}S_4(SPh)_{16}]$ was added and the solution heated to 250°C for approximately 30mins. The UV-Vis spectrum was monitored constantly and the reaction stopped by cooling down when the desired core size was reached. 2.5nm ZnS cores were grown in a similar way except the temperature was raised to 300 °C to grow the particles.

Synthesis of ZnS/CdSe composite quantum dots

The required size of ZnS cores made as detailed above was redispersed in 3-4ml chloroform. 75ml oleic acid was degassed at 100°C for 30mins before cooling to 30°C and injecting the ZnS cores. The solution was then put under vacuum to remove the chloroform and once complete was put under a nitrogen atmosphere. The solution was heated to 180°C and 0.08g (0.0249mmol) $Li_4[Cd_{10}Se_4(SPh)_{16}]$ was added slowly before slowly injecting 0.5ml TOPSe (0.5M soln). The solution was then heated to 270°C and monitored until the emission wavelength reached between 560-580nm. The solution was then cooled to room temperature and isolated using a mixture of chloroform and methanol. The precipitate was centrifuged, redispersed in chloroform and isolated once again with methanol. The dry powder was kept under nitrogen.

Synthesis of ZnS/CdSe/ZnS

The ZnS/CdSe made above was redispersed in 3-4ml chloroform. 75ml oleic acid containing 0.5g (2.7mmol) zinc acetate was degassed at 100°C for 30mins before cooling to 30°C and injecting the ZnS/CdSe composite dots. The solution was then put under vacuum to remove the chloroform and once complete was put under nitrogen then the solution was heated to 200°C. ZnS cladding layers of various thicknesses were grown by adding subsequent injections of a dilute solution of elemental sulfur (0.087g, 2.7mmol) dissolved in 10ml octadecene.

Growth of ZnS:CdS/CdSe/ZnS, ZnS/CdSe:CdS/ZnS, ZnS:CdS/CdSe:CdS/ZnS

Modifications of the above ZnS/CdSe/ZnS QDQW can be done by incorporating a CdS graded layer either between the ZnS core and CdSe shell, between the CdSe shell and ZnS cladding or both. Generally this is done by adding $Li_4[Cd_{10}S_4(SPh)_{16}]$ along with the precursor needed to grow the next layer in the ratio 1:8. For example, to grow ZnS:CdS/CdSe/ZnS the above experimental detail would be followed except 0.0095g $Li_4[Cd_{10}S_4(SPh)_{16}]$ would be added at the same time as $Li_4[Cd_{10}Se_4(SPh)_{16}]$.

Results and Discussion

Synthesis of ZnS Core

Two sizes of ZnS cores were grown. Both sizes were characterized by TEM and UV-Vis to grow the required size. Figure 1 shows the PL and UV-Vis spectrum of 2.5nm ZnS cores. The large peak in the photoluminescence spectrum at 385nm and tailing into the visible region comes from surface state traps. This gives a blue appearance and is quite common in ZnS nanocrystals.

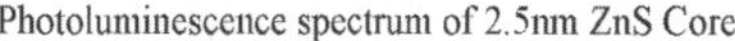

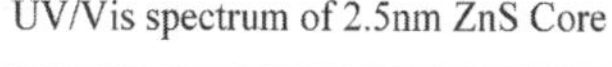

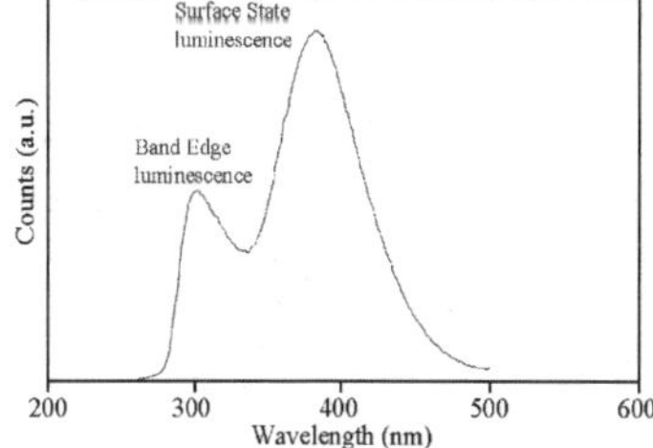

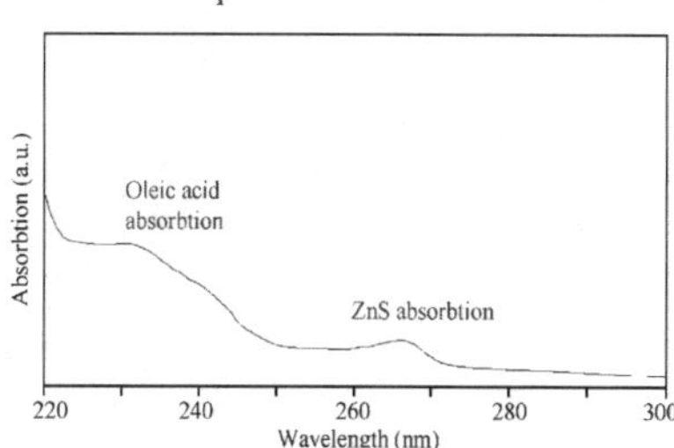

Figure 1 PL (left) and UV-Vis spectra (right) of 2.5nm ZnS core

Synthesis of ZnS/CdSe composite quantum dots.

Initially the particles were characterized by photoluminescence studies. If separate CdSe particles are formed as well as the desired ZnS/CdSe dots then at least two peaks would be seem in the PL. One arising from CdSe and the other from ZnS/CdSe. If the two emitting species happen to be close in wavelength, this could be mistaken as one peak but this is somewhat unlikely. Close wavelengths usually appear as broad odd shape PL spectra as seen in Figure 2 and can be resolved into two different peaks by exciting at different wavelengths. The large lattice mismatch between

ZnS and CdSe of 12% prevents anything beyond a monolayer of CdSe being grown onto the ZnS core. Attempts to increase the CdSe shell thickness by modifying the reaction conditions (growth temperature, injection temperature, heating rate, addition rate, addition amount etc) all proved unsuccessful.

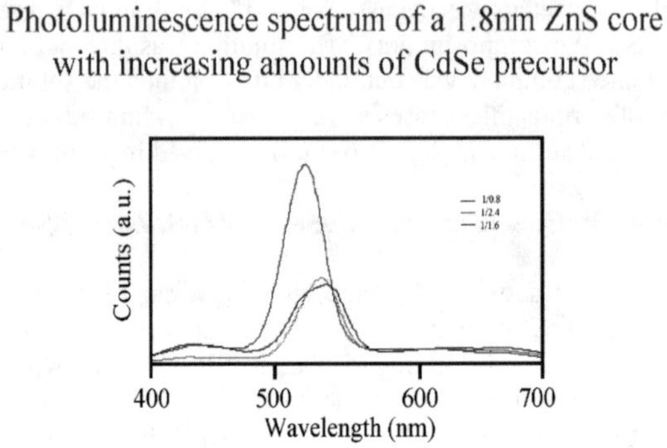

Photoluminescence spectrum of a 1.8nm ZnS core
with increasing amounts of CdSe precursor
Counts (a.u.)
1/0.8
1/2.4
1/1.6
400
500
600
700
Wavelength (nm)

Figure 2 PL of ZnS/CdSe with increasing CdSe precursor concentration

Some success was observed by trying to 'bridge' the lattice mismatch between ZnS and CdSe by incorporating a graded layer of CdS. Figure 3 shows the luminescence intensities of ZnS/CdSe and ZnS:CdS/CdSe, an improvement in excess of 15% was seen in the quantum yield when a graded CdS layer is used. This improvement in the quantum yield was only observed for a single monolayer of CdSe. Thicker CdSe layers, even when incorporating CdS couldn't be grown.

Photoluminescence spectrum of a 1.8nm ZnS core
with a graded CdS layer and increasing amounts of CdSe precursor
Counts (a.u.)
1/0.8
1/1.2
400
500
600
700
Wavelength (nm)

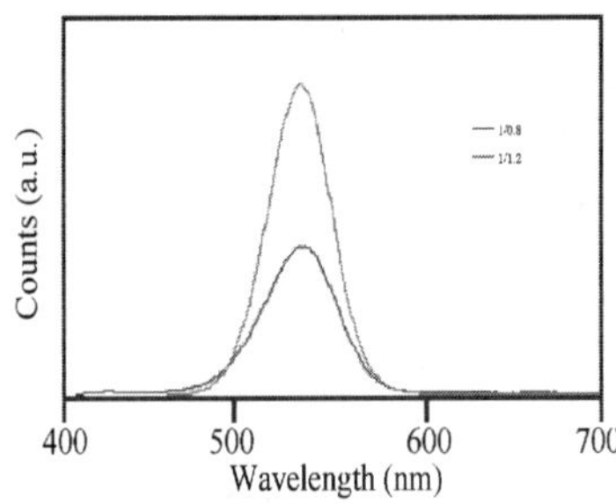

Figure 3 PL of ZnS:CdS/CdSe with increasing CdSe concentration

In an attempt to lower the lattice strain at the interface between ZnS and CdSe, a 2.5nm ZnS core was grown. The larger core has a lower surface curvature and therefore will possess lower surface energy compared to the smaller core. For these reasons a thicker CdSe shell was stable enough to be grown as can be seen in Figure 4. There is some evidence however of separately nucleated CdSe particles indicated by an 'edge' on the left hand side of the PL peak. TEM studies agreed with this observation showing the presence of two clearly different sized nanoparticles.

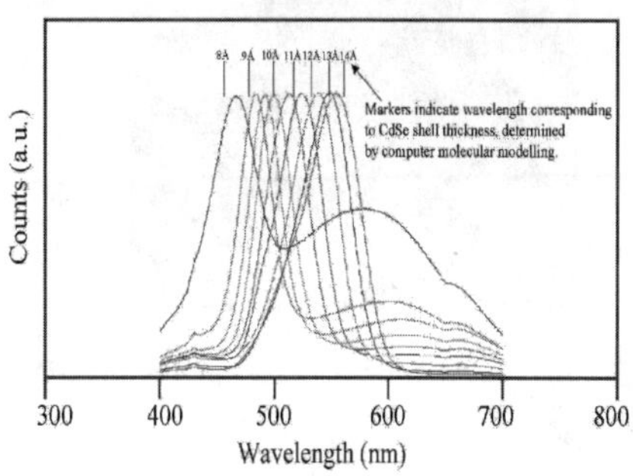

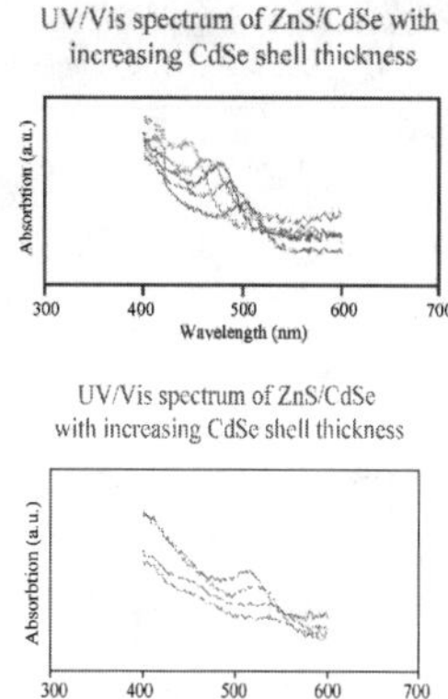

Figure 4 PL of ZnS/CdSe and right UV-Vis of ZnS/CdSe

Synthesis of ZnS/CdSe/ZnS QDQW's

The outer ZnS cladding shell was grown by adding elemental sulfur dissolved in octadecene. Figure 5 shows the trend in quantum yield as more sulfur is injected. The initial ZnS/CdSe quantum yield before any added sulfur is 20%. Upon addition of a small quantity of sulfur the quantum yield increases to 24% before appearing to decrease to the original value of 20%. Further addition of sulfur causes the quantum yield to increase to a maximum of 34% before steadily declining to a near constant 3%.

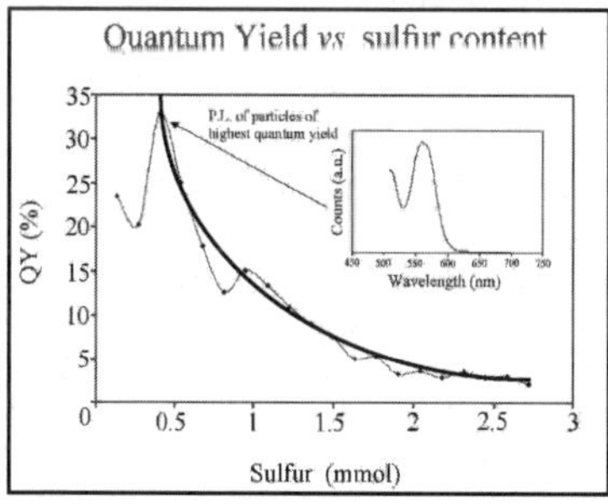

Figure 5 QY vs. sulfur added of ZnS/CdSe/ZnS

Very recently it has been found that by the slow initial addition of the CdSe cluster precursor, followed by rapid heating, ZnS/CdSe particles free from separately nucleated CdSe particles can be formed. The PL UV-Vis spectrum and TEM of the resulting nanocrystals are shown in Figure 6.

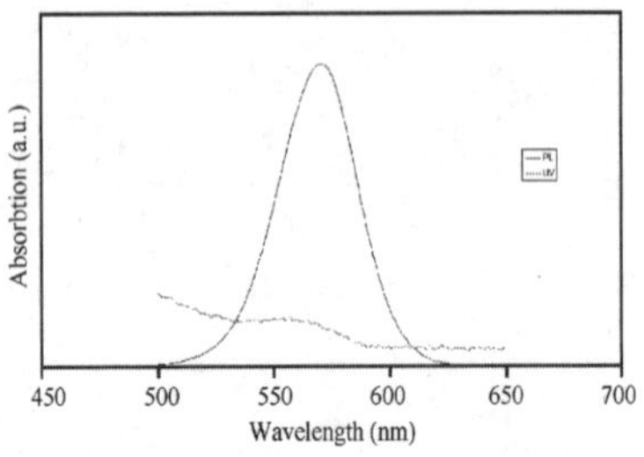

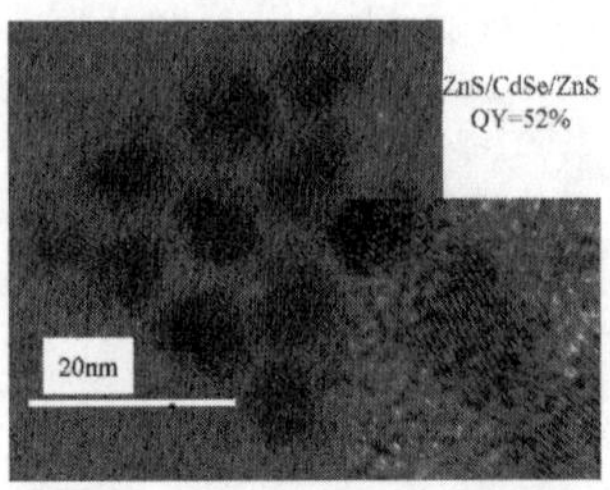

Figure 6 PL and UV-Vis (left) of CdSe free ZnS/CdSe/ZnS and the corresponding TEM (right)

Conclusion

In conclusion it has been shown that the thermolysis of precursors of the type $Li_4[M_{10}E_4(SPh)_{16}]$ have been used to grow ZnS/CdSe/ZnS QDQW's. Although the large lattice mismatch between ZnS and CdSe make the synthesis of ZnS/CdSe/ZnS QDQW's very difficult, we report the first growth of such a system. The size of the ZnS core seems to be an important factor in getting several monolayers of a stable CdSe layer to grow epitaxially. Another problem often encountered was the separate nucleation and subsequent growth of CdSe nanoparticles that required carefully controlled conditions to prevent, this has only recently been solved. The resulting nanocrystals have yet to be fully characterized.

References

1. C. B. Murray, D. J. Norris, M. G. Bawendi, *J. Am. Chem. Soc.,* **115**, 8706, (1993)
2. A. Eychmüller, A. Mews and H. Weller, *Chem. Phys. Lett.,* **208**, 59 (1993).
3. A. Mews A. Eychmüller, M. Giersig, D. Schooss, H. Weller, *J. Phys. Chem.,* **98**, 934, (1994)
4. A. Mews, A. V. Kadavanich, U. Banin, A. P. Alivisatos, *Phys. Rev. B.,* **53**, R13 242, (1996)
5. G. W Bryant, *Phys. Rev. B.,* **52**, R16 997 (1995).
6. W. Jaskolski and G. W. Bryant, *Phys. Rev. B.,* **57**, R4237 (1998).
7. R.-H. Xie, G. W. Bryant, S. Lee and W. Jaskolski, *Phys. Rev. B.,* **65**, 235306 (2002).
8. B. O. Dabbousi, J. Rodriguez-Viejo, F. V. Mikulec, J. R. Heine, H. Mattoussi, R. Ober, K. F. Jensen and M. G. Bawendi, J. Phys. Chem. B., **101**, 9463 (1997).
9. M. A. Hines and P. Guyot- Sionnest, J. Phys. Chem., **100**, 468 (1996).
10. L. Qu and X. Peng, J. Am. Chem. Soc. **124**, 2049 (2002).
11. S.A. Crooker, T. Barrick, J. A. Hollingsworth and V. I. Klimov., *App. Phys. Lett.,* **82**, 279 (2003)
12. K. T. Shimizu, R. G. Neuhauser, C. A. Leatherdale, S. A. Empedocles, W. K. Woo and M. G. Bawendi., *Phys. Rev. B,* **63**, 205316 (2001)
13. R. G. Neuhauser, K. T. Shimizu, W. K. Woo, S. A. Empedocles and M. G. Bawendi., *Phys. Rev. Lett.,* **85**, 3301 (2000).
14. M. Kuno, D. P. Fromm, H. F. Hamann, A. Gallagher and D. J. Nesbitt, J. Chem Phys., **112** 3117 (2000).
15. Ian G. Dance, Anna Choy and Marcia L. Scudder, J. Am. Chem. Soc., **106**, 6285 (1984).

Mat. Res. Soc. Symp. Proc. Vol. 789 © 2004 Materials Research Society N3.6

Quantum Calculations of Carbon Nanotube Charging and Capacitance

Pawel Pomorski[1], Lars Pastewka[1], Christopher Roland[1], Hong Guo[2] and Jian Wang[3]
[1] Department of Physics, NC State University, Raleigh, NC USA 27695-8202.
[2] Department of Physics, McGill University, Montreal PQ, Canada H3A 2T8
[3] Department of Physics, The University of Hong Kong, Pokfulam Road, Hong Kong

ABSTRACT

Although it has long been known that the classical notions of capacitance are altered at the nanoscale, few first principles calculations of these properties exist for real material systems. With a recently developed *ab initio* formalism, which combines nonequilibrium Greens function techniques with real-space density functional calculations, we have investigated charging effects for carbon nanotube systems, which are described by the capacitance coefficients. Specifically, the capacitance matrix of two nested nanotube armchair nanotubes, the insertion of one nanotube into another, and the properties of a nanotube acting as a probe over a flat aluminum surface were considered.

INTRODUCTION

Within classical electrostatics, the ability of a conductor to store charge is quantified by its capacitance. Essentially, the classical capacitance coefficients $C_{\alpha\beta}$ for a set of conductors gives us the charge accumulation Q_α on conductor α in response to a change in the electrostatic potential V_β on conductor β via $Q_\alpha=\Sigma_\beta C_{\alpha\beta}V_\beta$. Critical to the the notion of capacitance is that of a set of well-defined, equipotential conductors, with zero electric field in their bulk. This assumption typically breaks down at the nanoscale, where the screening length of the material becomes comparable to the dimensions of the systems [1]. In this case, there is only a finite density of states available to the system and conductors may no longer be equipotential surfaces. Hence, the classical notion of capacitance must be generalized to that of the electrochemical capacitance, where each conductor is connected to an electron reservoir with electrochemical potential μ [2]. The self-consistent charge variation dQ_α on conductor α in response to a variation in the electrochemical potential $d\mu_\beta$ in reservoir β is then given by:

$$dQ_\alpha = \sum_\beta C_{\alpha\beta}(d\mu_\beta / e).$$

These capacitance coefficients now take quantum effects into account, and may differ considerably from their classical counterparts. Note that the self-consistent induced rearrangement of charge in response to a potential change is the key quantity that needs to be measured.

Although the theory of quantum capacitance is almost a decade old [2], there have been few first principles calculation of these properties for real material systems. However, the recent advent of molecular electronic systems has given new urgency towards understanding these kinds of problems. As a example, we therefore have calculated the quantum capacitance of prototypical carbon nanotube systems [3]. We have focused on carbon nanotube systems because of the very prominant role that this material plays in field of nanotechnology [4]. Depending on their helicity, carbon nanotubes are either one-dimensional metals or semiconductors, which makes them an ideal system for exploring quantum transport at the nanometer length scale. Indeed, a number of prototypical nanotube-based devices have already been produced in the laboratory and their properties explored both theoretically and experimentally. To date, most of the theoretical investigations of these systems have focused on the conductance and their current-voltage characteristics. Here, we investigate the induced rearrangement of charge and the capacitance coefficients. A good understanding of these properties is of course important both from a fundamental and technological viewpoint. Capacitance properties are central to the workings of nanotubes as scanning probes, memory devices, the dynamic response of nanotubes, and their ability to store and retain charges.

METHODOLOGY

The numerical calculations are based on a recently developed *ab initio* formalism [5], which combines the Keldysh nonequilibrium Greens function theory with the power of real-space density functional simulations. As the details of this method are somewhat technical, and have already been outlined extensively elsewhere, we restrict ourselves here a short discussion of the most pertinent details only. Roughly speaking, the main advantages of this approach include: (i) the proper treatment of the open-boundary conditions for a quantum mechanical system under a finite bias voltage; (ii) a fully atomistic treatment of the electrodes; (iii) a self-consistent calculation of the charge density via the nonequilibrium Greens functions that includes both the effects of the scattering and bound states of the system; and (iv) the extensive use of real-space grids which enables parallelization and the treatment of large systems.

Although the code has generally been used to calculate the I-V characteristics of two-probe devicesn [5], it is the self-consistent charge density that is the most important quantity here. Essentially, the electronic states are modeled using a linear combination of atomic orbitals (LCAO) method with s,p, and d orbitals [5]. With nonequilibrium Greens function methods, one is then able to calculate the change in the charge $\Delta Q = Q(V+\Delta V)-Q(V)$, as a finite bias $\Delta V = ed\mu$ is applied to the reservoirs. Another feature is that we have used Dirichlet boundary conditions for the electrostatic potential at the walls of our finite-sized computational box [6]. This corresponds to the situation of surrounding the system with a metal container. This serves to terminate any field-lines that otherwise may escape from the nanotubes. When the box becomes infinitely large, the results reduce to that of a system in free space. A further advantage of this approach is that it allows for the treatment of charged nanotube systems. Since we will be dealing primarily with two-probe systems, the main quantities to be calculated are:

$$\begin{aligned}\Delta Q_1 &= C_{11}\Delta V_1 + C_{12}\Delta V_2 \\ \Delta Q_2 &= C_{21}\Delta V_1 + C_{22}\Delta V_2,\end{aligned}$$

with coefficients $C_{\alpha\alpha}$ representing the "self-charging" and $C_{\alpha\beta}$ the mutual-charging terms between the conductors, respectively.

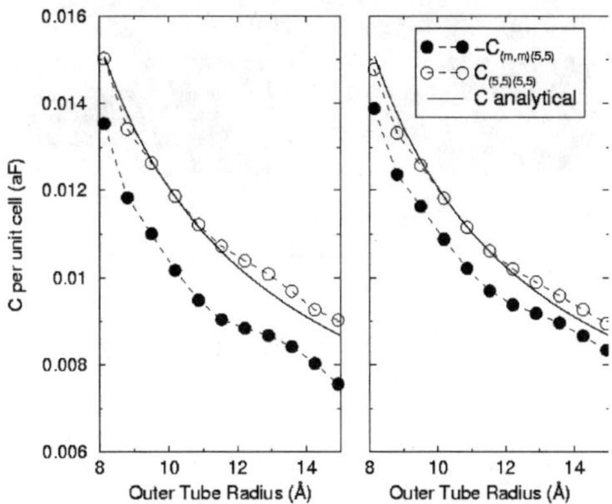

Figure 1: Capacitance versus tube radius for a (5,5) shell inside (m,m) tubes. The left (right) panel shows data for 40x40 Å (80x80 Å) metal container enclosing the respective system. The analytical result obtained is marked by a solid line.

RESULTS AND DISCUSSIONS

As a first example, we calculated the capacitance per unit length of two nested, armchair nanotubes, which corresponds to the classical case of a concentric, cylindrical capacitator. Such a system is reminiscent of multiwalled nanotubes, and has been realized experimentally [7], or alternatively may be constructed by attaching metal leads to two nanotubes [8]. Specifically, we focused on the case of (5,5) nanotube (conductor 1) inside larger (m,m) nanotubes (conductor 2), with m ranging from 12-22. Note that for the smallest (12,12) tube, the closest distance between atom in the inner and outer tubes is 9.1 Å, so that there is negligible overlap between the atomic orbitals on the different nanotubes. Hence, there is no current flow between the tubes. The tubes are, however, coupled by means of the electrostatic potential. The simulations were carried out in a periodic configuration, with one unit cell of the (5,5) tube with 40 atoms and 216 atoms for the largest (22,22) nanotube. Box sizes of (40x40x4.8) Å and (80x80x4.8) Å were considered. Each calculation used 200 k-points in the sampling. These systems are relatively large, and hence the simulations were carried out on 8 processors of a small Linux Beowulf cluster. The calculated capacitance response for this system – when they coupled to separate reservoirs, is linear to better than 1% up to about 5 V, so that a well-defined voltage-independent capacitance is obtained. This reflects the large energy window of a constant density of states around the Fermi level characteristic of armchair nanotubes.

For the (5,5)/(12,12) system, the calculated capacitance matrix coefficients are: $C_{11} = C_{(5,5)(5,5)} = 0.0150$ aF; $C_{21} = C_{(12,12)(5,5)} = -0.0135$ aF; $C_{12} = C_{(5,5)(12,12)} = -0.0134$ aF; and $C_{22} = C_{(12,12)(12,12)} = 0.0243$ aF. If the metal box surrounding the system were to be infinitely large, all four capacitance coefficients would have equal magnitude,. The value of $C_{(12,12)(12,12)}$ is expected to be larger because of the increased interaction of the outside

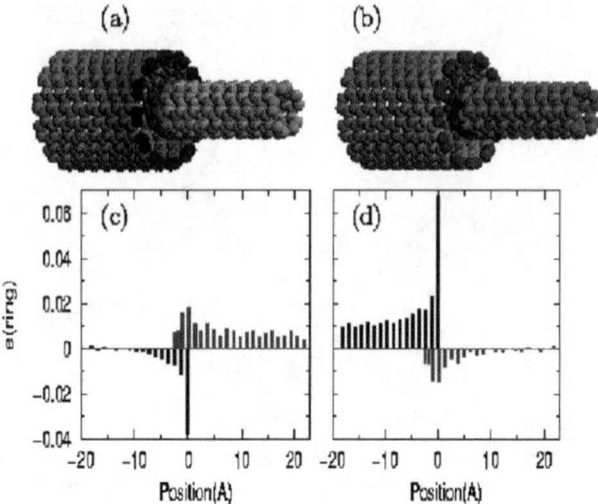

Figure 2: Tube charging of the (12,12)/(5,5) biased junction. Here, a +0.27V bias is applied to the right (5,5) tube in (a,c), and to the left (12,12) in (b,d). Note that large accumulation of charge on the outer ring of the (12,12) tubes, which is characterized by dangling bonds, and that the tube to which the bias voltage is applied acquires charge all along its length.

tube with the metal box. Thus, when the bias voltage is applied to the outside nanotube, the inner (5,5) tube holds only about 55% of the induced charge; the rest is on the metal box surrounding the system. By contrast, when the charge is applied to the inner tube, the outside (5,5) tube holds up to 89% of the induced charge, showing that the tubes act as efficient screens. Note also that the mutual charging coefficients agree with each other to within 1%, which is an important check of the goodness of the calculation. The capacitance coefficients for the different box sizes are shown in Figure 1. As the metal container gets larger, the outer tube screens more of the charge on the inner tube, so that the numerical value of the two becomes closer. As expected, the numerical values of the $C_{(5,5)(5,5)}$ coefficient changes only slightly, giving us an estimated asymptotic value of 0.015 aF per unit length.

It is interesting to probe for the classical limit for this nanotube system. Following the Buttiker formalism [2,3], we have modeled the capacitance of the system in terms of two concentric cylindrical tubes of length l carrying uniform sheets of two-dimensional charge. Within this model, all capacitance coefficients C are equal and are given by:

$$\frac{2\pi\varepsilon_o l}{C} = \log\left(\frac{R_2}{R_1}\right) + \frac{\lambda_1}{R_1} + \frac{\lambda_2}{R_2},$$

where $R_{1,2}$ are the radii of the inner and outer tubes and $\lambda_{1,2}=4\pi e^2(d\sigma/dE)$ is the screening length, and $d\sigma/dE$ the density of states per unit area of the cylindrical tubes. These parameters may all be estimated from our *ab initio* calculations. As shown in Figure 1, the agreement between our *ab initio* data and this simple model is quite good. Note also that the quantum correction terms – the λ/R terms – are comparable in magnitude to the

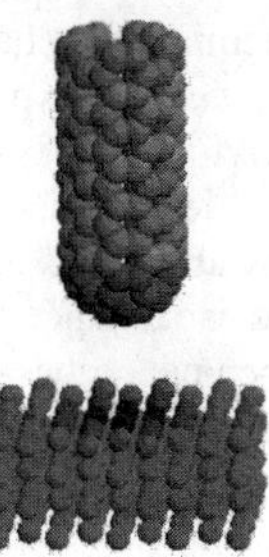

Figure 3: Accumulated charge on a (9x9) Al electrode and the capped (5,5) nanotube when a positive bias is applied to the nanotube. Note that the charge on the electrode in response to the positive charging of the nanotube is concentrated in the middle of the top electrode layer.

classical logarithmic term. Classical behavior is expected for $R_1 >> 1.35R_2$, *i.e.,* when the first term of the above equation dominates. Such values are never reached for our largest (22,22)/(5,5) system, and serves to emphasize the importance of keeping the quantum corrections when considering nanoscale capacitance.

As a second example, we briefly consider the insertion of a (5,5) capped nanotube into an open (12,12) tube with the center axis of the two tubes coinciding. Figure 2 shows plots of charging of the tubes, along with a histogram of the charge accumulated on each ring. In particular, note that the (12,12) tube acquires a very large amount of charge on its terminal ring, which is due to the presence of the dangling bonds on the open tube. Figure 2 also shows that the tube to which the bias voltage is applied acquires charge along its entire length. This "self-charging" is due to the capacitive coupling between the nanotube and the surrounding metal container. Again, if one could consider an infinite-sized box, this feature would disappear, and all capacitance coefficients would acquire an equal value. We have also measured the capacitance per unit length as more and more of the (5,5) nanotube is inserted into the (12,12) tube. Here, we calculate a capacitance of 0.012 aF, which is in good agreement with the previous results.

As a final example, we consider the efficacy of nanotubes as future capacitance probes for which first experiments are starting to emerge [9]. As a prototypical example, we have therefore looked at the capacitance properties of a capped (5,5) nanotube over a flat (100) Al surface, as shown in Figure 3. To this end, we calculated the charging coefficients of the nanotube as a function of the distance between the nanotube tip and the Al surface. As a function of this distance (d), all of our data is well-described by the fitted formula:

$$C_{(\infty,\infty)(5,5)}(d) = (\frac{0.197}{d} - 0.0069)aF\ .$$

This formula is only valid for d greater than about 5Å, so that no current flows between the nanotube and the infinitely-large Al surface. Inserting typical distances into the above formula, one can readily see that these capacitance coefficients are actually quite low.

This reflects the fact that most of the accumulated charge is localized near the nanotube tips, as is evident from Figure 3. Hence, we have explored ways of increasing the capacitive response of the tubes. One obvious way is to actually functionalize the nanotube tip with metallic or organic species. Placing a single Al atom outside the cluster was found to increase the capacitance by about 14%, indicating that when small metal clusters are placed outside the tubes, the capacitance is likely to increase substantially. By constrast, placing atoms inside the capped nanotubes did not lead to any significant changes in the capacitance, reflecting the screening of these species by the nanotube walls.

SUMMARY

In summary, we have computed the finite-bias capacitance coefficients of prototypical carbon nanotube systems with a recently developed transport formalism that combines the power of the nonequilibrium Greens function approach with the utility of density functional simulations. Here, we considered the capacitance of two-shelled nanotube systems, the insertion of a capped nanotube into an open nanotube, and the use of the a nanotube as a capacitance probe. While the capacitance for these systems is generally quite low (aF range), its value may be boosted by nesting and/or inserting a nanotubes into each other, or by means of functionalizing the ends of nanotubes with small metal clusters.

ACKNOWLEDGEMENTS

This work has been supported both by NSF and DOE. We also thank NCSA for extensive computer support.

REFERENCES

[1.] T.P. Smith, B.B. Goldberg, P.J. Stiles, and M. Heiblum, Phys. Rev. B **32**, 2696 (1985); T.P. Smith, W.J. Wang, and P.J. Stiles, Phys. Rev. B **34**, 2995 (1986).
[2.] M. Buttiker, H. Thomas, and A. Petre, Phys. Lett. A **100,** 364 (1993); M. Buttiker, J. Cond. Matter **5**, 9361 (1993); T. Christen and M. Buttiker, Phys. Rev. Lett. **77**, 143 (1996).
[3.] P. Pomorski, C. Roland, H. Guo, and J. Wang, Phys. Rev. B **67**, RC16140 (2003); P. Pomorski, L. Pastewka, C. Roland, H. Guo, and J. Wang, unpublished.
[4.] For a recent review on nanotubes, see for example: J. Bernholc, D. Brenner, M. Buongiorno Nardelli, V. Meunier, and C. Roland, Annu. Rev. Mater. Res. **32**, 347 (2002).
[5.] J. Taylor, H. Guo, and J. Wang, Phys. Rev. B **63**, RC121104 (2001); *ibid* 245407 (2001).
[6.] R. Tanura, Phys. Rev. B **64**, 201404 (2001).
[7.] J. Cumings, and A. Zettl, Science **289**, 602 (2000).
[8.] See for example: T. Nakanishi et al, Phys. Rev. B **66**, 073307 (2002); P. Poncharal et al, J. Phys Chem. B **106**, 12104 (2002).
[9.] See, for example: S.V. Kalinin, D.A. Bonnell, M. Freitag, and A.T. Johnsom, Appl. Phys. Lett. **81**, 754 (2002); *ibid* **81**, 5219 (2002).

Mat. Res. Soc. Symp. Proc. Vol. 789 © 2004 Materials Research Society N3.9

Observation of Size Confinement Effects of Excitons in AgX Nanocrystals by Cryo-Energy-Filtering TEM/EELS

Vladimir P. Oleshko

University of Virginia, Department of Materials Science & Engineering, Charlottesville, VA 22904-4745, USA

ABSTRACT

The non-uniform size-dependent contrast of $AgBr_{0.95}I_{0.05}$ nanocrystals (NCs) ranging from 22 to 80 nm in equivalent diameter (d_c) observed by cryo-energy-filtering TEM is referred to predominant excitations at the surfaces and near the edges. When the fields due to surface losses reach throughout the structure, they couple and the probability for their generation becomes periodic in the NC size. Since electronic sum rules must be satisfied, the surface excitations reduce the strength of the bulk excitations. Coupling of surface and volume losses may cause oscillations of the image intensities with the NC size. The appearance of such oscillations demonstrates a size confinement of excitations of valence electrons due to contributions to the energy-level structure from carrier confinement and surface states. The imaginary part of relative dielectric permittivity derived from electron energy-loss spectra shows an enhanced intensity of the band at 4 eV for NCs with $d_c = 50\pm4$ nm as compared to those of 109 ± 7 nm in size, while the bands at 7 eV and at 10 eV appear to be suppressed. An increase of the intensity of exciton-assisted direct interband transition at 4 eV (Γ_8-, Γ_6- $\rightarrow \Gamma_6$) correlates with the size-dependent enhancement of free exciton luminescence from AgBr NCs, when their size is less than 100 nm.

INTRODUCTION

Considerable complexity in the valence-band structure caused by hybridization of halogen p-states and Ag 4d-states and occurrence of a low energy indirect gap between the valence and conduction bands in AgX (X = Cl, Br, I) lead to unique size restriction effects on the electronic properties of nanocrystalline silver halides, when the particle size decreases below 100 nm [1-4]. The confinement occurs when the spatial extent of a material (boundary conditions) begins to affect the eigenenergies of the electron wave function, causing the electronic properties to differ from those of bulk material. Considering the relaxation properties of excitons in AgBr, size effects may be expected for crystals much larger than the Bohr radius of the indirect exciton ~3 nm. This is because the restricted geometry leads to the occurrence of surface recombination that becomes a dominant relaxation channel in crystals with sizes less than 500 nm [5]. Quantum size effects of excitons in AgX nanocrystals (NCs) have been demonstrated by the appearance of the indirect exciton luminescence forbidden by selection rules in large crystals and by an increase of the quantum yield by several orders of magnitude likely affected by surface states when the particle size is decreased from 80 nm to 7.9 nm [6,7]. A high energy shift of luminescence band maxima of free excitons in the case of Ag(Br,I) NCs with sizes of 40 nm and less was also observed [7,8]. In order to determine the critical size for the onset of free exciton emission in luminescence spectra, a series of NCs with sizes from 50 nm to 120 nm was examined. The indirect excitonic luminescence was observed starting with crystals of 80 nm in cubic edge length and below and its intensity increased as size decreased [7,9]. The enhancement of the free exciton emission is thought to be due to the quantum size effect in conjunction with contribution of the spatial electron and hole confinement and diminishing iodide bound exciton (IBE) emission [5,9]. In the 40 nm average edge size AgBr NCs (ranging from 20 nm to 58 nm) the position of the phonon-assisted emission lines was shifted by 1.8 meV to high energies and the zero-phonon emission was increased in intensity due to band restructuring induced by the size effect [7]. In this range of sizes deviations away from the cubic shape are beginning to occur, with the smaller NCs in the population converging on a spherical shape [5,7,9,10].

For a better understanding of unusual photophysical properties of AgX NCs, it is necessary t evaluate their electronic structure in the size range where deviations from the bulk crysta selection rules occur. Due to an opportunity to probe electron excitations caused by inelasti scattering with high spatial (0.1-1.0 nm) and spectral (0.1-1.5 eV) resolution, energy-filterin transmission electron microscopy and electron energy-loss spectroscopy (EFTEM/EELS) ma give new insights into the AgX nanocrystalline matter and provide information on the ban structure, bonding, electron densities and dielectric response. Tuning energy loss has bee employed to observe exciton states in Ag(Br,I) NCs [10-12]. The low-loss fine structure in EE spectra of a tabular Ag(Br,I) microcrystal (a AgBr core) was found to be in fair agreement wit first principles quantum mechanical calculations by the linear muffin-tin orbital method in it atom sphere approximation (LMTO-ASA) [4] that is capable of describing the exciton interban transitions. This paper reports size confinement effects on excitations of valence 4d (Ag^+) and 4 (Br^-) electrons in small (3-5 particles) arrays and in individual Ag(Br,I) NCs investigated b cryo-EFTEM/EELS.

EXPERIMENTAL PROCEDURES

Ag(Br,I) NCs containing ~5 mol.% of AgI have been synthesized by a double jet method at pBr= =3.5 and T=+35°C, as described elsewhere [10,11]. Two populations of NCs with d_c = 50±4 nm (**I**) and d_c = 109±7 nm (**II**) had sizes varying from 22 nm to 87 nm and from 34 nm to 185 nm, respectively. For the population **I**, the shape transformed from cubic (0.8%) to preferentially spherical (99.2%), as more stable at the given precipitation conditions. Larger NCs **II** (with a mean edge length of 54±4 nm) exhibited less shape uniformity (truncated cubes, 67.9%, sphere-like particles, 30.8% and a minor fraction of twins, 1.3%). The measured edge lengths were less than estimated on perimeters because cubes were usually truncated. For the population **I**, an effective mean "edge length" estimated from the mean perimeter was 39.5 nm and varied from 18.7 nm to 70.3 nm. All preparations and handling were done under non-actinic light to avoid the formation of printout silver. EFTEM/EELS were performed on a ZEISS CEM902 computerized electron microscope operating at accelerating voltage 80 kV and equipped with an integrated energy filter. The KRAKRO program [13] has been used to calculate the energy-loss function and the relative dielectric permittivity. Radiation damage of the NCs was reduced using a top-entry cryo-stage at T=-193°C supplied with a modified evacuated cooling trap.

RESULTS AND DISCUSSIONS

Cryo-EFTEM/EELS of NCs reveal variations of the electron intensity in the 0-100 eV range of energy losses due to an intersection with a low-loss fine structure (4-18 eV), a bulk plasmon (22-23 eV) and a background decrease and intersection with inner-shell excitations (56-70 eV) [10-12]. The intensity changes lead to contrast reversal when the energy is tuned over the range (Fig. 1), although the Bragg contrast of the NCs is preserved in inelastic scattering processes

[10,11]. Tuning the energy loss allows visualizing local excitations concerned with certain energy losses. At first, electrons hit the crystal directly causing internal ionization with the generation of electron-hole pairs. Swift electrons may also not strike the NC directly, but cause polarization inside it by their Coulomb field. Polarization waves formed by Mott-Wannier excitons (weakly bound electron-hole pairs with a radius exceeding several lattice parameters) are a prominent type of excitation for AgBr [10]. Iodine in AgBr acts as an isoelectronic trap for holes, producing a positively charged center. The latter can be neutralized by binding an electron, thus forming IBEs. Selection of proposed exciton losses at E=16±5 eV (Fig. 1) resulted in a weak rim around the NCs, because such losses can even extend at distances of a few nanometer away from the particle, which correspond satisfactorily to the expected exciton size [1]. No rims were observed in the zero-loss mode. Positioning the energy window at E=25±5 eV enables to visualize volume plasmons created inside the NC [11]. The image at E=50±5 eV demonstrates a size-dependent contrast reversal. Because of background decay, both surface and volume excitations are visible. The excitation probability decreases exponentially with distance from the surface. Fringes around NCs caused by surface excitations can be distinguished at distances of 10-20 nm. The non-uniform contrast and dark areas inside the NCs are referred to predominant excitations at the surfaces and near the edges; this can decrease the probability of volume losses because of the sum rule [14]. When the fields due to surface losses reach throughout the structure, they couple and the probability for their generation becomes periodic in the size of the system [14,15]. Numerical calculations of the bulk loss probability reveal its damped oscillations as a function of the nanoparticle radius [16]. Both plasmons and excitons are considered to be coherent superposition of electron-hole excitations corresponding to transitions from occupied to unoccupied states. Since electronic sum rules must be satisfied [14,15], the surface excitations necessarily reduce the strength of the bulk excitations. Consequently, the volume losses should exhibit similar periodic behavior. Coupling of surface and volume losses can then cause periodic oscillations of their intensities with the NC size and induced charge (Fig. 1, the profile across the NC at 25±5 eV). The lower intensity volume losses are localized within a 5 nm-thick surface layer and inside the NC in two regions (5 nm x 10 nm) surrounded by areas of the enhanced volume losses. The appearance of such oscillations obviously demonstrates a size confinement of

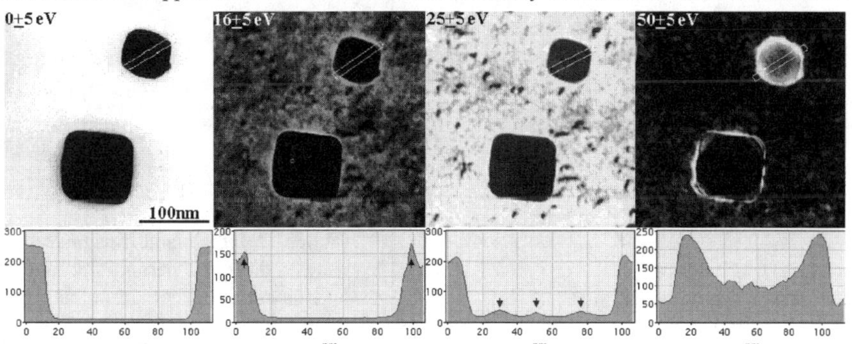

Figure 1. Cryo-EFTEM of Ag(Br,I) NCs with a 10 eV window at 0, 16, 25 and 50 eV, respectively, and intensity profiles across the right upper NC as marked by rectangular boxes, a 30 pixel integration width. Up arrows point to exciton states observed at 16±5 eV, down arrows indicate periodical intensity oscillations due to size confinement observed at the volume plasmon loss.

beam-induced excitations of valence 4d (Ag^+) and 4p (Br^-) electrons due to contributions to the energy-level structure from carrier confinement and surface states.

The single scattering EEL intensity (SSI) for low energy losses expressed as the differential cross-section $d^2\sigma/d\Omega dE$ is related to the imaginary part of the reciprocal of the complex relative permittivity $\varepsilon(\vec{q},E)=\varepsilon_1(\vec{q},E)+i\varepsilon_2(\vec{q},E)$ as a function of wave vector $\vec{q}$ and energy E, thus reflecting the local dielectric response of the media to a longitudinal field [13]:

$$\frac{d^2\sigma}{d\Omega dE} \propto \mathrm{Im}\left[-\frac{1}{\varepsilon(\vec{q},E)}\right]\ln\left(1+\frac{\beta^2}{\theta_E^2}\right), \qquad (1)$$

where β is the collecting semiangle ($\alpha<\beta<(E/E_0)^{1/2}$) and $\theta_E=\frac{E}{2T}(1-v^2/c^2)^{0.5}$ is the characteristic scattering angle; $T=\frac{1}{2}m_0v^2=\frac{E_0(1+E_0/2m_0c^2)}{(1+E_0/2m_0c^2)^2}$; m_0 is the free electron mass; v is the velocity of incident electrons; c is the light velocity and E_0 is the initial energy of electrons. Starting from the SSI, Kramers-Kronig analysis (KKA) enables the energy dependence of $\varepsilon(\vec{q},E)$ to be calculated. In principle, it should be carried out for known momentum transfer between initial and final states. However, it is possible to obtain ε_1 and ε_2 for transitions confined to small momentum transfers only through the KKA [13]. For each EEL spectrum, contributions from the zero-loss peak and due to multiple losses have been subtracted and corrections for the angular weighting were made. Ideally, thin uniform specimens are required that are inaccessible in particular case. Therefore EEL spectra were acquired on closely packed arrays of uniform cubic or sphere-like NCs [10]. The KKA is sensitive to the intensity inputs, because Im ($-1/\varepsilon$) is overwhelmed by the zero-loss peak, which should be subtracted. This may introduce uncertainties at energies <3-4 eV To overcome the problem, a smooth extrapolation of the intensities to zero energy was applied. Curves in Figs. 2a and 2c fall to 0 at E<3 eV taking into account the indirect exciton band gap, E^i_g, at 2.68 eV [4]. ε_1 describes the polarizability of the specimen (local minima at 5, 8, 12-14 eV

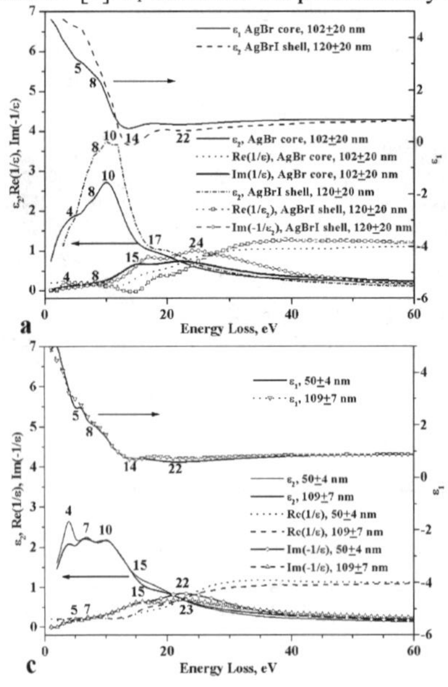

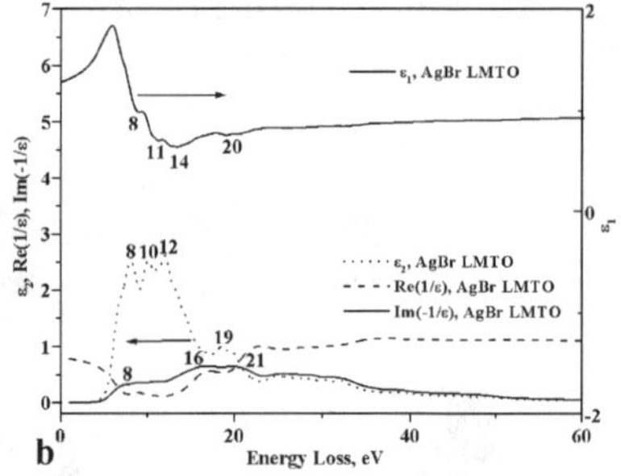

Figure 2. Real and imaginary part of relative dielectric permittivity and energy-loss function: (a) composite core (AgBr) - shell [Ag(Br,I)] tabular microcrystal [4,12]; (b) LMTO-ASA calculations for AgBr [4]; (c) Ag(Br,I) NCs **I** (d_c= 50 ± 4 nm) and **II** (d_c = 109 ± 7 nm), respectively.

and 22 eV), while ε_2 is related to absorption in the specimen (local maxima at 4, 7-8, 10, 15-17 eV). The energy-loss function is expressed as [13]:

$$\mathrm{Im}\left[-\frac{1}{\varepsilon(\vec{q},E)}\right]=\frac{\varepsilon_2}{\varepsilon_1^2+\varepsilon_2^2}=\frac{\omega\Gamma\omega_p^2}{(\omega^2-\omega_p^2)^2+(\omega\Gamma)^2}, \qquad (2)$$

where Γ is the damping constant, ω_p is the plasma resonance frequency. The energy of the volume plasma resonance, $\hbar\omega_p=[(\hbar\omega_p^f)^2+E_g^{i\,2}]^{1/2}$, where $\omega_p^f=[ne^2/(\varepsilon_0 m)]^{1/2}$ is the free electron plasma frequency, n is the electron density, e is the electron charge, ε_0 is the permittivity of vacuum, and m is the electron mass. For $E\leq\hbar\omega_p=21.6$ eV, Im (-1/ε) describes collective excitations against the ionic background caused by oscillations of bound electrons (interband transitions). When ε_1 has a local minimum (this points to the instability of the system against external perturbations leading to excitations) and ε_2 is small (this indicates small damping of oscillations due to absorption), Im (-1/ε) exhibits maxima corresponding to interband transitions at 4-5, 7, 15-17 eV and volume plasmon at 22-23 eV. ε_1 and ε_2 are connected by integral Kramers -Kronig relationships because of causality being involved in dielectric response. The peak at 4 eV in ε_2 was attributed to unresolved direct exciton transitions involving the spin-orbit split valence and conduction band states at Γ point, i.e., $(\Gamma_8^-,\Gamma_6^-\rightarrow\Gamma_6^+)$. The bands at 7 eV and 10 eV were assigned to exciton transitions at X point (unresolved $(X_6^-,X_6^-,X_7^-)\rightarrow X_6^+$) [4,10,12]. The band at 15-17 eV may be referred to excitonic transitions with $L_{4,5}^-$ and Γ_6^+ symmetries or to higher excited states at points different from Γ and L in the Brillouin zone.

Self-consistent *ab initio* LMTO-ASA calculations of dielectric parameters for AgBr (Fig. 2b) have been performed taking into account only direct transitions. Assuming that all the transition matrix (|*M*|) elements are equal and independent of the k-point and the band number of the initial and final electron states, the following expressions for ε_1 and ε_2 were derived [4]:

$$\varepsilon_1=1+\frac{V_1\hbar^2}{\pi}\sum P\int_{BZ}\frac{1}{[E_{j'}(\vec{k})-[E_j(\vec{k})]^2[E_{j'}(\vec{k})-E_j(\vec{k})-\hbar\omega]}d^3k \qquad (3)$$

$$\varepsilon_2=\frac{V_2}{\omega^2}\sum_{jj'}\int_{BZ}\delta(E_{j'}(\vec{k})-E_j(\vec{k})-\hbar\omega)d^3k \qquad (4)$$

Here fitting parameters $V_1=\frac{2h^4e^2}{\pi^2m^2}|M|^2=0.04$ and $V_2=\frac{2h^2e^2}{\pi m^2\omega^2}|M|^2=0.01$, respectively. The needed energies have been taken from the LMTO-ASA energy band calculations [4]. Although ε_2 and Im (-1/ε) fall to zero at 4.3 eV (the direct band gap of AgBr) due to limitations of the model, the maxima of ε_2 at 8 eV, 10 eV and 12 eV fit to the composite band at 7-14 cV (with maxima at 8 eV and 10 eV) assigned to the exciton transitions at X point. The minimum of computed ε_1 at 13 eV fits to the minimum of experimental ε_1 curve, similarly to the minimum of Re (1/ε) at 13 eV. The crossings of Re (1/ε) and Im (-1/ε) at 8 eV coincide as well as of Re (1/ε) and ε_2 at 22 eV, where Im (-1/ε) has its maximum and ε_1 has a minimum. Furthermore, there is the shoulder of ε_2 at 15-16 eV superimposed with the previous band and the maximum of Im (-1/ε). For E>25 eV, both computed and experimental curves are close. Although Re (1/ε) and Im (-1/ε) reveal general resemblance, ε_1 and ε_2 exhibit differences. An increase in the intensities of ε_2 bands at 8 eV and 10 eV for the Ag(Br,I) shell as compared to the AgBr core (Fig. 2a) could be assigned to the IBE. For NCs **I** (Fig. 2c), ε_2 shows reverse-order changes in relative intensities of the bands, i.e., the increased intensity of the band at 4 eV, while ones at 7 eV and at 10 eV are suppressed with no evidence of the IBEs possibly due to impurity exclusion [1,6]. This correlates with a size confined enhancement of exciton luminescence from AgBr NCs in this size range [1,2,7,9].

CONCLUSIONS

The non-uniform size-dependent contrast of individual $AgBr_{0.95}I_{0.05}$ NCs observed by cryo-EFTEM is referred to predominant excitations at the surfaces and near the edges. When the field due to surface losses reach throughout the structure, they couple and the probability for their generation becomes periodic in the NC size. Since electronic sum rules must be satisfied, the surface excitations reduce the strength of the bulk excitations. So, coupling of surface and volume losses may cause oscillations of the image intensities with the NC size. The appearance of such oscillations demonstrates a beam-induced size confinement effect on valence electron excitations due to the contributions to the energy-level structure from carrier confinement and surface states. For NCs **I** ($d_c = 50\pm4$ nm), ε_2 shows an enhanced intensity of the band at 4 eV as compared to NCs **II** ($d_c = 109\pm7$ nm), while the bands at 7 eV and at 10 eV are suppressed possibly due to impurity exclusion. An enhancement of exciton-assisted direct interband transition at 4 eV (Γ_8-, Γ_6-$\rightarrow \Gamma_6$) correlates with the size-dependent enhancement of free exciton luminescence, when the NC sizes are less than 100 nm. This effect does not appear in tabular Ag(Br,I) microcrystals of about the same thickness, where the intensity of the band at 4 eV is les than of the composite band at 7-14 eV satisfactorily described by LMTO-ASA calculations for bulk AgBr taking into account only direct transitions above 4.3 eV.

REFERENCES

[1] A. P. Marchetti, R. P. Johansson and G. L. McLendon, *Phys. Rev. B* **47**, 4268-4275 (1993).
[2] M. I. Freedhoff, A. P. Marchetti and G. L. McLendon, *J. Luminesc.* **70**, 400-413 (1996).
[3] P. J. Rodney, A. P. Marchetti and P. M. Fauchet, *Phys. Rev. B* **62**(7), 4215-4217 (2000).
[4] V. Oleshko, M. Amkreutz and H. Overhof, *Phys. Rev. B* **67,** 115409-1-7 (2003).
[5] M. Timme, E. Schreiber, H. Stolz and W. von der Osten, *J. Luminesc.* **55**, 79-86 (1993).
[6] H. Kanzaki and Y. Tadekuma, *Solid. State Comm.* **80**, 33 (1991).
[7] K. P. Johansson, A. P. Marchetti and G. L. McLendon, *J. Phys. Chem.* **96**, 2873 (1992).
[8] U. Scholle, H. Stolz and W. von der Osten, *Solid State Comm.* **86**, 657 (1993).
[9] K. P. Johansson, G. McLendon and A. Marchetti, *Chem. Phys. Lett.* **179** (4), 321-324 (1991),
[10] V. P. Oleshko, A. Van Daele, R. H. Gijbels and W. A. Jacob, *Nanostruct. Mater.* **10**(8) 1225-1246 (1998); **11**(5), 687-688 (1999).
[11] V. P. Oleshko, S. B. Brichkin, R. H. Gijbels, W. A. Jacob and V. F. Razumov, *Mendeleev Comm.* **7**, 213-215 (1997).
[12] V. P. Oleshko and J. M. Howe, *J. Soc. Photogr. Sci. Technol. Japan* **65**(5), 332-341 (2002).
[13] R. F. Egerton, *Electron Energy-Loss Spectroscopy in the Electron Microscope*, 2nd ed., (Plenum Press, New York, 1996), pp. 256-262, 414-416.
[14] L. Reimer, in *Energy-Filtering Transmission Electron Microscopy*. Springer series in optica sciences, vol.71, (Springer-Verlag, Berlin, 1995) pp. 370-373.
[15] P. E. Batson, *Surf. Sci.* **156**, 720-734 (1985).
[16] D. B. T. Thoai and E. Zeitler, *Phys. Stat. Sol.* (a) **107**, 791-797 (1988).

Mat. Res. Soc. Symp. Proc. Vol. 789 © 2004 Materials Research Society N3.10

Quantum Conductivity of Spatially Inhomogeneous Systems

Liudmila A. Pozhar
Air Force Research Laboratory, Materials and Manufacturing Directorate,
Sensor Materials Branch and Polymer Materials Branch (AFRL/MLPS/MLBP),
2941 P Street, Wright-Patterson Air Force Base, OH 45433, U.S.A.

ABSTRACT

A fundamental quantum theory of conductivity of spatially inhomogeneous systems in weak electro-magnetic fields has been derived using a two-time Green function (TTGF)-based technique that generalizes the original method due to Zubarev and Tserkovnikov (ZT). Quantum current and charge density evolution equations are derived in a linear approximation with regard to the field potentials. Explicit expressions for the longitudinal and transverse conductivity, and dielectric and magnetic susceptibilities have been derived in terms of charge density - charge density and microcurrent - microcurrent TTGFs. The obtained theoretical description and formulae are applicable to any inhomogeneous system, such as artificial molecules, atomic and molecular clusters, thin films, interfacial systems, etc. In particular, the theory is designed to predict charge transport properties of small semiconductor quantum dots (QDs) and wells (QWs), and is a significant step toward realization of a concept of virtual (i.e., theory-based, computational) synthesis of electronic nanomaterials of prescribed electronic properties.

INTRODUCTION

Recent progress in synthesis and characterization of electronic material components from tens to hundreds of nanometers in linear dimensions facilitated fabrication of integrated circuits, electronic devices and components built of sub-micron size units. Further advances in bringing the scale down to tens of Angstroms to realize integrated nanoprocessor systems capable of manipulating massive amounts of data, while minimizing size, weight and power consumption encounter fundamental challenges that have to be overcome. This includes development of theoretical and computational means to predict charge transport properties of nanoclusters, nanostructures, thin films, artificial atoms, etc. that perform as QDs and QWs in microelectronic devices. Such methods comprise a foundation of a qualitatively new approach to synthesis of such nanomaterials and devices, called virtual (or fundamental theory-based, computational) synthesis. The major advantage of this approach is its ability to "fabricate" numerically entirely new systems and manipulate their structure, composition and processing technology to achieve desirable electronic properties of these materials/systems. In particular, such virtual synthesis is necessary to rationalize and use rich, quantum-effect driven opportunities provided by the natural inhomogeneity of structure and composition in small atomic clusters and interfacial systems.

Existing theoretical approaches to description of charge transport in electronic materials rely primarily on fundamental theoretical and computational methods developed for bulk systems, or their half-heuristic modifications designed to describe available and emerging experimental data [1]. As a result, obtained theoretical results do not allow reliable evaluation and prediction of the charge transport properties of electronic sub-nanoscale materials, such as small semiconductor QDs composed of a few tenths of atoms. Novel theoretical techniques that originate in quantum statistical mechanics and quantum field theory of finite systems (and specifically designed to treat quantum spatially inhomogeneous systems) are emerging, and have to be developed further and linked to the correspondingly quantum computation methods to provide reliable description of charge transport in realistic electronic sub-nanomaterials. In this paper one of such fundamental theoretical methods is discussed.

Among existing fundamental quantum field theoretical and quantum statistical approaches (such as Green function method, random-matrix theory, supersymmetry, quantum kinetic approach, etc.) a projection operator methods originated by Zwanzig and Mori and developed further by other researchers have been brilliantly generalized by Zubarev and Tserkovnikov [2] to establish a powerful and at the same time, tractable technique capable of handling realistic problems. It allows for formulation of the quantum transport problem in terms of a system of coupled evolution equations for TTGFs of microscopic operators specific to observables, and further reduction of this system (the simplest of such equations is the Dyson equation). A generalization of this approach specific to inhomogeneous systems has also been developed [3] and successfully used to predict transport properties of nanosystems. A similar TTGF-based method described below is used to derive a fundamental quantum theoretical description of electronic transport in sub-nanosystems in weak electro-magnetic fields. In particular, evolution equations for the charge and quantum current densities (linear in the applied fields), and explicit expressions for the conductivity and dielectric susceptibility tensors applicable to semiconductor QDs, artificial molecules, etc. have been obtained in terms of the equilibrium TTGFs of the charge density and microscopic current. These Green functions are related to the electronic energy spectrum, and can be calculated theoretically or computationally using quantum chemistry software, such as GAMESS, GAUSSIAN-98 or NWCHEM. Such computations can be realized at present for small clusters of several to few hundreds of atoms, such as small semiconductor QDs.

QUANTUM STATISTICAL FUNDAMENTALS

In a weak electro-magnetic field when interaction between particle spins and the field can be neglected, and the Hamiltonian of a system of N particles of mass m and charge e occupying a volume V of space can be written in the form

$$H = \frac{1}{2m}\int d\mathbf{r}\,\psi^{+}(\mathbf{r})\left(\frac{\hbar}{i}\nabla_{\mathbf{r}} - \frac{e}{c}\mathbf{A}(\mathbf{r},t)\right)^{2}\psi(\mathbf{r}) + H_{\text{int}}(\mathbf{r}_1,\ldots,\mathbf{r}_N) + e\int d\mathbf{r}\,\psi^{+}(\mathbf{r})\psi(\mathbf{r})\varphi(\mathbf{r},t), \quad (1)$$

where ψ^{+} and ψ are the quantum field operators at a position **r**, defined in terms of the charged particle creation and annihilation operators, ${}^{+}{}_{k\sigma}$and ${}_{k\sigma}$, respectively,

$$\psi^{+}(\mathbf{r}) = \frac{1}{\sqrt{V}}\sum_{\mathbf{k},\sigma} e^{-i\mathbf{kr}}\delta_{\sigma S_z} a^{+}_{\mathbf{k}\sigma}, \quad \psi(\mathbf{r}) = \frac{1}{\sqrt{V}}\sum_{\mathbf{k},\sigma} e^{i\mathbf{kr}}\delta_{\sigma S_z} a_{\mathbf{k}\sigma}, \quad (2)$$

and where $\hbar$ is the reduced Planck's constant; $\nabla_{\mathbf{r}}$ is the momentum operator; **A** and φ are the vector and scalar potentials of the electromagnetic field, respectively; H_{int} consists of operators specific to interparticle interactions and interactions with the environment; c is the speed of light, t is time, **k** denotes the wave vector, S_z is the value of the z-component of the charged particle spin, and $\delta_{\sigma S_z}$ is the Kronecker delta. A sum of the term proportional to the squared gradient operator and H_{int} in the right hand side of equation 1 represents the Hamiltonian H_0 of unperturbed system (further assumed to be in its equilibrium state). Linearization of the Hamiltonian H with regard to the vector-potential **A** produces the expression

$$H = H_0 + H_1, \tag{3}$$

$$H_1(t) = -\frac{1}{c}\int d\mathbf{r}\,\mathbf{j}(\mathbf{r})\cdot\mathbf{A}(\mathbf{r},t) + \int d\mathbf{r}\,\rho(\mathbf{r})\varphi(\mathbf{r},t), \tag{4}$$

where $H_1(t)$ is the perturbation Hamiltonian (the dot "·"denotes the inner product),

$$\mathbf{j}(\mathbf{r}) = \frac{e\hbar}{2mi}\left\{\psi^+(\mathbf{r})\nabla_{\mathbf{r}}\psi(\mathbf{r}) - [\nabla_{\mathbf{r}}\psi^+(\mathbf{r})]\psi(\mathbf{r})\right\} \tag{5}$$

denotes the microscopic current density operators in the absence of the field,

$$\rho(\mathbf{r}) = e\psi^+(\mathbf{r})\psi(\mathbf{r}) = en(\mathbf{r}) \tag{6}$$

is the charge density operator, and n(**r**) denotes the operator of the number of particles at a position **r**. The total microscopic current density operator $\mathbf{j}_{\mathrm{tot}}(\mathbf{r},t)$ in the system in an external electromagnetic field is defined as $\mathbf{j}_{tot}(\mathbf{r},t) \equiv -e\dfrac{\delta\mathbf{H}}{\delta\mathbf{A}(\mathbf{r},t)}$, where $\delta/\delta\mathbf{A}(\mathbf{r},t)$ denotes the Frečhet derivative. Thus, from equations (3) and (4) one derives:

$$\mathbf{j}_{tot}(\mathbf{r},t) = \mathbf{j}(\mathbf{r}) + \frac{e^2}{mc}\mathbf{A}(\mathbf{r},t)\psi^+(\mathbf{r})\nabla_{\mathbf{r}}\psi(\mathbf{r}), \tag{7}$$

This operator satisfies the microscopic charge conservation (or continuity) equation,

$$\frac{\partial\rho(\mathbf{r},t)}{\partial t} = -div\,\mathbf{j}_{tot}(\mathbf{r},t), \quad \text{where} \quad \rho(\mathbf{r},t) = e^{iH_0t/\hbar}\rho(\mathbf{r})e^{-iH_0t/\hbar}. \tag{8}$$

Further considerations evolve according to the theory of linear reaction of a system to a small perturbation (that is the most general physics theory applicable to systems of any nature and is an example of the simplest projection operator-type method). According to this theory, the average value $<B>$ of an observable B specific to a non-equilibrium state of the system described by the Hamiltonian of equation 3 can be expressed in terms of the equilibrium average value of B, $<B(\mathbf{r})>_0$ [with the Hamiltonian H_0 of equation 7], and the equilibrium retarded TTGF (or ERTTGF) $<<BH_1>>_0$ of the observable B and the perturbation Hamiltonian (such as H_1 of equation 4). This theorem can be rigorously extended to include inhomogeneous systems:

$$< B(\mathbf{r},t) >=< B(\mathbf{r}) >_0 + \int_{-\infty}^{\infty} dt' << B(\mathbf{r},t)H_1(t') >>_0, \tag{9}$$

where $< B(\mathbf{r},t) >= Tr(\wp B)$ is the non-equilibrium average of B. Here the statistical operator $\wp$ is the solution of the quantum Liouville equation with the Hamiltonian of equation 3 and the equilibrium initial condition $\wp(t=-\infty) \equiv \wp_0 = e^{-\beta H_0} / Tre^{-\beta H_0}$, with β=1/k_BT being the reciprocal equilibrium temperature. The equilibrium average value $< B(\mathbf{r}) >_0 = Tr(\wp_0 B)$ is defined with respect to the equilibrium statistical operator $\wp_0$, and the Bogoliubov-Tyablikov's ERTTGF of two operators B and C is defined by the expression

$$<< B(\mathbf{r},t)C(\mathbf{r}_1,t_1) >>_0 = \vartheta(t-t_1)\frac{1}{i\hbar} < [B(\mathbf{r},t), C(\mathbf{r}_1,t_1)] >_0 \tag{10}$$

where $\vartheta(t-t_1)$ is the step function and [...,...] denotes the commutator of these operators.

EVOLUTION OF SPACE-TIME FOURIER TRANSFORMS OF THE CHARGE AND TOTAL CURRENT DENSITIES

Using equation (10) for the operators $\rho(\mathbf{r},t)$ and $\boldsymbol{j}_{tot}(\mathbf{r},t)$ and noticing that in equilibrium the current density is zero, $< j(\mathbf{r}) >_0 = 0$, one can derive the evolution equations for the charge and current density operators that correspond to a generalization of the Kubo-type description of system evolution to spatially inhomogeneous systems. Expending all of the averages in these equations over plane waves and then using the time-Fourier transformation, $< B(\mathbf{r},t) >= \frac{1}{V}\sum_{\mathbf{k}} \int_{-\infty}^{\infty} d\omega B(\mathbf{k},\omega) e^{i\mathbf{k}\cdot\mathbf{r}-i\omega t}$ (B symbolizes any operator and ω is the frequency), one can derive the following equations for the space-time Fourier transforms of the charge and current densities $\rho(\mathbf{k},\omega)$ and $\mathbf{j}_{tot}(\mathbf{k},\omega)$, respectively:

$$\begin{aligned}\rho(\mathbf{k},\omega) &= e\sum_{\mathbf{l}} n_0(\mathbf{k})\delta(\mathbf{k}-\mathbf{l})\delta(\omega) - \frac{1}{c}\sum_{\mathbf{l}} << \rho_{\mathbf{k}} | (\mathbf{l}\cdot\mathbf{j}_{-\mathbf{l}}) >>_{0,\omega} \frac{1}{|\mathbf{l}|^2 c}\mathbf{l}\cdot\mathbf{A}(\mathbf{l},\omega) \\ &+ \sum_{\mathbf{l}} << \rho_{\mathbf{k}} | \rho_{-\mathbf{l}} >>_{0,\omega} \varphi(\mathbf{l},\omega),\end{aligned} \tag{11}$$

$$\begin{aligned}\mathbf{j}_{tot}(\mathbf{k},\omega) &= -\frac{e^2}{mc}\sum_{l} n_0(\mathbf{k}-\mathbf{l})\mathbf{A}(\mathbf{l},\omega) - \frac{1}{c}\sum_{\mathbf{l}}\left\{\chi^{lon}(\mathbf{k},\mathbf{l},\omega)\frac{\mathbf{kl}}{|\mathbf{k}||\mathbf{l}|}\right. \\ &\left. + \chi^{tr}(\mathbf{k},\mathbf{l},\omega)\left\{\frac{\mathbf{k}\cdot\mathbf{l}}{|\mathbf{k}||\mathbf{l}|}\mathbf{I} - \frac{\mathbf{kl}}{|\mathbf{k}||\mathbf{l}|}\right\}\right\}\frac{1}{c}\cdot\mathbf{A}(\mathbf{l},\omega) + \sum_{\mathbf{l}} << (\mathbf{k}\cdot\mathbf{j}_{\mathbf{k}}) | \rho_{-\mathbf{l}} >>_{0,\omega} \varphi(\mathbf{l},\omega),\end{aligned} \tag{12}$$

In these equations $\rho_0(\mathbf{r})$ is the equilibrium charge density, $<\rho(\mathbf{r})>_0 = en_0(\mathbf{r})$, with $n_0(\mathbf{r})$ being the particle number density; $\delta(\omega)$ denote Dirac's delta function, the double bracket symbol $<<...|...>>_{0,\omega}$ denotes the space-time Fourier transforms of the corresponding ERTTGFs, the symbol |...| denotes the absolute value of a vector, **I** is the unit matrix, and the summation runs over all of the wave vectors (denoted by **l**). Quantities $\rho_{\mathbf{k}}$ and $\mathbf{j}_{\mathbf{k}}$ are momentum representations of the charge and total current density operators, $\rho_{\mathbf{k}} = \frac{e}{\sqrt{V}}\sum_{\mathbf{q},\sigma} a^+_{\mathbf{q}-\mathbf{k},\sigma} a_{\mathbf{q},\sigma}$ and

$\mathbf{j}_\mathbf{k} = \frac{e\hbar}{m\sqrt{V}}\sum_{\mathbf{q},\sigma}(\mathbf{q}-\frac{\mathbf{k}}{2})a^+_{\mathbf{q}-\mathbf{k},\sigma}a_{\mathbf{q},\sigma}$, respectively, where the summations run over all the wave vectors **q** and the z-components of the spins σ. The generalized longitudinal and transverse susceptibilities, χ^{lon} and χ^{tr}, respectively, are:

$$\chi^{lon}(\mathbf{k},\mathbf{l},\omega) = \frac{1}{|\mathbf{k}||\mathbf{l}|} << (\mathbf{k}\cdot\mathbf{j}_\mathbf{k}) | (\mathbf{l}\cdot\mathbf{j}_{-\mathbf{l}}) >>_{0,\omega}, \tag{13}$$

$$\chi^{tr}(\mathbf{k},\mathbf{l},\omega) = \frac{|\mathbf{k}||\mathbf{l}|}{2(\mathbf{k}\cdot\mathbf{l})^2} << [\mathbf{k}\times\mathbf{j}_\mathbf{k}]\cdot | [\mathbf{l}\times\mathbf{j}_{-\mathbf{l}}] >>_{0,\omega}, \tag{14}$$

where the symbol [...×...] denotes the vector cross-product.

Equations 12 and 13 are generalizations of their counterparts for the homogeneous system case, and reduce to the latter with the only term in the sums corresponding to **l**=-**k** and the number density becoming constant, $n_0(\mathbf{k})=n_0\delta(\mathbf{k})$, $n_0(\mathbf{l})= n_0\delta(\mathbf{l})$.

Taking into account that the electromagnetic field potentials define only the longitudinal component of the electric field intensity and the transversal component of the magnetic field intensity inside the system, one can write the field equations for the Fourier-components of the electric and magnetic fields intensities in the matter, $\mathbf{D}(\mathbf{l},\omega)$ and $\mathbf{B}(\mathbf{l},\omega)$, respectively. This, together with the Lorentz condition (that ensures that the field equations are gauge-invariant), permits re-writing of equations 12 and 13 in terms of the fields **D** and **B** in the matter.

ELECTRONIC PROPERTIES

Following Nozieres and Pines, one can introduce the scalar dielectric susceptibility [it's space-time Fourier transform is denoted by $\varepsilon(\mathbf{k},\omega)$], and using the standard definition of the polarization vector and its relation to the induced charge density of equation 12, to express this susceptibility in terms of the charge density-charge density ERTTGFs,

$$\varepsilon^{-1}(\mathbf{k},\omega) = 1 + 4\pi\sum_{\mathbf{l}}\frac{1}{l^2} << \rho_\mathbf{k} | \rho_{-\mathbf{l}} >>_{0,\omega}. \tag{15}$$

Similar considerations permit to introduce the magnetic susceptibility $\mu(\mathbf{k},\omega)$, to relate it to the magnetic momentum vector and to derive an approximation for this quantity:

$$\mu^{-1}(\mathbf{k},\omega) = 1 + \frac{4\pi}{c^2k^2}\sum_{\mathbf{l}}\{\chi^{tr}(\mathbf{k},\mathbf{l},\omega)\frac{(\mathbf{k}\cdot\mathbf{l})}{|\mathbf{k}||\mathbf{l}|} + \frac{e^2}{m}n_0(\mathbf{k}-\mathbf{l})\}. \tag{16}$$

Using these expressions in equation 12 one can finally re-arrange this equation and identify the longitudinal and transverse components of the conductivity tensor, respectively:

$$\sigma^{lon}(k,\omega)\frac{\mathbf{kk}}{k^2} = \frac{i\omega}{4\pi}\{1-\varepsilon(\mathbf{k},\omega)\}\frac{\mathbf{kk}}{k^2}, \tag{17}$$

$$\sigma^{tr}(k,\omega) = \frac{c^2k^2}{4\pi i\omega}\{\ 1-\mu^{-1}(\mathbf{k},\omega)\ \} \tag{18}$$

Note, that the diagonal form of the longitudinal conductivity and a possibility to obtained an explicit expression for the magnetic susceptibility (and the transverse conductivity in the diagonal form) is a consequence of weakness of the electromagnetic field. In particular, the dielectric susceptibility, and therefore, the "linear" longitudinal conductivity are defined completely by the charge density - charge density ERTTGFs.

SUMMARY

The above consideration confirms that in the case of weak electromagnetic fields the dielectric and magnetic susceptibilities and the conductivity tensors of inhomogeneous systems (such as artificial atoms, atomic/molecular clusters, QDs, interfacial systems, etc.) in the linear approximation with regard to the field intensities are entirely defined by the equilibrium retarded two-time Green functions (ERTTGFs) of microscopic operators corresponding to charge and total current densities. These functions can be related [2] to the energy spectrum of the charge carriers specific to the equilibrium state of the inhomogeneous system. This energy spectrum can be derived theoretically, but more accurate data in practical cases can be obtained numerically. In particular, in the case of atomic clusters of semiconductor atoms (such as artificial molecules, quantum wells or QDs) such calculations can be routinely performed using GAMESS, GAUSSIAN-98 or NWCHEM quantum chemistry software packages. The results discussed in this study can be used together with such computations to predict charge transport properties of such nanosystems and therefore, to provide a fundamental tool for virtual development of electronic nanomaterials of pre-designed charge transport properties. Several small semiconductor cluster QDs have already been designed virtually by the author and her colleagues.

ACKNOWLEDGMENTS

The author thanks the National Research Council, the Air Force Research Laboratory, and the Air Force Office of Scientific Research of the U.S. Air Force for support of this work.

REFERENCES

1. *Quantum Theory of Real Materials*, J.R. Chelikowsky and S.G. Louie eds: Kluwer, Boston, 1996; Ivchenko and G.E. Pikus, *Superlattices and Other Heterostructures*, Springer, Berlin, 1997, etc.
2. See D.N. Zubarev and Yu.A. Tserkovnikov, *Proc. Steklov Ins. Math.* **175**, 139 (1986), and references therein.
3. L.A. Pozhar, *Phys. Rev.* ***E*** **61**, 1432 (2000); L.A. Pozhar and K.E. Gubbins, *Phys. Rev.* ***E*** **56**, 5367 (1997); L.A. Pozhar and K.E. Gubbins, *J. Chem. Phys.* **94**, 1367 (1991), etc.

Mat. Res. Soc. Symp. Proc. Vol. 789 © 2004 Materials Research Society N3.16

Photoluminescence of $CuInS_2$ nanowires

Kazuki Wakita, Yoshihiro Miyoshi, Masaya Iwai, Hideto Fujibuchi, and Atsushi Ashida,
Department of Physics and Electronics, Graduate School of Engineering, Osaka Prefecture University, 1-1 Gakuencho, Sakai, Osaka 599-8531, Japan

ABSTRACT

$CuInS_2$ nanowires synthesized by a chemical treatment method were examined using X-ray diffraction and field emission-scanning electron microscopy (FE-SEM). The nanowires, which were identified to have a chalcopyrite structure of $CuInS_2$, were 30 – 100 nm in diameter and several micrometers in length. A remarkable correlation between the color of a sample with nanowires and their FE-SEM image was disclosed. Photoluminescence spectra of the obtained nanowires were also studied. At low temperatures (~ 10K) a broad peak centered at photon energy of 2.05 eV was observed. This energy is by 0.5 eV larger than the energy gap of the well-studied crystalline bulk samples of $CuInS_2$. The observed rise in energy can be ascribed to quantum size effects expectedly developing in $CuInS_2$ samples with nanosize dimensions.

INTRODUCTION

Nanotubes and nanowires have attracted much interest for their applications in various fields such as nanotechnology, biotechnology, electronics, and energy industry. Although studies on carbon nanotubes outnumber investigations of other materials with nanosize dimensions, the latter may also end up with the results important for devise application.

$CuInS_2$ is one of the most promising candidates for application in photovoltaic devises. So far, we have examined bulk crystals of this material to study photoluminescence, Raman scattering and other optical properties [1-4]. On the other hand, the $CuInS_2$ nanorods and nanotubes synthesized by a chemical treatment method were reported and comprehensively examined by x-ray diffraction (XRD), transmission electron microscopy (TEM), and other structural techniques [5]. In this work we make a report regarding relation between the conditions of syntheses and resultant nanowire structures, as well give and discuss a photoluminescence (PL) spectrum of the obtained products.

EXPERIMENTAL DETAILS

$CuInS_2$ nanowire samples were synthesized using a chemical treatment process similar to the one reported previously [5]. Pure copper, indium, and sulfur powders were placed in a stainless steer reactor together with ethylendiamine (80 % of the reactor volume) as solvent. The sealed reactor was kept in the electric furnace at the temperature between 280 and 300 °C for 24 h. After washing in water and alcohol, the product was keep at 60 °C for 6 h in vacuum. The structure of the nanowires was examined by XRD and field emission-scanning electron microscopy (FE-SEM). PL measurements at 10 K were carried out using a He-Cd laser (wavelength: 441.6 nm).

RESULTS AND DISCUSSION

The product obtained at 300 °C (sample A) was chocolate brown, while the powder obtained at 280 °C (sample B) was black. FE-SEM images of both are shown in figure 1, (a) sample A and (b) sample B. Most nanowires from sample A [Fig. 1 (a)] are 30 – 100 nm in diameter and several micrometers in length. On the other hand, sample B displays long (at a micrometer scale) belt structures with thickness at 30-50nm, and width at 200 nm – 500 nm. Other samples (not shown here) synthesized at 280 °C have displyed similar belt-like wires with thickness of 30 – 50 nm and width from several hundred nm to few μm. There is a strong correlation between the color and lateral dimensions of resultant structure. The smaller dimensions the brighter is the color of the powder. This relationship reflects absorption properties of the products.

Figure 2 shows a XRD pattern of sample A. Sharp diffraction peaks in the pattern can be indexed to almost chalcopyrite structure, in agreement with the reported data on bulk $CuInS_2$ (JCPDS Card No. 27-0159; a = b = 5.523 Å, c = 11.141 Å). The broadened diffraction peaks we attribute to the nanostructure of the obtained products with small diameters of the nanowires.

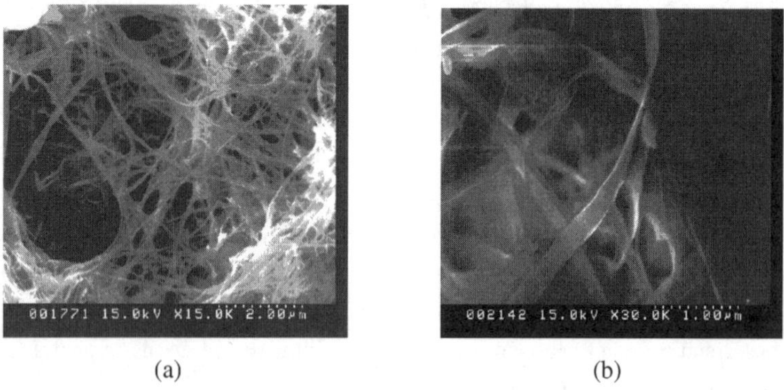

(a) (b)

Fig. 1 FE-SEM image of the products: (a) sample A and (b) sample B.

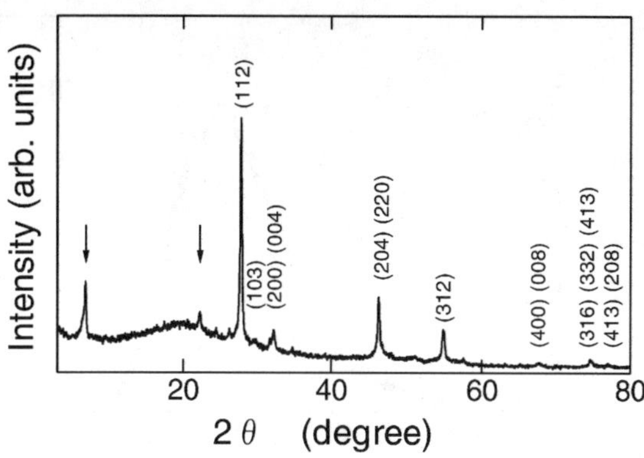

Fig. 2 XRD patterns with CuKα radiation for sample A.

Vertical arrows indicate unknown peaks apparently belonging to by-products.

Figure 3 shows PL spectrum of sample A at 10 K under excitation energy of 2.81 eV. We have observed a broad PL peak centered around 2.05 eV, with full width at half-maximum (FWHM) at 0.24 eV. Measurements of PL spectra of sample B with belt structures are not completed and comparisons with PL spectrum of sample A, which are important for tracing through the relationship between the structure of $CuInS_2$ nanowires and their PL spectra are not possible at this time. Here we discuss only PL spectrum of sample A.

The origin of the observed emission from sample A can be ascribed to the $CuInS_2$ nanowires or/and the by-products formed during the synthesis, or/and an oxide formed on the surface of the nanowires exposed to air ambience after synthesis. Among the just-listed possibilities, $CuInS_2$ nanowires look like the most plausible source of the observed emission for the following reasons. If we consider that at low temperatures the reported bandgap of bulky $CuInS_2$ with chalcopyrite structure is 1.555 eV [6], and that the photon energy corresponding to the observed PL peak of sample A is by 0.5 eV larger we can assume that the difference is due to quantum effects, which usually develop in nanostructures, with a result of elevation of the optical bandgap. Such assumption is consistent with the fact that only sample A with clear nanowires displayed chocolate brown color and an optical band increase that was disclosed by our preliminary ellipsometric measurements [7]. Indexed, for the most part, to $CuInS_2$ with chalcopyrite structure, the XRD pattern of sample A (Fig.2) is also in favour of correctness of the above assumption.

CONCLUSIONS

$CuInS_2$ nanowires synthesized from Cu, In, and S powders added to ethylendiamine solvent at 280 - 300 °C for 24 h have been examined using XRD and FE-SEM. As turned out,examined

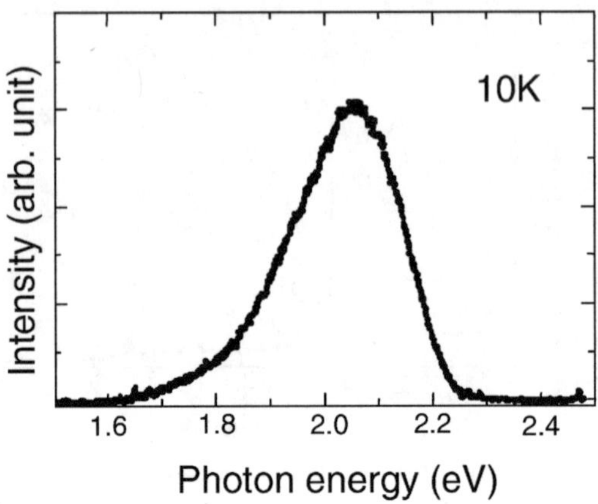

Fig. 3 PL spectrum at 10 K for sample A

nanowires are displayed through a remarkable broadened diffraction peaks in almost chalcopyrite XRD pattern of the obtained product. They are 30 – 100 nm in diameter and their lengths are about several μm. It is found that fine nanowires result in brighter color of the powder. A broad PL band at 2.05 eV, which is 0.5 eV larger than band gap of bulk samples, has been observed. We think that increased bandgap is a result of the quantum effects in the nanowires.

ACKNOWLEDGMENTS

This work was supported by a Grant-in-Aid for Scientific Research (C), No. 15510106, from the Ministry of Education, Culture, Sports, Science and Technology of Japan.

REFERENCES

1. K. Wakita, H. Hirooka, S. Yasuda, and N. Yamamoto, *J. Appl. Phys.* **83**, 443 (1998).
2. K. Wakita, F. Fujita, and N. Yamamoto, *J. Appl. Phys.* **90**, 1292 (2001).
3. K. Wakita, K. Nishi, Y. Ohta, and N. Nakayama, *Appl. Phys. Lett.* **80**, 3316 (2002).
4. K. Wakita, G. Hu, N. Nakayama, and D. Shoji, *Jpn. J. Appl. Phys.* **41**, 3356 (2002).
5. Y. Jiang, Y. Wu, S. Yuan, B. Xie, S. Zhang, and Y. Qian, *J. Mater. Res.* **16**, 2805 (2001).
6. J. J. M. Binsma, L. J. Giling, and J. Bloem, *J. Lumin.* **27**, 55 (1982).
7. K. Wakita *et al.*, will be published elsewhere.

Mat. Res. Soc. Symp. Proc. Vol. 789 © 2004 Materials Research Society N3.20

Si-N nanowire formation from Silicon nano and microparticles.

Chandana Rath[1], A. Pinyol[2] J. Farjas[1], P. Roura[1] and E. Bertran[2]

[1]*GRM, Department de Fisica, Universitat de Girona, Campus Montilivi, E17071 Girona, Catalonia, Spain*
[2]*FEMAN, Department de Fisica Aplicada i Òptica, Universitat de Barcelona, Av. Diagonal 647, E08028 Barcelona, Catalonia, Spain*
Corresponding author: E-mail: jordi.farjas@udg.es

ABSTRACT
We report silicon nitride whisker formation from hydrogenated amorphous silicon (a-Si:H) nanoparticles grown by PECVD for the first time. We compared the results with the kinetics of whisker formation from ball milled crystalline silicon (c-Si) microparticles. Whisker formation is analyzed at different temperatures (900-1440 °C) and oxygen partial pressures. At temperatures equal or above 1350 C and at low oxygen partial pressure we observe monocrystalline α-Si_3N_4 whiskers having 30-100 nm diameter and several microns length. By increasing the oxygen partial pressure, the structure of whiskers is completely changed, as shown by electron microscopy. In this case we observe α-Si_3N_4 whiskers covered by an amorphous silica layer at 1350 C. Finally, when the precursor material is silicon microparticles, thicker (170-330 nm) and longer whiskers are formed.

INTRODUCTION
Silicon nitride is an advanced ceramic and has the potential to be used in various applications thanks to its enhanced mechanical, chemical, electronic and thermal properties. They include, high mechanical strength with good flexibility, light weight, high fracture toughness, enhanced resistance to thermal shock and to oxidation [1-3]. These properties are related to the highly covalent chemical bonding between silicon and nitrogen atoms [4]. Therefore, studies of the synthesis of silicon nitride whiskers have received extensive attention and various synthesis methods have been developed to improve their properties. Among them we can find carbothermal reduction of silica-containing compound [2,5-7], chemical vapor deposition process [8-11], nitridation of silicon compacts (RBSN) [12-13] and combustion synthesis [14-16]. Some of these processes are very slow and energy consuming [12], and in some cases a very high temperature (1900°C) is required [16]. Besides, Si_3N_4 can not be formed from silicon suboxide prepared by carbothermal reduction [6].

In the present paper we report the formation of silicon nitride from a-Si:H nanoparticles and c-Si microparticles. As far as we know, there is not any publication related to whisker formation from amorphous silicon nanoparticles. We distinguish two competing processes such as oxidation and nitridation over the temperature range taken (900-1450 °C). By varying the temperature, the oxygen partial pressure and the specific surface area of Si particles we could produce α-Si_3N_4 whiskers in a wide range of diameters and lengths.

EXPERIMENTAL

a-Si:H nanoparticles have been prepared by plasma-enhanced chemical vapour deposition (PECVD) method from a RF glow-discharge of silane. c-Si microparticles have been produced from silicon wafers by ball-milling and have an average size of about 0.5 microns.

Nitridation experiments have been carried out in a Mettler Toledo thermobalance, model TG-851. Processes have been performed under a continuous flow of N_2 by heating the sample at 100K/min, in an alumina crucible, from room temperature up to a given temperature is reached and then it is kept at this temperature during one hour. Experiments have been performed using different constant temperatures (900-1440°C) and N_2 flows. The higher the N_2 flow is, the lower the O_2 partial pressure is. The highest flow rate is 300 ml/min which corresponds to an O_2 partial pressure of 2 10^{-3} atm. Mass-gain versus temperature curves have been obtained from thermograms by subtracting a blank curve. Temperature and heat flow have been corrected by proper calibration of the experimental conditions.

The chemical composition of all the samples has been determined by elementary analysis (EA), Fourier-Transform Infra red (FTIR) spectroscopy and Electron Energy Loss Spectroscopy (EELS). In addition, morphology and structure have been imaged by High Resolution and conventional Transmission Electron Microscopy (HRTEM and TEM), Scanning Electron Microscopy (SEM) and Selected Area Electron Diffraction (SAED) techniques. These measurements have been performed using a Phillips CM-30 (HRTEM) and Zeiss DSM 960A (SEM).

RESULTS AND DISCUSSION

Chemical composition of the original a-Si:H nanoparticles has been obtained by EA and mass-gain curves after complete oxidation. The initial stoichiometry is $H_{0.28}Si_1O_{0.134}$. TEM micrographs show that these particles are spherical with sizes between 40 and 100 nm, while electron diffraction (ED) reveals an amorphous structure. During constant heating at 100 K/min under a continuous N_2 flow, thermograms exhibit an exothermic peak around 800°C which corresponds to crystallization of a-Si. From EA and mass gain after isothermal heating, we have obtained the final stoichiometry. N and O content per silicon atom vs. temperature at a N_2 flow of 300 ml/min are plotted in Fig. 1. One can clearly point out that there are two competing processes, oxidation and nitridation. Up to 1200 °C oxidation plays a dominant role whereas above 1200 °C nitridation process becomes more prominent. The initial oxygen content (0.134) in the starting material increases above 900 °C and shows a maximum value at 1200 °C. Oxygen is incorporated forming an amorphous phase. This fact has been confirmed by SAED because the only crystalline phase observed corresponds to c-Si. Moreover, oxidation of a-Si:H nanoparticles above 800°C results mainly in the formation of a-SiO_2 [17].

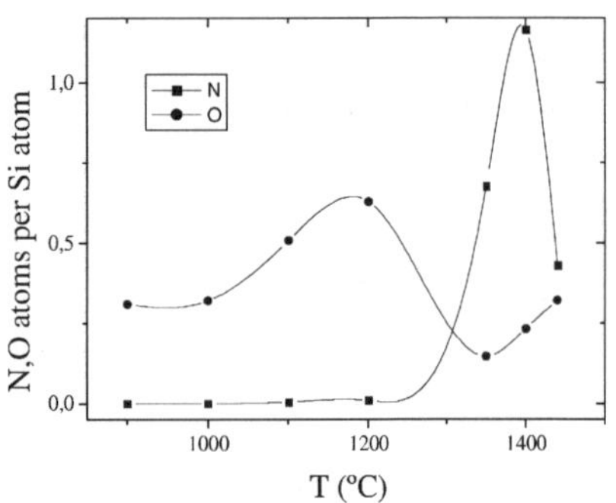

Fig 1. N and O content per silicon atom after treating a:Si:H nanoparticles at different constant temperatures and under a constant N_2 flow of 300 ml/min during 1 hour.

On the other hand, nitridation dominates over oxidation above 1200 °C. In general, nitrogen uptake involves the formation of Si_2N_2O and Si_3N_4. However, SAED analysis has ruled out the formation of Si_2N_2O in our samples. The absence of droplets at the tip of crystals on TEM micrograhs rules out the vapor-liquid-solid mechanism. Therefore, the most suitable mechanism leading to Si_3N_4 whiskers formation from Si and SiO_2 in the presence of N_2 is a two step vapor solid (VS) mechanism involving the formation of SiO in gas phase as an intermediate product [2]. This reaction only involves Si, SiO_2 and N_2 and it has been stated that silicon suboxides do not contribute to Si_3N_4 formation [6]. Then the reaction kinetics depends mainly on SiO partial pressure and this fact accounts for the lower onset on Nitrogen uptake [6]. So we tentatively propose these scenario:

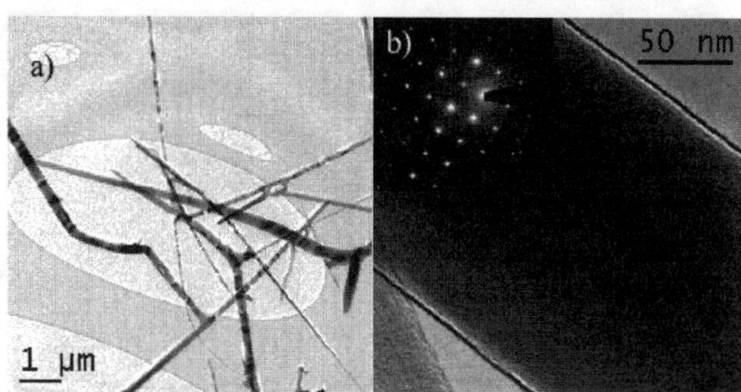

Fig 2. (a), (b) TEM micrographs of whiskers synthesized at 1400°C during 1 hour from a:Si:H nanoparticles. Inset is a SAED of a whisker (corresponding monocrystalline α-Si_3N_4).

1) Under 1200 °C, no SiO gas phase is formed and the only reaction that takes place is SiO_2 formation.
2) Above 1200 °C SiO gas partial pressure is high enough to form Si_3N_4 [5]. This reaction involves SiO_2 consumption, and consequently oxygen incorporation is significantly reduced.

At 1400°C we reach the highest nitrogen incorporation which is very close to Si_3N_4 stoichiometry. At this point, the remaining oxygen could be related to the initial oxygen content which is mainly bonded in a suboxide form, thus it does not participate in the nitruration process.

The proposed scenario is supported by FTIR and TEM measurements. At 1400°C, the TO_4 mode (1220 cm^{-1}) of silica is absent in the FTIR spectra while the peak related to the Si-O-Si stretching bond mode (1090 cm^{-1}) is present [17, 18]. Hence, O is mainly bonded to Si in a suboxide form. This fact is quite surprising because oxygen uptake above 800°C is in the SiO_2 form. Consequently at 1400°C all SiO_2 is consumed to form Si_3N_4 while the initial suboxide present in the particles remains.

Fig 3. TEM and HRTEM micrographs of whiskers formed from a:Si:H nanoparticles treated at 1350°C under low N_2 flow during 1 hour. Inset is a SAED of a whisker (indicating monocrystalline α-Si_3N_4).

Fig 2 are TEM (a,b) and SAED (c) micrographs obtained from whiskers formed at 1400°C treated during 1 hour. From SAED the crystalline phase has been identified as monocrystalline α-Si_3N_4. Although, in general α-Si_3N_4 from Si/SiO_2 precursors

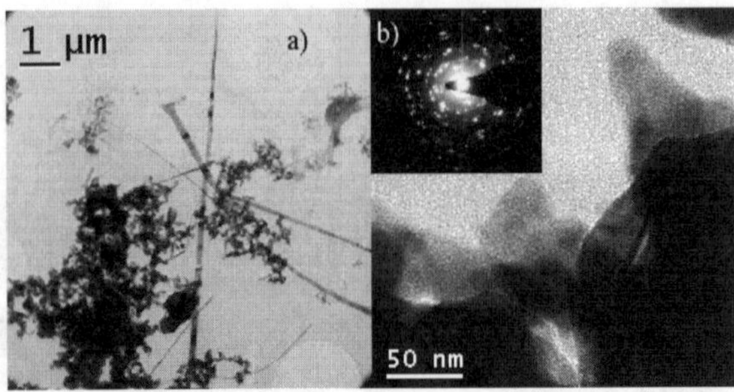

Fig 4. (a), (b) TEM of a region containing powders and whiskers synthesized from a:Si:H nanoparticles treated at 1350ºC under a constant N_2 flow of 300 ml/min during 1 hour. Powders are polycrystalline α-Si_3N_4 nanoparticles as confirmed by SAED (inset).

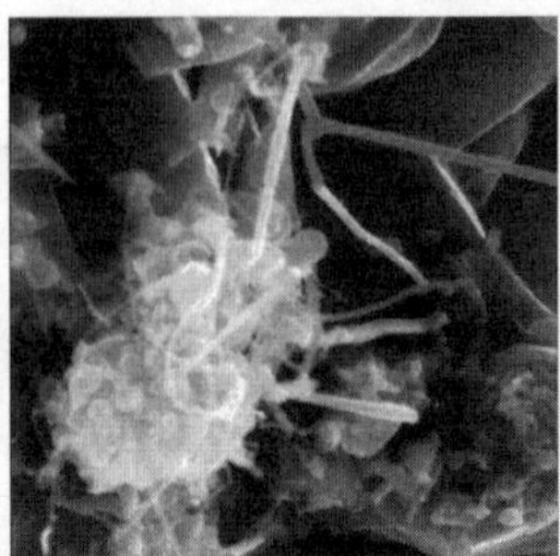

Fig 5. (a), (b) SEM and TEM images of c-Si treated at 1400 ºC during 1 hour under a constant N_2 flow of 300 ml/min.

involves the formation of a SiO_2 covering layer [2], HRTEM micrograh exhibits a single crysta without such a covering layer. This fact agrees with FTIR measurements. Whiskers are about 5C to 200 nm in width and several microns in length. At higher temperatures the whisker yield is reduced and whiskers are thicker.

However at higher O_2 partial pressure a SiO_2 covering layer appears. This can be statec from Fig 3 which corresponds to a HRTEM micrographs (3.a and b) obtained from whiskers formed at 1350ºC and at low N_2 flow. From SAED (fig 3.c) the core has been identified as α-Si_3N_4 monocrystal and no other crystalline phase has been noticed. EELS points out that the surrounding layer is composed by O and Si. Finally FTIR measurements confirmed the presence of SiO_2 through the TO_4 mode (1220 cm^{-1}) peak. Therefore the covering layer is a-SiO_2 and the core is α-Si_3N_4 monocrystal. Furthermore, the higher the O_2 partial pressure is, the lower the N uptake is.

For VS mechanism, it is generally believed that the growth form is closely related to the super saturation ratio which is a combination of the concentrations of the reactant gases [19]. Ir general, a higher super saturation ratio favors the formation of powders and lower value woulc proceed to form whiskers. This may be the case in 1350 ºC where we obtained very few whiskers compared to the silicon nitride powders. From fig 4, it is clear that the powders are polycrystalline in nature and α-Si_3N_4.

Finally, treating the c-Si micron particles at 1400 ºC at 300 ml/min N_2 flow, we obtainec thicker (150-350 nm) and longer whiskers (~100 µm) compared to the whiskers obtained from nanoparticles (Fig. 5). From SAED measurements whiskers are found to be single crystalline α-Si_3N_4. No covering SiO_2 layer is formed. Hence, in principle, the mechanism proposed for a-Si:H particles is also valid for c-Si microparticles.

CONCLUSIONS

For the first time, we have studied Si_3N_4 whisker formation from a-Si:H nanoparticles producec by PECVD. α-Si_3N_4 monocrystalline whiskers have been prepared from a-Si:H and c-S particles by nitridation processes at different temperatures from 900 to 1440 ºC and various O_2 partial pressures. The reaction involves two competing processes, oxidation and nitridation. Up to 1200 °C oxidation is the leading process which significantly diminishes when nitridation starts.

The mechanism responsible of whisker formation is a two-step VS reaction which agrees with published studies related to whisker formation when precursors are a mixture of c-Si and SiO_2. However, authors report the formation of a covering SiO_2 layer around whiskers that is not present in our case when O_2 partial pressure is low enough. Hence, we observe the formation of pure α-Si_3N_4 whiskers. Nevertheless, at higher O_2 partial pressure, an amorphous covering SiO_2 is formed. By treating the micro particles under the same conditions we have observed thicker as well as longer pure α-Si_3N_4 whiskers.

Finally, for both a-Si:H and c-Si ,whisker yield depends on temperature. The maximum yield is located around 1400°C. At 1400°C and for a-Si:H Nitrogen uptake is very close to Si_3N_4 stoichiometry and whisker yield is very high. Consequently, a-Si:H is a good precursor for α-Si_3N_4 whiskers formation because its high yield, chemical purity and moderate synthesis temperature (1400°C).

ACKNOWLEDGEMENT
This work has been supported by the Spanish Programa Nacional de Materiales under contract numbers MAT-2002-04236-C04--02. One of the authors (C. Rath) wishes to acknowledge the Ministerio de Ciencia y Tecnología, Government of Spain, for support. The authors also want to thank Drs J. Arbiol, J. Portillo, and also J. Mendoza for assistance in TEM observations.

REFERENCES

1. T. Sekine, H. Hongliang, T. Koboyashi, M. Zhang and F. Xu, Appl. Phys. Lett. **76**(25), 3706 (2000).
2. Y. Zhang, N. Wang, R. He. J. Liu, X. Zhang and J. Xhu, J. Crystal Growth, **233**, 803 (2001).
3. Y. Zhang, N. Wang, R. He. Q. Zhang, J. Xhu and Y. Yan, J. Mater. Res. **15**(5), 1048 (2000).
4. W. Dressler and R. Riedel, Int. J. Refract. Met. Hand Mater 15, 13 (1997).
5. M.J. Wang and H. Wada, J. Mater. Sci. 25, 1690 (1994).
6. P. D. Ramesh and K.J. Rao, *J. Mater. Res*, **9**, 2330 (1994).
7. P.S. Gopalakrishnan and P.S. Lakshminarsimham, *J. Mater. Sci. Lett*, **12**, 1422 (1993).
8. K. Kijima, N. Setaka and H. Tanaka, *J. Crystal Growth*, **24/25**, 183 (1974).
9. S. Motojima, T. Yamana, T. Araki, H. Iwanaga, *J. Electrochem. Soc*, **142**, 3141 (1995).
10. C. Kawai and A. Yamakawa, J. Mater. Sci. Lett. 14, 192 (1995).
11. B. Stannowski, C.H.M Van der Werf and R.E.I Schropp, *Proc. Of 3rd Intem. Conf. on Coatings on Glass*, Oct 29-Nov. 2, 2000) pp. 387-394.
12. L. Bingqiang, *The Silicon nitride Reaction-bonding process*, Doctoral Thesis, Lulea Tekniska University, 1999.
13. D.R. Messier and P. Wang, J. Am. Ceram. Soc., **56**(9), 480 (1973).
14. Y.G. Cao, C.C. Ge, Z.J Zhou and J.T Li, J. Mater. Res. **14**, 876(1999).
15. M.A. Rodiguez, N.S. Makhonin, J.A. Escrina et. Al. , Adv. Mater. **7**, 745(1995).
16. H. Chen, Y. Cao, X. Xiang, J. Li and C. Ge, J. Alloys and Compounds **325**, L1(2001).
17. D. Das, J. Farjas, P. Roura, G. Viera and E. Bertran, Appl. Phys. Lett. **79** (22), 3705 (2001).
18. M.I. Alayo, I. Pereyra, W.L. Scopel, M.C.A Fantini, Thin Solid Films **402** (2002) 154-161
19. W.B. Campbell, in *Whiskers Technology* edited by A.P. Levitt (Wiley, New York, 1970).

Mat. Res. Soc. Symp. Proc. Vol. 789 © 2004 Materials Research Society N3.25

A Novel Tick Film Nanocrystalline $Y_2O_3:Eu^{3+}$ Phosphor Synthesized by Post-Dispersion Treatment at Low Temperature

Sung-Jei Hong, Min-Gi Kwak, Dae-Gyu Moon, Won-Keun Kim, and Jeong-In Han
Information Display Research Center, Korea Electronics Technology Institute
#455-6, MaSan, JinWi, PyungTack, KyungGi, 451-865, Korea

ABSTRACT

In this study, a novel thick film nanocrystalline phosphor is synthesized by using post-dispersion treatment at low temperature. The post-dispersion treatment is to prevent the agglomeration between the precursors by mixing the organic dispersing agent with them before heat-treatment. The mean size of the particle heat-treated at 500°C is 4nm. Also, europium is uniformly distributed in the yttrium oxide with deviation below 0.5%. The nanocrystalline phosphor is principally composed of cubic structure having preferred orientation of <222>, <440>, and <400>. The photoluminescence properties of the nanocrystalline phosphor characterized with monochrometric systems reveals that the main PL peak is detected at 611 nm, and thick film nanocrystalline phosphor with 15wt% europium exhibits better properties.

INTRODUCTION

A nanocrystalline phosphor improves optical properties by controlling the recombination behavior between electron and hole efficiently. The phenomenon is achieved from the ultrafine size effect owing to the change of energy level by restricting the electron and hole [1]. However, particle size of nanocrystalline phosphor increases easily under even the small changes of heat-treating temperature because the wider surface is very unstable and tends to reduce the surface area [2, 3]. The increase of particle size gives rise to a reduction of the surface area of the nanoprticles inevitably, and the optical properties of nanocrystalline phosphor come to be degraded. In this study, we attempt to prevent the agglomeration of the nanocrystalline phosphor by mixing the precursor nanoparticle with organic solvent before heat-treatment. The attempted process is called as post-dispersion process.

EXPERIMENTAL DETAILS

Yttrium acetylacetonate and europium acetate are used as raw materials. To synthesize the thick film, $Y_xO_y:Eu^{3+}$ precursors with various concentration of europium are prepared. Then, post-dispersion treatment of the precursors is done under low temperature. Synthesized precursor nanoparticles are dispersed into organic solvent to block one nanoparitcle from others. The mixture of the precursor and the organic dispersing agent is printed and heat-treated on silicon wafer. Thermal analyses such as differential thermal analysis (DTA) and thermogravimetric analysis (TGA) of the precursor dispersed into organic solvent are done to determine the minimal limit of heat-treating temperature. Then, we attempt to optimize the size of the nanocrystalline phosphor by varying temperature of the post treatment ranging from 500°C to 900°C. The fabricated thick films of nanocrystalline phosphors are investigated by analyzing the size and crystal structure with high resolution transmission electron microscope (HRTEM) and X-ray diffractometer (XRD). Also, photoluminescence (PL) spectra are measured at room temperature by exciting samples with a xenon lamp combination with an interference filter. The wavelengths range is between 200nm and 800nm.

DISCUSSION

Physical properties of thick film $Y_2O_3:Eu^{3+}$ nanophosphor

Thermal analyses of the precursor dispersed into organic solvent, shown in figure 1, indicate that heat-treating temperature over 450°C is required to burn out the organic component, and to crystallize the nanophosphor. Change of heat flow is found at the temperature ranging from 100°C to 450°C as shown in figure 1 (a). Also, change of weight is observed at similar temperature range as seen in figure 1 (b). From the results, it is assumed that the acetate components consisting of C, H and O react with O_2 in air above 100°C, and it is oxidized and separated from yttrium and europium by reaching 450°C, respectively. Also, during the reaction, yttrium is supposed to oxidized with blocking an Eu^{3+} ion in a Y_2O_3 nanoparticle. Based on the thermal behaviors, the precursor requires at least heat-treatment at 500°C in order to crystallize the nanophosphor. So, the heat-treating temperatures are determined as 500, 700, and 900°C.

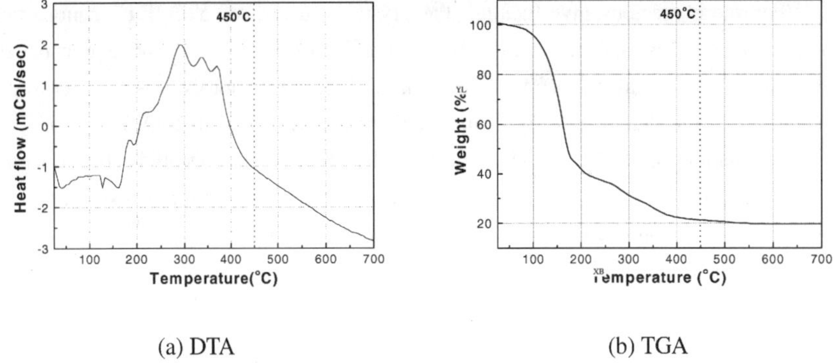

(a) DTA (b) TGA

Figure 1. Thermal analyses (TGA: thermogravimetric analysis, DTA: differential thermal analysis) of precursor dispersed into organic solvent

Size of the nanocrystalline phosphor is proportional to the heat-treating temperature. As seen in figure 2, the size of the nanocrystalline phosphor heat-treated at 700°C is about 20 nm. As the temperature is raised to 900°C, the size is increased to 30 nm. However, the nanocrystalline phosphor can be synthesized by suppressing the temperature of the post treatment below 500°C. That is, the nanocrystalline $Y_xO_y:Eu^{3+}$ phosphor of which mean particle size is 4 nm can be obtained applying the post-dispersion treatment. Investigating the nanocrystalline $Y_xO_y:Eu^{3+}$ phosphor in detail, crystalline layer structure is obviously observed. So, we assume that the precursor of the Eu^{3+} doped yttrium oxide is crystallized under the synthesis condition.

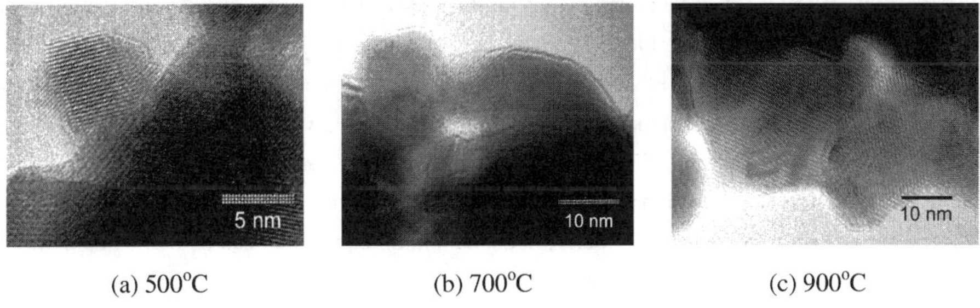

(a) 500°C (b) 700°C (c) 900°C

Figure 2. HRTEM observation of thick film nanocrystalline $Y_xO_y:Eu^{3+}$ phosphors according to heat-treating temperature

For more precious investigation, the crystal structure of $Y_xO_y:Eu^{3+}$ nanoparticles heat-treated at 500°C is analyzed with XRD. Figure 3 shows the X-ray diffraction pattern of the nanophosphor. Crystalline diffraction pattern is observed obviously. Also, the crystal structure according to the detected X-ray peak is coincident with that of $Y_2O_3:Eu^{3+}$ [4]. That is, the detected main peak is <222>, <440>, <400> indicating that the crystal is cubic structure. Only a small amount of peaks of monoclinic structure are observed from the nanocrystalline phosphor. So, we certify that all the particles are crystallized to $Y_2O_3:Eu^{3+}$.

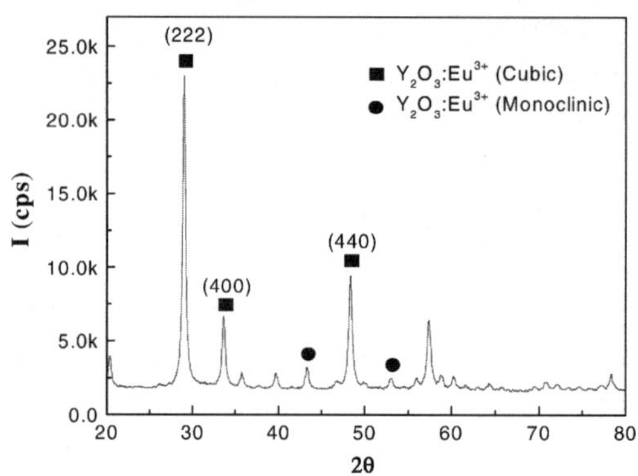

Figure 3. XRD analysis of nanocrystalline phosphor

From the results, we suppose that the thick film nanocrystalline phosphor is consistent of nano scaled Y_2O_3 particle blocking Eu^{3+} ion in. Because of the nano-dimensional scale, uniform distribution of europium is very important factor. So, we analyzed composition ratio of several points of the thick film nanocrystalline phosphor with energy dispersion spectrometer (EDS). As a sample, concentration of europium is designed to 7wt%. As seen in table 1, the analysis indicates a slight deviation within 0.5wt%. So, using the post-dispersion treatment, uniform composition of the $Y_2O_3:Eu^{3+}$ thick film nanocrystalline phosphor can be fabricated in this study.

Table 1. Distribution of the components in the nanocrystalline phosphor

	Y	Eu	O
Position 1	56.8 wt%	6.7 wt%	36.5 wt%
Position 2	60.1 wt%	7.5 wt%	32.4 wt%
Position 3	68.3 wt%	7.2 wt%	24.5 wt%

Optical properties of thick film $Y_2O_3:Eu^{3+}$ nanophosphor

It is important to optimize the concentration since the more exact combination between an Y_2O_3 nanoparticle and a Eu^{3+} ion enhances the optical properties of the nanocrystalline phosphor. So, using the synthetic method, three types of thick film nanocrystalline phosphors are fabricated of which europium concentration are 5, 10, 15 wt% respectively. Then, photoluminescence (PL) properties are observed after synthesizing the nanocrystalline phosphor. The spectra are shown in figure 4. In case of excitation, seen in figure 4 (a), wavelength range of the excitation is narrower as the europium concentration is increased. Also, the intensity of the excitation is increased as the europium concentration is increased. It is thought to be owing to the ultrafine size effect [5]. So, the more intense emission is expected as the europium concentration is increased. In fact, seen in figure 4 (b), the intensity of the nanocrystalline phosphor is enhanced as the europium concentration is increased. Also, the main PL peak is observed at 611nm, and the phenomenon indicates typical characteristics of red color phosphor. The PL peak is owing to the cubic structure of the synthesized $Y_2O_3:Eu^{3+}$ nanocrystalline phosphor based on the report that the emission wavelength is originated from the electron transfer of $^5D_0$ $Eu^{3+} \rightarrow {}^7F_2$ Eu^{3+} state [5]. So, we certify that optical properties of the nanocrystalline phosphor are enhanced applying a new novel post-dispersion treatment.

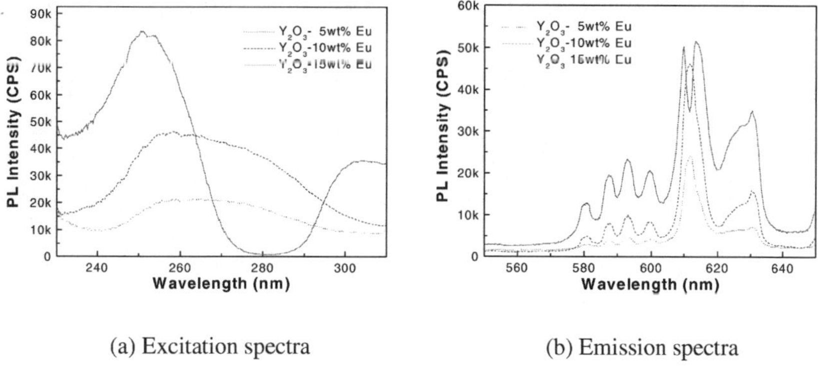

(a) Excitation spectra (b) Emission spectra

Figure 4. Photoluminescence properties of nanocrystalline $Y_2O_3:Eu^{3+}$ phosphors

CONCLUSIONS

In this study, we developed a post-dispersion treatment for thick film nanocrystalline $Y_2O_3:Eu^{3+}$ phosphor. Applying the post-dispersion treatment at 500°C, a thick film nanocrystalline $Y_2O_3:Eu^{3+}$ phosphor could be fabricated of which particle size and crystal structure are 4nm and mainly cubic, respectively. The uniform distribution of europium concentration within 0.5wt% could be achieved. The photoluminescence properties are enhanced as the europium concentration is increased. Also, the main peak detected at 611 nm properties exhibited a red phosphor. So, a novel thick film nanocrystalline phosphor with red color could be fabricated in this study.

REFERENCES

1. R. Schmechel, H. Winkler, Li Xaomao, M. Kennedy, M. Kolbe, A. Benker, M. Winterer, R.A Fischer, H. Hahn, and H. von Seggern, *Scripta mater.*, **44**, 1213, (2001)
2. J. Zhang, Z. Tang, Z. Zhang, W. Fu, J. Wang, and Y. Lin, *Mater. Sci. Eng. A*, **334**, 246 (2002)
3. S.J. Hong, J.I. Han, H.J. Kim, H.K. Chang, and C. Kim, *Proc. Pacific Rim 2002*, 187 (2002)
4. J. McKittrick, C.F. Bacalski, G.A. Hirata, K.M. Hubbard, S.G. Pattillo, K.V. Salaza, and M. Trkula, *J. Am. Ceram. Soc.*, **83**, 5, 1241 (2000)
5. R. Schmechel, M. Kennedy, von Seggern, H. Winkler, M. Kolbe, R. A. Fischer, Li Xaomao, A. Benker, M. Winterer, and H. Hahn, *J. Appl. Phys.*, **89**, 1679 (2001)

Mat. Res. Soc. Symp. Proc. Vol. 789 © 2004 Materials Research Society N3.28

The Effect of Gate Geometry on the Charging Characteristics of Metal Nanocrystal Memories

Authors: *Anirudh Gorur-Seetharam*, *Chungho Lee* and *Edwin C. Kan*
Organization: School of Electrical and Computer Engineering, Cornell University

Abstract

This study presents the effect of gate geometry on the charging characteristics of metal nanocrystal memories. The effect is studied by varying the perimeter to area ratio, number of convex corners and concave corners of the gate of a metal-oxide-semiconductor (MOS) capacitor with embedded gold nanocrystals. It can be observed that the nanocrystal charging rate increases for a smaller perimeter to area ratio. The presence of concave and convex corners increases the nanocrystal charging rate. Based on this study it is expected that gate geometries with low perimeter to area ratio and with selected convex and concave corners would increase the nanocrystal charging rate.

Introduction

The difference between the two main commercially available memory systems, dynamic RAM (DRAM) and Flash memories, is that DRAM provides fast read/write capability, whereas Flash memories have the advantage of nonvolatile data retention. Metal nanocrystal memories can be designed to have a good combination of both advantageous features. In metal nanocrystal memories the effective potential well designed by metal work function engineering enables the use of an oxide in the direct tunneling regime, where the oxide endurance in repeated write/erase cycles dramatically improves [2]. Also the discrete storage nodes of metal nanocrystal memories provide advantages in the design tradeoffs of write/erase voltage and retention time [1].

The effect of the control gate geometry on the charging characteristics of metal nanocrystal memories was studied by varying the control gate geometry of a metal-oxide-semiconductor (MOS) capacitor with embedded gold nanocrystals. We have considered three types of gate geometries: a circle, a square and a quasi-fractal pattern [3] with the same area. The geometric differences between the three structures are the perimeter to area ratio, the number of convex and concave corners. In this paper, we will present our device design, fabrication process, and structural and electrical characterization.

Device design and fabrication

An ultra thin (~2 nm) layer of silicon dioxide (tunnel oxide) is first grown thermally on a boron doped (10^{17} cm^{-3}) p-type silicon substrate. On this tunnel oxide gold nanocrystals are formed by a self-assembly process [2]. A 30nm PECVD oxide (control oxide) is then deposited followed by a 200nm chromium layer deposition by e-beam evaporation. The device structure is schematically shown in Fig 1. This chromium layer is patterned by photolithography into three geometric shapes: circle, square and quasi-fractal (Fig. 2). The area charge density Q in the oxide consists of charge stored in the nanocrystal, fixed oxide charge, mobile oxide charge, and silicon-oxide interface state trapped charge. The amount of charges stored in the nanocrystal scales with the writing voltage V_W till all of the energy states in the potential well of the nanocrystal are filled. From the Gauss law, the flat band voltage shift ΔV_{FB} in the high frequency capacitance-voltage (HFCV) measurement directly scales with Q (Eq. (1)). From Eq. (2) we observe that the slope of ΔV_{FB} vs. V_W indicates the nanocrystal charging rate with respect to the writing voltage V_W.

$$\Delta V_{FB} = \frac{xQ}{\varepsilon_{ox}} \tag{1}$$

where x is the distance of the nanocrystal layer from the gate and ε_{ox} is the permittivity of oxide.

$$\frac{d(\Delta V_{FB})}{d(V_W)} = \frac{x}{\varepsilon_{ox}} \frac{dQ}{d(V_W)} \tag{2}$$

Structural and electrical characterization results

From scanning transmission electron microscopy (STEM) images in Fig 3(a), the nanocrystals have formed on a 2 nm thin tunnel oxide. Convergent beam electron diffraction (CBED) images in Fig 3(b) confirm the crystallinity of the embedded metal dots. Scanning electron micrographs (SEM) in fig 3(c) show that the nanocrystal number density and the size distribution are quite uniform.

HFCV measurements were made on devices with different gate geometries. To estimate statistical fluctuations, measurements were repeated on three to five identical devices. Figures 4(a-c) show HFCV plots for capacitors with a control-gate area of 8464 sq. microns in different gate geometries. The flat band voltage shift ΔV_{FB} is estimated by defining a flat-band capacitance as a parameter λ times the strong inversion capacitance value, because the voltage sweep to MOSC accumulation may change the nanocrystal charge state. For larger areas, the value of λ was taken as 1.6 while at smaller areas 1.3 is used. The minor difference results from the fact that the maximum capacitance value reached in some smaller devices was less than 1.6 times of the strong inversion capacitance since the maximum gate voltage is limited due to concerns of changing the nanocrystal charge state.

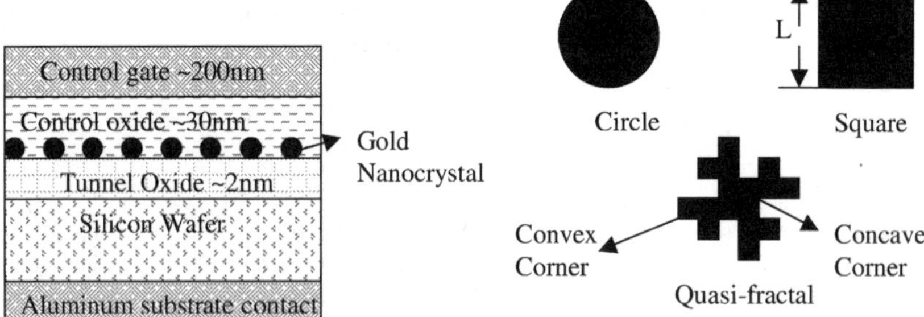

Fig. 1-Device Structure

Fig. 2- Control Gate shapes considered.

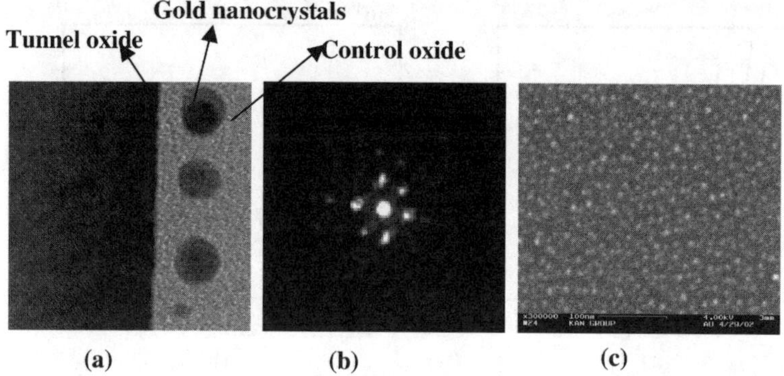

Fig. 3(a) An STEM image of gold nanocrystals on the ultra thin tunneling oxide; (b) A CBED image of the gold nanocrystal; (c) A SEM image showing the distribution and number density of the gold nanocrystals.

This extraction was repeated for identical devices and a least-square-fit technique was used to obtain the ΔV_{FB} vs. V_W plot in Fig. 4(d). The procedure was carried out for devices with different gate areas and the slope of ΔV_{FB} vs. V_W is plotted in Fig. 5.

3D device simulations [4] were made to find the magnitude of the electric field perpendicular to the control gate at various locations (Figs. 4(e) and 4(f)). From Fig. 4(e) for the case of a concave corner, the electric field remains the same until the corner is reached and then begins to decrease. Also, the field at the center of the gate is larger than that along the edge. In Fig. 4(f), the field begins to decrease in magnitude as a convex corner is reached and becomes zero some distance away from the corner. Comparing the magnitude of the field, E, at these locations we see the following relation holds.

$$(E)_{\text{center}} \cong (E)_{\text{concave corner}} > (E)_{edge} > (E)_{\text{convex corner}} \quad (3)$$

With consideration of the scattered data in Fig 5, we can make the following observations:

a) For large device areas, the nanocrystal charging rate with respect to the writing voltage is largest for the circular gate.
b) For smaller device areas, the devices with square and circular gates show similar nanocrystal charging rates.
c) Devices with quasi-fractal gates show lower nanocrystal charging rates than those with circular and square gates.

From Table 1, the geometric difference between the circle and square is that the circle has a smaller perimeter to area ratio than the square, and the square has 4 convex corners. At large device areas, the effect of corners is expected to be negligible and the effect on the nanocrystal charging rate should arise primarily from the perimeter to area ratio. The nanocrystal charging rate increases as the perimeter to area ratio decreases according to observation (a). This result is corroborated by the simulation result that the

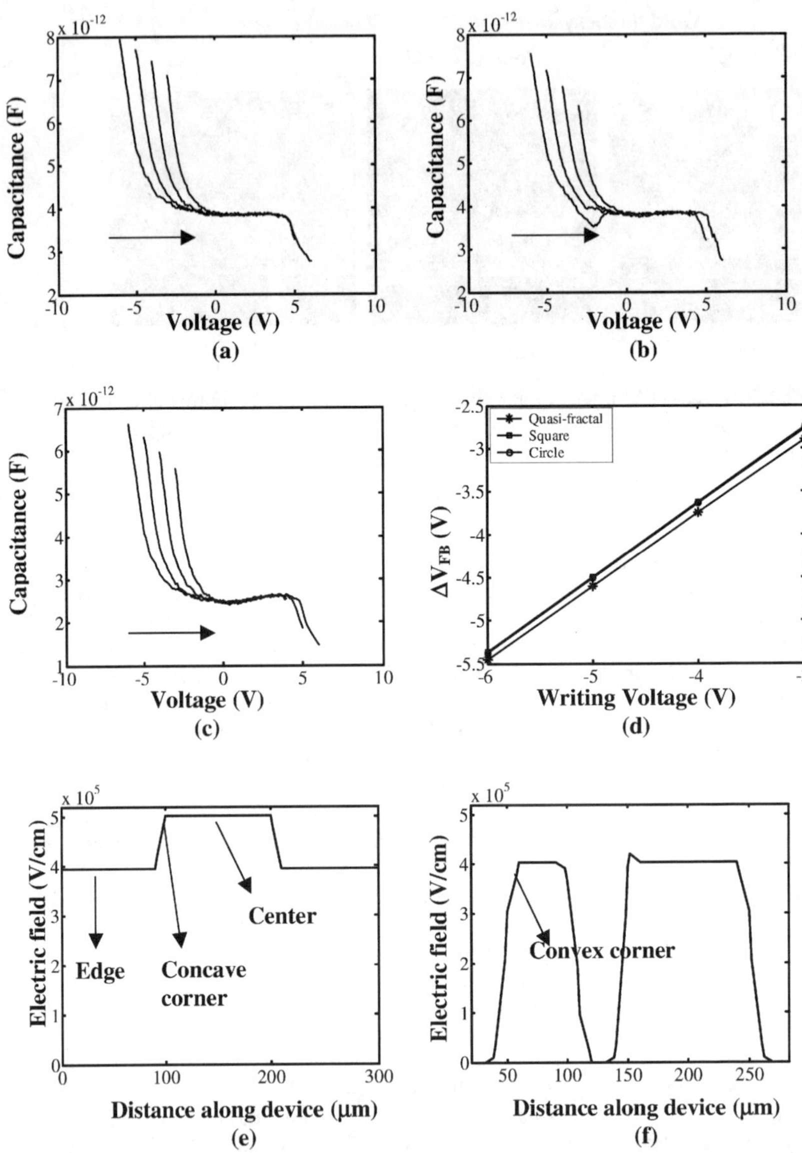

Fig. 4(a), (b) and (c) are CV plots of devices with a circular gate, a square gate and a quasi-fractal gate of area 8464 sq microns. Fig 4(d) is the extracted flat band voltage shift vs. writing voltage for the 3 different gate geometries. Fig. 4 (e) and (f) are the electric field data obtained by 3D device simulations [4].

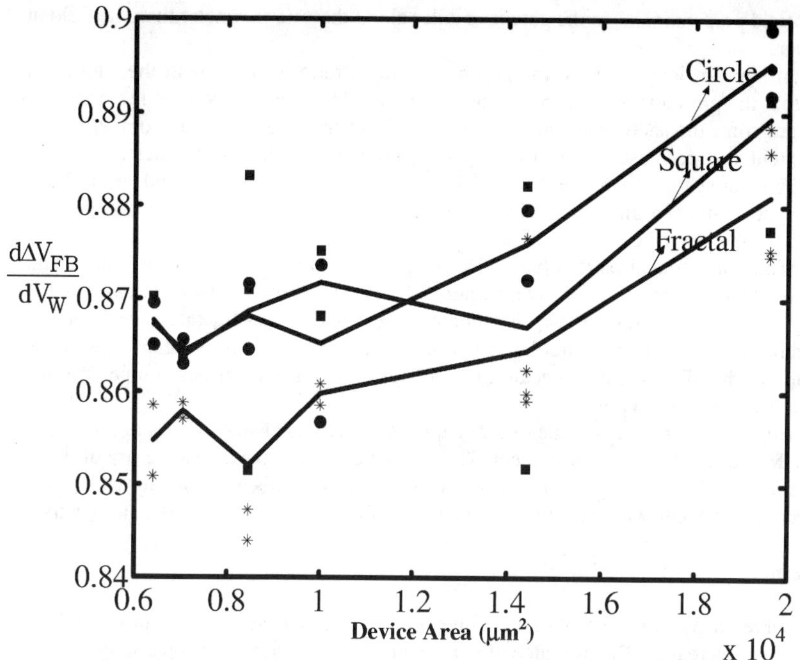

Fig. 5: Plot of the slope of ΔV_{FB} vs. V_W for various device areas. Legend: ● - Circular device data; ■ - Square device data; * - Quasi-Fractal device data. The solid lines represent the mean value of the data.

Shape of the control gate	Area (A)	Perimeter (P)	P/A	Number of Convex corners	Number of Concave Corners
Circle	L^2	$(2\sqrt{\pi})L$	$\frac{2\sqrt{\pi}}{L}$	0	0
Square	L^2	4L	$\frac{4}{L}$	4	0
Quasi-Fractal pattern	L^2	8L	$\frac{8}{L}$	16	12

Table 1: Geometric details of the three different shapes. L is the value of the side of the square. The area of the three gate geometries is made equal to L^2.

electric field perpendicular to the gate is smaller along the gate perimeter than the field at the center.

At smaller devices areas, the perimeter to area ratio is larger with the additional effect from the presence of the convex corners. From 3D simulation, we see that the field diminishes some distance away from the corner. These fringing fields can compensate to some extent the decrease in the nanocrystal charging rate due to an increase in the perimeter to area ratio .We see from observation (b) that the square and circle have similar nanocrystal charging rates. This shows that the convex corners indeed compensate the effect of the perimeter to area ratio.

The quasi-fractal pattern has twice the perimeter to area ratio of the square and 16 convex corners and 12 concave corners (Table 1). However, the decrease in the nanocrystal charging rate between square gates and quasi-fractal gates appears comparable to that between circular gates and square gates. This is again reasonable according to the 3D simulation because the convex and concave corners compensate the effect of the perimeter to area ratio.

However we notice that the difference in the electrical characteristics of the various devices is at most a few percent. This could be because the gate areas are of the order of several thousand square microns. We hope to clarify this issue by fabricating and characterizing devices with gate areas of the order of a few thousand square nanometers.

Conclusion

It is observed that as the control gate perimeter to area ratio decreases the nanocrystal charging rate increases. The presence of convex and concave corners can partially compensate for the decrease in the nanocrystal charging rate. From 3D simulation, it is expected that the concave corners should be more effective than convex corners. However from the present setup we cannot decouple the effect of concave corners as the quasi-fractal has both types of corners. In summary, a control gate geometry design which would maximize the nanocrystal charging rate and thus favorably reduce the writing/erasing voltage would have the smallest possible perimeter to area ratio along with a few convex and concave corners.

Acknowledgements

This project is supported by the ECS division of NSF. We acknowledge helpful discussions with the technical staff of the Cornell Nanoscale Facilities (CNF).

References

[1] Z. Liu, C. Lee, V Narayanan, G. Pei, E.C. Kan, IEEE Trans. Electron Devices, vol. 49, no. 9, Sept. 2002.

[2] C. Lee, Z. Liu and E.C. Kan, (MRS) Materials Research Symposium, Boston, MA, proc. vol. 737, F8.18, Dec. 2002.

[3] H. Samavati, A. Hajimiri, A.R. Shahani, G.N. Nasserbakht and T.H. Lee, IEEE J. Solid State Circuits, vol. 33, no.12, Dec 1998.

[4] ATLAS Device Simulation, Silvaco international, 2003.

Mat. Res. Soc. Symp. Proc. Vol. 789 © 2004 Materials Research Society N3.30

Tunable Optical Absorption of Composites of Nanocrystalline Copper Prepared by *in situ* Chemical Reduction within a Cu^{2+}-Polymer Complex

Cheng Huang,[1] Gang Huang,[2] and C. Z. Yang[3]
[1]Materials Research Institute and Electrical Engineering Department, The Pennsylvania State University, University Park, PA 16802, cxh57@psu.edu
[2]Electronic Materials & Thin Film Devices Division, Physics Dept., Suzhou University, 215006, China
[3]Department of Polymer Science and Engineering, Nanjing University, Nanjing 210093, China

ABSTRACT

Research on nanocrystalline materials and the physics behind their properties have attracted considerable attention. A number of physical and chemical techniques have been used to synthesize different nanomaterials and nanocomposites. Optical absorption characteristics of composites containing nanosized metals or semiconductors have been investigated for potential applications in nonlinear optics and photonic crystals and also to understand the effect of particle size on the band gap of the material concerned. These materials show a large third-order nonlinear susceptibility. A polymer-matrix nanocomposite containing copper particles has been prepared by *in situ* chemical reduction within a polymer-metal complex solid film. The copper particle size in the order of 10 nm is controlled by the initial content of the metal ions in the complex. Their fractal pattern and the value of the fractal dimension indicate that there exists a cluster-cluster aggregation (CCA) process in the present system. Optical absorption spectra of copper-polymer nanocomposites show distinct plasma absorption bands and quantum size effect in the samples. More studies on optical properties of composites containing nanosized metals are within the Drudeframe on the basis of Mie theory, but the electrons behave in a wavelike rather than a particlelike way as the particle size decreases to below 10 nm, and the classical Drude model should be modified considering the quantum confinement effect. In this paper, the calculated blueshift of the resonance peak based on a quantum-sphere model (QSM) proposed by Huang and Lue, gives remarkable agreement with the experimental data as the size of copper particles embedded in the polymer becomes smaller.

INTRODUCTION

Research on nanocomposites and their properties has attracted considerable attention in recent years [1,2]. A wide variety of physical and chemical synthetic approaches have been applied to the preparation of nanocomposites [3]. The association of subunits to form large clusters in metrices is still a phenomenon central to many synthetic processes of nanocompoaites. The Witten-Sander model [4] provides a basis for obtaining a better understanding of a variety of diffusion-limited processes, e.g., an metal electrodeposition process [3], however the features of this model are unrealistic for many other real colloidal system, such as metal colloids, soot, and coagulated aerosols [5], and later Meakin further developed a model for diffusion-controlled aggregation in which growing clusters as well as individual particles are mobile [6]. Optical absorption characteristics of composites containing nanosized metals or semiconductor have been investigated for potential applications in nonlinear optics and also to understand the effect of particle size on the band gap of the material concerned [1,2,7]. More studies on optical properties of composites containing nanosized metals [3,8] are within the Drudeframe on the basis of Mie theory [9], but the electrons behave in a wavelike rather than a particlelike way as the particle size decreases to below 10 nm, and the classical Drude model should be modified considering the quantum confinement effect [10,11]. In this paper a new method of making a metal-polymer nanocomposite via *in situ* chemical reduction within a Cu^{2+}-poly(itaconic acid-co-acrylic acid) complex solid film will be described. Fractal aggregation of these particles within the polymer matrix was observed, and tunable optical properties of the copper-polymer nanocomposite have also been investigated.

EXPERIMENTAL DETAILS

The random copolymer of itaconic acid and acrylic acid in the molar ratio about 2:3 synthesized by solution polymerization was used in the present experiment. A 5wt% solution of the polymer with the number-average molecular weight above 50 000 in N,N'- Dimethyl formamide was mixed with a series of stoichiometric amount of $Cu(CH_3CO_2)_2 \cdot H_2O$. After about 4h of stirring at 80^0C, the mixture was poured into a Teflon-casting plate and the transparent film ~20μm thickness, was formed by evaporating the solvent. Then the Cu^{2+}-polymer complex film was reduced by the dropwise addition of 25ml of 50.0wt% hydrazine hydrate aqueous solution with stirring at room temperature for a suitable time to produce the ultrafine metallic particles, dispersed in polymer matrix. For microstructural studies, the film (~80nm) was spin cast on a carbon-coated copper grid, then the copper grid with the chemically reduced film was mounted mainly on a JEOL TEM-200CX transmission electron microscope and the microstructure was investigated. The accelerating voltage is 200 kV. The optical absorption spectra were recorded a room temperature on a Shimadzu UV-31000 UV-VIS-NIR spectrophotometer over the wavelength range from 400nm to 900nm, and the films were spin coated on quartz substances. The typical thickness of the chemically reduced films is of the order of 2μm.

RESULTS AND DISCUSSION

Fractal Aggregation of Copper Nanoparticles Prepared by *in situ* Chemical Reduction within a Cu^{2+}-Polymer Complex

It has been proved that the metal ions in some macromolecule-metal complexes have a tendency to aggregate, forming nanoscale ionic domains owing to their electrostatic or dipole-dipole interactions [11]. Figure 1(a) shows the typical electron micrograph of one of the Cu^{2+}-poly(itaconic acid-co-acrylic acid) complex solid films, and there are black areas surrounded by gray areas, which is believed to be loose-packed ionic domains close up to each other. Selected area electron diffraction in Fig.1(a) indicates that there was no crystal phase existing in the polymer matrix. Figures 1(b) shows the electron micrograph of one of the copper-polymer nanocomposites at a magnificent of 100 000. The distinct diffraction rings in Fig 1(b) have been identified to correspond to the copper crystal, which indicates the copper ions can be transferred into copper nanoparticles in the polymer matrix by controlling the exposure of the metallic ionic sites to the reducing agent. Moreover, the ultrafine particles stick together to form chain-like aggregates, and it is evident that a ramified network fractal pattern is present in this system.

The fractal dimensions of these copper clusters in quasi-two-dimensional films have been estimated by the method delineated by Forrest and Witten [12]. The micrograph is first of all digitized by computer treatment to produce a matrix of 1's or blanks corresponding to the presence or absence of a particle. Different squares are then chosen in the image and the numbe

N of 1's in each square is counted. According to the power law: $N \propto L^D$, where L is the size of the square chosen and D is the value of the fractal dimension. In Fig. 2 the plot is shown well corresponding to the micrographs given in Fig. 1(b). The value of D estimated from the slope is of 1.75±0.02 over the length scale 100-10 000nm, and this value is in reasonable agreement with the D value predicted on the basis of a Cluster-Cluster Aggregation(CCA) model [5,6]. It is indicated that ionic domains can be acted as heterogeneous nucleation sites of nanoparticles during the chemical reduction process. Upon reduction, the metal atoms in the local structure aggregate to form the close-packed particles, meanwhile, drastic movement of the polymer chains and the change of interaction between the polymer chains and the particles, which ascribed to the change of volume occupied by the particles and the migration of copper atoms, results in the movement and aggregation of the growing particles in this system.

The particle size histograms of the nanocomposites synthesized as above have been fitted by a log-normal distribution function. Figure 3 shows the relationship between molar ratio of Cu^{2+} to carboxyl of the polymer and the copper particle size. It is noted that the particle size increases with the loading of metal in the polymer matrix, and the nanocomposites with different particle size can be prepared by starting with polymer precursors having different ion contents.

The evaluation of the particle size is also determined from the line width of the diffraction peaks in the X-ray diffraction patterns (Fig.4) using Scherrer's equation [13]:

$$D_{hkl} = {k \bullet \lambda}/{\beta \cos\theta}, \qquad (1)$$

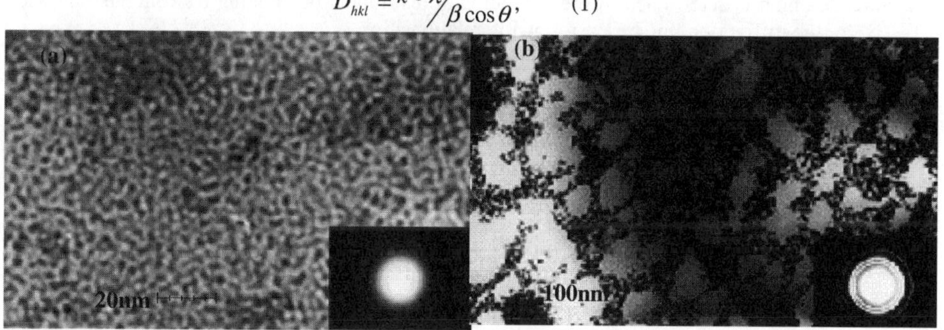

Figure 1. (a) Electron micrograph and selected area electron diffraction pattern of the Cu^{2+}-polymmer complex with the initial molar ratio of Cu^{2+} to carboxyl 1: 1. (b) Electron micrographs of the same copper-polymer nanocomposite at a magnificent of 100 000, and selected area electron diffraction pattern of (b).

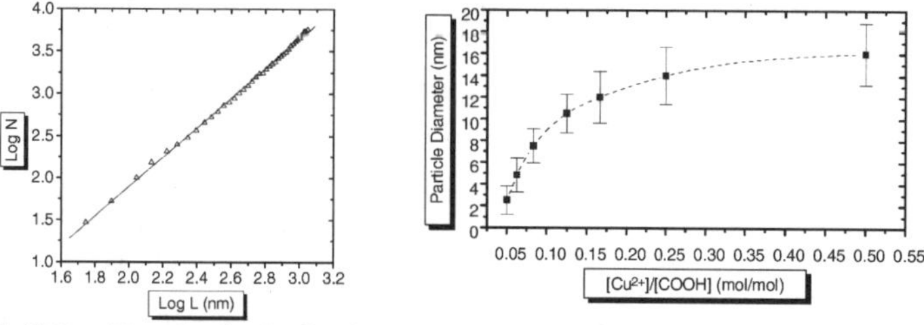

Figure 2. (left) Log N vs. log L plot for the sample shown in Fig.1 (b).
Figure 3. (right) Relationship between molar ratio of Cu^{2+} to carboxyl of the polymer ($[Cu^{2+}]/[COOH]$) and the copper particle size.

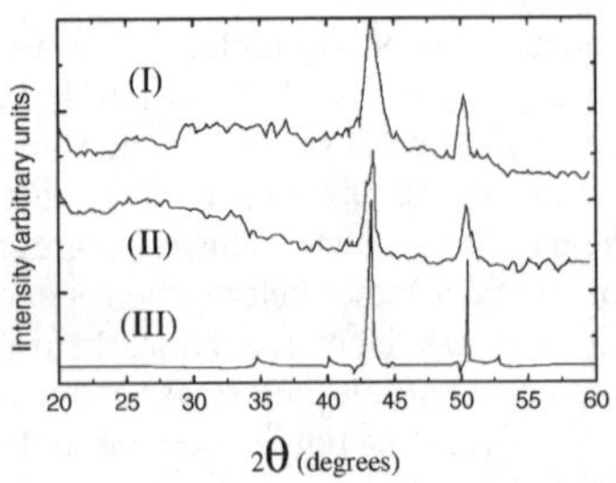

Figure 4. XRD patterns of the complex films reduced by $NH_2{\cdot}NH_2$ (I) and $NaBH_4$ (II). The pattern for bulk copper metal (III) is shown for comparison.

where D_{hkl} is the linear dimension of coherent areas in different crystallographic planes (hkl), which dominantly influence the broadening of diffraction lines neglecting the influence of microstresses on the basis of our preparation method of the experimental materials. λ is the X-ray wavelength (λ= 0.1542nm). θ is the Bragg diffraction angle of the diffraction peak, and k is geometric factor taken to be 0.9 if β is the half width of the diffraction peak. The physical broadening β is determined by the separation of instrument broadening B_S from the total width B_M of the diffraction line by the following relationship:

$$\beta = B_M - B_S, \qquad (2)$$

where B_S, the instrumental broadening was determined by a standard sample.

It is shown that the particle sizes calculated from XRD correspond well with the data obtained by TEM. In general, the particles reduced by hydrazine are typically smaller than those reduced by $NaBH_4$[14], which may be related to the reductibility or nature of the reducing agents[15].

Quantum-Sphere Model (QSM) of Metallic Nanoparticles

The quantum-sphere model (QSM) [10] was proposed for the evaluation of the dielectric function of small metallic particles. The QSM infers *N* independent electrons confined in sphere of diameter *D*. Based on Schrödinger equation

$$H\psi = [-\frac{\hbar^2}{2m}\nabla^2 + V(r)]\psi = E\psi, \qquad (3)$$

and the asymptotic approximations of the roots of the spherical Bessel functions

$$\alpha_{nl} \cong (2n+l)\tfrac{\pi}{2} \qquad (4)$$

for large n. The spacing of neighboring energy level can be simplified to

$$\Delta E_l = E_l - E_{l-1} \cong E_0[(2n+l)^2][\tfrac{\pi}{2}]^2 - E_0[(2\pi+l-1)^2][\tfrac{\pi}{2}]^2 \cong (2l+4n)E_0[\tfrac{\pi}{2}]^2 = \pi(E_0E_l)^{1/2}. \quad (5)$$

The dielectric function can be written as

$$\varepsilon^{total}(\omega,R) = \varepsilon_\infty + \varepsilon^f + \varepsilon^i + \varepsilon^j \cong [\varepsilon_1^f(\omega) + i\varepsilon_2^f(\omega)] + [\varepsilon_1^{QM}(\omega,R) + i\varepsilon_2^{QM}(\omega,R)]$$

$$= 1 - \frac{4\pi ne^2}{m\omega(\omega+i\tau^{-1})} + \frac{4\pi}{3V}\frac{1}{\hbar\omega^2}\sum_{m,n}\rho_{mn}^{(0)}|\pi_{mn}|^2[\frac{1}{\omega_{mn}-\omega-i\tau^{-1}} + \frac{1}{\omega_{mn}+\omega+i\tau^{-1}}], \quad (6$$

where V is the volume of the spherical particles ($= \frac{1}{6}\pi D^3$), the QM presents the parts contributed from the quantum-sphere model.

Tunable Optical Absorption and Quantum Size Effect of Copper Nanoparticle/Polymer Composites

In order to investigate the optical properties of such particles embedded in the polymer matrix, the optical absorption spectra in the region of plasmon resonance absorption were measured. Figure 5 shows the absorbance as the function of wavelength for all the samples mentioned in Fig. 3. in the visible region. The measured positions of the plasmon peak of various samples are compiled in Figure 6. An evident quantum size effect of the shift of the plasma resonance has been observed in the present experiment. It is clear that there exists a blue shifting of the peak with the decrease of the particle size. Also, the theoretical curves of position of the plasmon peak of copper particles versus the particle diameters were calculated. To find the resonance peak λ_D, we derive the wavelength λ_D from the differential equation

$$\partial\alpha/\partial\lambda\big|_{\lambda=\lambda_D} = 0, \qquad (7)$$

where the absorption coefficient α of the nanocomposite system can be derived from Mie's scattering [9] based on the dielectric function of the metal particles with diameter D, and the dielectric function is calculated based on the Drude expression from free-electron classical theory [16] and quantum state transitions with wave function solved from a quantum-sphere model (QSM) proposed by Huang and Lue [10], respectively. It appears that there exists a distinction between the classical conduction electron theory(free path effect) (curve B) and the quantum size calculations (curve A), and the latter are favored by the experimental results, while the Drude expression is in contradiction to the results in smaller range of the particle size, which indicated that the metal particles become so smaller that the conduction band breaks up into discrete levels separated by energies large compared to thermal energies, and the Drude expression is no longer valid. It must be replaced by a more realistic, quantum-mechanically derived model. In general, the classical size effect presumes the correction to the dielectric function with a reduction of the

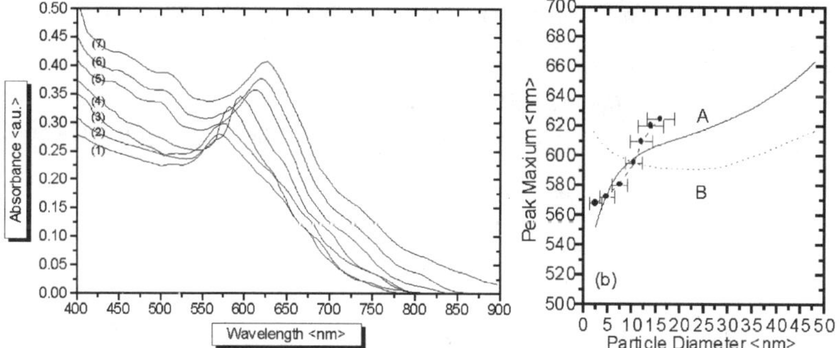

Figure 5. (left) Absorption spectra of copper particles in polymer matrix. The average size ranges from 2.5nm (1) to 16.0nm (7) shown in Fig.3.

Figure 6. (right) Position of the plasmon peak of copper particles vs. the particle diameter. Points: experimental values; curve A: calculated from the dielectric function based on the QSM; curve B: calculated from the Drude free-electron theory.

effective relaxation by the Drude expression to be only due to the scattering of electrons with a spherical boundary, while the blueshift from the QSM is mainly due to the choice of an infinite barrier, by which the electron density drops suddenly from its bulk value to zero at the spherical boundary, which results in the quantum confinement effect of electrons.

CONCLUSIONS

In conclusion, the metal ions in a polymer matrix can be reduced to form nanoparticles by controlling the exposure of the metallic ionic sites to a reducing agent and a copper-polymer nanocomposite was obtained. There exists a CCA process in the present system. The embedded copper particles exhibit quantum confinement effect with the enhanced optical resonance at visible region, and the measured size dependence of the shift of the plasma absorption was shown to be satisfactorily coincident with results calculated on a quantum mechanically derived dielectric function as the particle size decreases to below 10 nm. It should be mentioned here that the interparticle coupling effects on plasmon resonances of metal nanoparticles become very important when a cluster of metal nanoparticles are placed in close proximity to one another [17]. The resonant wavelength peak of two interacting particles is also red-shifted from that of a single particle because of near-field coupling. As shown in Figure 6, the difference between the experimental values and the calculated results as the particles become bigger maybe contributed to nanoparticle plasmon coupling as the interparticle spacing is reduced with the increasing of the metal loading. The collaborative oscillation of conductive electrons in metal nanoparticles which results in a surface plasmon resonance makes them useful for various applications including biolabeling. Synthesis and tunable optical absorption of metal nanoparticles in polymer matrix also offers potential for nonlinear optics and producing metallo-dielectric photonic crystals.

REFERENCES

1. K. Lee and S.A. Asher, *J. Am. Chem. Soc.* **122**, 9534 (2000).
2. S. Xu, J. Zhang, C. Paquet, Y. Lin and E. Kumacheva, *Adv. Funct. Mater.* **13**, 468 (2003).
3. S. Roy and D. Chakravorty, *Appl. Phys. Lett.* **59**, 1415 (1991).
4. T.A. Witten and L.M. Sander, *Phys. Rev. Lett.* **47,** 1400 (1981).
5. D.A. Weitz and M. Oliveria, *Phys. Rev. Lett.* **52**, 1433 (1984).
6. P. Meakin, *Phys. Rev. Lett.* **51,** 1119 (1983).
7. H.Z. Wang, F.L. Zhao, Y.J. He, X.G. Zheng, X.G. Huang and M.M. Wu, *Opt. Lett.* **23**, 777 (1998).
8. S. Banerjee and D. Chakravorty, *Appl. Phys. Lett.* **72,** 1027 (1998).
9. G. Mie, *Ann. Phys.* **25,** 377(1908).
10. W.C. Huang and J.T. Lue, *Phys. Rev. B* **49**, 17279 (1994).
11. C. Huang and C.Z. Yang, *Appl. Phys. Lett.* **74**, 1692 (1999).
12. S.R. Forrest and T.A. Witten, Jr., *J. Phys. A* **12,** L109 (1979).
13. H.P. Klug and L.E. Alexander, *X-ray diffraction procedures* (John Wiley, NY, 1954).
14. C. Huang, L. Chen and C.Z. Yang, *Polym. Bull.* **41**, 585 (1998).
15. I. Lisiecki and M.P. Pileni, *J. Am. Chem. Soc.* **115**, 3887 (1993).
16. U. Kreibig, *J. Phys. F: Metal Phys.* **4**, 999 (1974).
17. K.H. Su, Q.-H. Wei, X. Zhang, J.J. Mock, D.R. Smith, and S. Schultz, *Nano Letters* **3**(8), 1087 (2003).

Mat. Res. Soc. Symp. Proc. Vol. 789 © 2004 Materials Research Society N3.34

Synthesis of boron nitride nanolayers encapsulating iron fine particles and boron nitride nanotubes

Hisato Tokoro[1], Shigeo Fujii[1], Takeo Oku[2] and Shunsuke Muto[3]

[1]Hitachi Metals, Ltd., Advanced Electronics Research Laboratory, Mikajiri 5200, Kumagaya, Saitama, 360-0843, Japan
[2]Osaka University, Nanoscience and Nanotechnology Center, Institute of Scientific and Industrial Research, Mihogaoka 8-1, Ibaraki, Osaka, 567-0047, Japan
[3]Nagoya University, Department of Nuclear Engineering Graduate School of Engineering, Furo-Cho, Chikusa-ku, Nagoya, 464-8603, Japan

ABSTRACT

Boron nitride (BN) nanolayers encapsulating iron (Fe) fine particles have been synthesized by annealing mixtures of hematite (α-Fe_2O_3) and boron powders at 1373 K for 2 hours in nitrogen atmosphere. The Fe particles had an average diameter of ~300 nm with BN nanolayers coating of ~10 nm. The α-Fe_2O_3 was transformed into Fe and then Fe-B on a process of annealing. The Fe-B was decomposed into Fe and BN, and consequently Fe particles coated with BN nanolayers were synthesized. They showed soft magnetic properties with coercivity of 1.5 kA/m. The BN nanolayers encapsulation was effective on improving oxidation resistance. BN nanotubes with diameter of ~100 nm were also synthesized as a resultant product by this method.

INTRODUCTION

Magnetic metal nanoparticles are suitable for potential applications like high-density magnetic recording media [1], magnetic fluids [2], magnetic carrier in clinical cure [3] and other novel magnetic devices. Iron (Fe) or Fe-based alloy nanoparticles have an advantage of high saturation magnetization for these applications, whereas deterioration by oxidation has kept them away from practical usage.

To overcome this problem, nanocoating techniques for the metal nanoparticles have been reported. An arc discharge method has been proposed to coat metal nanoparticles with graphite carbon [4 – 6] or boron nitride (BN) [7, 8]. Surface oxidization of Fe and Co nanoparticles has also been studied [9, 10]. However, the arc discharge method is not considered to be suitable for a mass production because of low yield. The surface oxidation takes long times and causes decrease of saturation magnetization.

We have discovered a simple method that can synthesize BN nanolayers encapsulating Fe fine particles massively for industry by employing α-Fe_2O_3 and B powders as starting materials. A reaction process of this method, morphology of resultant products and magnetic properties are discussed in this paper.

EXPERIMENTAL

Commercially available α-Fe_2O_3 (hematite, 99.7%) and crystalline boron (99%) powders were used as starting materials. Their particle sizes were 30 nm and 20 μm (325 mesh under), respectively. Ratio in weight of α-Fe_2O_3 to B was 1:1. A differential thermal analysis (DTA-TG) was carried out from 300 K to 1773 K at temperature rise of 10 K/min in N_2 atmosphere in order to study reaction process between Fe_2O_3 and B from a thermodynamic point of view. Samples were prepared by annealing mixtures of the α-Fe_2O_3 and B powders at 773 K, 1273 K and 1773 K for 15 minutes and 1373 K for 2 hours in a furnace under nitrogen gas flow at ambient pressure. A temperature rising rate up to each temperature was 3 K/min. They were denoted as sample A, B, C and D, respectively.

An X-ray diffracto-meter (Rigaku RINT-2500) was used to detect metallurgical phases under an applied power of 50 kV and 250 mA with a Cu Kα irradiation. A goniometer was scanned at a step of 0.02 ° from 2θ of 20 to 80 ° in a 2θ/θ mode. A scanning speed was 1 ° per minute. A HRTEM was employed under an acceleration voltage of 300 kV in order to reveal microstructures of samples. Composition was also analyzed by an energy dispersive X-ray analysis (EDX) or an electron energy loss spectroscopy (EELS) equipped with TEM. Magnetic properties of the samples were measured by a VSM. Acrylic capsules (diameter: 6 mm, thickness: 8 mm) were used to pack the samples for this measurements. A DC magnetic field of 0.8 MA/m was applied to take hysteresis loops. The field was also applied in order to investigate variation of saturation magnetization due to heating from room temperature to 1073 K at temperature rise of 10 K/min in the air.

RESULTS

DTA-TG was taken as shown in Fig. 1 in order to reveal how the reaction between α-Fe_2O_3 and B proceed in nitrogen atmosphere. Exothermal peaks are clearly appeared at 1130 K and 1580 K in the DTA curve ('P1', 'P2' in Fig. 1). The thermal gravity (TG), which is normalized by the gravity before heating and represented '100 %' at 273 K, keeps constant value below 1473 K and abruptly increases above the temperature.

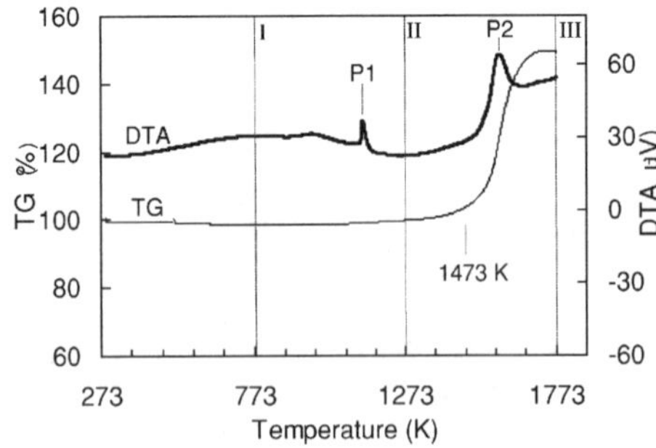

Figure 1. DTA-TG curve showing a reaction process of α-Fe_2O_3 and B.

To make the reactions in Fig. 1 clear, X-ray diffraction measurement and SEM observation were carried out. Figure 2 (a), (b) and (c) show X-ray diffraction patterns of the sample A, B and C, respectively, which were annealed at 773 K, 1273 K and 1773 K ('I', 'II', 'III' in Fig. 1). In Fig. 2(a), there are peaks assigned to Fe_3O_4 and starting materials, that is α-$Fe_2O_3$and B. This means some amount of α-Fe_2O_3 reduces to Fe_3O_4 at 773 K. The reduction of these iron oxides is completed at 1273 K and Fe-B compounds such as FeB, Fe_2B and FeB_{49} are formed as shown in Fig. 2(b). B_2O_3 is considered as a product by reaction of B with oxygen in iron oxides. The peak P1 in Fig. 1 would represent the formation of the Fe-B compounds after the reduction of Fe_2O_3. The Fe-B compounds change to α-Fe and BN at 1773 K as shown in (c). The BN consists of hexagonal (h-BN) phase and rhombohedral (r-BN) one. It is obvious that the peak P2 results from the synthesis of BN through nitridation of Fe-B compounds. The TG also increases above 1473 K. The temperature of BN generation is higher than the reported one, which is 1373 K [11]. SEM images of these samples are also shown in Fig. 2 (d), (e) and (f) at the right side of the X-ray patterns. Particles with various diameters can be seen in (d). They grow more than 10 μm in diameter at 1273 K as shown in (e) probably because of a sintering effect. Particles become fine, less than diameter of several micrometers, at 1773 K as shown in (f). This could be attributed to decomposition of Fe-B compounds and generation of BN.

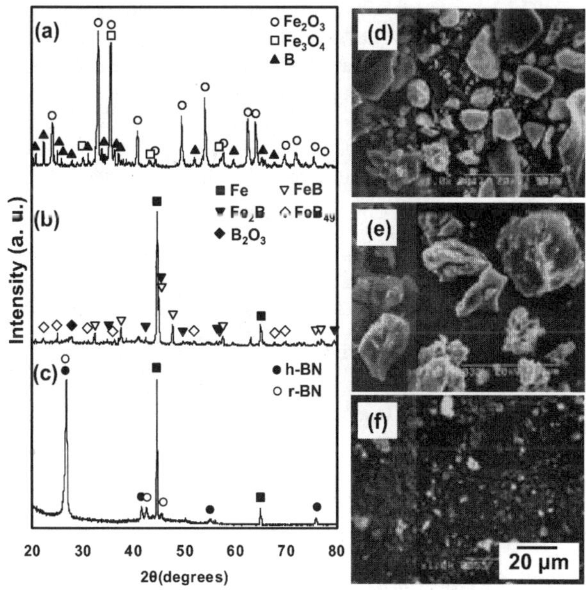

Figure 2. X-ray diffraction patterns and SEM images of sample A (a, d), sample B (b, e) and sample C (c, f).

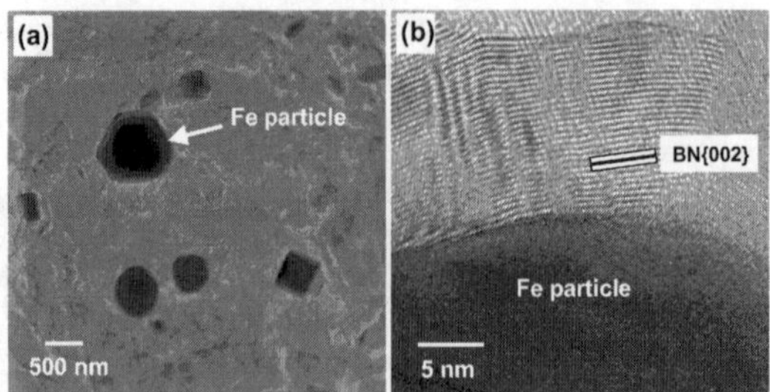

Figure 3. TEM images of sample D; (a) Fe particles, (b) detailed morphology on the surface of the Fe particle.

Figure 3 shows HRTEM images of the sample D annealed at 1373 K for 2 hours. The sample D contained same products as the sample C. Dark contrast particles can be seen in Fig. 3(a). The diameters of about 100 particles have been measured giving an average diameter of ~300 nm. EDX analysis revealed that they were mainly composed of Fe. A detailed morphology on a surface of the Fe particle is shown in (b). Nanolayers with ~10 nm in thickness can be seen on the surface. Spacing of the lattice fringes is measured as 0.33 nm, which is almost equal to that of h-BN (002) planes. So, each Fe fine particle was coated with h-BN nanolayers.

Magnetic properties of the sample D were measured by VSM. A M-H hysteresis loop shown in Fig. 4 exhibits soft magnetic properties. The saturation magnetization (Ms) and coercivity is 47.3 $A{\cdot}m^2/kg$ and 1.50 kA/m, respectively. The value of Ms is smaller than that of commercially available carbonyl Fe, that is 210 $A{\cdot}m^2/kg$, because the sample contains not only α-Fe but also h-(or r-) BN and a little amount of derivatives. The Fe particles are good soft magnetic materials due to the small coercivity.

In order to investigate oxidation resistance of the sample D, variation of the Ms as a function of temperature were measured when it was heated from 300 K to 1073 K in air (Fig. 5). The carbonyl Fe powders with an average diameter of 3 μm were measured as a control sample. Ms

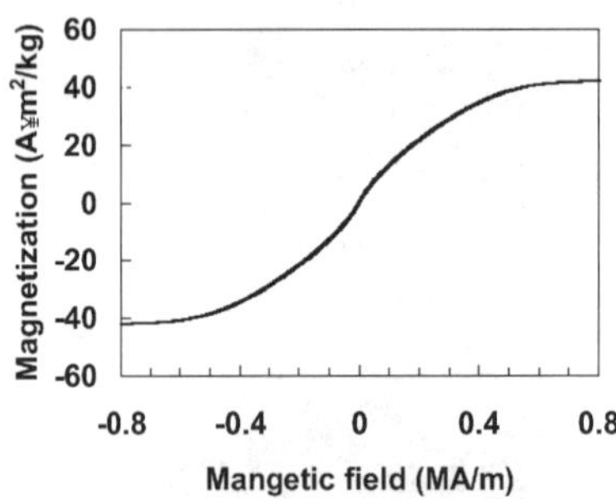

Figure 4. Magnetic hysteresis loop of the sample D at room temperature.

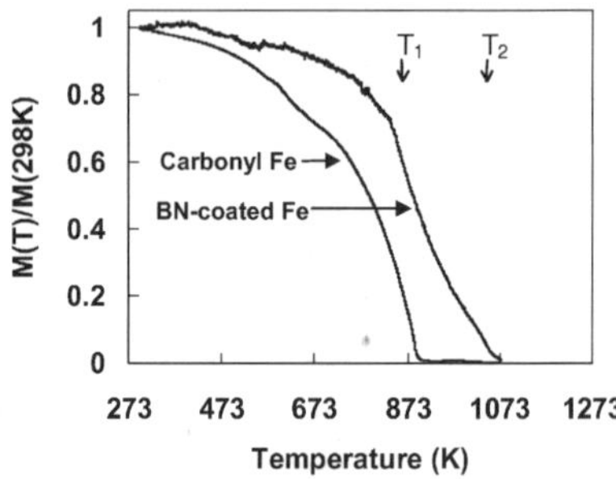

Figure 5. Variation of saturation magnetization as a function of temperature; BN-coated Fe (sample D) and carbonyl Fe (control sample).

measured at each temperature in this experiment is normalized by that at room temperature and is expressed as M(T)/M(298K). The normalized Ms of both samples decreases gradually as temperature increases and become falling to zero abruptly at a temperature. The Ms of the carbonyl Fe is zero value at ~873 K, while that of the sample D falls down to zero at ~1073 K. A Curie temperature of magnetite (Fe_3O_4) is known as 860 K ('T_1' in Fig. 5), which suggests that the carbonyl Fe has been oxidized and transformed into Fe_3O_4. The ~1073 K coincides with the Curie temperature of iron (1040 K, indexed as 'T_2' in the figure). Accordingly, Fe particles coated with BN nanolayers exhibit good oxidation resistance.

A kind of tubes can also be found in the sample D as shown in Fig. 6(a). The tube is folded, and has some nodes and something like joints inside. Most of tubes observed in our experiment had multi-walls with diameters of ~100 nm and lengths of a few micron meters. Composition of the nanotubes was analyzed by means of an EELS and its typical spectrum is shown in Fig. 6(b). There are peaks assigned to B, N and O atoms. The energy-loss near edge structures (ELNES) of B-K and N-K edges are very similar to those obtained from sp^2-hybridized BN samples [12]. Therefore, the nanotube is identified with BN. An oxygen peak is very weak and probably not assigned to the tubes. The diameters of the BNNTs are also larger than other BNNTs [13 – 16].

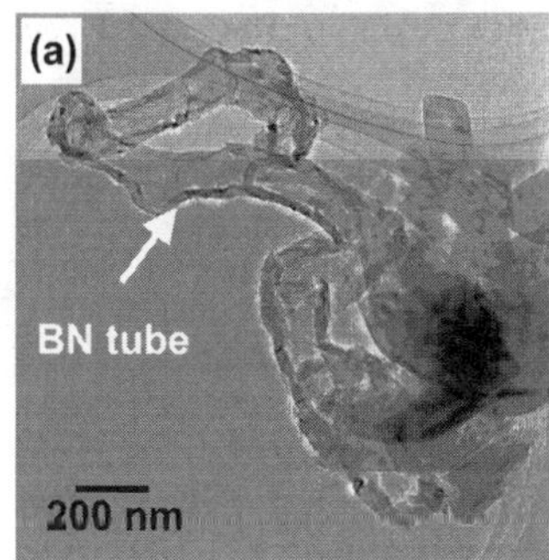

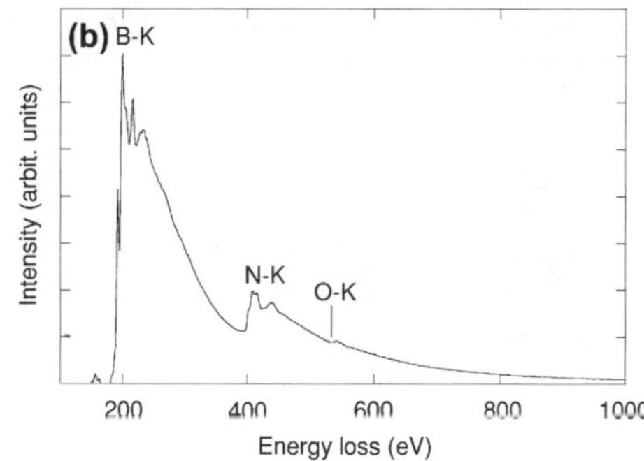

Figure 6. A TEM image of a BN nanotube in the sample D (a), and EELS spectrum of the tube (b).

DISCUSSIONS

It is revealed that BN nanolayers encapsulating Fe fine particles and BN nanotubes have been synthesized by annealing α-Fe_2O_3 and B powders in nitrogen atmosphere. The reaction formula in this synthesis can be proposed as follows:

$$Fe_2O_3 + xB \rightarrow Fe + Fe\text{-}B + B_2O_3 \qquad (x>2) \qquad (1)$$

$$Fe\text{-}B + yN_2 \rightarrow Fe + 2yBN \qquad (2)$$

B acts like a reducing agent and generates B_2O_3 by reducing α-Fe_2O_3 to Fe. The reduced Fe reacts with excess B and produce Fe-B compounds (Fig. 2). The Fe-B reacts with nitrogen and decompose into Fe and BN at above 1473 K (Fig. 1). The decomposition occurred even at 1373

K by annealing for 2 hours. The Fe-B would play an important role to generate BN. It would be considered that B in the Fe-B diffuses to the surface of particles in order to react with nitrogen, so that the BN nanolayers on the surface would be generated simultaneously with breaking large grains into pieces (~500 nm) as shown in Fig. 2(e) and (f).

BNNTs have been successfully synthesized in NH_3 atmosphere by using α-Fe_2O_3 powders as a catalyst [13]. It was also reported that BNNTs are generated from B_2O_3 and N_2 [17]. These reports suggest that α-Fe_2O_3 nanoparticles play a role as a catalyst for BNNTs, which would be attributed to the decomposition of the Fe-B, and that the BNNTs also can be fabricated from boron oxide produced by the reduction of α-Fe_2O_3. A detailed mechanism of the generation of the nanolayers and the BNNTs are under investigation.

CONCLUSIONS

The BN nanolayers encapsulating Fe fine particles have been synthesized by simply annealing the mixture of α-Fe_2O_3 and B powders in nitrogen atmosphere, namely by means of solid phase reaction. The Fe particles exhibited soft magnetic properties and good oxidation resistance. Multi-walled BNNTs with diameter of ~100 nm have also been synthesized. This method could be a promising candidate for a mass production of those BN derivatives.

REFERENCES

1. F. E. Kruis, H. Fissan, A. Peled, *J. Aerosol Sci.*, **29**, 511 (1998).
2. J. Ding, W. F. Miao, P. G. McCormick, R. Street, *Appl. Phys. Lett.*, **67**, 3804 (1995).
3. Q. Liu, Z. Xu, J. A. Finch, R. Egerton, *Chem. Mater.*, **10**, 3936 (1998).
4. Y. Yoshida, S. Shida, T. Ohsuna, N. Shiraga, *J. Appl. Phys.*, **76**, 4533 (1994).
5. Z. D. Zhang, J. G. Zheng, I. Skorvanek, G. H. Wen, J. Kovac, F. W. Wang, J. L. Yu, Z. J. Li, X. L. Dong, S. R. Liu, X. X. Zhang, *J. Phys. Condens. Matter.*, **13**, 1921 (2001).
6. S. Subramoney, *Adv. Mater.*, **10**, 1157 (1998).
7. T. Hirano, T. Oku, K. Suganuma, *Diamond and Related Mater.*, **9**, 476 (2000).
8. T. Oku, M. Kuno, H. Kitahara, I. Narita, *Int. J. Inorg. Mater.*, **3**, 597 (2001).
9. S. Sun, C. B. Murray, *J. Appl. Phys.*, **85**, 4325 (1999).
10. K. K. Fung, B. Qin, X. X. Zhang, *Mater. Sci. Eng. A*, **286**, 135 (2000).
11. K. F. Huo, Z. Hu, J. J. Hu, H. Xu, X. Z. Wang, Y. Chen, Y. N. Lu, *J. Phys. Chem. B*, **107**, 11316 (2003).
12. C. C. Ahn and O. L. Krivanek, EELS Atlas (1983), distributed by Gatan Inc.
13. C. C. Tang, M. Chapelle, P. Li, Y. M. Liu, H. Y. Dang, S. S. Fan, *Chem. Phys. Lett.*, **342**, 492 (2001).
14. Y. Saito, M. Maida, T. Matsumoto, *Jpn. J. Appl. Phys.*, **38**, 159 (1999).
15. O. Lourie, C. Jones, B. Bartlett, P. Gibbons, R. Ruoff, W. Buhro, *Chem. Mater.*, **12**, 1808 (2000).
16. L. Bourgeois, Y. Bando, T. Sato, *J. Phys. D: Appl. Phys.*, **33**, 1902 (2000).
17. Y. Bando, K. Ogawa, D. Golberg, *Chem. Phys. Lett.*, **347**, 349 (2001).

Mat. Res. Soc. Symp. Proc. Vol. 789 © 2004 Materials Research Society N3.35

Spectroscopy Studies of InP Nanocrystals Synthesized Through a Fast Reaction

Madalina Furis, David J. MacRae[1], D. W. Lucey[1] Yudhisthira Sahoo [1], Alexander N. Cartwright, Paras N. Prasad[1]
Department of Electrical Engineering, University at Buffalo,
Buffalo, NY, 14260, USA
[1]Institute for Lasers, Photonics and Biophotonics, University at Buffalo,
Buffalo, NY, 14260, USA

ABSTRACT

We present spectroscopic characterization of InP nanocrystals grown through a fast reaction in a non-coordinating solvent. The photoluminescence (PL) spectra collected from these nanocrystals exhibit a sharp feature associated with the band-edge emission and a broad infrared feature associated with deep level surface trap emission. The emission efficiencies of the as-grown nanocrystals vary between 0.3% and 1% from sample to sample. After undergoing an HF etching process, the emission efficiency increases to 18% and the emission associated with surface states is eliminated from the PL spectrum. Time-resolved photoluminescence (TRPL) experiments conducted at room temperature on the as-grown and HF-etched nanocrystals show that before etching the PL intensity decay is multi-exponential, with a fast (3ns) component independent of wavelength, associated with the non-radiative recombination processes. The etching process effectively eliminates the non-radiative component and the post-etching PL decay can be fitted with a single exponential decay characterized by long (45ns) lifetimes. We tentatively associate these long lifetimes with the recombination of carriers from spin-forbidden states. This assignment is supported by the observation of a significant redshift of the feature associated with band-edge recombination in the PL spectrum with respect to the lowest energy feature in the photoluminescence excitation (PLE) spectrum.

INTRODUCTION

Semiconductor nanocrystals are very important for potential photonic and optoelectronic applications due to the possibility of tuning the wavelength of the emission by varying the nanocrystal size [1] and because of the high emission efficiencies measured in most II-VI and III-V nanocrystals [2,3]. Historically, II-VI nanocrystals were the first ones to be synthesized due to the less restrictive conditions imposed on the growth process and are now the only ones readily available in commercial quantities. By comparison, the synthesis of III-V nanocrystals is more challenging since the reaction must take place under very strict vacuum and humidity conditions. The traditional method of preparing II-VI as well as III-V nanocrystals consists of heating the cation and anion precursors in the presence of a coordinating solvent, such as trioctylphosphine oxide (TOPO) or dodecylamine (DDA) at high temperatures (~ 150-200°C for II-VI nanocrystals and 300°C for III-V nanocrystals) for several (2-5) days [4-8]. The role of the coordinating solvent is to arrest the growth of nanocrystals while they are still only a few nanometers in size and passivate the dangling bonds on the surface. The products of these reactions are solutions containing nanocrystals with large size distributions and a post-growth size-selective precipitation is typically necessary in order to obtain samples with narrow size distributions (+/-10%) [9,10] . A complication resulting from this growth method is the possible presence in the PL spectrum of features associated with emission from heated impurities present in the coordinating solvent [11]. These features can mask the emission from nanocrystals in the blue and visible region of the spectrum, if the nanocrystals are characterized by low emission efficiencies.

A recent alternative to the traditional synthesis of InP nanocrystals, proposed by Battaglia et al. [12], eliminates the TOPO or DDA from the synthesis and cuts down the reaction time from a few days to a few hours. However, a coordinating solvent and/or surfactant is still added to the solution in order to achieve surface passivation.

In this paper we report on the optical properties of InP nanocrystals grown through a fast reaction which does not involve any coordinating solvent or added surfactant. The precursors are heated in the presence of a non-coordinating solvent and the surfactant is formed "in-situ". Transmission Electron Microscopy (TEM) and standard UV-VIS absorption measurements indicate the nanocrystals grown through this method are characterized by diameters smaller than 50Å and fairly narrow size-distributions. The PL spectrum of the "as-grown" nanocrystals, passivated by the "in-situ" surfactant contains contributions from the band-edge as well as surface state-related recombination. After undergoing an HF etching processes the surface state recombination contribution to the PL spectrum is eliminated and the emission efficiencies increase by a factor of 40. Time-resolved photoluminescence studies were employed to determine the nature of the recombination mechanism responsible for the band-edge luminescence.

EXPERIMENTAL DETAILS

Indium Phosphide nanocrystals were prepared by heating 0.199 g (0.25 x 10^{-3} moles) of $In(O_2C_{13}H_{28})_3$ ($In(MYR)_3$) in 40 mL of dry octadecene to 300 °C followed by the rapid injection of 0.031 mg (0.125 x 10^{-3} moles) of $P(SiMe_3)_3$. The surfactant is formed "in-situ" and no surfactant is added from outside. Immediately upon injection the solution became orange in color. The monitoring of the reaction was performed by drawing several 0.5ml aliquots at several time intervals during the growth and dispersing them in hexane. Standard UV-VIS absorption measurements were performed on these aliquots using a Shimadzu spectrophotometer and the reaction was stopped when the wavelength corresponding to the absorption edge of the nanocrystals no longer increased as a function of growth time. The reaction was typically stopped after 2 to 3 hrs.

After isolation by precipitation from a hexane solution with acetone, a colloidal solution of the nanocrystals was prepared by dissolving 10 mg of nanocrystals in a solution of hexane, 1-butanol and acetonitrile in a 1:0.1:1 ratio and the solution sonicated for 1 min to completely solubilize the nanocrystals. Next, the 0.1 mL of an HF solution of 5% HF, 10% H_20 and 85 % 1-butanol was added to the colloidal nanocrystal solution and sonicated 1 min. The colloidal nanocrystal/HF solution was allowed to sit for 24 hrs under ambient conditions. After 24 hrs the nanocrystals were isolated by addition of excess acetone to precipitate the nanocrystals. A single fraction was then isolated by centrifugation and redispersed in hexane.

Various precursor ratios and temperature sequences were explored in order to optimize the optical properties. The two samples presented in this study were both grown with an In:P precursors ratio of 2:1. After the precursors were injected at 300°C, for sample 1 the temperature was set to 245°C and the solution was left to gradually cool down. In the case of sample 2 the reaction was quenched immediately after the 300°C injection with entire reaction volume equivalent of octadecene and then the temperature was set to 245°C.

Preliminary continuous wave (CW) photoluminescence (PL) and photoluminescence excitation (PLE) optical characterization measurements have been conducted using a Jobin-Yvon 3-11 Fluorolog spectrofluorometer equipped with a 400W Xenon lamp as an excitation source and a Hamamatsu R928 photomultiplier tube as a detector. The PL and PLE spectra were measured for 0.5 ml aliquots drawn at different times during the growth process, dispersed in hexane and placed in a 1 cm optical path UV transparent quartz cuvette. Time-resolved photoluminescence (TRPL) measurements were performed in a right angle geometry using 400 nm, 200 fs pulses with a repetition rate of 250kHz obtained by frequency doubling the 800 nm pulsed output of a Coherent Rega 9000 regenerative amplifier. The PL spectra were spectrally and temporally resolved using a Chromex 250IS monochromator (spectral resolution 0.15 nm) equipped with a Hamamatsu C4334 streak camera with typical jitter of 50 ps. All the time-resolved measurements were carried out at room temperature with the samples placed in the same 1cm optical path quartz cuvettes used in the CW studies.

RESULTS AND DISCUSSION

A summary of the absorption spectra from the two samples under study is presented in Figure 1. The feature associated with the InP nanocrystals is present in the absorption spectrum measured on the aliquot drawn 5 minutes after the reaction started, indicating the reaction takes place on a very fast scale. As time passes, the wavelength associated with the InP nanocrystals absorption feature shifts towards longer wavelengths, indicating the particles grow in size. For sample 1 (Figure 1(a)) the wavelength remains approximately constant after the first 30 minutes of the growth process indicating the particles grow very fast and their size is not controlled by the time of the reaction, but rather by the precursors ratio and the temperature conditions. After two hours the wavelength associated with the absorption feature in sample 1 corresponds to a nanocrystal diameter of approximately 26 Å [13,14], a number which is in very good agreement with the results of the X-ray experiments performed on the same sample. The crystalline domain size estimated from the Debye-Scherrer formula was found to be 26.3 Å.

In order to investigate if several aliquots characterized by different average nanocrystal sizes could be obtained from the same reaction, the growth rate was slowed down using the

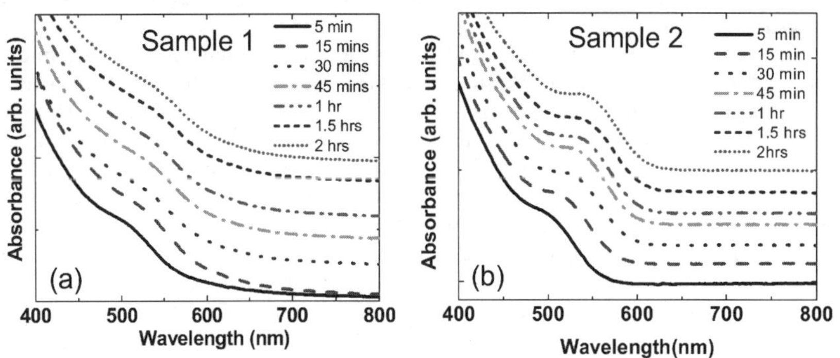

Figure 1 Absorption spectra from (a) sample 1 and (b) sample 2 measured on several .5 ml aliquots drawn at different times after the precursors injection.

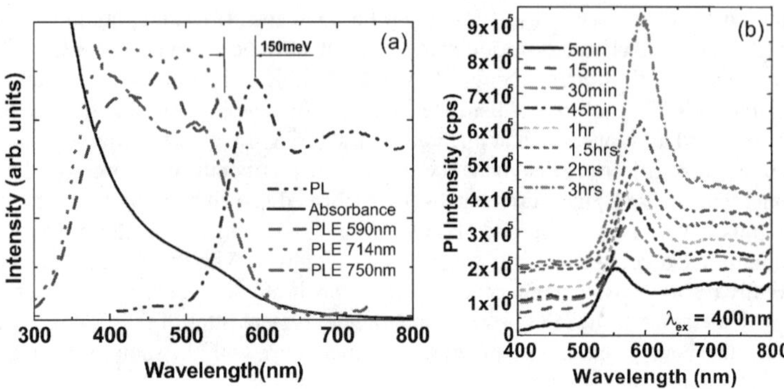

Figure 2 (a) PL, PLE and absorption spectra from sample 1, measured after the reaction was stopped and (b) PL spectra of sample 2 measured as a function of growth time. All measurements were performed on the "as-grown" nanocrystals, coated with the surfactant formed "in-situ".

procedure described for sample 2 in the previous section. In contrast to sample 1, the wavelength associated with the InP nanocrystals absorption for sample 2 increases continuously as a function of growth time until the reaction is stopped. The particle diameter is equal to approximately 18 Å after the first 5 minutes of growth time and increases to 30 Å two hours later. In addition, the absorption features of sample 2 are considerably sharper than those of sample 1, indicating sample 2 is characterized by a sharper size distribution [12].

The results of the CW photoluminescence experiments are summarized in Figure 2. Panel (a) depicts the PL, PLE and absorption spectra from sample 1 measured after the reaction was stopped, on the "as–grown" sample. The sharp feature present in the PL spectrum, associated with band-edge recombination is red-shifted by 150meV in comparison to the lowest energy feature in the PLE spectrum. A similar red-shift has been observed in all PL experiments performed on InP nanocrystals [3,13] and its origins can be related not only to the existence of a size distribution inside the sample, but possibly to the dark nature of the ground state in such nanocrystals. [14,15].

In order to investigate the nature of the broad feature extending into the near infrared region, which is also present in the PL spectrum of the "as-grown" nanocrystals, we investigated the PLE spectrum at several wavelengths across the PL spectrum. As shown in Figure 2(a), the PLE spectrum at long wavelengths (714nm and 750nm) contains no features in the 600 to 800nm region. This result implies that the broad emission is not due to larger particles that might be present in the solution and indicates that the broad emission is most likely associated with surface state recombination._The presence of surface states is expected in these nanocystals since they are passivated with weakly bound "in-situ" formed surfactant. In addition, the relative PL intensity between the sharp and broad features increases in the favor of the former as a function of growth time, as shown in Figure 2(b). This observation is consistent with the assignment of the broad feature to surface state recombination. As the particles grow in size, the surface to volume ratio decreases and the contribution of the surface state recombination diminishes as well. At the end of the growth process the feature associated with band-edge recombination dominates the spectrum. Its full width at half-maximum (FWHM) (approximately 60 nm) is comparable to that of the PL spectrum of narrow size distributions obtained as a result of size selective precipitation processes [5]. This result indicates that the size distribution can be controlled by tuning the growth parameters and a post-growth size-selection is not necessary with this method.

The emission efficiencies measured on the "as–grown" nanocrystals (for example the 3hrs spectrum in Figure 2(b)) varies between 0.3% and 1% from sample to sample. In order to increase the emission efficiency we subjected the samples under study to an HF etching process which has been proven to significantly increase the emission efficiencies by passivating the surface. The time-integrated PL spectrum of sample 2, taken using the 400 nm pulsed excitation described in the experimental section, before and after the sample underwent the etching process are shown in Figure 3(a). The emission efficiency increased to 18% after etching, similar to reported efficiencies from InP nanocrystals grown by other methods. [3] In addition the broad feature associated with surface state emission is no longer present in the spectrum of the etched nanocrystals , indicating the surface was passivated through this process.

The differences in the photoluminescence decay times measured before and after etching at the wavelength corresponding to the peak of the band-edge emission reflect the passivation of the surface states as a result of etching. The PL decay as a function of time measured for sample 2 is plotted in Figure 3(b). Before etching, the decay is multi-exponential and dominated by a long-live component characterized by a lifetime equal to 100 ns. Since the decay is measured at the peak of the sharp feature associated with band-edge recombination, where the spectrum contains a contribution from the surface state emission, the long lived component of the decay can be associated with the surface state recombination. After etching, the long-lived component is absent from the PL decay, and the longest lifetimes measured are of the order of tens of nanoseconds (45 ns for sample 2). These lifetimes are too long to be associated with band-to-band or excitonic recombination. Our measurements of the PL decay times at different growth times indicated these lifetimes are independent of particle size and therefore, the recombination cannot be associated with states created by the In-dangling bonds present on the nanocrystal surface but rather with spin-forbidden intrinsic states [14]. This result is very similar to the one obtained by Micic et. al [13] who studied the PL decay times in InP nanocrystals grown by the traditional TOPO method.

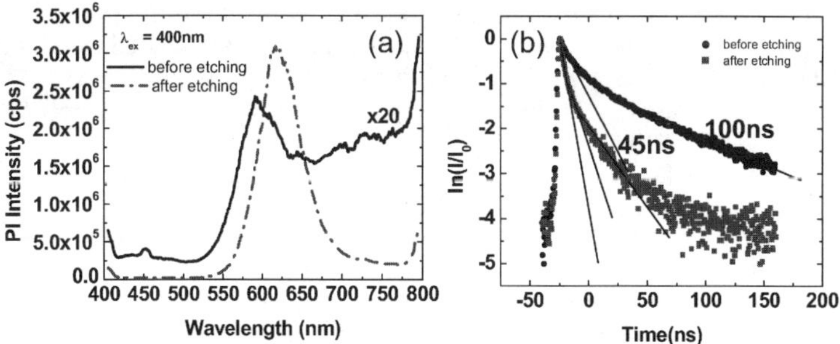

Figure 3 (a) PL spectrum from sample 2 measured before and after the sample underwent an HF etching process and (b) PL decay time measured before and after etching at a wavelength corresponding to the peak of the PL spectrum of sample 2.

CONCLUSIONS

CW and time-resolved spectroscopy studies of InP nanocrystals grown through a fast reaction have shown that nanocrystals with narrow size distributions can be obtained through this method without undergoing a size-selective precipitation process. The reaction rate can be controlled such that several sizes can be obtained during a given growth sequence. The PL spectrum of "as-grown" nanocrystals exhibits a feature associated with surface state recombination. This surface state recombination feature can be eliminated by subjecting the samples to an HF etching process. The emission efficiency increases by a factor of 40 as a result of the same etching process. The long carrier lifetimes indicate that the band-edge recombination is associated with a spin-forbidden state. The emission efficiencies, FWHM of the PL feature and carrier lifetimes are similar to the ones measured on InP nanocrystals grown through different methods, indicating all these parameters are controlled by the intrinsic properties of the nanocrystals rather than the surfactants or other parameters of the growth process.

ACKNOWLEDGEMENTS

This work was supported in part by ANC's NSF CAREER Grant #9733720, ONR YIP Grant #N00014-00-1-0508,and a Defense University Research Initiative on Nanotechnology Grant #F496200110358 through the Air Force Office of Scientific Research.

REFERENCES

1. A. P. Alivisatos, *Science* **271**, 933 (1996).
2. M. Kuno, J. K. Lee, B. O. Dabbousi, F. V. Mikulec, and M. G. Bawendi, *Journal of Chemical Physics* **106**, 9869 (1997).
3. O. I. Micic, J. Sprague, Z. H. Lu, and A. J. Nozik, *Applied Physics Letters* **68**, 3150 (1996).
4. O. I. Micic and A. J. Nozik, *Journal of Luminescence* **70**, 95 (1996).
5. A. A. Guzelian, J. E. B. Katari, A. V. Kadavanich, U. Banin, K. Hamad, E. Juban, A. P. Alivisatos, R. H. Wolters, C. C. Arnold, and J. R. Heath, *Journal of Physical Chemistry* **100**, 7212 (1996).
6. A. P. Alivisatos and M. A. Olshavsky, Patent No. 5,505,928 (9 April 1996).
7. A. P. Alivisatos, X. G. Peng, and M. Liberato, Patent No. 6,306,736 (23 October 2001).
8. U. Banin and Y. -W. Cao, USA Patent No. 20030010987 (16 January 2003).
9. O. I. Micic, C. J. Curtis, K. M. Jones, J. R. Sprague, and A. J. Nozik, *Journal of Physical Chemistry* **98**, 4966 (1994).
10. R. J. Ellingson, J. L. Blackburn, P. R. Yu, G. Rumbles, O. I. Micic, and A. J. Nozik, *Journal of Physical Chemistry B* **106**, 7758 (2002).
11. M. Furis, Y. Sahoo, D. J. MacRae, F. S. Manciu, A. N. Cartwright, and P. N. Prasad, *Journal of Physical Chemistry B* **107**, 11622 (2003).
12. D. Battaglia and X. G. Peng, *Nano Letters* **2**, 1027 (2002).
13. O. I. Micic, H. M. Cheong, H. Fu, A. Zunger, J. R. Sprague, A. Mascarenhas, and A. J. Nozik, *Journal of Physical Chemistry B* **101**, 4904 (1997).
14. H. X. Fu and A. Zunger, *Physical Review B* **56**, 1496 (1997).
15. M. Nirmal, D. J. Norris, M. Kuno, M. G. Bawendi, A. L. Efros, and M. Rosen, *Physical Review Letters* **75**, 3728 (1995).

Synthesis, Shape Control and Characterization

Synthesis and Characterization of $Cd_{1-x}Cu_xSe$ Quantum Dots

Khalid M. Hanif and Geoffrey F. Strouse
Department of Chemistry & Biochemistry
University of California
Santa Barbara, CA 93106-9510, U.S.A.

ABSTRACT

Cu:CdSe nanoparticle alloys were synthesized using inorganic cluster precursor via a lyothermal method. The dots synthesized were approximately 4 nm in size and exhibited doping levels as high as 20% for Cu(I). Cu(I) cations are randomly doped on the Cd-T_d site based on a Vegard's law analysis, suggesting that doping arises from a random ion substitution mechanism. The resulting dots exhibited essentially no band edge photoluminescence due to a high level of trap site inclusions in these materials resulting from Se vacancy formation in order to compensate for the Cu^+ replacing a Cd^{2+} ion. Se vacancy formation is evident in the XPS analysis of the Cd:Cu:Se ratios for these materials.

INTRODUCTION

Transition metal doped quantum dots are excellent candidates for applications such as phosphor displays and photovoltaics.[1,2] Cu(I) doping of bulk II-VI semiconductors has been extensively studied due to the observation of improved performance in electrochemical solar cells. In these studies, it has been shown that Cu(I) ion substitution of a Cd site in CdSe is advantageous to the solar cell performance. The improved performance is attributable to an increase in carrier concentrations in these materials arising from p-doping. Cu also has been used to increase the carrier level in ZnO.[3] Along with p-doping, the appearance of Se vacancies in these materials arise due to self-compensating effects for charge balance.

While bulk-CdSe is readily doped, Cu(I) doping of lyothermally grown materials is more difficult due to the high propensity for self-annealing of the particles with loss of the internal dopant ion. Doping is readily achievable at low temperatures, however these materials are generally poorly formed. High temperature synthetic techniques are favored over room temperature methods because they tend to yield materials that are more crystalline and that have fewer defects. Only within the past few years has successful doping of semiconductor quantum dots been accomplished using high temperature techniques based on preformed inorganic clusters, in which the dopant ion is introduced into a single source molecular cluster containing Cd and Se.

One of the most important issues that exist when doping is the location of the impurity. The impurity can exist either on a lattice point in the core, at interstitial sites, or localized at the surface. To date studies have shown that it has been very difficult to reliably dope the core of nanoparticle with impurities. In most cases only one or two ions were doped in the core of each nanoparticle which equates to a core doping level of approximately $<0.2\%$.[4] In the bulk the most common way of proving doping in the core of particles is by observing a shift in the pXRD data using Vegard's law. As smaller ions replace existing Cd^{2+} ions the lattice is expected to contract. The lattice contraction should be linear with the doping level. If the dopant ion is smaller than the ion it is replacing then a contraction of the lattice as a function of doping should be observed in the XRD spectra.[5] The only time this is not observed is when the dopant is either on the surface or if the doping level is low enough that it has no net effect upon the structural properties of the host lattice. . Application of Vegard's law analysis of core-doping has been used by several researchers to establish structural evidence using powder-XRD data to prove doping within the material core.[5,6]

In this manuscript we will provide pXRD proof of core doping of 4nm Cu:CdSe nanoparticles with Cu(I) on a Cd-T_d site. A linear shift in the lattice parameters obtained from pXRD is observed, thus providing direct evidence for statistical doping of the core with Cu(I) ions. Repetitive particle stripping provides no change in the observed concentrations measured either by pXRD or Atomic emission. Cu(I) doping instead of Cu(II) can be verified by inspection of the XPS data that provides evidence for only Cu(I), and a correlated increase in Se vacancies with increasing Cu(I) doping concentration.

EXPERIMENTAL DETAILS

Synthesis of $[Cu_4(SPh)_6]$ $(TMA)_2$. $[Cu_4(SPh)_6]$ $(TMA)_2$ was prepared by adding a solution of (15.8 g, 68 mmols) of $[Cu(NO_3)_2 \cdot H_2O]$ in 70 ml methanol to a stirred, room temperature solution of (20.0 g, 182 mmols) of benzenethiol and (18.5g, 182 mmols) triethylamine in 40 mL methanol. To this mixture, a solution of (8.4g, 77 mmols) tetramethylamonium chloride (TMACl) in 40mL methanol was added, and the product was then allowed to crystallize at 0°C producing a yellow solid. The solution was filtered, washed with cold methanol and vacuum dried. The sample was characterized by ESI-MS with a parent ion peak for $[Cu_4(SPh)_6]^{2-}$ observed at 454.7 m/z.

Synthesis of $(Li_4)[Cd_{10}Se_4(SPh)_{16}]$ $(TMA)_2$. $(Li_4)[Se_4Cd_{10}(SPh)_{16}]$ was synthesized according to literature methods.[7] ESI-MS analysis of the cluster was performed to verify the cluster formation.

Synthesis of $Cd_{1-x}Cu_xSe$ Quantum Dots. Synthesis of $Cd_{1-x}Cu_xSe$ nanoparticles was performed using a parallel plate reactor by adding ~1.200 g of $(Li_4)[Se_4Cd_{10}(SPh)_{16}]$ and various concentrations of $(TMA_2)[Cu_4(SPh)_6]$, [0.100g, 0.200 g, 0.300 g, 0.400g, 0.500g, 0.600g, 0.700g, and 0.900g] in 9 different sealed reactor vials. To the reactor vials, 35mL aliquots of n-Hexadecylamine (HDA) at ~100°C were added. Growth of the doped nanomaterials was accomplished by heating the solution to ~250°C at a rate of ~ 20°C/hour on a parallel heating block. The parallel plate methodology allows nine separate reactions to be carried out under identical reaction conditions. This is important for controlled doping of the semiconductor alloys for a given nanoparticle size. Nanomaterial growth is followed by monitoring the absorption spectra to ensure that that the band edge absorption stays narrow. Growth of the doped quantum dots was halted when they approached ~ 4 nm in size. The quantum dots were isolated from the HDA and recapped with pyridine 2 X according to literature methods.[6]

Materials Characterization. Powder X-ray diffraction measurements were taken on a Scintag X2 diffractometer with a Cu Kα source. Si powder was used as an internal standard, which gave sharp peaks in the p-XRD spectra at 1.6, 1.9 and 3.1 Å. Absorption measurements were taken on a CARY 50 Bio UV-VIS spectrophotometer. Photoluminescence was performed on a CARY Eclipse Fluorescence Spectrophotometer using a 490 nm excitation source. Elemental analysis was performed on a Thermo-Jarrell Ash IRIS inductively coupled plasma atomic emission spectrometer (ICP-AE). The Cd and Cu concentrations were measured against Cd and Cu standards purchased from High Purity Standards. Transmission electron microscopy (TEM) was performed on a JEOL 2010 FX high resolution TEM using Ni holey carbon grids.

DISCUSSION

The Cu:CdSe semiconductor alloys prepared on the parallel plate allow nine doping levels to be studies for materials grown under identical reactor conditions. The isolated nanomaterials are ~4 nm in size, with a roughly spherical morphology in a zinc blende lattice (Figure 3). ICP-AE was used to measure the Cu alloy concentration in the nine Cu:CdSe. Elemental analysis of these materials was performed after the surface of the quantum dots had been stripped and recapped with pyridine. This step ensures removal of any Cu that may be electrostatically attached to the surface of the particle or that may not have been removed during the isolation of the material from the hexadecylamine. The observed alloy concentration was constant following the first pyridine stripping of the particle. The measured concentration range for the Cu ion in the alloy was from 0.033 to 0.203 mole percent based on ICP-AE analysis of the nanoparticles. The Cu:CdSe nanoparticles have a band edge absorption that ranges from ~540 – 560 nm consistent with the absorption of 3.8-4.2 nm particles as can be seen in figure 1. The band edge absorption was narrow, as expected for materials with a narrow size distribution. All of the Cu(I) doped quantum dots exhibited no band edge luminescence. These samples had very broad emission spectra in the red which is consistent with what is thought to be defect emission. Preliminary XPS data shows that these samples are Cu(I) doped rather than Cu(II), which leads to a charge

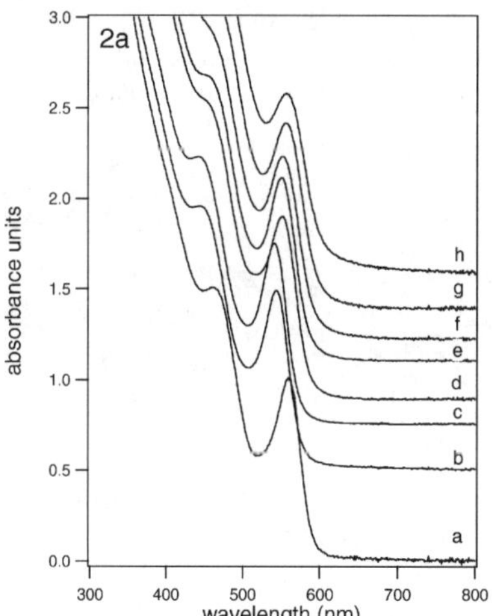

Figure 1. Absorption spectra of $Cd_{1-x}Cu_xSe$ Quantum dots. a)$Cd_{0.967}Cu_{0.033}Se$ b) $Cd_{0.945}Cu_{0.055}Se$ c) $Cd_{0.935}Cu_{0.065}Se$ d) $Cd_{0.901}Cu_{0.099}Se$ e) $Cd_{0.897}Cu_{0.113}Se$ f) $Cd_{0.868}Cu_{0.132}Se$ g) $Cd_{0.847}Cu_{0.153}Se$ h) $Cd_{0.777}Cu_{0.223}Se$

imbalance in the final particles. The charge imbalance produces Se vacancies, as shown in an analysis of the Cu:Cd:Se ratios in the materials. An increasing vacancy formation is observed with increasing Se concentration. This would create a large number of trap sites for electrons thereby hindering the recombination of electrons with holes in the quantum dots. Based on the TEM image of $Cd_{0.91}Cu_{0.09}Se$ shown in figure 2, it can be determined that these materials are fairly narrow in size distribution. The observation of lattice fringes in the TEM shows that these materials are crystalline. An ABCABC packing can be observed in these materials consistent with that of a zinc blende structure.

Random Ion Substitution in Cu:CdSe alloys. The formation of the alloy is believed to be achieved due to seeded growth process inherent in the use of a $[Cd_{10}Se_4(SPh)_{16})]^{4-}$ single source precursor growth technology. Substitution of the Cd ions in the cluster can be achieved by stoichiometric addition of Cu ions to a solution containing the cluster at 100°C. The facile metal substitution of the M_{10} core by the M_4 species in the above reaction has been studied by several researchers. Holm et al showed that cluster with the structure $[M_4(SPh)_{10}]^{2-}$ readily exchanged the following metal ions (Fe(II), Co(II), Zn(II), Cd(II)) to give the following structure $[M_{4-n}M'_n(SPh)_{10}]^{2-}$ when mixed together in CD_3CN.[8] Bowmaker, et al extended this study to M_{10} clusters and observed core substitution of the desired ionic species by ESI-MS analysis.[9] Although isolation of the Cu containing cluster in this study was not carried out, we believe core substitution of the cluster is critical for the substitution of the nanoparticle.

As previously stated the location of the impurity is very important. The dopant can exist either in the core of the particle or on the surface of the nanoparticle. While stripping of the particle removes adventitious Cu ions, it will not remove Cu ions located at or near the surface or doped throughout the particle by a random ion substitution (RIS) process. Powder X-ray diffraction (pXRD) provides an invaluable tool to distinguish between the two possibilities. Random ion doping is predicted to give rise to a linear shift in the lattice as predicted by Vegard's law. Near surface doping will produce strain, but the magnitude of the shifts is

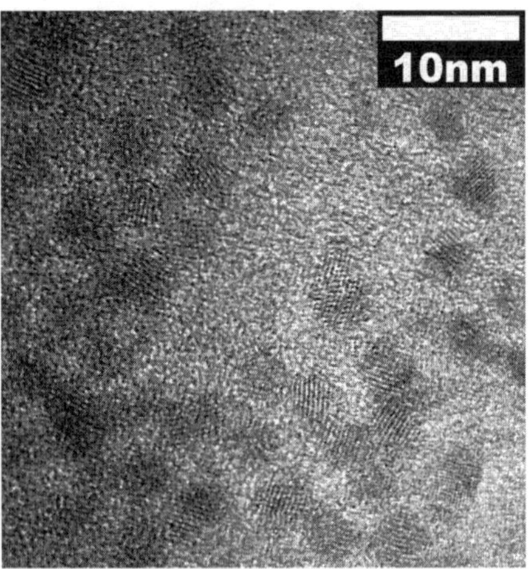

Figure 2. TEM image of $Cd_{0.901}Cu_{0.099}Se$.

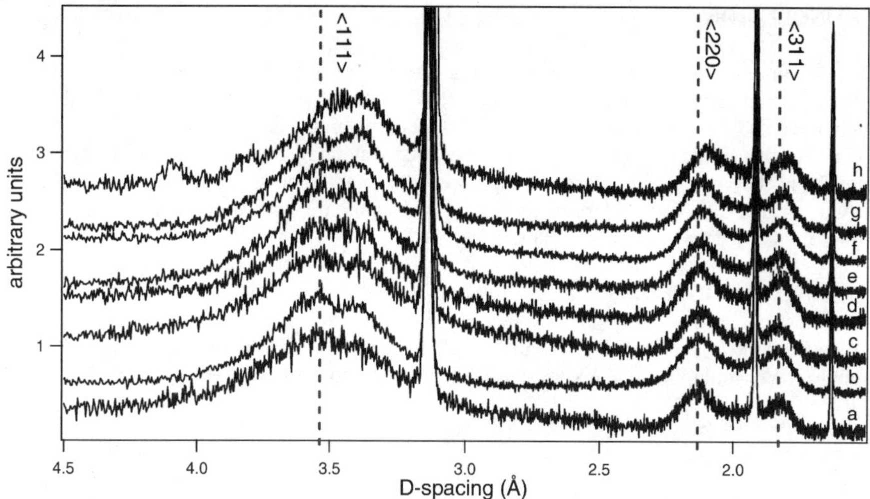

Figure 3. XRD of $Cd_{1-x}Cu_xSe$ Quantum dots. a) $Cd_{0.967}Cu_{0.033}Se$ b) $Cd_{0.945}Cu_{0.055}Se$ c) $Cd_{0.935}Cu_{0.065}Se$ d) $Cd_{0.901}Cu_{0.099}Se$ e) $Cd_{0.897}Cu_{0.113}Se$ f) $Cd_{0.868}Cu_{0.132}Se$ g) $Cd_{0.847}Cu_{0.153}Se$ h)$Cd_{0.807}Cu_{0.203}Se$

predicted to be much smaller than observed for a RIS mechanism. The structure of Cu:CdSe alloys are zinc blende (Figure 3). Although the hexagonal wurtzite phase is generally seen in lyothermally grown nanomaterials, the presence of Cu creates enough of a driving force to form a zinc blende structure. This is consistent with the observation that Cu_2Se exists as a cubic lattice. The powder diffraction of these materials exhibit three peaks that can be assigned as <111>, <220> and <311> reflections of the zinc blende structure. The peaks are broadened due to Scherrer broadening resulting from the small size of the quantum dots. The three sharp peaks in the XRD spectra are due to the addition of a Si internal standard to allow measurement of small lattice shifts. It is noticeable in the pXRD pattern of these compounds that the peaks shift to smaller d-spacing as the doping level is increased. This behavior is expected based on a Vegard's law treatment of the pXRD data, when the material core is doped with a smaller cation than the one it is replacing. Literature values of the tetrahedral covalent radii for both Cd and Cu are 1.48 and 1.38 Å respectively.[10] The pXRD data is consistent with increased levels of Cu^+ cations being incorporated within the lattice of the host material with increased amounts of dopant cluster being added to the reaction mixture. The magnitude of the shift provides evidence of a random ion substitution mechanism in the Cu:CdSe materials.

CONCLUSION

In this paper we have shown a simple and powerful synthetic technique that yields nanomaterial alloys with high levels of core doping. Structural evidence was provided indicating that the core of these materials is statistically doped through a random ion substitution process by Cu^{+} ions that reside on a Cd -T_d site. It is believed that the incorporation of Cu^{+} within the lattice results in Se vacancies in order to accommodate the charge imbalance that results from replacing a Cd^{2+} with a Cu^{+}. This is expected to create many more trap sites accounting for the lack of photoluminescence in these materials.

ACKNOWLEDGEMENTS

We would like to thank Frederic Diana and Prof. Pierre Petroff at UCSB for acquiring the TEM image. We would like to thank UC-CARE for funding to do this project.

REFERENCES

(1) Kokubun, Y.; Hatano, H.; Wada, M. *Jpn J Appl Phys* **1979**, *18*, 1947-1950.

(2) Bhargava, R. N. *J Lumin* **1996**, *70*, 85-94.

(3) Han, S. J.; Song, J. W.; Yang, C. H.; Park, S. H.; Park, J. H.; Jeong, Y. H.; Rhie, K. W. *Appl Phys Lett* **2002**, *81*, 4212-4214.

(4) Mikulec, F. V.; Kuno, M.; Bennati, M.; Hall, D. A.; Griffin, R. G.; Bawendi, M. G. *J Am Chem Soc* **2000**, *122*, 2532-2540.

(5) Jun, Y. W.; Jung, Y. Y.; Cheon, J. *J Am Chem Soc* **2002**, *124*, 615-619.

(6) Hanif, K. M.; Meulenberg, R. W.; Strouse, G. F. *J Am Chem Soc* **2002**, *124*, 11495-11502.

(7) Cumberland, S. L.; Hanif, K. M.; Javier, A.; Khitrov, G. A.; Strouse, G. F.; Woessner, S. M.; Yun, C. S. *Chem Mater* **2002**, *14*, 1576-1584.

(8) Hagen, K. S.; Stephan, D. W.; Holm, R. H. *Inorg Chem* **1982**, *21*, 3928-3936.

(9) Lover, T.; Henderson, W.; Bowmaker, G. A.; Seakins, J. M.; Cooney, R. P. *Inorg Chem* **1997**, *36*, 3711-3723.

(10) Suresh, C. H.; Koga, N. *J Phys Chem A* **2001**, *105*, 5940-5944.

QD-Bioconjugates: Design and Use in Biotechnological Applications

Mat. Res. Soc. Symp. Proc. Vol. 789 © 2004 Materials Research Society N5.4

Towards the Design and Implementation of Surface Tethered Quantum Dot-Based Nanosensors

Igor L. Medintz[1,*], Kim E. Sapsford[2], Joel P. Golden[1], Aaron R. Clapp[3], Ellen R.Goldman[1] and Hedi Mattoussi[3]

[1]Center for Bio/Molecular Science and Engineering, Code 6900, U.S. Naval Research Laboratory, Washington DC 20375
[2] George Mason University, 10910 University Blvd, MS 4E3, Manassas, VA 20110
[3] Division of Optical Sciences Code 5611, U.S. Naval Research Laboratory, Washington, DC 20375

* Electronic address: imedintz@cbmse.nrl.navy.mil

ABSTRACT

Considerable progress has been made towards creating quantum dot (QD) based nanosensors. The most promising developments have utilized QDs as energy donor in fluorescence resonance energy transfer (FRET) processes. Hybrid QD-protein-dye complexes have been assembled to study FRET, to prototype analyte sensing and even to control or modulate QD photoluminescence. In order to transition the benefits of this technology into the field, QD-based nanosensors will have to be integrated into microtiter wells, flow cells, portable arrays and other portable devices. This proceeding describes two examples of QD-protein-dye assemblies. The first investigates the concepts of FRET applied to QD energy donors and the second describes a prototype biosensor employing QDs. We also introduce the first steps towards implementing surface-tethered QD-bioconjugates, which could potentially serve in the design of solid-state QD-based sensing assemblies.

INTRODUCTION AND BACKGROUND

The first reports on the design and preparation of water-soluble biocompatible colloidal luminescent semiconductor nanocrystals (or quantum dots, QDs) were published almost six years ago. Since then, considerable progress has been made towards developing QD biotechnology-related applications. However, most investigations have focused on using the QDs as fluorescent probes to detect proteins or DNA, as fluoro-immunoreagents, and more recently as labels in cellular imaging. In all these applications, QDs offer significant advantages compared to small molecule organic dyes [1,2]. These include broad excitation and size-dependent tunable photoluminescence (PL) spectra with narrow emission bandwidths that span a wide region of the optical spectrum. One additional unique feature of QD fluorophores is the ability to simultaneously excite several color QD sizes with a single source and generate discrete emissions characteristics of each QD population. QDs also have exceptional photo- and chemical stability [3,4,5].

A very promising development in using these inorganic fluorophores has recently emerged: QDs as efficient energy donors in fluorescence resonance energy transfer (FRET) processes. Developing QD-based FRET assays moves the use of water-soluble QDs from a passive bio-label mode into a more active bio-sensing role. Protein-QD hybrid ensembles have been assembled to study FRET, to prototype analyte sensing and even to control or modulate QD photoluminescence. However to date, all the prototypes developed have been solution-based [6-8]. In order to transition the benefits of this technology into the field, QD based nanosensors will have to be integrated into microtiter wells, flow cells, portable arrays and other portable devices.

In this report we present example studies where QD conjugates have been used as efficient energy donors in FRET investigations. In particular, we show that control over the rate of energy transfer cross sections can be achieved by either tuning the spectral overlap between the donor and acceptor or by having a single donor interact with several acceptors simultaneously in the same assembly. We then describe a prototype of a solution-based nanoscale FRET sensing ensemble using QD bioconjugates. These findings motivated us to develop a scheme to attach QD conjugates onto a variety of surfaces, which could potentially allow for their integration into functional devices.

EXPERIMENTAL SECTION

Quantum dots were synthesized and made water compatible as described previously [9,10]. Briefly, CdSe-ZnS core-shell nanoparticles were prepared using organometallic precursors in a high temperature coordinating solution made of a mixture of trioctyl phosphine/trioctyl phosphine oxide (TOP/TOPO) mixed with additional amine terminated ligands. QDs were rendered water-soluble by replacing the native organic cap with bidentate dihydrolipoic acid (DHLA, a dithiol-alkyl-COOH) ligands, as described previously [4,5], providing water-soluble QD preparations with good quantum yields and which were exceptionally stable in basic pH buffer [4,5].

QDs conjugated to maltose binding protein appended with an oligohistidine attachment domain (MBP-his) and labeled with Cy3 dye (MBP-his-Cy3) were prepared by adding varying amounts of labeled and unlabeled proteins to 100 μL of 10 mM sodium tetraborate buffer solution (pH 9). The oligohistidine attachment allows for non-covalent self-assembly of the protein onto the QD surface [4,11,12]. In these preparations, the molar ratios of MBP-his-Cy3:QDs were discretely varied from 0 to 10 while the overall ratio of MBP-his to QD was maintained at 15. The solutions were placed in a quartz fluorescence cuvette (10 mm path length) and the emission spectra measured using a SPEX Fluorolog-2 fluorimeter (Jobin Yvon/SPEX, Edison, New Jersey). The excitation wavelength was fixed at 430 nm where absorption of the Cy3 dye is minimum to reduce effects of direct excitation of the acceptors. Moreover, emission from MBP-Cy3 only solutions was subtracted from the spectra of the QD-MBP-Cy3 conjugate solutions.

Substrate preparation and surface tethering of capture biomolecules to create surface-tethered QD-based structures in discrete locations was carried out following previously described methods developed for antibody immobilization [13,14]. An immuno-array biosensor was utilized to create QD assemblies on glass slides and to visualize the results. This array biosensor has been used extensively for pathogen detection. Glass slides were first activated and coated with a layer of NeutrAvidin using silane chemistry and heterofunctional linkers as described previously [13,14]. Using a PDMS flow cell, discrete portions of the substrate surface were exposed to a hyper-biotinylated MBP solution, which would serve a bridge between the NeutrAvidin functionalized surface and the QD-avidin conjugates [15,16]. A solution of 555 nm emitting QDs coated with ~1 avidin per QD and 15 MBP per nanocrystal was self-assembled in a small volume of 10 mM NaTetraborate buffer pH 9.55 for 30 mins. Since the MBP surrounding the QDs will bind to amylose resin, this solution was passed over and adhered to amylose resin to purify the QD-protein conjugate from excess free avidin [4,16]. The resulting purified QD-MBP/avidin conjugate solution was exposed for 3 hours to the biotin-MBP slides using a second PDMS flow cell aligned perpendicular to the previous one, followed by washing. The slides were visualized by illuminating with a laser derived evanescent wave at 488 nm and the resulting images were captured with a CCD camera. Imaging of the captured QDs was achieved using a variation of the array biosensor, described in refs 13 and 14, modified with a 488 nm excitation source and appropriate filters.

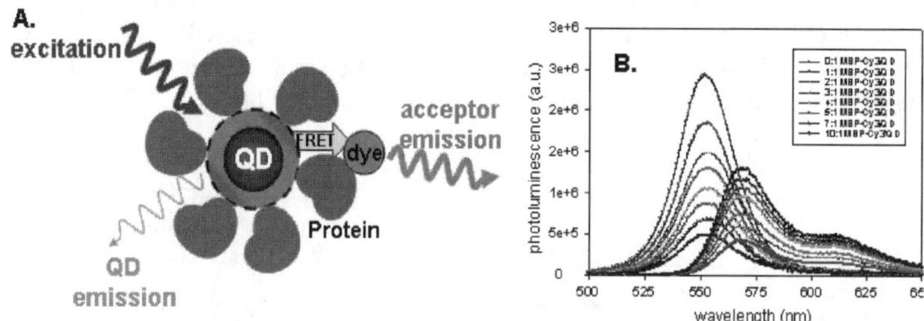

Figure 1. (A) Schematic depiction of a QD donor surrounded by surface conjugated proteins. In this scenario, one of the proteins is dye-labeled and acts as a FRET acceptor. **(B)** Corrected donor and acceptor photoluminescent changes from titrating a 555 nm emitting QD with an increasing ratio of Cy3-dye-labeled acceptor maltose binding protein (MBP-his), while keeping the total amount of protein on the QD fixed

RESULTS AND DISCUSSION

1. QD as Energy Donors in Fluorescence Resonance Energy Transfer (FRET)

Figure 1A shows a schematic representation of a QD conjugated to several proteins, with one of them dye-labeled and acting as the energy acceptor. Figure 1B shows the corrected donor and acceptor photoluminescent (PL) contributions resulting from titrating a 555 nm emitting QD conjugate solution with an increasing fraction of Cy3 dye-labeled MBPs. A progressive quenching of the QD emission, concomitant with a steady enhancement of the acceptor signal is recorded when the number of acceptors is increased. This set of spectra permitted us to investigate the role of protein conjugated QDs in active FRET. For instance, the data in Figure 1B show that by covering energy donating QDs with an increasing ratio of Cy3-labeled maltose binding proteins, symmetrically arrayed around the QD center, the rate of FRET can be substantially enhanced even for a system having a modest energy overlap integral [11]. Moreover, varying the donor acceptor energy overlap by changing the QD emission spectrum (via its size), for a given acceptor, could substantially improve FRET efficiency [11]. These two features combined with one's ability to pack several proteins around a single QD donor (in a QD conjugate) indicate that QD-based FRET could potentially prove to be a robust sensing tool for bio-oriented applications.

2. Prototype of QD-based FRET nanosensor

The above findings were further utilized to construct a prototype QD-based FRET nanosensor (schematically shown in Figure 2A) [12]. The prototype sensor consisted of 555 nm emitting QDs surrounded by ~ 10 MBP-his, each coordinated to the ZnS surface of the QD by the MBPs C-terminal penta-histidine sequence via Zn-histidine metal affinity. The central binding pocket of the MBP is pre-labeled with a beta-cyclodextrin-QSY9 quencher analog (BCD-QSY9), which results in FRET quenching of the QD emission, due to its proximity to the QSY 9 dyes. Addition of the targeted analyte-maltose results in displacement of the BCD-QSY9 analog, altering FRET in a concentration dependent manner as depicted in Figures 2A and 2B. Again, the ability to assemble multiple acceptors around a single QD donor results in a higher FRET absorption cross-section, which can improve sensing capabilities.

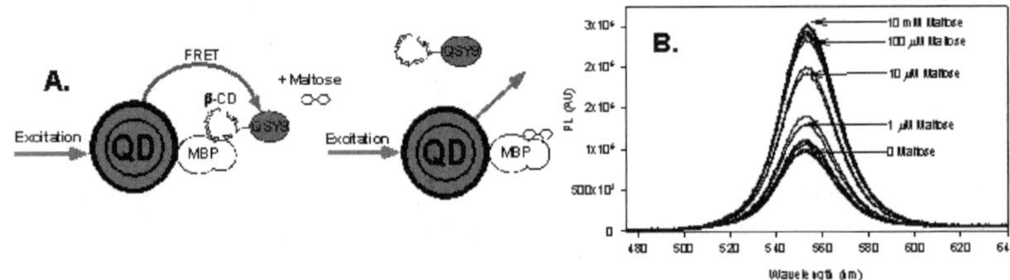

Figure 2. (A) Schematic depiction of a QD-MBP prototype nanosensor. In this scenario, each QD donor is surrounded by ~10 MBPs each prelabeled with a BCD-QSY9 analog in the MBP central binding pocket (for brevity only 1-MBP is shown). Close proximity of the QSY9 dark quencher to the QD donor results in FRET quenching. Addition of the maltose analyte results in displacement of the BCD-QSY9 analog, altering FRET in a concentration dependent manner. **(B)** Results from titrating the solution phase sensor shown in A with an increasing concentration of maltose, demonstrating recovery of QD emission.

3. Surface immobilization of QD bioconjugate assemblies

We now discuss the design of a surface anchoring system that could allow immobilizing a QD-protein-dye assembly onto a solid substrate. This should eventually permit one to realize a solid-state surface tethered sensor that has the benefit of being potentially regenerable. Figure 3A presents a schematic description of a three-layer assembly terminating with a QD-avidin conjugate, where we used two bridging proteins between the functionalized surface and the QD conjugates. We utilized a biotin functionalized MBP as a bridge between NeutrAvidin on the substrate and avidin conjugated QDs on the other end. Figure 3B shows the corresponding photoluminescent (PL) image captured on a CCD camera when exciting the structure at 488 nm with a Ar ion laser source; the PL signal result from the immobilized QD conjugates after optical excitation. The image clearly shows that assembly of discrete structures containing immobilized

QDs were formed. Moreover, the data shows that control of the fraction of immobilized QDs can be achieved by varying the concentrations of QD avidin conjugates in the initial solutions or the MBP-biotin on the surfaces. Very little non-specific binding or background was observed.

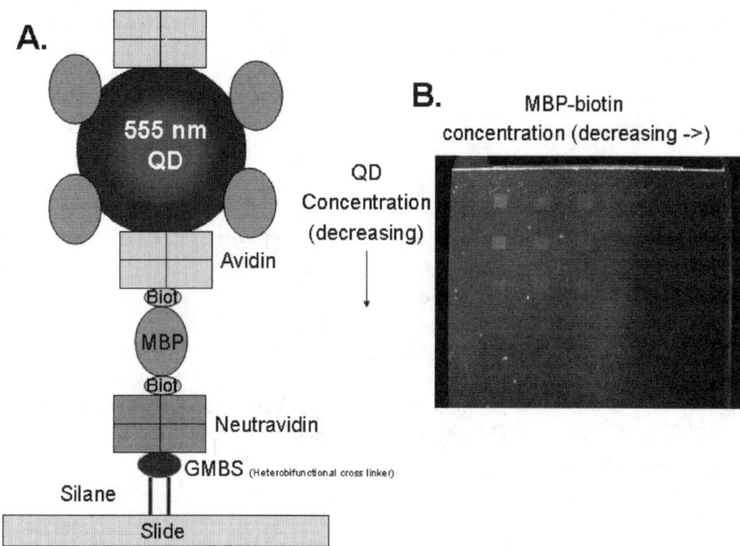

Figure 3. **(A)** Schematic depiction showing the structure of a surface immobilized QD. The slide surfaces are coated with silane and then a heterobifunctional linker is used to immobilize neutravidin on the surface. The QD is surrounded with MBP and avidin as described in reference 16. A highly biotinylated MBP acts as the linker between the NeutrAvidin on the surface of the slide and the MBP/avidin coated QD. **(B)** Image captured from a slide prepared as in A, illuminated with an evanescent 488 nm laser wave. The image demonstrates controlled assembly of discrete 555 nm emitting QD surface structures.

CONCLUSION

We have shown the use of QD bioconjugates as energy donors in FRET assays. In these investigations, QD-bioconjugates offered the unique configuration where multiple dye acceptors interacted with a single QD donor in a QD-protein assembly. Energy transfer efficiency increased with increasing surface coverage of dye-labeled proteins. These findings were further exploited to design a prototype FRET-based solution phase biosensor using QDs as energy donors. We then demonstrated controlled assembly of QD-avidin conjugates onto a fixed surface using a biological attachment strategy. This represents the first step towards creating a variety of QD based nano-devices. These devices will be able to exploit the numerous spectroscopic benefits offered by QDs. The surface assembled (anchored) QD-protein structures remain viable and can be rescanned anywhere from days to weeks after (data not shown). This can be attributed to the high affinity binding constants of the protein-complexes used. The avidin-biotin complex has a binding affinity (K_D) of 10^{-15} and the 5-histidine MBP-QD association is estimated to have a K_D of 10^{-10} to 10^{-12} [17]. Potential uses include QD powered nanosensors that are resistant to photo- and chemical degradation and light harvesting devices.

Acknowledgements: ILM and ARC are supported by National Research Council (NRC) fellowships through the U.S. Naval Research Laboratory.

REFERENCES

1. M. Bruchez-Jr, M. Moronne, P. Gin, S. Weiss, and A.P. Alivisatos, Science **281**, 2013-20 (1998).
2. W.C.W. Chan and S. Nie, Science **281**, 2016-2018 (1998).
3. C.J. Murphy, Anal. Chem. **74**, 520A-526A (2002).
4. H. Mattoussi, J.M. Mauro, E.R. Goldman, G.P. Anderson, V.C. Sundar, F.V. Mikulec, and M.G. Bawendi, J. Am. Chem. Soc. **122**, 12142-12150 (2000).
5. J.K. Jaiswal, H. Mattoussi, J.M. Mauro, and S.M. Simon, Nature Biotechnology **21**, 47- (2003).
6. D.M. Willard, L.L. Carillo, J. Jung, and A. Van Orden, Nano Lett. **1**, 469-474 (2001).
7. P.T. Tran, E.R. Goldman, G.P. Anderson, J.M. Mauro, and H. Mattoussi, Phys. Stat. Sol. (b) **229**, 427-432 (2002).
8. F. Patolsky, R. Gill, Y. Weizmann, T. Mokari, U. Banin, and I. Willner, J. Am. Chem. Soc. **125**, 13918 -13919 (2003).
9. M.A. Hines and P. Guyot-Sionnest, J. Phys. Chem. **B100**, 468-471 (1996).
10. B.O. Dabbousi, J. Rodriguez-Viejo, F.V. Mikelec, J.R. Heine, H. Mattoussi, R. Ober, K Jensen, and M.G. Bawendi, J. Phys. Chem. **101**, 9463-9475 (1997).
11. A.R. Clapp, I.L. Medintz, J.M. Mauro, B.R. Fisher, M.G. Bawendi, and H. Mattoussi, J. Am. Chem. Soc., 125, 301-310 (2004).
12. I.L. Medintz, A.R. Clapp, H. Mattoussi, E.R. Goldman, and J.M. Mauro, Nature Materials 630-638 (2003).
13. K.E. Sapsford, P.T. Charles, C.H. Patterson, and F.S. Ligler, Anal. Chem. **74**, 1061-1068 (2002).
14. C.A. Rowe-Taitt, J.W. Hazzard, K.E. Hoffman, J.J. Cras, J.P. Golden, and F.S. Ligler, Biosen. Bioelect. **15**, 579-589 (2000).
15. B.M. Lingerfelt, H. Mattoussi, E.R. Goldman, J.M. Mauro, and G.P. Anderson, Anal. Chem. **75**, 4043-4049 (2003).
16. E.R. Goldman, E.D. Balighian, H. Mattoussi, M.K. Kuno, J.M. Mauro, P.T. Tran, and G Anderson, J. Am. Chem. Soc. **124**, 6378-6382 (2002).
17. J.F. Hainfeld, W. Liu, C.M.R. Halsey, P. Freimuth, and R. D. Powell, J. Str. Bio. 1**27**, 185-1 (1999).

Mat. Res. Soc. Symp. Proc. Vol. 789 © 2004 Materials Research Society N5.8

Design of Water-Soluble Quantum Dots with Novel Surface Ligands for Biological Applications

H. Tetsuo Uyeda[1], Igor L. Medintz[2], and Hedi Mattoussi[1]
U.S. Naval Research Laboratory, Washington, DC 20375, [1]Division of Optical Sciences, Code 5611, [2]Center for Bio/Molecular Science and Engineering, Code 6910

ABSTRACT

We have designed a series of organic oligo- and polyethylene glycol (PEG) based surface capping ligands that allow for QD manipulation in aqueous media. We utilized readily available thioctic acid and various oligo- and polyethylene glycols in simple esterification schemes, followed by reduction of the dithiolane to produce multi-gram quantities of capping substrates. Cap exchange of the native trioctyl-phosphine and -phosphine oxide based ligands with the PEG-terminated dithiol-alkyl cap readily resulted in aqueous dispersions of QDs that were homogeneous and stable in various pH ranges over an extended period of time. Mixed surface capping strategies utilizing ratios of dihydrolipoic acid to the pegylated dihydrolipoic acid were also prepared. We anticipate that such systems should allow one to covalently attach amine containing biomolecules to nanoparticle systems bearing carboxylates, employing known coupling agents, such as (dimethylamino) propyl-3-ethyl-carbodiimide (EDC). This design and conjugation strategy may facilitate the development of a new generation of QD-bioconjugates, which can be directly utilized in bio-related applications such as sensing and cellular imaging.

INTRODUCTION

The use of fluorescent tags to label biological molecules is a general and practical method in biological research. Organic dyes have been employed in single and multiplex detection schemes [1,2]. However, this approach has several limitations. Organic fluorophores tend to have narrow excitation spectra with red tailing broad emission bands. Due to spectral overlap, simultaneous quantitative assessment of the relative amounts of different fluorophores present in the same sample becomes exceedingly difficult [3]. Moreover, variation of the absorptive and emissive properties of dye labeled conjugates requires the use of chemically distinct labels with non-uniform synthesis and conjugation methods.

Colloidal semiconductor nanocrystals, or quantum dots (QDs), with unique properties such as high photobleaching threshold, chemical stability, and readily tunable spectral properties have the potential to overcome many of the limitations in biological research encountered by conventional dye systems [4-7].

Recently, the utility of stable water-soluble luminescent colloidal quantum dots (QDs), either pure or conjugated to biomolecular receptors, has been demonstrated in biosensing and cellular imaging applications [4-14]. However, due to the process in which these systems are prepared, the surface properties of the QDs limit one's ability to manipulate them in aqueous environments and apply simple covalent conjugation techniques to prepare stable versatile QD-bioconjugates. Several examples of surface chemistry include the use of QDs coated with carboxylic acid terminated mono-thiol ligands, such as mercaptopropionic acid (MPA), many of which have been used in bioconjugation schemes; however, the temporal stability of these systems is transitory [8-11,15]. Schemes that employ dihydrolipoic acid (DHLA) as a capping ligand and bioconjugation based on non-covalent (e.g., electrostatic) binding motifs necessitate the use of basic solutions to preserve the QD sample integrity [8-11]. Recently, Kim and co-workers reported the synthesis a novel multidentate oligomeric phosphine-based ligand that provide for luminescent water-soluble and stable QD systems [16].

Here, we present the synthesis of a series of DHLA ligands modified with several oligo- and poly(ethylene glycol) molecules, detail the surface exchange of the native TOP/TOPO ligands and transfer into aqueous environments, describe their photophysical behavior, and present tests of such systems, including stability in acidic conditions.

EXPERIMENTAL DETAILS

Synthesis of the ligands used in this study is outlined in Figure 1.

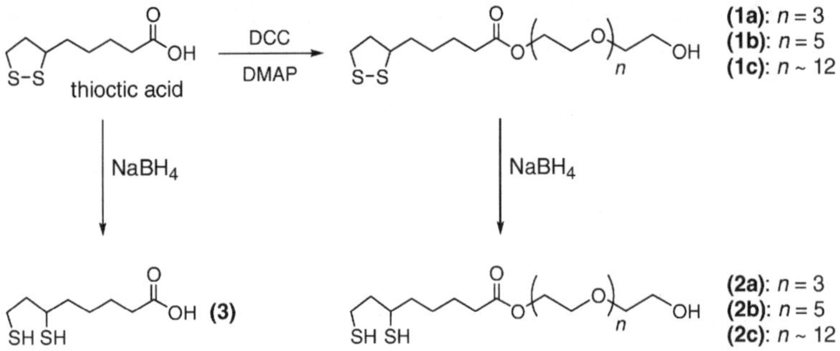

Figure 1. Synthetic scheme for QD surface ligands.

General procedure for the synthesis of compounds **1a-c**: A round bottomed flask was charged with thioctic acid (5.16 g, 25 mmol), 4-dimethylamino-pyridine (915 mg, 7.5 mmol), poly(ethylene glycol) (125 mmol) and CH_2Cl_2 (300 mL). The contents were stirred under nitrogen and cooled to 0 °C while a solution of 1,3-dicyclohexylcarbodiimide (5.67 g, 27.5 mmol) in CH_2Cl_2 (20 mL) was added drop wise over 0.5 h. The reaction was allowed to warm to room temperature and was stirred for 20 h. The solution was filtered over a bed of Celite and rinsed with chloroform. The organics were washed repeatedly with water and brine to remove the unreacted poly(ethylene glycol). The combined extracts were dried over $MgSO_4$, filtered, and evaporated to give a viscous yellow oil. The crude product was chromatographed on silica utilizing a solvent system consisting of $CHCl_3$, methanol, and acetone. The general procedure for the synthesis of compounds **2a-c**: A round bottomed flask was charged with the appropriate pegylated thioctic acid derivative (1 mmol), ethanol (10 mL), and water (2 mL). $NaBH_4$ (150 mg, 4 mmol) was added and the reaction was heated gently until the yellow color vanished to give a cloudy colorless solution. The contents were diluted with $CHCl_3$ (200 mL), dried over $MgSO_4$, filtered and evaporated to give a colorless oil.

Water-soluble CdSe-ZnS core-shell QDs were prepared by replacing the native trioctylphosphine/trioctylphosphine oxide (TOP/TOPO) organic growth solution with functionalized bidentate dihydrolipoic acid (DHLA) ligands in a modified procedure as described previously [8]. To form the water-soluble QD assembly, native trioctylphosphine and trioctylphosphine oxide (TOP/TOPO) capped nanocrystals were mixed with an excess of the desired surface ligand and incubated to displace the TOP/TOPO molecules. A small amount of the QD growth solution is precipitated with ethanol and centrifuged to remove additional TOP and TOPO. A large excess of compound **2c** is added to the QDs and the solution is heated while stirring at 90 °C for several hours [8]. The QDs are precipitated from the solution with a mixture of hexanes and chloroform and centrifuged. The QDs dissolved in water are filtered and concentrated using a membrane centrifugal device with a nominal molecular cut-off of 50,000 and a 0.45-μm filter disc.

RESULTS AND DISCUSSION

The identification of compounds **1a-c** and **2a-c** was confirmed through ^{1}H-NMR spectroscopy. Identification of compounds **1a-c** and **2a-c** were achieved by comparing the parent thioctic acid, dihydrolipoic acid, and poly(ethylene glycols) with the numbered compounds (Figure 2). In all cases, the summation of the individual units most closely resembled that of the proposed species. Yields for the coupling are typically good (higher than 80%) and employing an excess amount of the poly(ethylene glycols) in which two hydroxyl substituents are present ensures minimal formation of the bis-substituted ligand. The reduction of the dithiane ring system with sodium borohydride proceeded smoothly and gave near quantitative amounts of the desired product, as was observed with thioctic acid reduction [8]. This strategy has allowed one to quickly prepare multi-gram quantities of the desired capping ligands. This is convenient for the cap exchange process in which a large excess of the capping ligand is used relative to the amount of QD.

Surface ligand exchange of DHLA based ligands prepared with tetra- and hexa(ethylene glycol) (**2a**, **2b**) produced species that did not adequately disperse in aqueous solutions. Presumably, at those oligo(ethylene glycol) chain lengths favorable interactions of the PEG units with the water molecules is not sufficient enough to produce stable and aggregate-free water-soluble QDs. Such systems however, were readily soluble in polar organic solvents such as methanol and ethanol. Surface ligand exchange of QDs with compound **2c** produced systems that were readily soluble in water and free of aggregates, and which maintained their absorption and luminescence characteristics.

Confirmation of the cap exchange was also obtained by investigation of changes in the gel electrophoretic mobility shift of QDs upon capping the nanocrystals with mixtures of DHLA and DHLA-PEG ligands at various ratios. Due to a net negative charge from the carboxylic end groups of the DHLA molecules, QDs that were fully or partially capped with DHLA ligands migrate towards the positive electrode in conventional agarose gel electrophoresis. The data in Figure 3 clearly shows that for QDs coated with 100% of compound **2c** results in a little or no shift from the loading well under the applied current, due to the non-ionized nature of these QD systems. When 50% of the pegylated compound **2c** was replaced with the ionized DHLA (lane 2 in Figure 3), a substantial gel shift was measured, due to the finite electrophoretic mobility brought by the DHLA caps on the QD surface. Similarly, QDs coated with a mixture of DHLA and DHLA-PEG (**2c**) ligands but at lower DHLA ratios (30% in lanes 3 and 10% in lane 4) resulted in smaller shifts relative to the case shown in lane 2. This clearly reflects a lower electrophoretic mobility of the QDs, due a smaller density of surface charges when the fraction of DHLA ligands used is reduced.

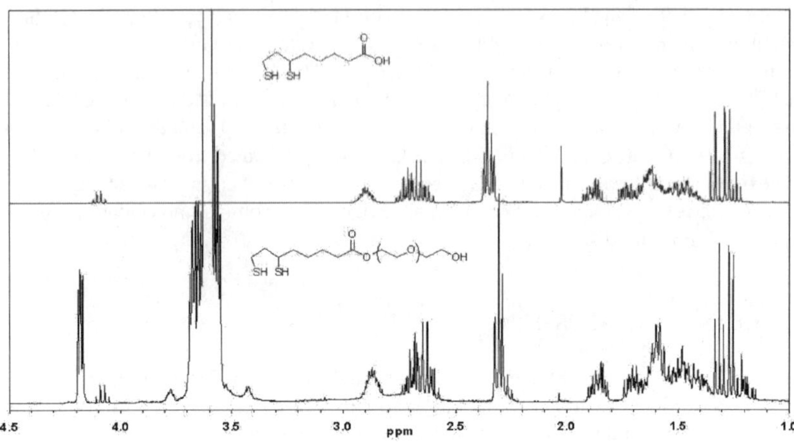

Figure 2. ^{1}H-NMR spectra of compounds **3** and **2c**.

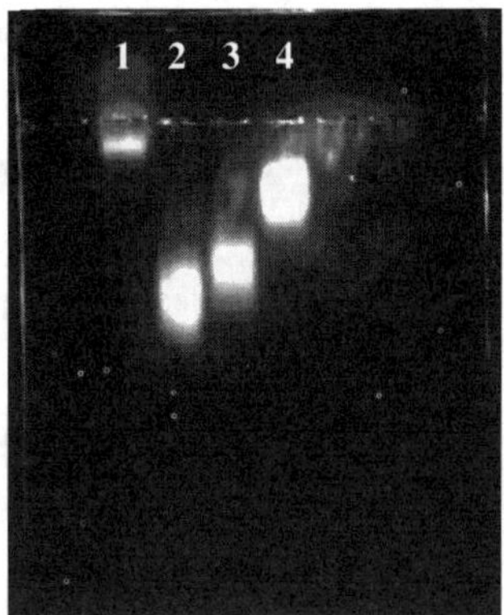

Figure 3. Electrophoresis of CdSe/ZnS QDs coated with various ratios of **2c** to **3** in 1% agarose gel buffered with Tris borate EDTA buffer. Lane 1: 1:0 of **2c** to **3**, lane 2: 1:1 of **2c** to **3**, lane 3: 3:1 of **2c** to **3**, lane 4: 10:1 of **2c** to **3**.

It has been well established that CdSe/ZnS based QDs coated with DHLA are soluble in aqueous solutions at basic pHs (7-12). This phenomenon is driven by the ionization of the surface carboxyl functional groups at the ends of the DHLA moieties. This situation however, precludes the use of these QDs in acidic media due to the aggregative effects of re-protonation of the carboxylic functionality. We hypothesized that by basing the solubility on conventional poly(ethylene glycol) substituents, rather than the ionization of a carboxyl group, we would not be limited to using basic solutions for more elaborate experiments. Preliminary experiments of incubating the QDs coated with ligand **2c** in buffers at various pHs suggests that the luminescence of such systems are quite stable and unaltered (as shown in Figure 4). Stability of DHLA-PEG-capped QDs is better and more reproducible in acidic buffer solutions at pH values larger than 3, however.

pH 3 4 5 6 9

Figure 4. Luminescence image set of CdSe-ZnS QD solutions coated with compound **2c** in buffers ranging from 3 – 9 after 2 weeks. Samples were excited with a UV lamp at 365 nm; QD emission is centered at 590 nm.

Quantum yield measurements of the native TOP/TOPO capped QDs and QDs that were cap-exchanged with compound **2c** were obtained by comparison to rhodamine 6G as a standard (QY = 0.95) [17]. Quantum yields for the native TOP/TOPO capped QDs in organic media were quite high (48 – 80%), however, as expected, a decrease in the QY for the pegylated-DHLA capped QDs in aqueous media (15 – 22%) was also observed (Table 1).

Efforts are currently underway to directly conjugate QDs that were cap exchanged with a mixture of DHLA and pegylated-DHLA molecules with amino-functionalized proteins and other biological molecules such as DNA.

Table 1. Optical properties of a few QD-complexes used in this study. $\lambda_{abs\ max}$ designates the location of the first absorption peak (or band edge absorption).

Complex	Ligand	$\lambda_{abs\ max}$ (nm)	$\lambda_{em\ max}$ (nm)	Quantum Yield (%)
QD-TOP/TOPO	-	522	534	80
QD-TOP/TOPO	-	515	531	76
QD-DHLA-PEG-600	2c	535	552	15
QD-DHLA	3	518	536	22

CONCLUSIONS

This study shows that a series of oligo- and poly (ethylene glycol) derivatized dihydrolipoic acid molecules were successfully synthesized. The use of these ligands in the surface ligand exchange process of the native TOP/TOPO capped QDs has also shown that the dihydrolipoic acid modified with the poly(ethylene glycol) (MW ~ 600) results in homogeneous aqueous dispersions. Furthermore, preliminary investigations of these systems have shown (increased drop) pH tolerance over an extended period of time. We are presently developing simple covalent conjugation schemes to attach these water-soluble QDs to a variety of biomolecular receptors.

ACKNOWLEDGMENTS

Research supported by the U.S. Naval Research Lab and Office of Naval Research. Authors (HTU and ILM) were supported through the NRL-NRC Postdoctoral Research Program.

REFERENCES

1. E. Schrock, S. duManoir, T. Veldman, B. Schoell, J. Wienberg, M. A. FergusonSmith, Y. Ning, D. H. Ledbetter, I. Bar-Am, D. Soenksen, Y. Garini, and T. Ried, Science (Washington, D.C.) **273,** 494-497 (1996).
2. M. Roederer, S. DeRosa, R. Gerstein, M. Anderson, M. Bigos, R. Stovel, T. Nozaki, D. Parks, and L. Herzenberg, Cytometry **29,** 328-339 (1997).
3. G. T. Hermanson, *Bioconjugate Techniques* (Academic, San Diego, 1996).
4. M. Bruchez, M. Moronne, P. Gin, S. Weiss, and A. P. Alivisatos, Science (Washington, D.C.) **281,** 2013-2016 (1998).
5. W. C. W. Chan and S. Nie, Science (Washington, D.C.) **5385,** 2016-2018 (1998).
6. J. K. Jaiswal, H. Mattoussi, J. M. Mauro, and S. M. Simon, Nature Biotech. **21,** 47-51 (2003).
7. X. Wu, H. Liu, J. Liu, K. N. Haley, J. A. Treadway, J. P. Larson, N. Ge, F. Peale, and M. P. Bruchez, Nature Biotech. **21,** 41-46 (2003).
8. H. Mattoussi, J. M. Mauro, E. R. Goldman, G. P. Anderson, V. C. Sundar, F. V. Mikulec, and M. G. Bawendi, J. Am. Chem. Soc. **122**, 12142-12150 (2000).
9. P. T. Tran, E. R. Goldman, G. P. Anderson, J. M. Mauro, and H. Mattoussi, Phys. Stat. Sol. B. **229,** 427-432 (2002).
10. M. Y. Han, X. H. Gao, J. Z. Su, and S. Nie, Nature Biotech. **19,** 631-635 (2001).
11. E. R. Goldman, G. P. Anderson, P. T. Tran, H. Mattoussi, P. T. Charles, and J. M. Mauro, Anal. Chem. **74,** 841-847 (2002).
12. H. Mattoussi, M. K. Kuno, E. R. Goldman, G. P. Anderson, and J. M. Mauro, in *Optical Biosensors: Present and Future*, Elsevier, Amsterdam, 2002, p. 537-569.
13. M. E. Akerman, W. C. W. Chan, P. Laakkonen, S. N. Bhatia, and E. Ruoslahti, P.N.A.S. **99,** 12617-12621 (2002).
14. B. Dubertret, P. Skourides, D. J. Norris, V. Noireaux, A. H. Brivanlou, and A. Libchaber, Science (Washington, D.C.) **298,** 1759-1762 (2002).
15. J. Aldana, A. Wang, and X. Peng, J. Am. Chem. Soc. **123,** 8844-8850 (2001).
16. S-J. Kim and M. G. Bawendi, J. Am. Chem. Soc., in press (2003).
17. J. N. Demas and G. A. Corsby, J. Phys. Chem. **75,** 991-1024 (1971).

Poster Session II: QDs and Nanoparticles for Biological Applications

Mat. Res. Soc. Symp. Proc. Vol. 789 © 2004 Materials Research Society N6.2

Optimization of Quantum Dot – Nerve Cell Interfaces

Jessica Winter[1*], Christine Schmidt[2], Brian Korgel[1]
[1]Department of Chemical Engineering and [2]Department of Biomedical Engineering
University of Texas at Austin, MC0400, Austin, TX, 78712, [*]Speaker

ABSTRACT

Nerve cells communicate by passing electrical signals through extensions from their cell bodies. Micron-scale devices such as microelectrode arrays and field effect transistors have already been used to externally manipulate these signals. However, these devices are almost as large as the cell body of the neuron. In order to make functional electrical connections to nerve extensions or their component ion channels that propagate signals, it will be necessary to utilize increasingly smaller components. We propose the use of semiconductor quantum dots, which can be optically activated, as a potential means of perturbing the nerve membrane potentials. As a first step, we have created qdot-neuron interfaces with cadmium sulfide quantum dots. The qdots may be attached to cells either non-specifically or through selected interactions exploiting biorecognition molecules attached to the qdot particle surface. We have investigated the effect of altered synthesis conditions including pH, concentration, reactant ratio, ligand length, ligand R group, and ligand concentration on the nature and quality of qdot-cell binding. We discuss the effect of these altered synthesis conditions on particle fluorescence intensity and color. Additionally, we studied the interaction of these particles with cells and determined that larger particles are more likely to bind non-specifically than smaller particles produced with the same amount of passivating ligand. It is possible this results from reduced surface area coverage of passivating chemical. Finally, we have produced particles passivated with the biorecognition peptide CGGGRGDS in the absence of other chemical stabilizers and characterized their surfaces using NMR and IR.

INTRODUCTION

Nanometer-sized colloidal particles, or quantum dots (qdots), exhibit sizes between that of bulk material and molecular clusters. This results in a unique mix of material properties. For example, qdots are large enough to retain semiconductor characteristics, yet small enough so that quantum confinement of electron-hole pairs[1] results in distinct spectroscopic traits. Optical excitation can produce dipole moments,[2] heat production,[3] fluorescent emission,[4] and electron transfer.[5,6] Further; many of these properties are size-dependant,[2,7] allowing for the development of devices with tunable characteristics. Several applications utilizing qdots have been explored, including LEDs,[8] solar cells,[9] and biological labels.[10-12] Although the latter application has led to the development of commercially available qdot dyes, it fails to take advantage of the potential for direct interaction between qdots and cells. Because cells communicate through nanometer-sized components (e.g.., biomolecules, receptors), the ability to interface nanomaterials and biological components affords many new opportunities to alter cell behavior. This technology may lead to the development of new prosthetic devices and bio-inorganic electronics.

We are designing a system to utilize qdot material properties (e.g., dipole moment, electron transfer, magnetism) to both visualize and interact with cellular receptors. For example, the dipole moment produced by qdots is much larger than that of bulk material.[3] Attachment of biorecognition molecules to the qdot surface can facilitate intimate contact between the dipole and neuronal ion channels. If the separation distance is small enough, it might be possible to induce a membrane potential change, producing an action potential. However, this will require optimization of the qdot-cell labeling system. Labeling efficiency needs to be high to increase signal, while qdot-cell separation distance must be minimized to reduce charge screening. Also, it may be necessary to decrease non-specific binding so that labeled tissue may be distinguished from surrounding tissue. Here, we present the initial development of this system, including the optimization of aqueous qdot syntheses, studies of qdot non-specific binding, and the development of bioconjugated labels. Additionally, we are currently using whole-cell clamping to examine the effect of qdot labeling on nerve cell membrane potential.

EXPERIMENTAL DETIALS

Aqueous arrested precipitation of cadmium sulfide qdots was examined using a model ligand, mercaptoacetic acid (Fluka, Cat. No. 88652), as described previously.[12] Several synthesis conditions were systematically altered, including pH, reactant ratios and concentrations, and ligand concentrations, carbon chain length, and R group. A second type of particle was also created, using the peptide CGGGRGDS as a capping ligand with no additional chemical ligands. The presence of the peptide was confirmed using FTIR and NMR. All particles were manufactured at least four hours before analysis, which consisted of measuring the resultant spectroscopic properties. A Beckman DU500 spectrophotometer (Beckman Coulter, Fullerton, CA) was used to measure particle absorbance, and PL spectroscopy was performed using a Quanta Master Model C Cuvette-based scanning spectrofluorometer (Photon Technology International, Lawrenceville, NJ).

Additionally, particle non-specific binding to biological substrates was investigated using SK-N-SH neuroblastoma cells (ATCC, HTB-11). Cells were cultured following ATCC recommendations in EMEM with 10% FBS and 2% Penicillin-Streptomycin. Then, cells were incubated with 2% BSA in Dulbecco's PBS for one hour to block inherent protein binding. The cells were rinsed with Dulbecco's PBS. Next the cells were incubated for 30 minutes with quantum dots which had been isolated using 1-propanol and centrifugation, resuspended in Dulbecco's PBS, and sterile filtered. The cells were then rinsed with Dulbecco's PBS three times, and stored in Dulbecco's PBS for observation.

RESULTS AND DISCUSSION

Previously, particles were examined while systematically varying the pH, reactant concentrations and ratios, and ligand concentration, carbon chain length, and R group.[13] Regardless of the variable examined, we observed a fluorescence emission intensity maximum at intermediate particle sizes (~ 2 nm). We theorized that fluorescent maxima result from thermodynamically stable surfaces, perhaps because of the formation of Cd-thiol precursor complexes. These qdots displayed an emission wavelength of ~ 530 nm (green-yellow) with a quantum yield of ~ 15%. This wavelength is easily distinguishable from blue autofluorescence produced by the cell. Next, the particles were examined for non-specific binding. Larger particles demonstrated the most non-specific binding, in some cases forming a thin film over both the cells and substrate (Figure 1A). Minimal non-specific binding appeared to occur at intermediate qdot sizes (Figure 1B). However, non-specific binding for small qdots was difficult to measure because the emission wavelength is similar to that of cellular autofluorescence (Figure 1C).

Given the high quantum yield and low non-specific binding shown at intermediate sizes, it was decided to develop biolabels in this size range.

Figure 1: Non-Specific Binding Produced by (A) Large (~ 5 nm) Qdots, (B) Intermediate Size (~ 2nm) Qdots, and (C) Small (~ 1 nm) Qdots. Qdots appear orange/yellow, cells autofluoresce blue. For large qdots (A), a film is formed on the surface of the qdots and substrate. Intermediate sizes (B) produce minimal non-specific binding. Smaller sizes (C) produce limited binding that is difficult to distinguish from background autofluorescence.

Thus, we attempted to construct particles passivated with peptide ligands (i.e., CGGGRGDS). These ligands are substantially larger than the model ligand examined previously. From our model experiments, we knew that increasing ligand length would blue shift emission fluorescence and decrease particle size (Figure 2A). However, we had also discovered that increasing reaction pH can compensate for this effect. Therefore, using altered synthesis conditions, we were able to produce qdots with sizes near 2 nm. However, the amount of peptide required for such syntheses is large. Given cost limitations, we also experimented with composite systems, incorporating mercaptoacetic acid as a "filler" ligand in addition to the peptide. The resultant qdots were studied using both IR and NMR to verify the presence of peptide, evaluate qdot-peptide binding characteristics, and roughly determine the amount of peptide present. Finally, the peptide-conjugated qdots were used to label cells (Figure 2B).[12] Ligands recognized and bound model cell surface receptors (i.e., integrins), allowing for nanometer scale (i.e., 3-30 nm) separation distances.[12] Currently, we are attempting to measure changes in the nerve cell membrane potential produced by optically excited qdots.

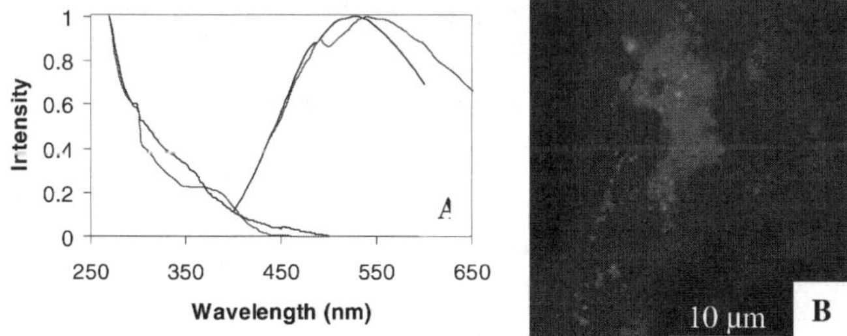

Figure 2: (A) Absorbance and PL Emission Spectra of MAA (Red) and Peptide (Blue) Passivated Nanocrystals. (A) Peptide conjugation blue shifts the absorbance (i.e., 370 nm vs. 350 nm) and emission peaks (i.e., 520 nm vs. 550 nm) when compared to the MAA control. (B)

Cells were incubated with peptide passivated qdots, and bound to integrins on the cell surface. Qdots appear green-yellow, the cell autofluorescence is blue.

CONCLUSIONS

The goal of our work is to demonstrate the ability of qdots to serve not only as labels, but also to manipulate cell function. In pursuit of this goal, we have examined the effect of synthesis conditions on the spectroscopic properties of the qdots produced. In particular, we optimized qdots for maximum quantum yield and minimum non-specific binding. Further, we produced peptide conjugated qdots that are capable of binding cell surface receptors. *These qdots will serve as a starting point for future work, where we hope to demonstrate the ability of optically excited qdots to produce nerve membrane potential changes.* Although our goal was to create biological labels, the ability to tune spectroscopic properties is relevant to several applications (e.g., LEDs). Ultimately, this work will improve the ability of nanometer-scale colloids to interact with cells and biomolecules. As the emphasis in materials science shifts from the micron-scale to nanometer-scale applications, the ability to interface technologies at this level becomes increasingly important. Additionally, cells communicate using nanometer-sized biomolecules, and the capability for integration of biological components and nanomaterials will provide intriguing opportunities to influence cell behavior. This work highlights many of the issues that may inhibit the construction of such devices, and describes some of the first steps taken to build novel nanoscale biomaterials.

ACKNOWLEDGEMENTS

The authors gratefully acknowledge discussions with Fred Mikulec and David Jurbergs. Funding for this work was provided by the Texas Higher Education Coordinating Board through their ARP program (CES, BAK), an NSF graduate research fellowship (JOW), and the NSF IGERT program (JOW). FTIR facilities were provided by the center for nano- and molecular materials sponsored by the Robert A. Welch foundation.

REFERENCES

1. R. Rossetti, J.L. Elison, J.M. Gibson, L.E. Brus. J. Chem. Phys. **80**, 4464,1984.
2. Y. Wang, N. Herron. J. Phys. Chem. **95**, 525, 1991.
3. S.R. Shershen, S.L. Westcott, N.J. Halas, J.L. West. J. Biomed. Mat. Res. **51**, 293, 2000.
4. A. Henglein. Ber. Bunsenges. Chem. **86**, 301, 1982.
5. A.P. Alivisatos. Science. **271**, 933, 1996.
6. C.B. Murray, C.R. Kagan, M.G. Bawendi. Ann. Rev. Mat. Sci. **30**, 545, 2000.
7. H. Weller, H.M. Schmidt, U. Koch, A. Fojtik, S. Baral, A. Henglein, et. al. Chem. Phys. Lett. **124**, 557, 1986.
8. V.L. Colvin, M.C. Schlamp, A.P. Alivisatos. Nature. **370**, 354, 1994.
9. N.C. Greenham, X. Peng, A.P. Alivisatos. Phys. Rev. B. **54,** 17628, 1996.
10. M. Bruchez, Jr., M. Maronne, P. Gin, S. Weiss, A.P. Alivisatos. Science. **281**, 2013, 1998.
11. W.C.W. Chan, S. Nie. Science. **281**, 2016, 1998.
12. J.O. Winter, T.Y. Liu, B.A. Korgel, C.E. Schmidt. Adv. Mats. **13**, 1673, 2001.
13. J.O. Winter, N. Gomez, S. Gatzert, C.E. Schmidt, B.A. Korgel. Submitted. 2003

Photoluminescence Properties and Zeta Potential of Water-Dispersible CdTe Nanocrystals

Masanori Ando, Chunliang Li and Norio Murase
Photonics Research Institute, National Institute of Advanced Industrial Science and Technology (AIST), Kansai Center, 1-8-31 Midorigaoka, Ikeda, Osaka 563-8577, JAPAN

ABSTRACT

The photoluminescent (PL) properties and zeta potential of green-emitting CdTe nanocrystals (diameter: 3 nm) capped with a stabilizing surfactant, thioglycolic acid (TGA), have been investigated as a function of pH of the aqueous solution. The green PL intensity reached the maximum at pH5.1 and was somewhat lower in the pH range of 6-10, which was similar to the previously reported result. However, when the pH was at and below 4, the green PL intensity decreased drastically. The relative ratio of the dissociation form of the carboxyl group of TGA showed a large diminution at and below pH5 accompanied by a significant decrease of the absolute value of zeta potential. Since the absolute value of zeta potential reflects the stability of nanocrystals, the results obtained shows that the TGA-capped CdTe nanocrystals are stable only in basic to neutral regions and that the agglomeration of the nanocrystals in acidic range reflects the transition from the dissociated (charged) form to the non-dissociated (non-charged) form of a carboxyl group in TGA. Encapsulation of nanocrystals in glass is a promising way to further improve the long-term photostability of nanocrystals. Therefore, we chose an alkoxide having an amino group for a matrix for the encapsulation. The amino group has a good affinity to TGA as well as promotes the sol-gel reaction. As the result, the CdTe nanocrystals have been dispersed finely in the glass matrix without a deterioration of PL intensity.

INTRODUCTION

Over the years, semiconductor nanocrystals (quantum dots (QDs)) attract much attention due to their strong, size-dependent photoluminescence (PL) [1-3]. Many possible applications are expected ranging from optical, optoelectronic to biological fields [4-6]. To date, two solution based syntheses of QDs have been reported. The first method is a non-aqueous solution based one which utilizes a thermal decomposition of organometallic compound, and trioctyl phosphine is used as a hot coordinating solvent [1,7]. Hydrophobic QDs are obtained by this method. The second method is an aqueous solution based one in which hydrophilic QDs are prepared by using organic capping surfactants such as thioglycolic acid (TGA) [2,8]. The QDs synthesized by the second method usually retain their PL for longer time in ambient than the QDs prepared by the first method. Since the PL intensity depends on the pH of the solution [2], we have measured zeta potential as an indicator of the stability of particles in solution. As the result, we have found that the transition between the dissociated form and non-dissociated form of carboxyl group of the stabilizing surfactant plays an important role in the diminution of PL intensity and stability of QDs in acidic region.

In order to further improve the long-term photostability of QDs they need to be encapsulated in a rigid matrix. One promising way is to incorporate the water-dispersible QDs into silica glass matrix by a sol-gel method. Because it has advantages over organic polymers for their chemical and thermal stability together with the robustness to ultraviolet light. A sol-gel incorporation of CdSe QDs prepared in a non-aqueous solution has been reported [9]. However, the final material obtained was a gelatinous one. As the TGA-capped CdTe QDs are most stable in basic region, we have chosen a silane coupling agent with an amino group for sol-gel fabrication of glass. By using this silane coupling agent, highly luminescent silica glass incorporating the CdTe QDs was obtained. The glass phosphor thus prepared is expected to be useful for practical applications.

EXPERIMENTAL DETAILS

Synthesis of water-dispersible, TGA-capped CdTe QDs

TGA-capped CdTe QDs were synthesized in aqueous solution by reaction of Cd^{2+} with H_2Te gas or with NaHTe in the presence of a stabilizing agent (TGA) as previously reported [7]. The size of CdTe QD increases with reflux time. We obtained green-emitting CdTe (peak wavelength: ~549 nm; diameter: 3 nm) by selecting the reflux time. Since the mercapto group in TGA is bound to CdTe, hydrophilic carboxyl group of TGA is assumed to be situated on the outer surface of the TGA-capped CdTe QDs. This makes the TGA-capped CdTe QDs dispersible in water.

Measurement of PL spectra and zeta potential of the pH-controlled QD solution

The PL spectra of the above solution were measured with a Hitachi F-4500 fluorescence spectrometer. The excitation wavelength was 365 nm. The PL quantum efficiency was estimated by comparing the PL intensity and absorbance with those of the standard dyes, rhodamine 6G (R6G) and quinine [10]. In order to investigate the pH dependence, the pH of the QD solution was controlled in the range of 3-10 using HCl and NaOH. The zeta potential of the pH-controlled QD solution was measured with an electrophoretic light scattering spectrometer (Otsuka Electronics ELS-8000).

Fabrication of CdTe-incorporated silica glass phosphor by sol-gel method

As the TGA-capped CdTe QDs are most stable in basic solution, a silane coupling agent having an amino group, 3-aminopropyltrimethoxysilane (APS) was selected as a silica precursor for the sol-gel synthesis. Methanol was used as co-solvent, and the molar ratio of APS to methanol was fixed to 1:50. The methanol solution was stirred, then transferred to a Teflon chalet. The pH-controlled aqueous CdTe QD solution was added to the solution in the chalet, followed by stirring of the mixture and drying in a dark place. The final concentration of CdTe QDs in glass was roughly estimated to be 1×10^{-5} M. The PL spectra of the resulting glass phosphor was measured with a Hitachi F-4500 spectrometer, and the PL quantum efficiency was estimated by comparing with the standard dyes.

RESULTS AND DISCUSSION

The pH dependence of PL properties of the TGA-capped CdTe QD solution

The PL quantum efficiency of the green-emitting CdTe QDs in basic aqueous solution was estimated to be approximately 5% at room temperature [10]. Figures 1 and 2 show the pH dependence of the PL peak wavelength and that of the PL intensity of the QD solution under various pH conditions, respectively. In the basic region (pH10-8), the PL intensity was

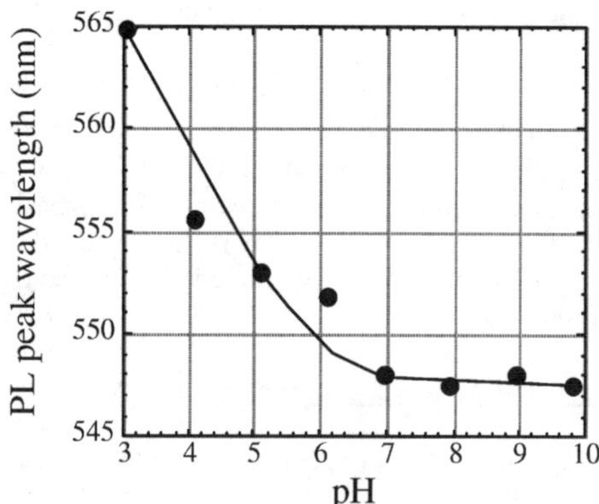

Figure 1. pH dependence of the PL peak wavelength of TGA-capped CdTe QDs.

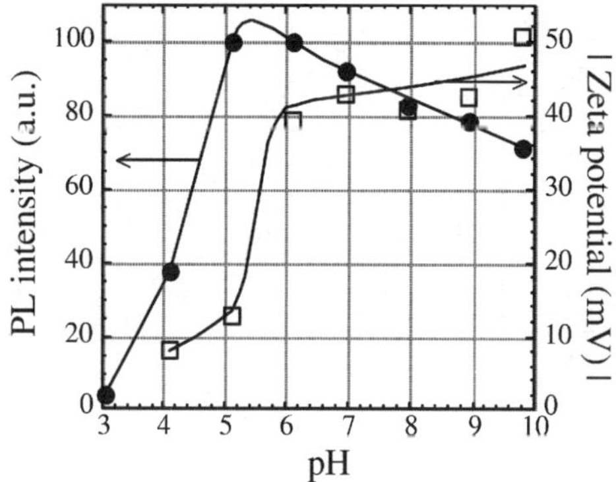

Figure 2. PL intensity and absolute value of zeta potential of TGA-capped CdTe QDs, as a function of pH.

relatively strong, and somewhat increased with decreasing pH. In this region, the PL peak wavelength remained unchanged. In the neutral to weak acidic region (pH7-5), the PL intensity was slightly larger than that in the basic region, and reached the maximum at pH5.1. The PL peak wavelength slightly shifted to the red when the solution became weakly acidic (pH6-5). This small red shift suggests that the QDs are slightly agglomerated in this pH region. On the other hand, in the region of pH4-3, the PL intensity drastically decreased and the PL peak wavelength was heavily red-shifted. Almost no PL was obtained at pH3. This shows that the degree of agglomeration is enhanced by the acidic environment of pH4-3. We assume that the significant decrease of the PL intensity at pH4-3 is mainly caused by defect centers created by agglomeration.

The pH dependence of zeta potential of the TGA-capped CdTe QD solution

In figure 2, the absolute value of zeta potential ($|\zeta|$) of the TGA-capped CdTe QDs is shown as a function of pH. In the basic, neutral to weak acidic region (pH10-6), the $|\zeta|$ value was large, and slightly decreased with decreasing pH. However, when the solution became more acidic (pH 5-3), the $|\zeta|$ value drastically decreased. The $|\zeta|$ value increases with increasing stability of particles in solution. Therefore, these results tells that the TGA-capped CdTe QDs in aqueous solution are most stable in the basic region, a little less stable in the neutral region (pH7-6), and unstable in the acidic region (pH5-3).

The pH dependence of the dissociation ratio of carboxyl group of TGA

As described in the Experimental section, the carboxyl group of TGA situates on the outer surface of the QD, and makes the QDs hydrophilic. Generally, in a basic solution, carboxyl group tends to have a dissociated form ($-COO^-$) due to deprotonation. Therefore, the surface of each TGA-capped QD has negative charge. This negatively charged surface prevents the QDs

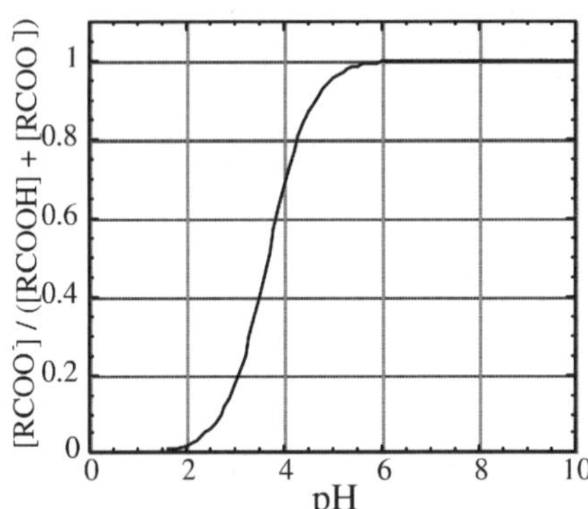

Figure 3. Dissociation curve of carboxyl group of TGA (pKa: 3.67).

from agglomeration. On the other hand, in an acidic solution, carboxyl group tends to be a non-dissociated form ($-COOH$) due to protonation. Since the non-dissociated carboxyl group does not have a charge, the TGA-capped QDs agglomerate easily. The relative ratio of the dissociation form and the non-dissociation form of the carboxyl group of TGA ($[RCOO^-]$ /

[RCOOH]) is determined by the dissociation constant (pKa: 3.67) of TGA, as a function of pH. Figure 3 shows the dissociation curve of carboxyl group of TGA. It is seen from the figure that in the basic, neutral to weakly acidic region (pH10-6), the relative ratio of the dissociation form of TGA is nearly unity. In such situation, the agglomeration of the QDs is effectively suppressed. Under a little more acidic condition (pH5), the dissociated form is still predominant. On the other hand, in the highly acidic region (pH4-1), the relative ratio of the dissociation form of TGA decreases rapidly with decreasing pH. Such low dissociation ratio at pH equal to or below 4 accelerates the agglomeration of QDs.

All the three factors, namely PL intensity, zeta potential and the dissociation ratio of a carboxyl group in TGA, showed drastic changes or transitions at pH4-5. This result shows that when the pH becomes below 5, the carboxyl group of TGA changes from the dissociated form with a negative charge to non-dissociated form without a charge, which makes the QDs unstable and promotes the agglomeration of the QDs in the solution. This agglomeration results in the strong suppression of PL intensity of the QDs.

The dissociation ratio of carboxyl group of a molecule in the acidic region increases with decreasing pKa of the molecule. Therefore, if the CdTe QDs are capped with another stabilizing surfactant whose pKa is lower than that of TGA, the resulting capped CdTe QDs are expected to be more stable and to show stronger PL intensity in acidic region.

<u>CdTe-incorporated sol-gel glass phosphor fabricated from APS</u>

In the process of sol-gel reaction, the amino group of APS reacted quickly with the carboxyl group of TGA. This reaction plays an important role for preventing the CdTe QDs from agglomeration during sol-gel synthesis. Furthermore, the amino group accelerates the sol-gel reaction drastically. Due to these two effects, glasses with good transparency were readily prepared from APS at room temperature [10]. The resultant glass also showed bright green PL. The PL quantum efficiency of CdTe QDs in glass depends on the pH of the sol-gel solution. Glass prepared in both acidic and basic ranges showed higher quantum efficiency than the glass prepared in neutral range. The maximum value of the PL efficiency of the glass was similar to that of the solution. It has been reported that the speed of the hydrolysis of APS is much faster in both acidic and basic ranges than in the neutral range [11]. When the reaction is fast, the possibility of degradation of the surface by exposure to the amino group during the reaction decreases. Probably this is the reason for the pH dependence of the PL efficiency. The glasses from APS retained their PL over at least several months.

CONCLUSIONS

CdTe QDs showing green PL were synthesized in aqueous solution in the presence of a stabilizing surfactant TGA. The PL quantum efficiency was estimated to be around 5%. PL intensity and zeta potential of the TGA-capped CdTe QDs were measured as a function of pH of the QD solution. The PL intensity and stability of the QDs are strong in the basic region and become weak in the acidic region. The comparison of the experimental results with the calculated dissociation ratio of carboxyl group of TGA showed that when the pH becomes below 5, the carboxyl group of TGA changes from a dissociated form with a negative charge to a non-dissociated form without a charge. This makes the QDs unstable and promotes the agglomeration, accompanied by the strong suppression of PL intensity.

In order to further improve the long-term photostability of QDs, silica glass phosphor incorporating CdTe QDs was fabricated by a sol-gel process using a silane coupling agent, aminopropyl silane (APS). Because the amino group in APS reacts with TGA and accelerates the sol-gel reaction, high quality glasses have been prepared from APS. The fast reaction in both acidic and basic regions resulted in the formation of glass phosphor showing bright green PL. The PL quantum efficiency of the glass was similar to that of the QD solution, indicating that the incorporation process of QDs in sol-gel glass did not cause a deterioration of PL intensity of the QDs.

ACKNOWLEDGMENTS

This study was supported by the Nanotechnology Glass Project as a part of the "Nanotechnology Materials" Program supported by the New Energy and Industrial Technology Development Organization (NEDO), Japan.

REFERENCES

1. B. O. Dabbousi, J. R.-Viejo, F. V. Mikulec, J. R. Heine, H. Mattoussi, R. Ober, K. F. Jensen and M. G. Bawendi, *J. Phys.Chem. B* **101**, 9463 (1997).
2. M. Gao, S. Kirstein, H. Möhwald, A. L. Rogach, A. Kornowski, A. Eychmüller and H. Weller, *J. Phys. Chem. B* **102**, 8360 (1998).
3. A. L. Rogach, L. Katsikas, A. Kornowski, D. Su, A. Eychmüller and H. Weller, *Ber. Bunsen-Ges. Phys. Chem.* **100**, 1772 (1996).
4. V. I. Klimov, A. A. Mikhailovsky, S. Xu, A. Malko, J. A. Hollingsworth, C. A. Leatherdale, H.-J. Eisler and M. G. Bawendi, *Science* **290**, 314 (2000).
5. N. P. Gaponik, D. V. Talapin, A. L. Rogoch and A. Eychmüller, *J. Mater. Chem.* **10**, 2163 (2000).
6. M. J. Bruchez, M. Moronne, P. Gin, S. Weiss and A. P. Alivisatos, *Science* **281**, 2013 (1998).
7. C. B. Murray, D. J. Norris and M. G. Bawendi, *J. Am. Chem. Soc.* **115**, 8706 (1993).
8. T. Rajh, O. I. Micic and A. J. Nozik, *J. Phys. Chem.* **97**, 11999 (1993).
9. S. T. Selvan, C. Bullen, M. Ashokkumar and P. Mulvaney, *Adv. Mater.* **13**, 985 (2001).
10. C. L. Li, M. Ando and N. Murase, *Phys. Status Solidi C* (4), 1250 (2003).
11. C. J. Brinker and G. W. Scherer, *Sol-Gel Science* (Academic Press, San Diego, 1990).

Mat. Res. Soc. Symp. Proc. Vol. 789 © 2004 Materials Research Society N6.6

Aqueous Ferrofluid of Citric Acid Coated Magnetite Particles

A. Goodarzi[1], Y. Sahoo[2], M.T. Swihart[1], P.N. Prasad[2]
[1]Department of Chemical and Biological Engineering and Institute for Lasers, Photonics and Biophotonics, University at Buffalo (SUNY), Buffalo, NY, 14260
[2]Department of Chemistry and Institute for Lasers, Photonics and Biophotonics, University at Buffalo (SUNY), Buffalo, NY, 14260

ABSTRACT

Magnetic nanoparticles have found application in medical diagnostics such as magnetic resonance imaging and therapies such as cancer treatment. In these applications, it is imperative to have a biocompatible solvent such as water at optimum pH for possible bio-ingestion. In the present work, a synthetic methodology has been developed to get a well-dispersed and homogeneous aqueous suspension of Fe_3O_4 nanoparticles in the size range of 8-10 nm. The surface functionalization of the particles is provided by citric acid. The particles have been characterized using transmission electron microscopy, magnetization measurements with a superconducting quantum interference device, FTIR spectroscopy (for surfactant binding sites), thermogravimetric studies (for strength of surfactant binding), and x-ray photoelectron spectroscopy and x-ray diffraction (for composition and phase information). The carboxylate functionality on the surface provides an avenue for further surface modification with fluorescent dyes, hormone linkers etc for possible cell-binding, bioimaging, tracking, and targeting.

INTRODUCTION

Magnetic nanoparticles have drawn much scientific interest for a variety of studies on topics including superparamagnetism[1], magnetic dipolar interactions[2], single electron transfer[3], magnetoresistance[4] and so on. One area that is particularly promising is the use of magnetic nanoparticle systems for probing and manipulating biological systems[5]. Magnetite (Fe_3O_4) nanoparticles have been widely studied, and their colloidal dispersion is a well-known ferrofluid that has many potential bio-medical applications. The particles, for this purpose, are subjected to suitable surface modifications by various coating agents such as dextran[6], dimercaptosuccinic acid [7,8], starch and methoxypoly(ethylene glycol)[9], protein[10], silica coating [11] etc.

It is a technological challenge to acquire control over the nanoparticles' sizes and dispersibility in desired solvents. Because of their large surface to volume ratio, nanoparticles possess high surface energies. The particles tend to aggregate to minimize total surface energy. In the case of metal oxide surfaces, such energies are in excess of 100 dyn/cm [12]. Because the particles are magnetic in nature, there is an additional contribution from inter-particle magnetic dipolar attraction that tends to destabilize the colloidal dispersion further. Suitable surface functionalization of the particles and choice of solvent are crucial to achieving sufficient repulsive interactions between particles to prevent aggregation and obtain a thermodynamically stable colloidal solution For magnetite particles, oleic acid acts as an efficient surfactant that binds through the carboxyl end leaving aliphatic chains extending out from the surface[13]. This makes the coated particles effectively hydrophobic and dispersible in non-polar solvents. Sahoo et al. [14] have shown that alkyl phosphonates and phosphates also bind to magnetite particles well and render the surfaces hydrophobic. There has been some work on the aqueous suspension of magnetic particles. For example, Fauconnier et al. have thiolated the particle surface[8] and Shen et al. [15] have created a bilayer structure of fatty acids to make the surface hydrophilic, enabling stable aqueous dispersion. In the present work, we have used citric acid (CA) as surfactant to form a stable aqueous dispersion of magnetite particles and simultaneously provide functional groups on the particle surface that can be used for further surface derivatization.

EXPERIMENTAL

Magnetite(Fe_3O_4) particles were prepared by co-precipitation of $FeCl_2$ and $FeCl_3$ (1:2 molar ratio) by the addition of NH_4OH. In a typical reaction, 0.86 g $FeCl_2$ and 2.35 g $FeCl_3$ were mixed in 40 ml water and heated to 80° C under argon in a three-necked flask. While vigorously stirring the reaction mixture, 5 ml of NH_4OH was introduced by syringe and the heating continued for thirty minutes. After that, 1g of citric acid (CA) in 2ml water was introduced, the temperature raised to 95° C and the stirring continued for an additional ninety minutes. A small aliquot of the reaction mixture was withdrawn, diluted to twice its volume and placed on a 0.5 Tesla magnet in a vial. At the point when the particles did not settle down under the influence of the magnet, the colloidal solution was stable. The as-formed reaction product contained an excess of citric acid. Therefore, the nanoparticle dispersion was subjected to dialysis against water in a 12-14 kD cut-off cellulose membrane (Spectrum Laboratories, Inc., USA) for 72 hours to remove the excess unbound CA. The citric acid coated magnetic particles are abbreviated as MP-CA.

The particles were characterized by Fourier transform infrared (FTIR) spectroscopy, X-ray photoelectron spectroscopy (XPS), thermogravimetric analysis (TGA), and transmission electron microscopy (TEM). FTIR spectra of neat citric acid and coated magnetic particles were taken in KBr pellets using a Perkin Elmer FTIR spectrometer model 1760X. XPS studies were performed on a Physical Electronics/PHI 5300 x-ray photoelectron spectrometer operated at 300W (15 kV and 20 mA) with a Mg K_α radiation (1253.6 eV) source. TGA was performed on a Perkin Elmer instrument model TGA7 on roughly 5mg samples heated to 800^0C. TEM images were obtained by employing a model JEOL 100 CX II microscope at an acceleration voltage of 80 kV in the bright field image mode. Magnetization measurements (DC) were made on aqueous samples hermetically sealed in non-magnetic ampoules, by using a superconducting quantum interference device (SQUID) MPMS C-151 magnetometer from Quantum Design.

RESULTS AND DISCUSSION

The formation of Fe_3O_4 particles by co-precipitation of Fe^{2+} and Fe^{3+} by an alkali is fairly well known and is widely used. XPS (Fig. 1(a)) and XRD (Fig. 1(b)) results are consistent with the expected composition of Fe_3O_4 particles. XPS shows a binding energy of the Fe $2p_{3/2}$ shell electron at slightly above 710 eV, which agrees with the oxidation state of Fe in Fe_3O_4 [16]. The XRD peaks can be indexed into the spinel cubic lattice type with a lattice parameter of 8.34 Å. It cannot be ascertained from the XRD whether the further oxidized Fe_2O_3 phase exists in the

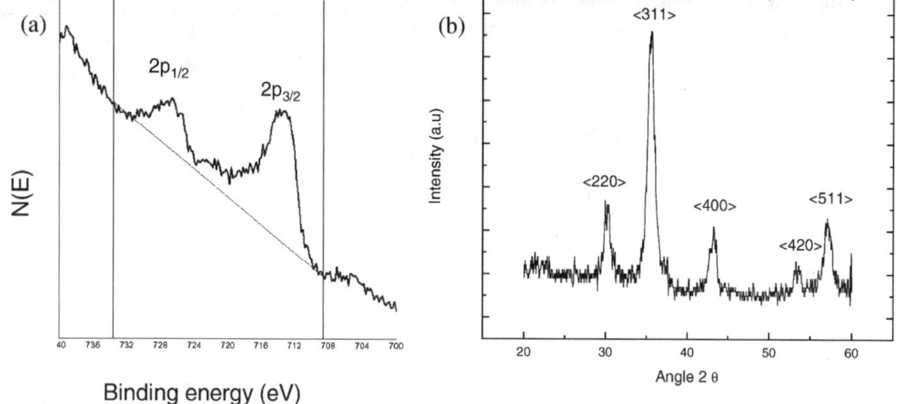

Figure 1. (a) XPS and (b) XRD of MP-CA show the composition and crystal structure of magnetite.

sample because of similar lattice type and lattice constant [11]. The TEM image of the citric acid coated magnetite particles without any size selection is shown in Fig. 2(a). The sizes range from 5 to 13 nm with a dominant population of 6-8 nm (44%) (Fig 2.b). A size selection can be carried out by adding incremental volumes of acetonitrile as non-solvent and carrying out sequential centrifugation. However, for the present paper we have characterized and present results for the moderately polydisperse sample. XRD shows broadening of the diffraction peaks, and the particle size calculated using the Scherrer formula is ~10 nm.

FTIR spectroscopy shows that the surface passivation of the particles occurs via the –COOH group. Fig. 3 shows the FTIR spectra of the neat citric acid (CA) and magnetite particles coated

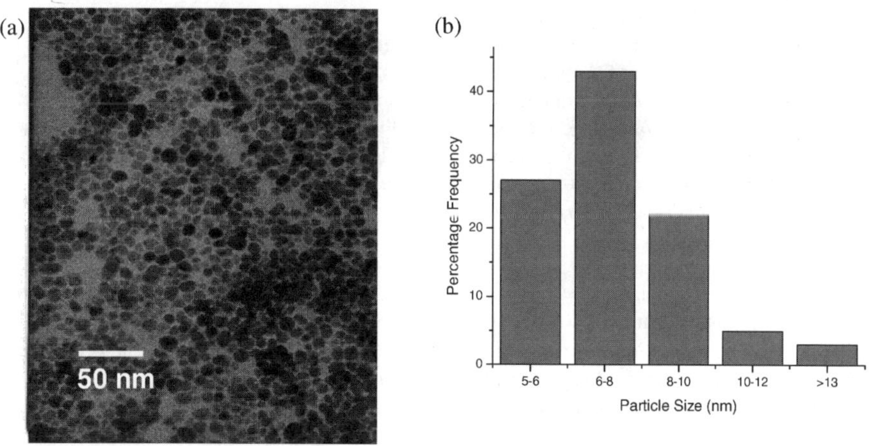

Figure 2. (a) TEM of MP-CA before any size sorting and (b) the corresponding size distribution.

with citric acid (MP-CA). The vibration bands for the CA are found to be rather broad, as expected due to the strong intra- and intermolecular hydrogen bonding. The 1715 cm^{-1} peak assignable to the C=O vibration in neat CA is present as a broad band. On binding of CA to the magnetite surface, this band shifts to 1618 cm^{-1} in MP-CA. The carboxylate end of CA may complex with the Fe of the magnetite surface and render the C=O bond partial single bond character. This observation is similar to the citrate complex in $YFeO_3$ studied by Todorovsky et al [17]. It is proposed that CA binds to the magnetite surface by chemisorption of the carboxylate i.e citrate ions. Earlier studies by Matijevic's group [18] have shown that oxalic acid and citric acid bind to the hematite surface through chemisorption that is highly pH dependent. The citric acid was inferred to be bound either as a bidentate or a tridentate ligand from their zeta potential measurement. TGA results (Fig.4) show a single step weight loss on the neat CA and MP-CA. The weight loss in the latter case starts at a higher temperature and occurs more gradually than for the neat CA. CA does not have a normal boiling point because it decomposes before boiling at atmospheric pressure. There is about 30°C difference (195° and 225°C for the neat and MP-CA respectively) in the weight loss onset temperatures. This is an indicator of the enthalpy of adsorption of CA molecules on the magnetite surface. Also, while the net weight loss in the neat sample is 100% as expected, the weight loss for MP-CA is 40%. We can attribute the weight loss to desorption of citric acid molecules from the surface of the magnetite particles. The mass of surfactant bound to the particle surface can be calculated as follows. If we assume a close-packed monolayer of the surfactant on the surface of a nanoparticle of diameter d_p then the total weight of the nanoparticle plus the monolayer is $(1/6)\pi d_p^3\rho + (\pi d_p^2/a)(M/N_0)$,
where d_p = the diameter of the particle, ρ = the density of the particle, a = the head area per molecule of the surfactant, M = the molecular weight of the surfactant, and N_0 = Avogadro's number. Assuming that the TGA heating causes weight loss of only the surface bound surfactant, the percentage weight loss from a particle of diameter d_p is $100 \times (\pi d_p^2/a)(M/N_0) / ((1/6)\pi d_p^3\rho + (\pi d_p^2/a)(M/N_0))$. For a polydisperse sample with discrete size distribution $f(d_p)$, the fractional weight loss is obtained by summing over the size distribution:

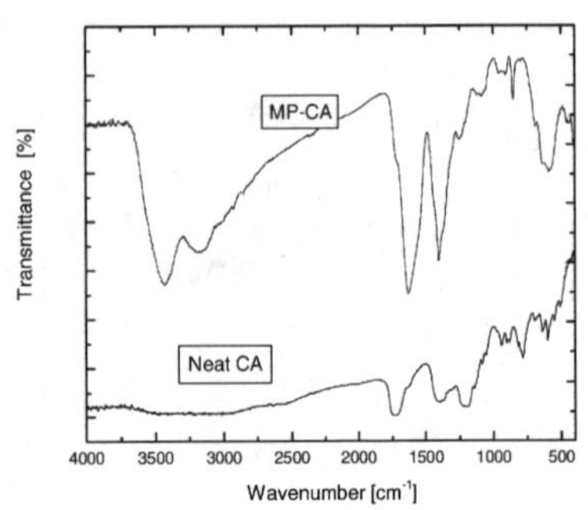

Figure 3. FTIR of neat CA and MP-CA showing binding through a –COOH group.

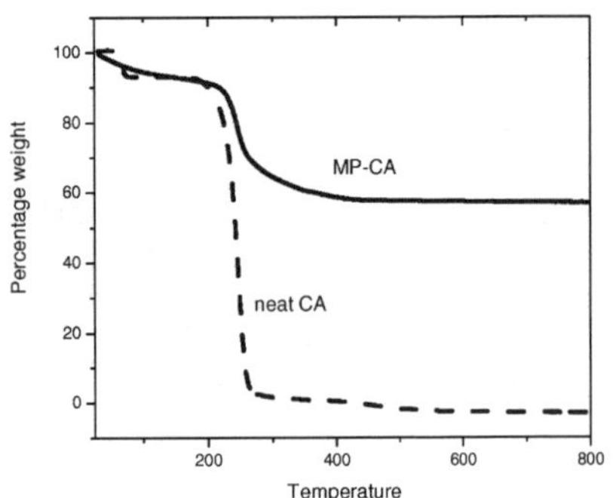

Figure 4. TGA of neat CA and CA coated Fe_3O_4.

$$\text{weight loss} = \frac{\sum_i \left(\pi d_{p,i}^2 / a \right) \left(M / N_0 \right) f(d_{p,i})}{\sum_i \left(\frac{1}{6} \pi d_{p,i}^3 \rho + \left(\pi d_{p,i}^2 / a \right) \left(M / N_0 \right) \right) f(d_{p,i})}$$

Summing over the discrete nanoparticle size distribution fractions as in the size distribution plots of Fig 3.b, with $\rho = 5.18$ g/cm^3, $a = 21$ Å^2, and $M = 192.12$ amu gives the percentage weight loss for this particle size distribution as 18 %. This value is about half the value of the weight loss observed from TGA (Fig. 5). This discrepancy can be explained as follows. In the calculation, the head area of 21 Å^2 is appropriate for close packed monodentate ligands [13]. This suggests that in spite of repeated washing, substantially more than a monolayer of surfactant is present, possibly due to potentially strong hydrogen bonding among the citric acid molecules. Surfactant not bound to the particle surface, but held by interdigitation with the first monolayer should show an extra step in the TGA curve [15]. However, because of the strong hydrogen bonding in CA, the molecules might still remain bound and leave *en masse* on desorption giving rise to a single step weight loss feature.

The CA coated particles forming the ferrofluid in this study are in their superparamagnetic regime. At 300K, magnetization measured up to a dc magnetic field of 10000G shows saturation of magnetization (Fig. 5), but no hysteresis loop. However, on lowering the temperature to 5K, the hysteresis appears, with a coercive field of ~430 G. This suggests that the blocking temperature of the particles is above 5K.

In this study, we have stabilized the surface of magnetite nanoparticles by adsorption of citric acid. The citric acid may be adsorbed on the surface of the magnetite nanoparticles by co-ordinating via one or two of the carboxylate functionalities depending on the steric necessity and the curvature of the surface. Yet, there will be at least one carboxylic acid group exposed to the solvent, that should be responsible for making the surface charged. Further, the presence of a terminal carboxylic group provides an avenue to extended bond formation with fluorescent dyes, proteins, and hormone linkers so that specific targeting within biological systems can be facilitated.

ACKNOWLEDGEMENTS

This work was partially supported by the Interdisciplinary Research and Creative Activities Fund of the University at Buffalo (SUNY). We are grateful to Shumin Wang and Hong Luo for performing the magnetization measurements reported here.

REFERENCES

1. C. P. Bean and J. D. Livingstone, J. Appl. Phys. **30**, 120S (1959); Neel, L. *Rev. Mod. Phys.* **25**, 293 (1953).
2. G. A. Held, G. Grinstein, H. Doyle, S. Sun and C. B. Murray, *Phys. Rev.* B **64**, 012408 (2001).

3. S. N. Molotkov, S. S. Nazin, *Phys. Low-Dimens. Struct.* **10**, 85 (1997); R. Weisendanger, MRS Bull. **22**, 31 (1997).
4. P. Poddar, T. Fried and G. Markovich, *Phys. Rev.* B **65**, 172405 (2002); C. T. Black, C. B. Murray, R. L. Sandstrom and S. Sun, Science **290**, 1131 (2000).
5. K. Raj, B. Moskowitz, R. Casciari, *J. Magn. Magn. Mater.* **149**, 174 (1995); R. F. Ziolo, E. P. Giannelis, B. A. Weinstein, M. P. O'Horo, B. N. Ganguly, V. Mehrotra, M. W. Russell and D. R. Huffman, *Science* **257**, 219 (1992).
6. D. K. Kim, W. Voit, W. Zapka, B. Bjelke, M. Muhammed , K. V Rao , *MRS Symp. Proc.* 676, (2002).
7. C. Wilhelm, C. Billotey, J. Roger, J.N Pons, J. C. Bacri and F. Gazeau, *Biomaterials*, **24**, 1001 (2003).
8. N. Fauconnier, J.N. Pons, J. Roger, A. Bee, J. *Coll. Interfac. Sci.* **194**, 427(1997).
9 D. K Kim, K. Do ; M. Mikhaylova , Y. Zhang, M. Muhammed, *Chem. Mater.* **15(8),** 1617 (2003).
10. K. Nishimura, M. Hasegawa, Y. Ogura, T. Nishi, K. Kataoka, H. Handa , M. Abe, *J. App. Phys.* **91(10, Pt. 3)**, 8555 (2002).
11. L. Levy, Y. Sahoo, K. Kim, E. J. Bergey, and P. N. Prasad, *Chem. Mater.* **14**, 3715 (2002).
12. L. Vayssieres, C. Chaneac, E. Troc, J.P Jolivet, *J. Coll. Interfac. Sci.* **205**, 205 (1998).
13. T. Fried, G. Shemer, and G. Markovich, *Adv. Mater.* 13, 1158, (2001).
14. Y. Sahoo, H. Pizem, T. Fried, D. Golodnitsky, L. Burstein, C. N. Sukenik, and G. Markovich, *Langmuir*, **17**, 7907 (2001).
15. L. Shen, A. Stachowiak, S. K. Fateen, P. E. Laibinis, T. A. Hatton, *Langmuir*, **17**, 288 (2001).
16. C. D. Wagner, R. M. Riggs, L. E. Davis, J. F. Moulder, *Handbook of Electron Spectroscopy*, edited by G. E. Muilenberg (Perkin Elmer Corp., Physical Electronics, Eden Prairie, MN, 1979.)
17. D. S. Todorovsky, D. G. Dumanova, R. V. Todorovska, M. M.Getsova, *Croatica Chemica Acta,* **75(1)**, 155 (2002).
18. N. Kallay and E. Matijevic, *Langmuir*, **1**, 195 (1985).

Quantum Dots and Wires: Structure, Spectroscopy, and Transport

Mat. Res. Soc. Symp. Proc. Vol. 789 © 2004 Materials Research Society N8.3/T6.3/Z6.3

Polarization Spectroscopy of Charged Single Self-Assembled Quantum Dots

Morgan E. Ware[1], Allan Bracker[1], Daniel Gammon[1], David Gershoni[2]
[1]Naval Research Laboratory, Washington, DC 20375
[2]Technion-Israel Institute of Technology, Haifa, 32000, Israel

ABSTRACT

We have demonstrated single dot spectroscopy of InAs/GaAs self-assembled quantum dots embedded in a bias controlled Schottky diode. The photoluminescence spectra exhibit discrete lines depending on bias, which we attribute to the recombination of positively charged, neutral, and negatively charged confined excitons. With excitation directly into the dot, large circular polarization memory is exhibited by the two charged exciton (trion) lines. This indicates long spin lifetimes for both the electron and the heavy hole in the quantum dots.

INTRODUCTION

Semiconductor quantum dots have been viewed in recent years as attractive components for implementing quantum computation schemes [1]. In particular, the unpaired spin of an electronically charged semiconductor quantum dot is especially exciting, because of its relatively long spin dephasing time, and because its spin can be controlled and measured optically [2]. Knowledge of the discrete energy spectrum of these dots and the ability to control the individual charges in the dots is vital for future progress in this field.

Several groups have succeeded in identifying charging in single dot photoluminescence (PL) by high power excitation [3,4] or by controlling the bias across a field effect device containing quantum dots [5,6,7]. In order to control the charge on a dot at lower excitation power, in the single exciton regime, it is more useful to embed the dots in a gated structure. Regelman, et. al., demonstrated both hole and electron charging in high power excitation of InAs dots [3]. Ashmore, et. al., were able to make separate devices on n and p-type substrates which formed negatively and positively charged excitons respectively in $In_{0.5}Ga_{0.5}As$ dots under optical excitation [6]. Both groups demonstrated that compared to the neutral exciton the positive exciton is blueshifted by ~2 meV and the negative exciton is redshifted by ~5 meV.

In addition to being able to controllably charge the dot, a requirement for quantum information processing is the ability to control and measure the spin of the charges in the dot [1]. This is accomplished with circularly polarized light which through selection rules couples to certain spin states. In particular, if one defines the quantization axis as that of the incident photon, a spin up heavy hole and spin down electron pair is photogenerated by a right circularly polarized photon (σ^+), and the opposite spins pair is photogenerated by a left circularly polarized photon, (σ^-) [8]. The reverse process, recombination, involving these same spin polarized pairs will in turn result in the creation of a σ^+ and σ^- photon, respectively. In isolated quantum dots, recombination of single pairs can be readily identified spectrally [9]. This allows one to probe the spin of the recombining charge carriers by detecting the polarization of the emitted photon. When a single electron (hole) is present in the dot, optical excitation of one electron-hole pair (exciton) results in a pair of electrons (holes) in the ground state and an unpaired hole (electron) in its ground state. By exciting with circularly polarized light and analyzing the degree of circular polarization of the emitted light, one can measure the polarization memory of the dot, which is related to the rate by which the photogenerated unpaired charge carriers flip their spin.

EXPERIMENTAL

Quantum dots are grown by molecular beam epitaxy and form by Stransky-Krastinov strain driven self-assembly. An indium flush technique is used to raise the emission energy from ~1eV to ~1.3 eV [10]. This involves growing ~2.4 monolayers of InAs at 485°C, which forms dots from 5 nm to 8 nm in height. Then a partial cap of 3.5 nm GaAs is deposited at the same temperature. Following a temperature ramp to 545°C and anneal for 120 s, the parts of the dots left exposed are desorbed, and growth of the GaAs cap is continued at that temperature. This allows for very precise control of the vertical thickness of the dot and thus their PL emission energy. The flush technique is described in detail in Ref. 10.

The quantum dot layer is embedded in a n+ - intrinsic-Schottky (NIS) diode in order to control the dot charging. We use a highly doped n-type GaAs(001) substrate and grow a 500 nm GaAs buffer layer doped with tellurium at ~$5*10^{17}$ cm^{-3}. On that we grow 80 nm of un-doped GaAs, followed by the dot layer as described above, then 230 nm undoped GaAs, 40 nm $Al_{0.3}Ga_{0.7}As$, then a final 10 nm GaAs cap. A schematic showing the structure and the approximate band diagram is given in Figure 1. All growth is performed at 545°C except the InAs and partial cap. We use electron-beam lithography to fabricate submicron apertures in an Al mask on the surface of the samples. A 5 nm semi-transparent titanium layer along with the Al mask forms the Schottky barrier for the front contact, and a AuGe layer on the backside forms an ohmic contact with substrate.

Using a tunable cw Ti-Sapphire laser PL was excited and detected through the Al apertures The light was collected and dispersed with a 0.5 m single grating spectrometer and a charge-coupled device array. The excitation energies for all data presented was less than 1.37eV which is well below the emission of the 2D InAs wetting layer. The exciting light was circularly polarized and the polarization of the emitted light was analyzed with variable liquid crystal retarders. The excitation power for the polarization memory measurements was chosen such that multiexciton spectral lines were not present unless otherwise stated.

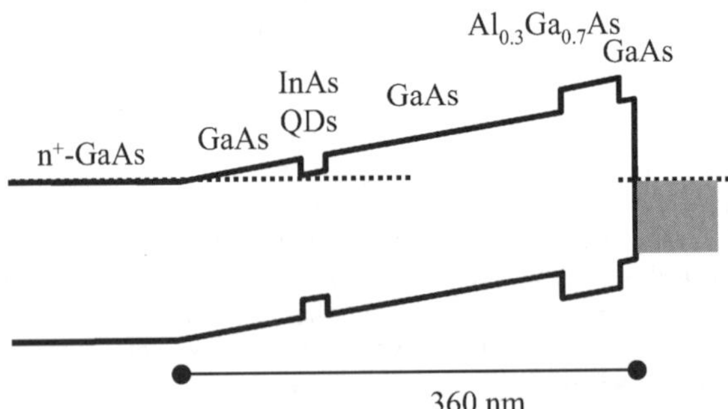

Figure 1 Band diagram showing the Schottky diode structure and the positions of the bands relative to the Fermi level.

RESULTS AND DISCUSSION

By changing the bias across the diode we vary the charge in the dot. As a result, different spectral lines appear in the PL spectra. In particular, as can be seen in Figure 2a, we were able to identify recombination of an electron-hole (e-h) pair in the absence of charge (exciton, X), in the presence of an electron (negative trion, X^-) and in the presence of a hole (positive trion, X^+). At higher excitation power, as shown in Figure 2b, the biexcitons and their respective charged complexes become visible. This pattern of spectral lines has been observed for several dots.

The observation of the positive trion in NIS structures, which is reported here for the first time, deserves explanation. At high reverse bias, with no external excitation, the dot is uncharged. For optical excitation at energies below the wetting layer emission, e-h pairs are photogenerated in the dot. Since the field induced tunneling rate of the electron out of the dot is much larger than that of the hole, the dot will become positively charged and may remain so for

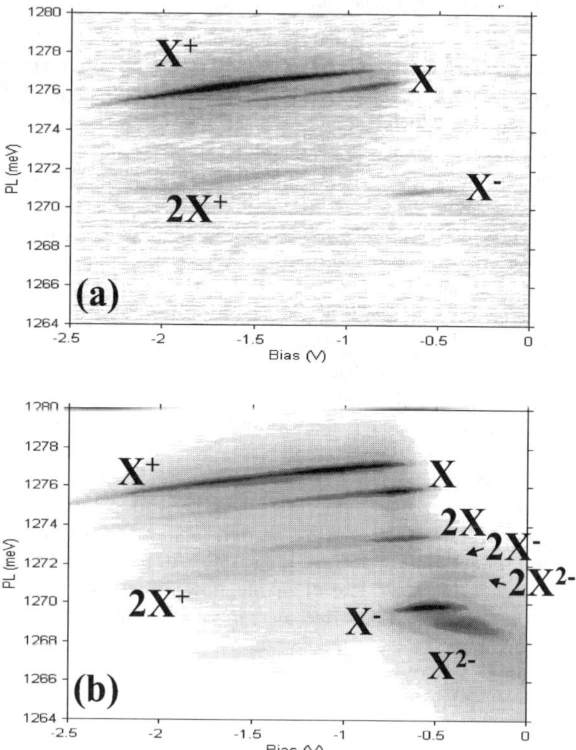

Figure 2 Typical bias dependent PL intensity map of two representative InAs/GaAs quantum dots used in this experiment. (a) shows one under low power excitation, and (b) shows the other at high power excitation where many biexciton complexes are found. Excitation in both cases was well below the energy of the wetting layer emission.

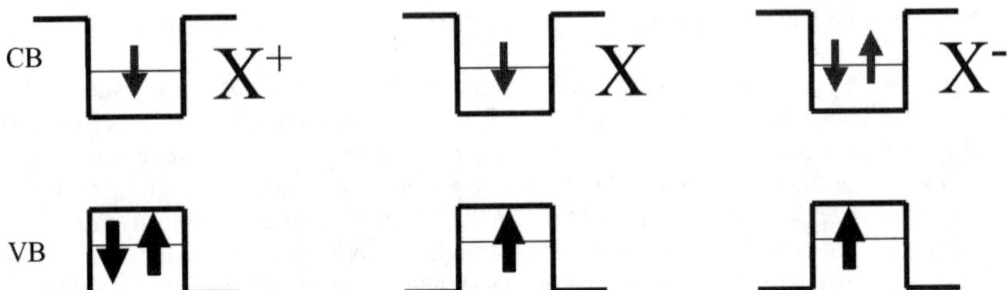

Figure 3 Diagram showing the positively charged exciton (X^+), the neutral exciton (X), and the negatively charged exciton (X^-) with σ^+ excitation. The unpaired charge in the trions dictates the polarization of the emitted light. The emitted light from the X is expected to be unpolarized due to the anisotropic electron-hole exchange.

a relatively long time. If a second e-h pair is photogenerated, the additional positive charge significantly reduces the electron tunneling rate, and thus e-h pair recombination in the presence of additional positive charge, becomes possible. It is unlikely that this positive trion is derived from a biexciton precursor which loses an electron since the X^+ is very prominent even at the lowest powers where the creation of a biexciton is extremely unlikely.

Reducing the reverse bias reduces the tunneling rate of the electrons making recombination from a neutral exciton more favorable. Further increase in the bias injects electrons from the substrate into the dot and, depending on the bias, recombination of e-h pairs in the presence of one or more electrons may be observed. This allows for fine control over the spectra as a result of charging in a single quantum dot. A brief examination of Figure 2 shows that compared to the neutral exciton, the positive exciton is blueshifted by 1-2 meV, and the negative exciton is redshifted by ~5-6 meV. This is in agreement with published experimental and theoretical results for the positive and negative trions [3,6].

In Figure 3 we schematically describe the arrangement of the spins of the charge carriers within the ground state of the exciton and trions. For excitation of spin polarized excitons in the presence or absence of one additional bound charge we must examine the three different configurations. The positive (negative) trion, in which the holes (electrons) are paired, leaves a single electron (hole) unpaired. In these cases, the polarization of the emitted photon is solely determined by the spin state of the unpaired carrier. Spin memory of the unpaired charge carrier, in these cases, will reflect itself in positive polarization of the emitted photon (copolarized with the exciting laser light). In the absence of charge, however, the situation is expected to be different. In this case, the anisotropic exchange interaction mixes the bright ground exciton states and upon recombination the emitted photon is expected to be linearly polarized [11,12]. This results in a complete loss of the circular polarization for the exciton.

In Figure 4a we display the bias dependence of the polarization, P, for the positive, neutral and negatively charged excitons. Here the polarization is given by $P=(I^+-I^-)/(I^++I^-)$, where the intensities I^+ and I^- correspond to the detected intensities of right and left circularly polarized light respectively. The bias dependence of the PL intensity of this particular dot is shown in Figure 4b. Quite noticeably, the exciton has a vanishingly small polarization at most biases with some large fluctuations as the PL intensity gets small. This, again, is exactly what we would expect for the exciton with no external applied magnetic field [12]. The trions in contrast both exhibit a very large polarization memory. It should be reiterated here that these

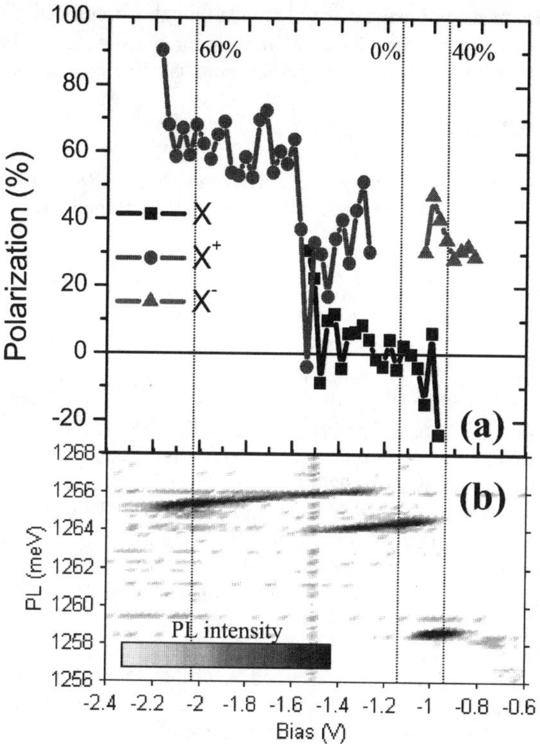

Figure 4 Polarization memory of each charged species is shown in (a). (b) shows the bias dependence of the intensity for this dot. Here we are exciting at 1305 meV which is near the first excited exciton state of the dot and well below the wetting layer emission. The dashed lines mark the approximate maximum in PL intensity for each of the three lines.

polarizations are dependent on the lifetimes of the unpaired charges. Thus, large positive polarization indicates long spin lifetimes of both the unpaired electron and hole, compared with the recombination time, at least for excitation directly into the dot. This agrees with the current understanding of the electron spin flip time [13], but it is in conflict with the common description of the hole spin flip time from higher dimensional systems, which is believed to be very short [8].

CONCLUSIONS

We have observed single exciton emission from InAs self-assembled quantum dots and have identified positive, neutral, and negative charged exciton complexes, X^+, X, X^-. These different charge states of the exciton are readily interchanged by varying the bias across an NIS diode structure. This charge tunability is coupled with large polarization memory for optical excitation of excitons directly into the dot. Positive polarizations as large as 60% for the X^+ and as large as

40% for the X^- are measured using σ^+ excitation, although with no externally applied magnetic field we would not expect the results to be different for σ^- excitation. This opens promising avenues to explore the physics of single particle spin using single dot spectroscopy.

ACKNOWLEDGEMENTS

This work was funded in part by DARPA/QUIST, NSA/ARDA, and ONR. MEW holds a National Research Council Research Associateship Award at the Naval Research Laboratory.

REFERENCES

1. D. Loss, D. P. DiVincenzo, Phys. Rev. A **57,** 120 (1998).
2. D. Gammon and D.G. Steel, Physics Today **55**, 36 (2002)
3. D.V. Regelman, E. Dekel, D. Gershoni, E. Ehrenfreund, A. J. Williamson, J. Shumway, and A. Zunger, W. V. Schoenfeld and P. M. Petroff, Phys. Rev. B **64**, 165301, (2001).
4. E. S. Moskalenko, K. F. Karlsson, P. O. Holtz, B. Monemar, W. V. Schoenfeld, J. M. Garcia and P. M. Petroff J. App. Phys. **92**(11) 6787 (2002).
5. R.J. Warburton, C. Schäflein, D. Haft, F. Bickel, A. Lorke, K. Karrai, J. M. Garcia, W. Schoenfeld and P. M. Petroff, Nature, **405,** 926, (2000).
6. A.D. Ashmore, J.J. Finley, R. Oulton, P.W. Fry, A. Lemaitre, D.J. Mowbray, M.S. Skolnick, M. Hopkinson, P.D. Buckle and P.A. Maksym, Physica E, **13**, 127 (2002).
7. A. Zrenner; F. Findeis, M. Baier, M. Bichler, G. Abstreiter, U. Hohenester and E. Molinari, Physica E **13**, 95 (2002).
8. *Optical Orientation*, edited by F. Meier and B. Zakharchenya, "Modern Problems in Condensed Matter Sciences," Vol. 8 (North-Holland, Amsterdam 1984).
9. E. Dekel, D. Gershoni, E. Ehrenfreund, D. Spektor, J.M. Garcia and P.M. Petroff, Phys. Rev. Lett. **80**, 4991-4994 (1998)
10. Z. R. Wasilewski, S. Fafard, and J. P. McCaffrey, J. Cryst. Growth **201/202**, 1131 (1999).
11. D. Gammon, E. S. Snow, B. V. Shanabrook, D. S. Katzer, and D. Park , Phys. Rev. Lett. **76,** 3005 (1996); D. Gammon, Al. L. Efros, T. A. Kennedy, M. Rosen, D. S. Katzer, and D. Park, S.W. Brown, V. L. Korenev and I. A. Merkulov Phys. Rev. Lett. **86**, 5176 (2001).
12. M Bayer, G Ortner, O Stern, A. Kuther, A. A. Gorbunov, A. Forchel, P. Hawrylak, S. Fafard, K. Hinzer, T. L. Reinecke S. N. Walck, J. P. Reithmaier, F. Klopf, and F. Schäfer, Phys. Rev. B **65** (19): 195315 (2002).

13 A. V. Khaetskii and Y. V. Nazarov, Phys. Rev. B **61**, 12639 (2000).

Mat. Res. Soc. Symp. Proc. Vol. 789 © 2004 Materials Research Society N8.5/T6.5/Z6.5

Growth Structure, and Optical Properties of III-Nitride Quantum Dots

Hadis Morkoç[1], Arup Neogi[2], Martin Kuball[3]

[1] *Virginia Commonwealth University, Richmond, VA 23284, USA*
[2] *North Texas University, Denton, Texas 76203*
[3] *University of Bristol, Bristol, BS8 1TL UK*

ABSTRACT

Quasi-zero-dimensional (0D) semiconductors have been the subject of considerable interest which is stemmed from their unique physical properties which in turn are conducive to devices such as low threshold lasers and light polarization insensitive detectors, in addition to exciting basic physical phenomena. A laboratory analogue of 0D systems is semiconductor quantum dots (QDs) wherein the electronic states are spatially localized and the energy is fully quantized, loosely similar to an atomic system, making it more stable against thermal perturbations. In addition, the electronic density of states near the band gap is higher than in 3D and 2D systems, leading to a higher probability for optical transitions. Furthermore, the electron localization may dramatically reduce the scattering of electrons by bulk defects and reduce the rate of non-radiative recombination. Semiconductor based and metal based dots have been produced, the former via self-assembly and also by lithographic methods in many II-VI, III-V, and group IV semiconductor. The aim of this paper is focused on III-Nitride based quantum dots covering their production and optical properties, as well as reporting on the GaN quantum dots produced by molecular beam epitaxy utilizing standard, ripening, metal spray followed by nitridation methods.

INTRODUCTION

Semiconductor nitrides such as aluminum (AlN), gallium nitride (GaN), and indium nitride (InN) are very promising materials for their potential use in optoelectronic devices and high-power/temperature electronic devices, as have been treated in length and reviewed recently (Strite and Morkoç[1]; Morkoç et al.[2]; Mohammad et al.[3]; Mohammad and Morkoç[4]; Ambacher[5]; Morkoç[6]; Pearton et al.[7]). These materials and their ternary and quaternary alloys cover an energy bandgap range of 0.7-6.1 eV, suitable for band-to-band light generation with colors ranging from red to ultraviolet (UV) wavelengths. Specially, nitrides are suitable for application such as UV detectors, UV and visible light emitting diodes (LEDs). GaN based LEDs and Laser diodes (LDs) incorporate quantum wells (QWs). In addition to quantization, the quantum dots have other benefits such as localization which lead to reduced internal quantum yield degradation. Gérard et al.[8] pointed out that once the carriers are captured by QDs, they become strongly localized and their migration toward nonradiative recombination centers is made difficult. Furthermore, the increased localization gives rise to increased radiative recombination rates which brings one to the expected low threshold for lasers that have

[1] hmorkoc@vcu.edu

already been experimentally observed in the InGaAs system. A flurry of interest in low-dimensional GaN and other III-nitrides is in part due to a desire to develop new optoelectronic devices with improved quality and wider applications. Development of the light emitters with QDs, for example, is expected to have a lower threshold current in LDs and a higher thermal stability.[9]

The size of QDs and of its distribution as well as density over the wafer are important parameters. The typical value of size is on the order of a few nm which necessitates a large assembly of QDs rather than a single one, although probes are continually developed to interrogate individual dots. The fluctuation in dot size produces an inhomogeneous broadening in quantized energy levels and may destroy the very fundamental properties expected from a single QD. The random rather than well-ordered distribution may destroy the coherence of the optical and electronic waves propagating through the structure. Similar to other semiconductor heterostructures, the surface or interface of the QDs must also be free of defects, see recent chapters for details.[10,11] Otherwise, the surface/interface may become the effective scattering center for electrons. In metal dots, minimization of the chemical potential lead to merging of small dots with large ones which leads to convergence to a more or less uniform dots on the surface. Initial variation of the dots size obviously is caused by disorder in the system. In semiconductor dots, however, the surface mobility en mass is not available, and other methods, such as strain and periodic topological features such atomic steps on vicinal substrates, must be used to the extent possible to accomplish this.

As in the case of a majority of quantum dots in more established systems inclusive of conventional III-Vs, Si/SiGe and metals, the III-nitride QDs are also synonymous with strain, as strain is often times used to form and assemble them. The bulk wurtzitic GaN lattice constants are 3.189 Å and 5.185 Å in-plane and c directions, respectively. For AlN which provides the strain, they are 3.11 Å and 4.98 Å. For InN, these values are 3.55 Å and 5.76 Å. Using the binary data, the lattice mismatch between GaN and AlN in basal plane is 2.5% and is much larger between GaN and InN (10%). Confinement due to size reduction causes a blue shift in optical transitions. However, wurtzitic III-nitrides are noncentrosymmetric and uniaxial crystals and exhibit large spontaneous and strain induced polarization effects which lead to red shift. In the end it is the combination of these two competing processes which determine the eventual transition energy. The piezoelectric coefficients are almost an order of magnitude larger than in traditional III-V compounds such as GaAs.[4,12] Consequently, unless non polar surfaces are used, a large amount of polarization charge may appear at the heterointerface(s) even in the absence of strain due to the difference in the spontaneous polarization across the interface. The intense internal electric field induced by the polarization charge and piezoelectricity is very unique to III-nitride heterostructures and has a dramatic effect on the properties of QDs. A great majority of III-nitride QDs is grown by self-assembly during growth by molecular beam epitaxy (MBE) or organo metallic vapor deposition (OMVPE).

Due to the lack of a suitable material, in terms of both lattice and thermal matching, The III-nitride heterostructures are commonly grown on foreign substrates. Sapphire (α-Al_2O_3) is the one most extensively used. Wurtzitic GaN QDs have also been grown on other substrates such as 6H-SiC(0001) and Si(111). All of these mean that suitable buffer layers must first be grown before quantum dot growth can be attempted.

In the case of GaN dots driven by strain, either AlGaN or AlN buffer layers must be used. The composition of the ternary and the thickness of these layers depend on the level of strain desired, as thin barrier layers tend not to relax fully. In the case of InGaN dots, GaN under-layer becomes an option also. In addition to strain driven dots, anti-surfactants such as Si (the exact mechanism of which has evolved quite substantially in the case of GaN and is now believed to be due to SiN_x formation[13]) and lithographic processes have been explored. In the latter, lithography in conjunction with anisotropic nature of growth associated with OMVPE has been employed to further reduce the dimensions to what is closer to the Bohr radius. There are many other methods such as colloidal forms and inorganic matrices. In this paper, however, only GaN dots driven by strain caused by underlying AlN layers will be discussed.

Growth of GaN self-assembled QDs on AlGaN, with the aid of a sub- to monolayer Si layer, which are then covered with AlGaN has been reported by Tanaka et al.[14], and Shen et al.[15] Other approaches have been reported by Widmann et al.[16,17] and Damilano et al.[18] who used AlN layers which provide a larger lattice mismatch to GaN than AlGaN, and in turn provide the impetus for a 3-D growth. In addition, the surface topology of AlN is smoother which removes the surface features from being the nucleation sites for dots. Dots have been demonstrated on 6H-SiC[14, 15] and sapphire (0001).[16,17] Blue-light emission has been reported from such QD structures.[18] By changing the size of the quantum dots, one can in fact tune the color of the emission due to large polarization induced band bending.[18] However, the wavelength would be dependent on the injection level, as injected carriers tend to screen the polarization charge.

Assembly of dots can be attained on exact or vicinal surfaces. On exactly oriented surfaces, strain, topology and to some extent imperfections are the driving mechanism for positioning the dots. Naturally, the growth conditions and layer dimensions in the growth directions are tuned to maximize the desired effects. The work reported in this paper relies mainly on this class of dots. An additional degree of ordering provided can be had by incorporating vicinal substrates with regularly spaced terraces. Obviously the substrate surface preparation must be optimum for exposing and retaining these terraces. In the case of GaN, being grown on non-native substrates with large lattice and other structural mismatches, it is nearly impossible to retain the terraced structure during and after the buffer layer growth. The terraces formed in GaN buffer layers in OMVPE and HVPE layers typically are distorted and terminate at structural defects. Nevertheless, representing the best, these are the sorts of surface that can be used to produce quantum dots relying on strain induced phenomena.

The typical growth modes in any epitaxial deposition are Frank–van der Merwe (FV) which results in 2D growth, desired for epitaxial growth with lateral uniformity. The other mode is Volmer-Weber (VW) mode which results in 3D growth from the get go and typically occurs in metals. The third mode is the Stranski-Krastanov (SK) mode for which strain is the driving force. After an initial 2D growth which is called the wetting layer, the built in strain drives the system to a 3D growth for strain minimization. Obviously, the wetting layer must be grown on a buffer whose lattice constant is smaller than that of the dot material. Schematic representation of the aforementioned processes as shown in **Figure 1**.

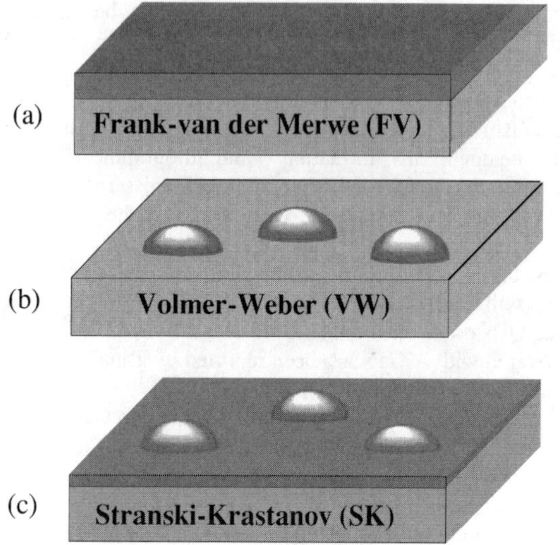

Figure 1. Various growth modes occurring in general on exactly oriented templates. (a) Frank– van der Merwe (FV) mode, layer-by-layer 2D growth mode, (b)Volmer-Weber mode applicable mostly to metals and leads to 3D growth, (c) Stranski-Krastanov mode which is driven by strain and is typical of semiconductors

On vicinal substrates which often consists of a staircase of equally spaced steps in the tilting direction, one may expect the approximate 2-D analogs of all the 3-D equilibrium growth modes, e.g. Frank–van der Merwe (FV), Stranski-Krastanov (SK), and Volmer-Weber (VW), to be as shown in **Figure. 2**. Which mode would be in effect is determined by the balance between the inter-step energy (laterally between the film and the substrate steps, equivalent to interfacial energy in 3D) and energies of film steps and substrate steps (equivalent to surface energies in 3D).[19]

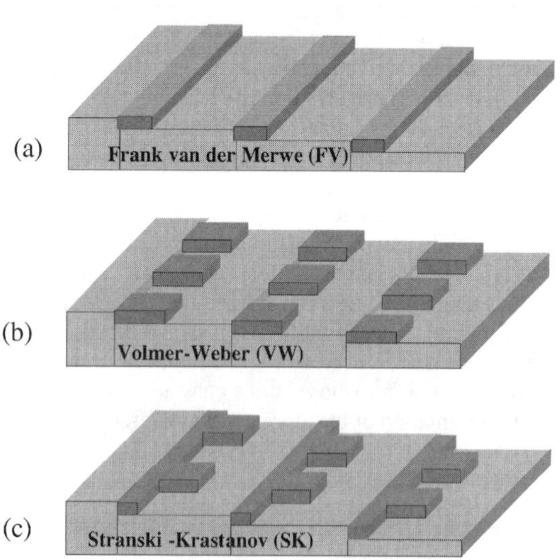

Figure 2. 2D analogs of 3D equilibrium growth mode, which occurs during step flow growth on vicinal substrates. (a) FV mode, row-by-row growth which is termed step flow growth. In the case of GaAs, there will be not RHEED oscillations observed in this mode. Continuous stripes formed under "ideal" step flow. (b) VM mode, island growth, (c) SK mode, row-by-row followed by island growth. The island being square in footprint is just an assumption for simplicity. Patterned after reference 19.

For homooepitaxy on a vicinal substrate [**Figure 2(a)],** one would expect "ideal step flow growth. In the case of heteroepitaxy, there is misfit strain, and depending on the extent of the strain, different modes occur. For example, with wetting layer (stripe in this case), **Figure 2(c)** the SK growth mode occurs, and without the wetting stripe the VW mode occurs, **Figure 2(b).** The wetting stripe corresponds to the wetting layer in the 3D case.

There have been several approaches towards the growth and synthesis of GaN QD structures. One of the common approaches utilizes GaN nanoparticles embedded in inorganic matrices.[20] GaN nanocrystals (≤ 2-3 nm) were synthesized by sequential implantation of Ga and N ions into either crystalline (quartz, sapphire) or amorphous (silica) dielectrics.[21] Recently GaN dots were grown on AlGaN using using Si atoms as an antisurfactant.[22] GaN QDs have also been grown on AlN taking advantage of the self-organization of QDs during growth employing Stranski-Krastanov (SK) growth mode[23] or by use of stressor islands.[24] The SK method has proven successful in achieving nanostructures with excellent optical properties. To reiterate, the present work reports on the growth and optical characteristics of self-organized GaN QD's on AlN layers. The employed methods were the Stranski-Krastanov (SK) growth mode and Ga spray mode followed by the nitridation of GaN dots either by an atmosphere of ammonia or by an RF nitrogen plasma source. One of the purposes of this work is to study the optical properties of the quantum dots or quantum boxes, particularly in terms of recombination dynamics of excitons. However, due to space limitation, the recombination dynamics will not be covered here. In GaN and Al(Ga)N QDs or QWs both piezoelectric and spontaneous polarization effects contribute to the electric field in the confined structure and in the surrounding barriers. The period or the barrier width in QDs or QWs therefore significantly alters the structural[25] and optical properties.[26] The stacking of multiple layers is also expected to influence the optical properties due the strain-induced alignment of dots in one level with respect to the adjacent level.

The holy grail of quantum dots is the nature of the confined states and the resultant density of states. If we consider a semiconductor whose constant energy surface for conduction band in k-space is a sphere, such as the case in GaN, the volume of that sphere in k-space is proportional to k^3 in terms of momentum and $E^{3/2}$ in terms of energy as shown in **Figure 3a**. The density of states associated with this system is proportional to $E^{1/2}$, again as shown in **Figure 3a**. The area in k-space in an ideal system confined in one direction only (representing quantum wells), which is often the z or the growth direction, is proportional to k^2 or E, as shown in **Figure 3b**. The density of states in this case is given by $m^*/\pi\hbar^2$ and forms a staircase as shown in **Figure 3b**. If we continue and place a confinement in the x-direction in addition to the z-direction, which represents quantum wires, the line length in k-space is proportional to k in terms of momentum and $E^{1/2}$ in terms of energy as shown in **Figure 3c**. The corresponding density of states takes the dependence of $E^{-1/2}$, again as shown in **Figure 3c**. If confinement is imposed in all three directions, which represent the pseudo atomic or quantum dot state, the energy is discretized in all directions and the resultant density of states takes on a delta-like function in energy, as shown in **Figure 3d**.

The optical properties of III-nitride QD-systems are influenced by the confinement effect and by the compressive strain, both producing a blue shift of the QDs emission compared with that of the GaN bulk, and by the internal electric fields (induced by spontaneous and piezoelectric polarization), producing a significant red shift in the wavelength of emission.[27,28]

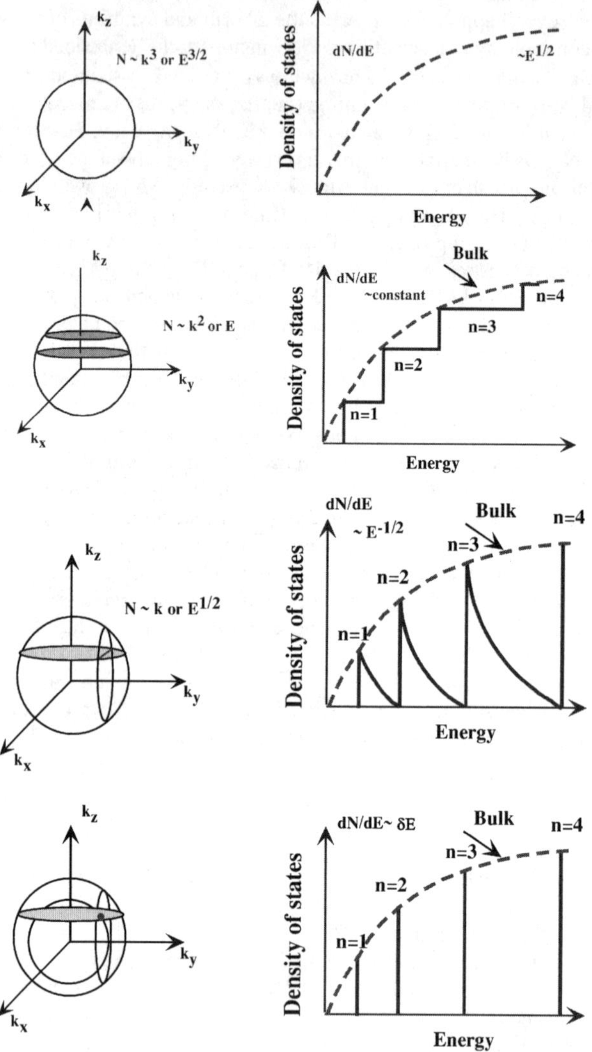

Figure 3. Constant energy surfaces (a line and a point in 1 and 0D systems, respectively) and density of states for 3D in (a), 2D in (b), 1D in (c), and 0D in (e) systems.

The quantum confinement effect shifts the band gap to a higher energy. This shift, called the confinement energy, depends on the size and shape as well as the material properties of both QDs and surrounding matrix. Here, we estimate the confinement energies for two simplified cases: a plate or disk, and a sphere. Assuming an infinite barrier, the confinement energy of the ground state for an electron in a rectangular box is given by

$$E = \frac{h^2}{8}\left(\frac{1}{m_x d_x^2} + \frac{1}{m_y d_y^2} + \frac{1}{m_z d_z^2}\right)$$

Equation 1

where d_j (j=x, y, and z) are the dimensions of the box and m_j are the electron mass in j directions. For a plate- or disk-like dot, in which the in-plane sizes d_x and d_y are much larger than the height $d_z = d$, the confinement energy is simplified to $h^2/(8m_z d^2)$. The shift in the band gap is calculated using $1/m_z = 1/m_{e,z} + 1/m_{h,z}$, where $m_{e,z}$ and $m_{h,z}$ are respectively the effective masses of the electrons and holes in III-Nitride along the z-direction. For a cubic box of size d, the confinement energy is given by the same expression, $h^2/(8m_z d^2)$, but with $1/m_z$ replaced by $(2/m_{e,t} + 2/m_{h,t} + 1/m_{e,z} + 1/m_{h,z})$, where $m_{e,t}$ and $m_{h,t}$ are the transverse effective masses of the electrons and holes. In **Figure 4**, the confinement energies as a function of *d* for a plate and a sphere have been plotted.[29] In both cases, the effective masses of $0.22m_0$ for electrons and $1.1m_0$ for holes were assumed. For small dots of a few nm in size, the confinement energy is very sensitive to the dot size in that when the size decreases from 10nm to 2nm, the confinement energy changes from 20meV to more than 1eV depending on the shape of the dots.

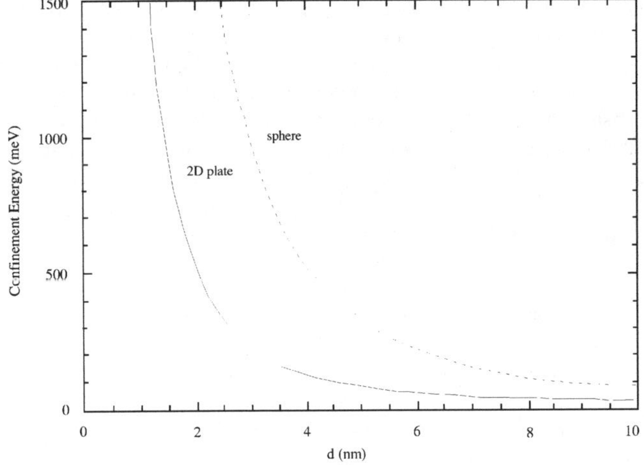

Figure 4. Shift in the band-gap energy from bulk GaN for a 2D GaN plate of thickness d (solid line) and a GaN sphere with a diameter d (dotted line) as a function of d. The shift in the band-gap energy in a cubic box of size d is between these two curves.

The curves shown in **Figure 4** represent good lower and upper bounds for the effective confinement energy in GaN QDs, both free and embedded in an AlN matrix, if the size is not too small. For electrons, the barrier height in GaN/AlN interface ($0.75\Delta E_g$ or 2.1 eV)[30] is high and the mass anisotropy is small. For holes, the barrier is lower (0.7eV) and the mass anisotropy is larger, and the effect on confinement energy is much reduced due to the large mass. For a specific GaN QD, the actual energy shift from the bulk value is expected to be within those two bounds mentioned earlier. In the case of self-assembled QDs, the plate-like results may be more suitable since the aspect ratio is very large. In other cases such as GaN nanocrystallines, sphere-like shape may be more appropriate.

The effect of strain in the heterostructures surrounding the dots must be taken into account, as it is the driving force for dot formation. For GaN QDs grown on $Al_xGa_{1-x}N$ and $In_xGa_{1-x}N$ QDs on GaN, the strain is commonly compressive which induces a blue shift in the band gap. This shift is proportional to the strain through the relevant deformation potentials. In the case of vanishing shear strain, the bandgap shift can be conveniently expressed as $\Delta E_g = (2\alpha+\beta\nu)\varepsilon_{xx}$ or $(2\alpha/\nu+\beta)\varepsilon_{zz}$,[31,32] where ν is the Poisson ratio and can be estimated as $-2C_{13}/C_{33}$, C_{13} and C_{33} are the elastic constants, ε_{xx} and ε_{zz} are the xx and zz components of the strain tensor, and α and β are the constants related to the in plane and out of plane deformation potentials of both conduction and valence bands. The reported value of $(2\alpha+\beta\nu)$ is from -6 to -12 eV,[31,32,33,34] depending on measurements and the choice of ν (from –0.51 to –0.60 which represents an overestimate of the Poisson ratio).[35] Assuming $(2\alpha+\beta\nu) = -10$ eV, we obtain $\Delta Eg = -10\varepsilon_{xx}$ eV. It is 0.25eV in the case of fully strained GaN on AlN. For a fully strained GaN on $Al_xGa_{1-x}N$, assuming a linear dependency of $Al_xGa_{1-x}N$ lattice constant on x, the energy shift $\Delta E_g = 0.25x$ eV.

As compared to the strain induced by lattice mismatch, the strain induced by thermal mismatch is negligible. For a rough estimation, we use a temperature difference of 1000 K between the growth and the measurements. Using the thermal expansion coefficient $\Delta a/a$ of $5.6\text{x}10^{-6}\,K^{-1}$ for GaN and $7.5\text{x}10^{-6}\,K^{-1}$ for sapphire,[1] a compressive strain of about 0.002 is obtained, assuming that there is no relaxation during cool down. This strain will produce only 20 meV blue shift in the band-gap. The extent of strain can be obtained from the shift of the Raman modes from their unstressed frequency position, as has been done for four different GaN/AlN QDs samples.[36] The lower than expected value of the residual strain of the QDs may be due to efficient relaxation by the large number of stacked layers. This has bearing on the piezoelectric polarization contribution which depends on the strain level of the QDs structures and barrier layers, and hence on the QDs size.[37]

Both spontaneous polarization and strain-induced polarization (piezoelectric effect) produce large electric fields which have a significant effect on the optical properties of QDs. From the symmetry arguments, it can be concluded that the wurtzitic (hexagonal) crystal has the highest symmetry showing spontaneous polarization.[38] Bernardini et al.[39] who updated their calculations later,[40] have calculated the spontaneous polarization P_{spon} and the piezoelectric constants e_{31} and e_{33} for III-Nitrides. By defining the spontaneous polarization as the difference of electronic polarization between the wurtzitic and zincblend structures, they obtained P_{spon} from first-principle calculations. As compared to GaAs, the piezoelectric constants of III-nitrides are about an order of magnitude larger.

The electric field associated with the polarization can be equivalently described by the bulk and interface polarization charges, $\rho_{p(r)} = \nabla.P(r)$ and $\sigma_p = (P_2 - P_1).n_{21}$. Here n_{21} is the directional vector of the interface pointing from medium 1 to 2 and $(P_2 - P_1)$ is the differential polarization vector across the interface. Let us consider a thick GaN layer grown on the AlN (0001) surface, the interface charge due to the difference in spontaneous polarization in two crystals is –0.056 C/m^2 corresponding to a carrier density of 3.5x10^{13}cm^{-2}. The electric field E_P created by this polarization charge is $\sigma_P/(2\varepsilon_r\varepsilon_0)$, here $\varepsilon_r\varepsilon_0$ is the static dielectric constant. This leads to -2.8x10^6 V/cm in GaN and 3.5x10^6 V/cm in AlN for the field, with the sign reference to the growth direction. The difference between the fields in the two semiconductors is due to difference in the dielectric constant, AlN one being lower. For a thin GaN plate straddled by AlN, as schematically shown in **Figure 5**, the polarization charges at both the bottom and top interfaces must be considered and the result is doubled (-5.6x10^{6}V/cm).

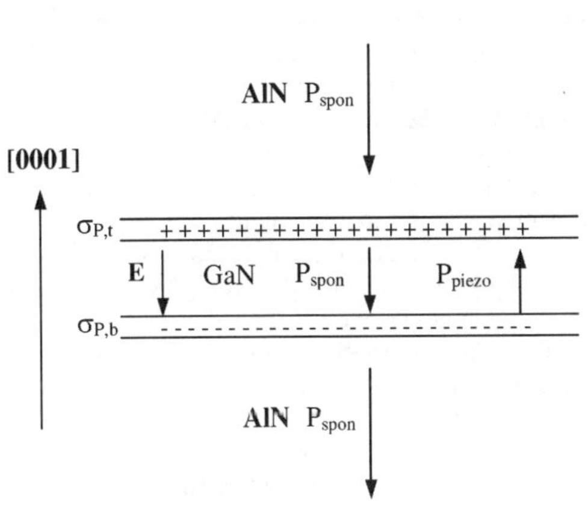

Figure 5. A schematic diagram of a GaN plate (fully strained) grown straddled by two thick (0001) AlN layers (relaxed), showing the polarization effect. Here $\mathbf{P}_{spon}$ is the spontaneous polarization that is– 0.034 C/m^2 in GaN and - 0.090 C/m^2 in AlN. $\mathbf{P}_{piezo}$ is the polarization induced by strain (2.5% in in-plane direction in GaN), which is 0.027 C/m^2 only appearing in GaN layer. $\sigma_{P,b}$ and $\sigma_{P,t}$ are the polarization charges at the bottom and top interfaces, which are -0.083 C/m^2 and 0.083C/m^2 respectively. **E** is the electric field, in opposite direction to **P**, which is about –9.0x10^6 V/cm. Note the directions of **P** and **E** with respect to the [0001] direction and the signs of the interface charges.

The strain induced polarization, on top of spontaneous polarization, can be calculated from the elastic and piezoelectric constants listed in reference 39, using $P_z=2e_{31}\varepsilon_{xx}+e_{33}\varepsilon_{zz}=2(e_{31}-e_{33}C_{13}/C_{33})\varepsilon_{xx}$. For a fully strained GaN plate on bulk AlN, P_z is calculated to be 0.027 C/m^2. It should be reminded that the effect of spontaneous and strain induced polarization in Ga polarity GaN/AlN heterostructures is additive. As a result, the total electric field induced by spontaneous polarization and piezoelectricity is raised to -4.6 x10^6 V/cm if only a single interface is considered. This field is doubled (about -9x10^{6}V/cm) in a thin GaN plate straddled by two AlN layers. This huge field will drive free electrons in the GaN toward the top interface and holes towards the bottom interface. As a result of the quantum confined stark effect (QCSE),[41,42] the electron-hole transition energy is greatly reduced despite the confinement effects unless the dots are very small. There is a limit as to the slope of the band caused by this field in that if the total band bending approaches the bandgap, free carrier screening would occur which is compensatory in nature.

The field-induced shift in the transition energy in quantum wells has been first observed by Mendez et al.[41] in the GaAs/AlGAAs system, and investigated extensively by Miller et al.[42] The shift is more significant for a wider well than a narrower one, particularly under the linear approximation which holds in narrow quantum wells. In much wider wells, a quadratic approach which reduces the red shift is used. In addition, as mentioned above if the Fermi levels come close to the conduction band on one side and the valence band on the other, the resultant free carriers would screen the induced field, reducing the shift. A similar result is expected for plate-like QDs. For a rough estimate, the shift is approximately $E_p d$ in the limit of large size, assuming the linear regime. This gives a value of 2 eV for a plate of 4nm thick under a field of $5x10^6$ V/cm. In the following, experiments dealing with GaN quantum dots fabricated in our laboratories. The approach is based on the SK mode of growth on AlN interlayers.

GROWTH AND STRUCTURES

GaN dots and associated templates discussed here were primarily grown on sapphire substrates by molecular beam epitaxy, equipped with both plasma activated N and ammonia. A typical structure is shown in **Figure 6**. Nitridation of substrates was achieved by exposing the sapphire to nitrogen plasma. The buffer layer consists of alternating layers of AlN and GaN grown on a thin layer of initiation AlN buffer. This buffer layer normally leads to an epilayer with Ga-polarity with a smoother and more impervious surface than that with the N-face. Quantum-dots were formed by growing a GaN layer just below a critical thickness, which allows it to maintain its coherence with the AlN lattice.[43] Just before the GaN relaxes, strain relaxation results in the formation of GaN QDs. The layer of GaN dots was often capped with a thin AlN layer. The procedure could be repeated 20-30 times in order to obtain multiple layers of GaN dots (MQDs). No attempt was made for vertical coupling of the dots in the mechanical sense (strain) for vertical alignment (stacking). The thickness of the GaN QD layer was varied to study the size dependency of the QDs growth and the resulting quantum confinement effects. It was shown by Widmann et al.[44] that the QD size varies significantly depending whether the QDs are allowed to evolve (ripening effect) under vacuum (before covering with AlN), or not, as a result of a ripening mechanism. This variation in size can lead to a large variation in the piezoelectric effect in the self-assembled GaN layers. Our experiments indicate that ripening leads to reduced footprint and increased height for a larger aspect ratio, as the dots are not spherical. In our work the modification of the optical properties due to the variation in the size of the 20 period quantum dots (MQDs) was achieved by ripening under RF nitrogen plasma source in addition to the length of time employed for growing the dots (sample 745, 747 and 748). These QD structures are represented as NP-MQDs in this paper. The single (sample number 650 denoted as AA-SQD) and one multiple period quantum dot structure (sample number 983 denoted as AA-MQD) studied in this paper were grown by spraying Ga on the AlN layer followed by conversion of the sprayed Ga to GaN by nitridation in an atmosphere of ammonia. The dot size and density depend on the growth conditions, deposition time as well as post growth treatment. The list of the samples characterized for this study is presented in **Table 1**. The growth rate for GaN layer is ~ 0.5 μm/hr and for AlN is 0.05 μm/hr.

Table 1. Process details of quantum dots investigated.

Sample	Period	Nitridation technique	Growth time	
			GaN	AlN
NP-MQD 733	20	Nitrogen Plasma	8"	120"
NP-MQD 734	20	Nitrogen Plasma	8"	120"
NP-MQD 745	20	"	8"	150"
NP-MQD 747	20	"	6"	150"
NP-MQD 748	20	"	4"	150"
NP-MQD 772	20	"	6"	180"
NP-MQD 785	20	"	4"	60"
AA-SQD 650	1	Ammonia	15"	75"
AA-MQD 983	20	Ammonia	15"	75"

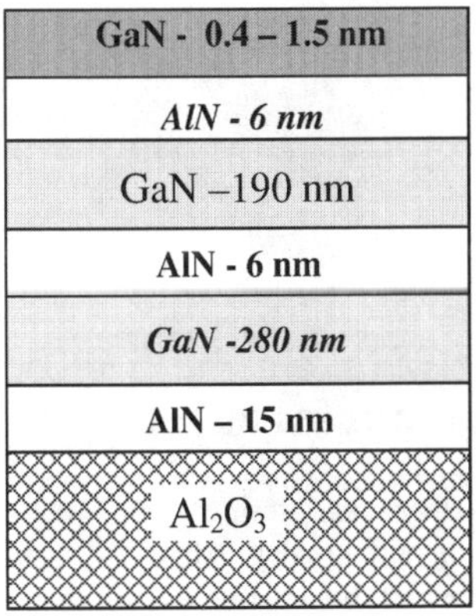

GaN - 0.4 – 1.5 nm

AlN - 6 nm

GaN –190 nm

AlN - 6 nm

GaN -280 nm

AlN – 15 nm

Al_2O_3

Figure 6a. Schematic of single quantum dot structure (multiple ones can be added on top by replicating the top two layers).

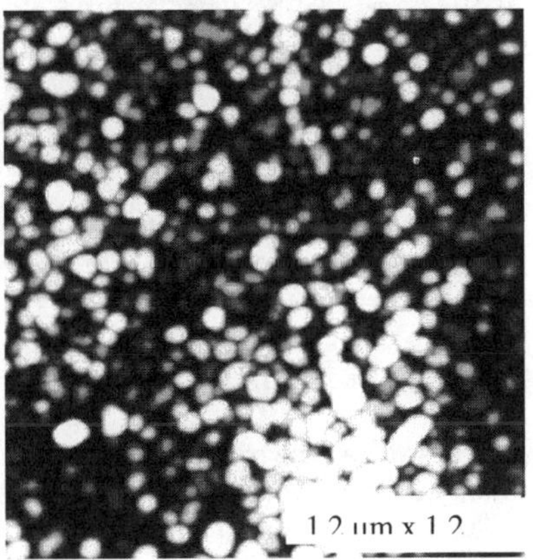

Figure 6b. AFM image of multiple period QD structure. The scale in z-direction is 12 nm.

The general surface morphology of the dots was investigated by atomic force microscopy. **Figure 6b** shows the typical AFM image of a 20 period NP-MQD (sample 747) grown at 1125 °C. The dot density is observed to be typically 3 x 10^{10} cm^{-2}, while the average size (height/diameter) is 5/50 nm.

In case of the single QD layer (650), the dot density is somewhat lower (**Figure 6c**) and in some cases, depending on the particulars of the growth, the size of the dots is larger by nearly four folds (24/174 nm). This is due to the length of time associated with the nitridation of Ga droplets. These structures are more similar to quantum boxes or clusters discussed by Talierco et al.[26] As discussed in the following section these structures have a large piezoelectric field as a signature of the induced strain. The gross morphological feature seen as associated with the surface morphology of the GaN/AlN stacked buffer layer grown by MBE. The buffer layers prepared by OMVPE and HVPE are free of such features and therefore are more conducive to being better templates for dot growth which is the approach that will soon be taken.

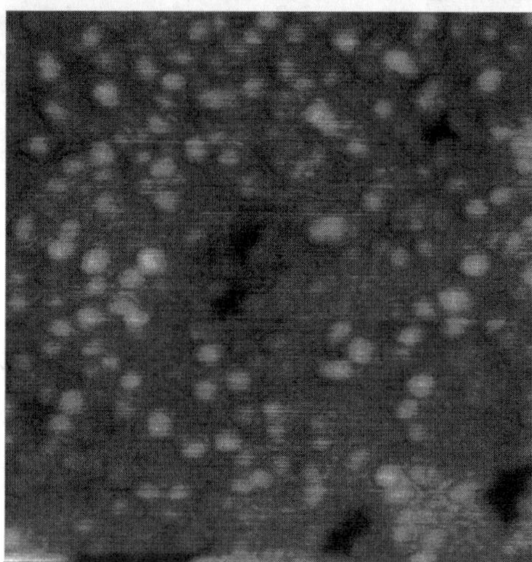

Figure 6c. AFM image of a single QD structure (500 nm x 500 nm, z=7 nm)

RAMAN SCATTERING IN GAN/ALN QUANTUM DOT STRUCTURES

Raman spectra were recorded at room temperature in the backscattering configuration with light propagating along the growth direction and unpolarized detection, i.e., in $Z(X,.)\overline{Z}$ configuration. The experiments were performed using a Dilor micro-Raman spectrometer with the Ar^+-laser line at 2.57 eV as excitation source using a 100x objective to focus and collect the laser light. **Figure 7** shows Raman spectra of three different quantum dot samples, grown using different growth techniques but over a similar buffer layer shown in **Figure 6a.** SVT 650 is a single layer quantum dot structure (AA-SQD), SVT 733 is an NP-MQD samples (with GaN (0.85 nm)/AlN(1.0 nm) layer thickness, without a cap layer, and (SVT-785) is a NP-MQD sample with a thin GaN cap layer (4 nm) over the 20 period quantum dot layers. The spectra are dominated by the E_2 (high) and A_1(LO) Raman modes of the GaN layer underneath the GaN/AlN quantum dot layer structure. For all three samples, the GaN E_2 (high) Raman mode is shifted significantly with respect to unstressed GaN with its E_2 (high) frequency of 567 cm^{-1} (determined using a thick GaN reference layer). The underlying thick buffer GaN layers in these structures are therefore under increased compressive stress. For most of the NP-MQD samples, including SVT-785 shown in **Figure 7**, the E_2 (high) Raman mode is observed at 573 cm^{-1} corresponding to a compressive stress of –2.1 GPa (strain: -0.4%). This Raman shift and the strain are identical for the AA-SQD sample (SVT-650). For SVT-733 we find an E_2 (high) frequency of 575 cm^{-1}, which corresponds to a compressive stress of –2.8 GPa (strain: 0.6%).[45]

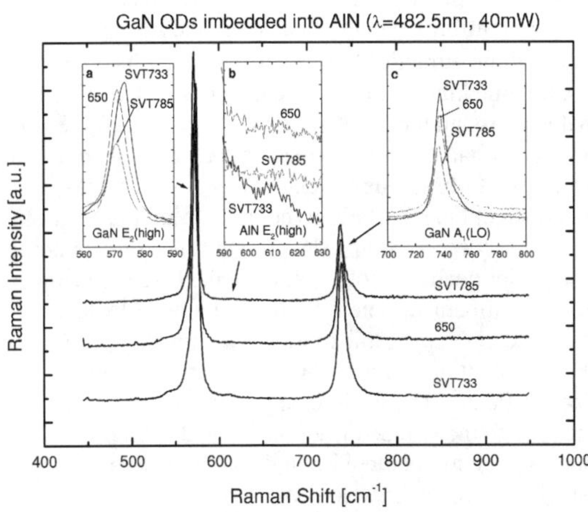

Figure 7. Raman spectra of GaN QDs imbedded into AlN, obtained under 2.57 eV excitation. The main point to be drawn is that the AlN layers are strained as well because of their small thickness.

As shown in the inset (**Figure 7-b**), there is a Raman peak at 612 cm^{-1}, which can be attributed to the E_2 (high) phonons of the AlN spacer layers in these structures. The AlN layers in the grown structures are therefore under high tensile stress as unstressed AlN[46] has an E_2 (high) phonon frequency of 657 cm^{-1}. This is in agreement with the presence of a high compressive stress in the GaN buffer layer in the samples. This stress in the AlN layers is significantly higher than the one commonly found in AlN (5 nm)/GaN (6 nm) superlattices grown on thick AlN layers [E_2 (high) frequency of 622 cm^{-1} or in a GaN QDs matrix grown on a thick AlN layer [E_2 (high) frequency of of 653 cm^{-1}, see references [47 48],

These results suggest that lattice mismatch between the GaN and the AlN spacers where the GaN is deposited is reduced due to the high tensile stress in the AlN layers. As a result, GaN quantum dots form despite the reduced difference in in-plane lattice parameters between the strained AlN spacers and the GaN. The formation of such quantum dots with reduced lattice mismatch between the GaN and AlN is therefore expected to show reduced piezoelectric polarization fields. This is in contrast to the QD structures investigated by Daudin el al.[17,44] which were grown on thick AlN spacer layers, and consequently exhibits giant built-in fields.

PHOTOLUMINESCENCE SPECTROSCOPY

The time-integrated photoluminescence experiments were performed with a frequency tripled Ti:Sapphire laser delivering pulses of 10 ps duration at 267 nm (photon energy 4.655 eV). The luminescence was analyzed with a 0.5 m grating spectrometer and a photon counter. The temperature was varied from 10 K – 300 K using a closed cycle cryostat. Figure 8 shows the photoluminescence spectra from NP-MQDs with various GaN and AlN barrier layer thickness. The PL peaks from the QD layers are shifted to a higher energy as compared to the underlying bulk GaN for the wurtzite phase (bandgap energy $E_g = 3.45$ eV). The blue shift is likely related to the exciton confinement effect. By assuming a spherical quantum dot with infinite potential barrier in the strong confinement regime (i.e. negligible Coulomb attraction), the energy shift ΔE is given by the contribution of three terms: the excitonic kinetic energy, the Coulomb interaction, and the correlation effect.[49] It is reasonable to assign the high energy PL spectrum (fig. 8) to the superposition of blue-shifted near-band-gap excitonic emissions arising from clusters of dots with size smaller than the excitonic Bohr radius for GaN ($a_B \sim 3$ nm), at least in the growth direction. There is also a distinct blue shift observed in the peak PL position with the reduction in the GaN growth time for a constant AlN growth time corresponding to a AlN layer thickness of 1.1 nm. Smaller quantum dots result and the blue shift results from the corresponding increase in the confinement potential. If the thinner AlN layer thickness is reduced (0.9 nm) with a fixed GaN layer thickness ~ 0.85 nm (sample SVT 734), the PL peak is blue shifted due to reduction of the strain in the barrier layers induced by the spontaneous polarization field.[50] The PL intensity decreases with decreasing GaN dimensions not only due to the reduced fraction of incident photons absorbed by the thinner layers,[51] but also due to the increase in the nonradiative recombination rate, the origin of which has not yet been investigated, but may be related to the dot density. In the NP-MQD structures, a higher dot density due to smaller dot size may result in additional relaxation channels due to lateral coupling of adjacent dots resulting in an overlap of wavefunctions that are not confined within the quasi-zero dimensional states.

The absorption edge of these NP-MQD structures is depicted in Figure 9. Photoluminescence emission spectra (PLE) were measured using a 1 kW Xe-lamp as the excitation source with the detection wavelength set around the peak PL wavelength (Figure 8) of the respective samples. A comparison of figures 8 and 9 shows a large Stokes shift exceeding 250 meV. The PLE spectrum measured at room temperature for sample 745 and 747 indicates absorption resonance at 3.97 eV and 4.03 eV respectively, likely corresponding to higher excitonic transitions[21] in the GaN dots. The dot size distribution throughout the layer is responsible for inhomogeneous broadening of this part of the emission band. However, exciting discrete levels does not seem to result in selective excitation of a family of QDs. Instead we obtain a broad luminescence spectra corresponding to the quasi-totality of the QD size distribution. This strongly suggests, that the higher energy (smaller) dots are coupled in NP-MQDs. This allows hot carriers to diffuse laterally before they relax and thus populate most of the QDs.

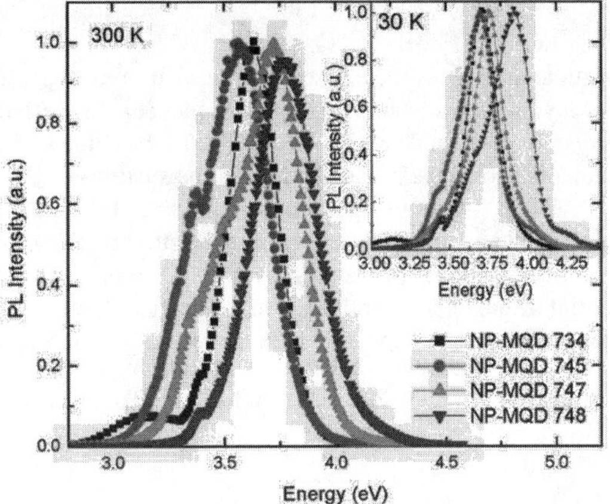

Figure 8. PL spectra of NP-MQDs with varying GaN and AlN layer thickness.

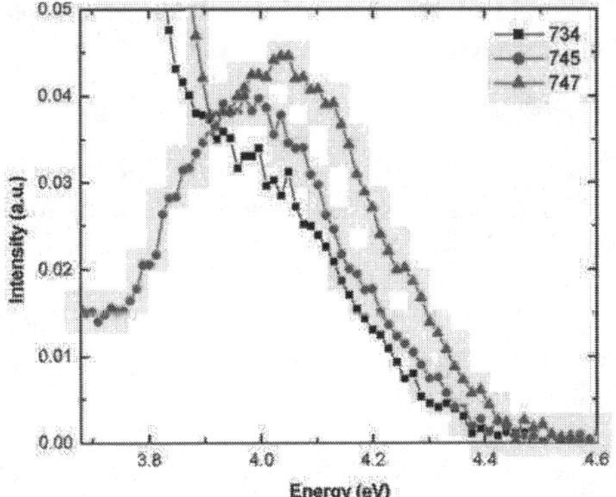

Figure 9. PLE spectra of NR-MQDs samples.

Finally, we should mention that the thickness of the AlN barrier layers is such that while providing the necessary strain for S-K growth, it is not relaxed which leads to GaN dots to vertically stack as dots prefer to situate where the strain between GaN and AlN below is minimum. We believe that vertical stacking is evident in our NP type multiple quantum dots which resulted in the reduction of the non-radiative recombination channels due to vertical stacking. Additional supporting argument is the observation of a relatively small change in the PL intensity with temperature (the PL intensity drops a mere 3 times from 10 K to 300 K) without any change in the PL linewidth.

CONCLUSIONS

We have investigated the optical properties of self-organized quantum dots grown using relatively thin AlN spacer layers which results in a reduced piezoelectric effect relative to highly strained QD system with thicker AlN layers. The highly blue shifted PL along with the Raman measurements shows the reduced strain in the system of QDs grown with thinner AlN layer. The QDs grown using ammonia (larger dots) appear to have a larger built-in polarization field and are red-shifted compared to the QDs grown using nitrogen plasma. The dots are smaller in rf-plasma grown MQD samples resulting in a stronger confinement potential of the exciton and leads to a significant blue shift in the PL emission edge. The extent of blue shift depends on the growth time, thus size, of the GaN layers. The PL efficiency is considerably enhanced for multiple period QD structures, and an intense PL is observed at room temperature for the NR-MQD samples.

ACKNOWLEDGMENTS

The VCU portion of this work was funded by grants from NSF (Drs U. Varshney, and L. Hess), and ONR (Dr. Y. S. Park), and AFOSR (Dr. G. L. Witt and T. Steiner). A. N. was supported by the Army Research Office and the National Research Council. The authors thank Prof. D. Huang and Dr. M. Reshchikov for useful discussions and contributions to the topic. M.K. would like to thank M. A. Renucci, J. Frandon and F. Demangeot (Toulouse) for use of their facility"

REFERENCES

[1] S.T. Strite and H. Morkoç, "GaN, AlN, and InN: A Review," J. Vacuum Science and Technology **B10**, 1237-1266 (1992).

[2] H. Morkoç, S. Strite, G. B. Gao, M.E. Lin, B. Sverdlov, and M. Burns, "A Review of Large Bandgap SiC, III-V Nitrides, and ZnSe Based II-VI Semiconductor Structures and Devices," J. Appl. Phys. Reviews **76**(3), 1363 (1994).

[3] S. N. Mohammad, A. Salvador, and H. Morkoç, "Emerging GaN Based Devices," Proc. IEEE **83**, 1306 (1995).

[4] S. N. Mohammad and H. Morkoç, "Progress and Prospects of Group III-V Nitride Semiconductors," Progress in Quantum Electronics **20**(5 and 6), 361-525 (1996).

[5] O. Ambacher, "Growth and Applications of Group III-Nitrides", J. Phys. D: Appl. Phys. **31**, 2653, (1998).

[6] Hadis Morkoç, Aldo Di Carlo and R. Cingolani, "GaN-Based Modulation Doped FETs and UV Detectors", Solid State Electronics, **46**(2), 157, (2002).

[7] S. J. Pearton, J. C. Zolper, R. J. Shul, and F. Ren, "GaN: Processing, Defects, and Devices", J. Appl. Phys. **86**,1 (1999).

[8] J. M. Gérard, O. Cabrol, and B. Sermage, Appl. Phys. Lett. , **68**, 3123 (1996).

[9] Y. Arakawa and H. Sakaki, Appl. Phys. Lett., **40**, 939 (1982).

[10] D. Huang, M. A. Reshchikov and H. Morkoç, "Growth, Structure, and Optical Properties of III-Nitride Quantum Dots", in "Quantum Dots", International Journal of High Speed Electronics Vol. **25**, No. 1 pp.79-110. (March 2002), Eds. E. Borovitskaya and M. S. Shur.

[11] D. Huang, Y. Fu, and H. Morkoç, "Preparation, Structural and Optical Properties of GaN based quantum dots", in " " Ed. T. Steiner, Artech House.

[12] H. Morkoç, *Nitride Semiconductors and Devices* (Springer Verlag, Heidelberg, 1999, the second edition is in process); S. Nakamura and G. Fasol, *The Blue Laser Diode* (Springer-Verlag, Heidelberg, 1997).

[13] M. Kuball, J. Gleize, Satoru Tanaka, and Yoshinobu Aoyagi, Appl. Phys. Lett. **78**, 987 (2001).

[14] S. Tanaka, S. Iwai, and Y. Aoyagi, Appl. Phys. Lett., **69**, 4096 (1996).

[15] X. Q. Shen, S. Tanaka, S. Iwai, and Y. Aoyagi, Appl. Phys. Lett., **72**, 344 (1998).

[16] F. Widmann, B. Daudin, G. Feuillet, Y. Samson, J. L. Rouvière, and N. Pelekanos, J. Appl. Phys., **83**, 7618 (1998).

[17] F. Widmann, J. Simon, B. Daudin, G. Feuillet, J. L. Rouvière, N. T. Pelekanos, and G. Fishman, Phys. Rev. B, **58**, R15989 (1998).

[18] B. Damilano, N. Grandjean, F. Semond, J. Massies, and M. Leroux , Appl. Phys. Lett., **75**, 962 (1999).

[19] Adam Li, Feng Liu, D.Y. Petrovykh, J.-L. Lin, J. Viernow, F. J. Himpsel, and M. G. Lagally,", Phys. Rev. Lett, **85**, 5380 (2000).

[20] Kenneth E. Gonsalves, Sri Prakash Rangarajan,Greg Carlson, Jayant Kumar and Ke Yang, Appl. Phys. Lett, **71**, 2175 (1997).

[21] E. Borsella, M.A. Garcia, G. Mattei, C. Maurizio, P. Mazzoldi, E. Cattaruzza, F. Gonella,G. Battaglin, A. Quaranta, and F. D'Acapito, J. Appl. Phys, **90** (9), 4467 (2001).

[22] S. Tanaka, M. Takeuchi, Y. Aoyagi, *Jpn J. Appl. Phys*, **39**, (L831), (2000).

[23] U. Woggon, Optical Properties of Semiconductor QD, Springer, Berlin, Heidelberg, NY (1997).

[24] H. Lipsanen, M. Sopanen, J. Ahopelto, Phys. Rev. B, **51**, 13868 (1995).

[25] V. Chamard, T.H. Metzger, B. Daudin, C. Adelmann, H. Mariette and G. Mula, Appl. Phys Lett, **79**, 1971 (2001).

[26] T. Talierco, P. Lefebvre, M. Gallart and A. Morel, J. Condensed Matter Phys., **13**, 7027 (2001).

[27] F. Bernardini, V. Fiorentini, and D. Vanderbilt, Phys. Rev. B **56**, R10024 (1997).

[28] A. D. Andreev and E. P. O'Reilly, Physica E **10**, 553 (2001).

[29] D. Das and A. C. Melissinos, *Quantum Mechanics*, (Gordon and Breach Science Publishers, New York, 1986).

[30] G. Martin, A. Botchkarev, A. Rockett, and H. Morkoç, Appl. Phys. Lett., **68**, 2541 (1996).

[31] V. Y. Davydov, N. S. Averkiev, I. N. Goncharuk, D. K. Delson, I. P. Nikitina, A. S. Polkovnokov, A. N. Smirnov, and M. A. Jacobson, J. Appl. Phys. **82**, 5097(1997).

[32] A. Shikanai, T. Azuhata, T. Sota, S. Chichibu, A. Kuramata, K. Horino, and S. Nakamura, J. Appl. Phys. **81**, 417 (1997).

[33] H. Amano, K. Hiramatsu, and I. Akasaki, Jpn. J. Appl. Phys. 2 **27**, L1384 (1998).

[34] W. Shan, R. J. Hauenstein, A. J. Fischer, J. J. Song, W. G. Perry, M. D. Bremser, R. F. Davis, and B. Goldenber, Phys. Rev. B**54**, 13460 (1996).

[35] K. Shimada, T. Sota, and K. Suzuki, J. Appl. Phys. **84**, 4951(1998).

[36] O. Martinez, M. Mazzoni, F. Rossi, N. Armani, G. Salviati, P. P. Lottici, and D. Bersani, Phys. Stat. Sol. (a), **195**, pp.26-31, (2003).

[37] P. Ramvall, P. Riblet, S. Nomura, Y. Aoyagi, and S. Tanaka, J. Appl. Phys. **87**, 3883 (2000).

[38] J. F. Nye, *Physical Properties of Crystals*, (Oxford University Press, Oxford, 1985).

[39] F. Bernardini, V. Fiorentini, and D. Vanderbilt, Phys. Rev. B**56**, R10024 (1997).

[40] F. Bernardini, V. Fiorentini Physical Review B, **63**, 193201 (2001).

[41] E. E. Mendez, G. Bastard, L. L. Chang, L. Esaki, H. Morkoç, and R. Fischer, Phys. Rev. B**26**, 7101 (1982).

[42] D. A. B. Miller, D. S. Chemla, T. C. Damen, A. C. Gossard, W. Wiegmann, T. H. Wood, and C. A. Burrus, Phys. Rev. B**32**, 1043 (1985).

[43] M. A.Reshchikov, J. Cui, F.Yun, P.Visconti, M. I. Nathan, R. Molnar, and H. Morkoç Fall MRS, 2000, Mat. Res. Soc. Symp. Proc. **639**, G11.2 (2001).

[44] F. Widmann, J. Simon, N.T. Pelekanos, B. Daudin, G. Feuillet, J.L. Rouviere, G. Fishman, Microelectronic Journal, **30**, 353 (1999).

[45] J.M. Hayes, M. Kuball, A. Bell, I. Harrison, D. Korakakis, and C.T. Foxon, Appl. Phys. Lett. **75**, 2097 (1999).

[46] J.M. Hayes, M. Kuball, Ying Shi, and J.H. Edgar, Jpn. J. Appl. Phys. **39**, L710 (2000).

[47] J. Gleize, F. Demangeot, J. Frandon, and M. A. Renucci, F. Widmann and B. Daudin, Appl. Phys. Lett., **74**, 703 (1999).

[48] J. Gleize, F. Demangeot, J. Frandon, M.A. Renucci, M. Kuball, B. Damilano, N. Grandjean, and J. Massies, Appl. Phys. Lett. **79**, 686 (2001).

[49] A.D. Yoffe, Adv. Phys., **42**, 173 (1993).

[50] N. Iizuka, N. Suzuk, Jpn J. Appl. Phys -1 **39** (4B), 2376 (2000).

[51] J.C. Harris, T. Someya, S. Kako, K. Hoshino and Y. Arakawa, Appl. Phys. Lett. 77, 1005 (2000).

Mat. Res. Soc. Symp. Proc. Vol. 789 © 2004 Materials Research Society N8.6/T6.6/Z6.6

Diffuse x-ray scattering from InGaAs/GaAs quantum dots

Rolf Köhler, Daniil Grigoriev, Michael Hanke, Martin Schmidbauer, Peter Schäfer, Stanislav Besedin, Udo W. Pohl[1], Roman L. Sellin[1], Dieter Bimberg[1], Nikolai D. Zakharov[2], and Peter Werner[2]
Institut für Physik, Humboldt-Universität zu Berlin, Newtonstr. 15, D-12489 Berlin, Germany
[1]Institut für Festkörperphysik, Technische Universität, Hardenbergstr. 36, D-10623 Berlin, Germany
[2]Max-Planck-Institut für Mikrostrukturphysik, Weinberg 2, D-06120 Halle, Germany

ABSTRACT

Multi-fold stacks of $In_{0.6}Ga_{0.4}As$ quantum dots embedded into a GaAs matrix were investigated by means of x-ray diffuse scattering. The measurements were done with synchrotron radiation using different diffraction geometries. Data evaluation was based on comparison with simulated distributions of x-ray diffuse scattering. For the samples under consideration ((001) surface) there is no difference in dot extension along [110] and [-110] and no directional ordering. The measurements easily allow the determination of the average indium amount in the wetting layers. Data evaluation by simulation of x-ray diffuse scattering gives an increase of In-content from the dot bottom to the dot top.

INTRODUCTION

Based on a wealth of research on semiconductor quantum dot system (see [1] for a review) there are ongoing efforts to further improve the quality of InGaAs/GaAs quantum dot systems. This system is of special interest not least because of its potential for the application for lasers emitting at 1.3 μm and above. Therefore, appropriate techniques are required for the characterization of the structural properties of such systems. Transmission electron microscopy (TEM) in cross section or plane view is used in a substantial part of the papers on the subject (see [2] for a review). A comparatively new approach is the investigation by means of cross-sectional tunneling microscopy (XSTM). In case of InGaAs/GaAs several different indium distributions within the dots were revealed as reversed trapezoidal [3] or reversed pyramidal [4] indium rich cores – both of them without any wetting layer underneath – or a reversed pyramidal indium core on a comparatively thick wetting layer [5].

X-ray diffuse scattering investigations are complementary to those 'direct space' techniques as they are non-destructive and characterize a comparatively large sample area, thus providing a rather good statistical measure of sample quality. Not least, there is the advantage of the very high strain sensitivity of x-ray scattering. This was already used in order to investigate InGaAs/GaAs dot structures in the case of free standing dots [6,7] and of dots in a layer stack[8,9]. This paper is intended to demonstrate the capabilities of x-ray diffuse scattering in conjunction with an appropriate evaluation approach.

EXPERIMENTAL DETAILS

The sample presented here was grown on a (001) GaAs substrate by means of metalorganic chemical vapor deposition and consists of five layers of $In_{0.6}Ga_{0.4}As$ separated by 20 nm GaAs spacer layers, capped by a further 20 nm thick topmost layer. As the transmission electron micrograph in figure 1 shows, dots form in the InGaAs layers and there are thin remaining wetting layers. There is a rather good vertical correlation of dots, but there is no indication of lateral ordering as obvious by TEM in plane view (not shown here).

We have used TEM and x-ray diffuse scattering in the complementary way indicated. However, in the present paper we will mainly focus on x-ray diffraction. Because of the small amount of material in the dots synchrotron radiation (beamline BW2 at HASYLAB/DESY) was used. It has proven advantageous to investigate the samples in different scattering geometries as high resolution x-ray diffraction (HRXRD) at large glancing angles, grazing incidence diffraction (GID) and grazing incidence small angle scattering (not considered in this paper). HRXRD is especially sensitive to out-of-plane strain components and probes the layer stack, whereas GID is sensitive to in-plane strain components only and can be tuned to penetration depths from less than 30 Å to about 1 μm.

Our evaluation of x-ray diffuse scattering is based on simulation using a three step approach: first a structural model is constructed taking into account some prior information e.g. from crystal growth, secondly the strain within the model is calculated in the framework of linear elasticity theory using the finite element method (FEM) and thirdly, the diffuse scattering is simulated using either a kinematical approximation in the case of HRXRD or the distorted-wave Born approximation (DWBA) in the case of GID. By trial and error and taking into account as much external information as possible the model is adapted until satisfactory agreement with measurements is achieved. As we will show, x-ray diffuse scattering is quite sensitive to comparatively small changes of the model.

For more details of experimental set-up and data evaluation see [10].

DISCUSSION

Fig. 2 presents the diffuse scattering measured in the vicinity of the (224)-reciprocal lattice point in comparison with three simulations. q_{001} and q_{110} are the components of the diffraction vector in [001]- and [110]-directions, respectively. The simulations show exclusively the diffuse scattering, i.e. only the perturbation with respect to a laterally homogeneous layer system is

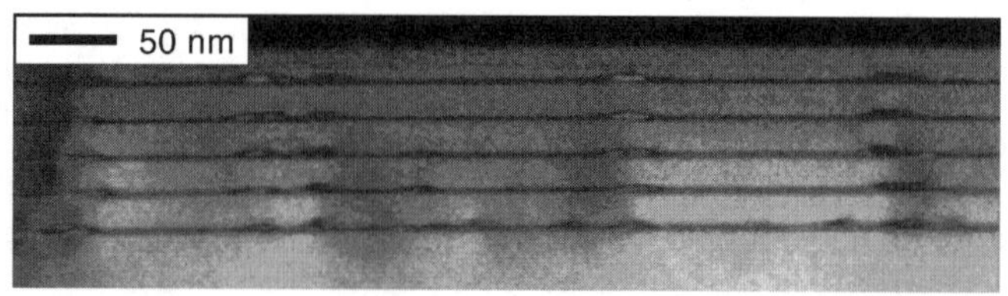

Figure 1. Transmission electron micrograph (dark field, (020)-reflection) of the InGaAs dot layer stack

taken into account. These simulations are a small selection of a large variety calculated for different models. The corresponding models are shown schematically in figure 3. The (224)-reflection proved useful in that it is sensitive to vertical and lateral normal components of the strain tensor.

A first point worth noticing is that the q_{001}-position of the 0th order satellite reflection depends on the amount of indium in the wetting layer only – at least for the comparatively large in-plane distances of dots in the investigated sample. More precisely: in a rather good approximation the q_{001}-position of the 0th order satellite is given by the average indium concentration in the entire periodic layer stack outside the vicinity of the dot stack. We have checked this with models containing dots of different sizes within the size range indicated by TEM. This is interesting since it allows for an independent determination of information on the wetting layers. The reason is that the dot strain field is confined to a rather close vicinity of the dot whereas within the main area of a layer the strain field is solely due to the wetting layer. This is demonstrated in figure 4 by a FEM calculation of ε_{33}, the normal component in [001]-direction of the tensor of total strain (i.e. strain as compared to the substrate lattice). This means that already scans along the truncation rod (parallel to the surface normal) give reliable values regarding the indium content of the wetting layer. Besides: our measurements give a stacking

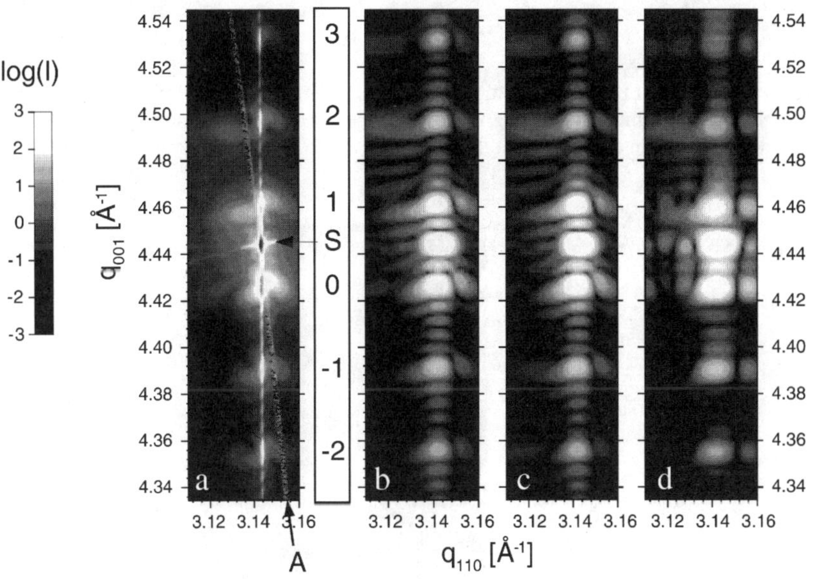

Figure 2. Comparison of measured (a) x-ray diffuse scattering in the (224)- reflection with three different simulations: (b) lens shaped, (c) inverted cone, (d) lens shaped. (b) and (c) were calculated with an indium content increasing from dot bottom to top, (d) with a homogeneous content. For further details of the models see schemes in figure 3. 'S' marks the substrate reflection (peak intensity marked by black spot), numbers '-2' to '3' indicate the orders of satellite reflections. 'A' is an artifact due to crosstalk from the very high substrate peak intensity.

period of 18.5 nm, i.e. less than the nominal spacer layer thickness.

However, only diffuse scattering provides information on indium content and strain field of the dots. There is also 'chemical' information due to difference in the structure factor on InAs and GaAs, which becomes accessible by anomalous scattering close to absorption edges. As a rule for small dots the 'chemical' contribution as well as strain has to be taken into account. In the given case of figure 2 the influence of strain is dominating.

Actually, the simulations for the (224)-reflection proved rather insensitive to the specific shape of dots for the system under investigation. This can be inspected by comparing the simulations (b) for lens shaped dots and (c) for inverted cone geometry (such indium rich

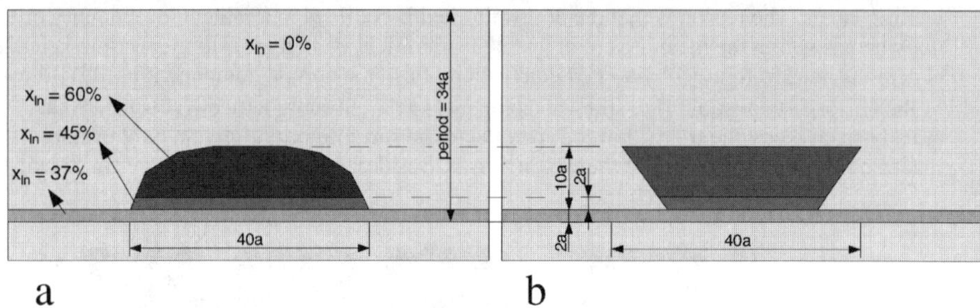

Figure 3. Scheme of the structural models used for simulations in figure 2: (a) lens shaped, (b) inverted cone. The dimensions are given in multiples of the lattice parameter a. The schemes show only the area in the vicinity of the dots – the total width used in the simulation is 240a.

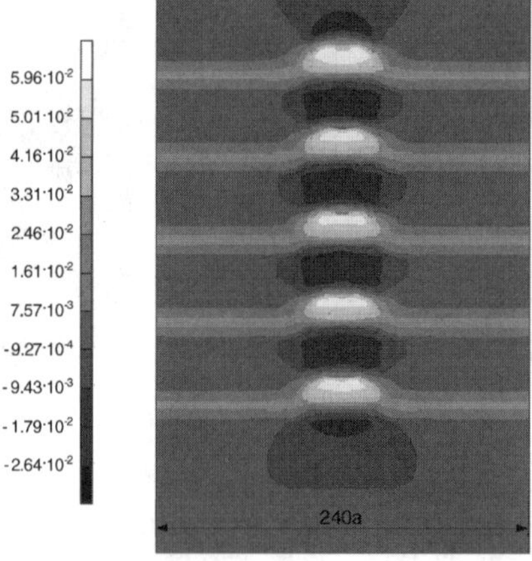

Figure 4. FEM calculation of ε_{33}, the normal component in [001]-direction of the tensor of total strain for a five-fold stack of dot layers. The dot model based on the scheme in figure 3a was used. As explained in the text the entire model width is 240a = 135.7 nm.

inverted cones were detected e.g. by [5]). However, there is a considerable difference between simulations (b) and (d), both calculated for lens shaped dots. This difference is due to the different indium distributions inside the dots (see figure 3 and corresponding explanations in the text). As a comparison of simulations shows, the typical shape of the intensity distributions around satellite reflections can only be achieved by means of an indium distribution increasing from bottom to top within each single dot. Such segregation effects were already observed by means of x-ray diffuse scattering [6] and by TEM [11] at free standing InGaAs dots.

Figure 3 shows in detail the most significant section of the model in the vicinity of the dots. The actual models used for FEM and scattering calculation of this paper have dimensions of 240*a* in [110]-direction, 160*a* in [1-10]-direction and 230*a* in [001]-direction. In view of subsequent scattering calculations (these are based on a quasi atomistic approach – see [10, 12]) all model dimensions are chosen as multiples of the lattice parameter *a*. The following correspondences hold: simulation fig.2b is based on model fig.3a, simulation fig.2c on model fig.3b and simulation fig.2d on model fig.3a, but with an indium concentration of 60% in the transition layer (instead of 45%) as in the dot top.

We have checked different thicknesses and indium concentrations of the transition layer. There is no essential change in the diffuse scattering if the transition layer is slightly thicker than shown here. However, without this transition layer with an indium content close to the mean value of wetting layer and dot top it is obviously impossible to achieve an intensity distribution close to the experiment.

Figure 5 presents the in-plane intensity measured in the (220)-reflection (GID). The experimental intensity distribution (figure 5a) is nicely reproduced by simulation (figure 5b) without taking into account any lateral directional correlation of dots. This is in accordance with

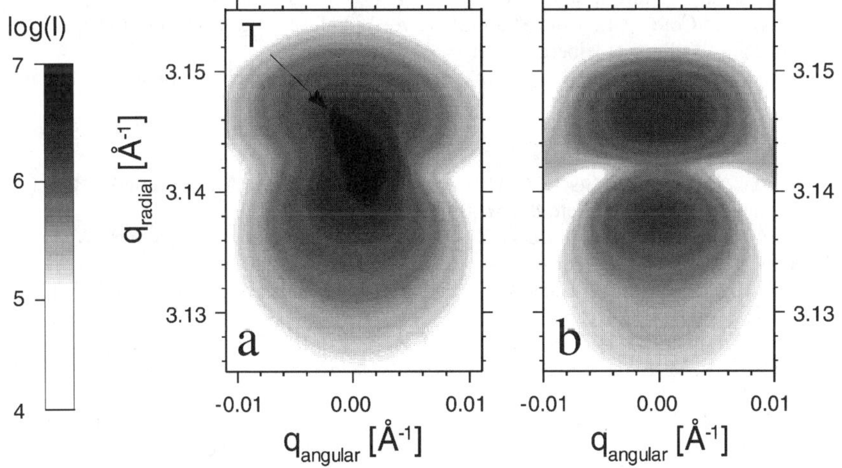

Figure 5. In-plane x-ray diffuse scattering, measured (a) in the (220)-reflection. (b) is the corresponding simulation. In order to reduce noise, (a) is the diffuse scattering signal integrated over some q_{001}-range. The intensity streak 'T' is due to the truncation rod which is not taken into account in the simulation.

TEM results, which also do not indicate lateral directional correlation. However, the distance between the two dark areas in figure 5 is determined by the average distance between dots, because this distance coincides with the average main harmonic of lateral strain modulation. This is simulated by a periodical boundary condition for FEM. The size of the optimized model – 240a ≈140 nm – is an appropriate estimate of the average dot stack distance (see also figure 1).

It should be also noted that we did not observe significant differences between HRXRD maps for the (110)-plane and the (1-10)-plane or, correspondingly, for GID in-plane intensity distributions taken with (220)- and (2-20)-reflections. This indicates that the investigated dots do not show any elongation along either [110]- or [1-10]-directions.

CONCLUSIONS

It is shown that x-ray diffuse scattering in conjunction with scattering simulation is an appropriate means for the evaluation of essential data of InGaAs/GaAs quantum dot systems such as indium content in the wetting layer and in the dots, precise stacking period, symmetry and vertical and lateral positional correlation of dots.

We acknowledge the financial support by the Deutsche Forschungsgemeinschaft (Sfb296).

REFERENCES

1. D. Bimberg, M. Grundmann, and N.N. Ledentsov, Quantum Dot Heterostructures (Wiley, Chichester, NY, 1999)
2. A. Rosenauer, Transmission Electron Microscopy of Semiconductor Nanostructures: Analysis of Composition and Strain State, vo. 182 of Springer Tracts in Modern Physics (Springer, Berlin, Heidelberg, 2003)
3. N. Liu, H. Lyeo, C. Shih, M. Oshima, T. Mano, and N. Koguchi, Appl. Phys. Lett. **80**, 4345 (2002).
4. N. Liu, J. Tersoff, O. Baklenov, J. A. L. Holmes, and C. K. Shih, Phys. Rev. Lett. **84**, 334 (2000).
5. A. Lenz, R. Timm, H. Eisele, C. Hennig, S. Becker, R. Sellin, U. W. Pohl, D. Bimberg, and M. Dähne, Appl. Phys. Lett. **81**, 5150 (2002).
6. I. Kegel, T. H. Metzger, A. Lorke, J. Peisl, J. Stangl, G. Bauer, K. Nordlund, W. V. Schoenfeld, and P. M. Petroff, Phys. Rev. B **63**, 035318 (2001).
7. K. Zhang, Ch. Heyn, W. Hansen, Th. Schmidt, and J. Falta, Appl. Surf. Sci. **175-176**, 606 (2001).
8. A. A. Darhuber, V. Holý, J. Stangl, G. Bauer, A. Krost, F. Heinrichsdorff, M. Grundmann, D. Bimberg, V. M. Usinov, P. S. Kop'ev, A. O. Kosogov, and P. Werner, Appl. Phys. Lett. **70**, 955 (1997).
9. M.Hanke, D.Grigoriev, M.Schmidbauer, P.Schäfer, R.Köhler, U.W. Pohl, R.L. Sellin, D. Bimberg, N.D. Zakharov, and P. Werner, Physica E (2003), to be published.
10. M. Schmidbauer, M. Hanke and R. Köhler, Cryst. Res. Techn. **37**, 3 (2002).
11. A. Rosenauer, D. Gerthsen, D. Van Dyck, M. Arzberger, G. Böhm, and G. Abstreiter, Phys. Rev. B **64** (2001) 245334
12. D.Grigoriev, M.Hanke, M.Schmidbauer, P.Schäfer, O.Konovalov, and R.Köhler , J. Phys. D **36**, A225 (2003)

Mat. Res. Soc. Symp. Proc. Vol. 789 © 2004 Materials Research Society N8.8/T6.8/Z6.8

Near-Field Magneto-Photoluminescence of Singe Self-Organized Quantum Dots.

A. M. Mintairov, A. S. Vlasov and J. L.Merz.
Department of Electrical Engineering, University of Notre Dame, Notre Dame, IN 46556, USA

ABSTRACT

We present results obtained using low temperature near-field scanning optical microscopy for the measurements of Zeeman splitting and the diamagnetic shift of single self-organized InAs/AlAs, InAs/GaAs and InP/GaInP quantum dots. The measurements allow us to relate the bimodal size distribution of InAs dots with variations in In content. For single InP QDs we observed a strong circular polarization at zero magnetic field accompanied with a negative energy shift, suggesting that strong internal magnetic fields exist in these QDs.

INTRODUCTION

Near-field scanning optical microscopy (NSOM) allows one to extend the spatial resolution of optical experiments far beyond the light diffraction limit, which opens new possibilities for nanoscale characterization. Combining this with high magnetic fields allows the use of NSOM to study structural parameters and spin fine structure of exciton states.[1,2,3,4] In the present paper we study the observed bimodal size distribution of InAs/GaAs and InAs/AlAs QDs by addressing the size and In content of individual dots using near-field magneto-photoluminescence (NMPL). We also report for the first time the NMPL measurements of single InP/GaInP QDs.

EXPERIMENTAL DETAILS

We used 50-300 nm diameter tapered fiber tips with an Al coating of 50-200 nm thickness and transmission 10^{-4}-10^{-2} (see Fig.1). The taper of the single mode fibers were prepared using a pulling technique, which provides apertures 150-300 nm. To make smaller apertures (down to 50 nm) we etch the tips after pulling.

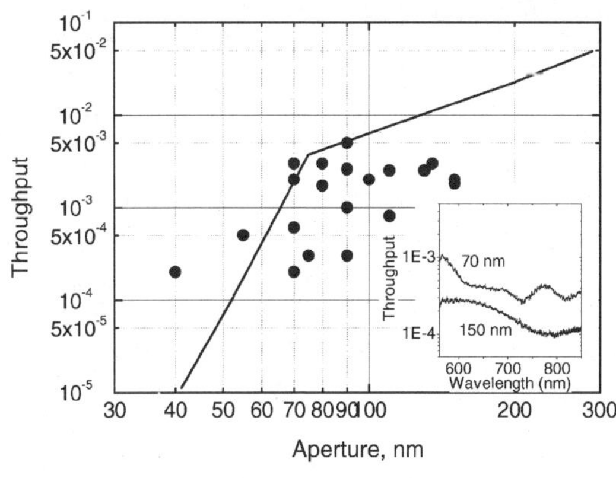

Fig. 1. Transmission of the near-field fiber probes (solid circles) at 631 nm. Straight lines show best results reported in literature. Insert shows transmission spectra of 150 and 70 nm diameter tips taken using 250W Halogen lamp as the source.

An Oxford CryoSXP cryogenic positioning system was used for 3D scanning.

The near-field photoluminescence (NPL) spectra were taken in collection-illumination mode at 10 K under excitation of the 514.5 nm Ar-laser line. Spectra were measured using a CCD detector together with a 270 mm focal length monochromator. The excitation power was 1-50 μW, which provided power densities generally below the emission threshold of QD excited states (<20 W/cm^2). The spectral resolution of the system is 0.2-0.4 meV. A quarter wave-plate and linear polarizer were used for separation of right- and left-hand circular polarization components of the QD emission spectra in a magnetic field up to 10 T.

The InAs/GaAs and InAs/AlAs QD samples described in ref[5] were used to study the bimodal size distribution. The QDs were formed within a short-period superlattice consisting of two/eight monolayers (ML) of AlAs/GaAs. For the InAs/AlAs sample the InAs QDs were grown between 2ML of AlAs, whereas for the InAs/GaAs sample the dots were grown between 8 ML of GaAs. The InAs/AlAs and InAs/GaAs dots have base 15 nm and density 300 and 100 μm^{-2}, respectively. The height of the dots is ~5 nm. The bimodal distribution of dot sizes results in the existence of smaller dots having base, height and density half of the larger ones (see below).

We also measure near-field magneto-photoluminescence spectra of InP QDs embedded in GaInP, which were studied by us previously.[6] The dots have a base ~100 nm, height 10 nm and density 20 μm^{-2}.

RESULTS AND DISCUSSION.

Near-field spectra of individual QDs.

In Fig. 2 we present the near-field photoluminescence (NPL) spectra of InAs/AlAs QDs; insert shows the spectra of InAs/GaAs QDs. The spectra consist of a number of sharp lines (up to ten) corresponding to ground state emission of single QDs. The halfwidth of the sharp lines is equal to the spectral resolution of our setup (0.2-0.4 meV).

The lines are observed at energies 1.50-1.7 eV for InAs/AlAs and 1.3-1.45 eV for InAs/GaAs

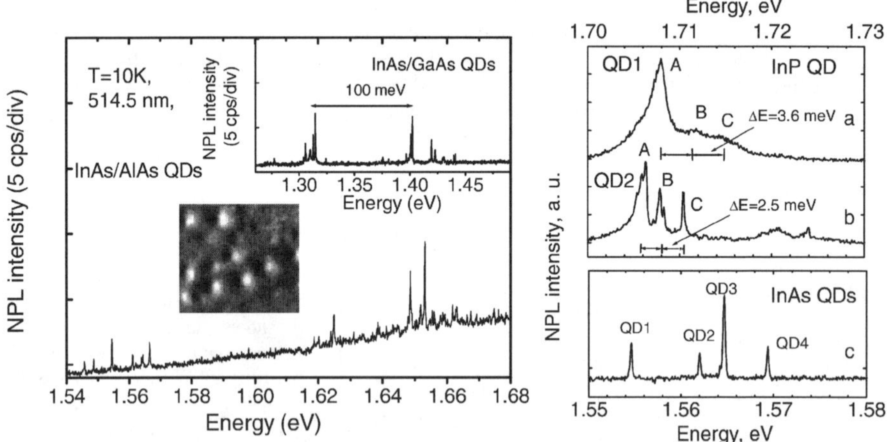

Fig.2. NPL spectra of InAs/AlAs and InAs/GaAs (upper right insert) QDs. Insert at the center shows TEM image of InAs/AlAs QDs.

Fig.3. Comparison of NPL spectra of individual InP/GaInP (a and b) and InAs/AlAs QDs .

QDs. We attribute this difference in the emission energy of InAs/AlAs and InAs/GaAs QDs to the difference of the band-gap energy of the matrix material. We can also see that for each sample the lines are grouped into two distinct energy ranges separated by 100 meV (1.55 and 1.64 eV for InAs/AlAs and 1.31 and 1.42 eV for InAs/GaAs QDs), which reflects the bimodal distribution of dot sizes (see transmission electron microscopy image in Fig. 2).

The spectra of single InP QDs consist of an emission manifold having energy at 1.7-1.79 eV.[6] Inspecting the spectra of ~40 QDs we found that in most cases the emission manifold of the single InP QD has three components, denoted as A, B, C, having halfwidth (γ) of 3-5 meV and energy splitting of 4-15 meV (Fig. 3a QD1). For some dots we found a fine structure of the components consisting of a several ultranarrow lines with γ=0.2 meV (Fig. 3b QD2). For these dots, the splitting between the A, B and C components reduces to values as small as 1-3 meV. Several meV halfwidth and multiple peak structure with splitting up to 20 meV is the intrinsic feature of the low-temperature emission of single InP/GaInP QDs independent of their sizes,[7] which is quite different compared with InAs QDs, for which individual dots have a single narrow line at low excitation power (Fig. 3c). In the ref. [8] it was shown that the multiple peak structure arises in InP QDs due to residual doping (10^{16} cm^{-1}) of the GaInP matrix and accumulation of up to 20 electrons in the QD at low temperature.

Bimodal size distribution of InAs QDs.

In Fig. 4 we present the observation of the Zeeman splitting and diamagnetic shift for InAs/GaAs QDs in a magnetic field. Each line observed in the spectra splits into σ+ and σ- circularly polarized components in the magnetic field, and thus corresponds to a different QD.

We found that the diamagnetic coefficient (β) has different values for two sets of emission

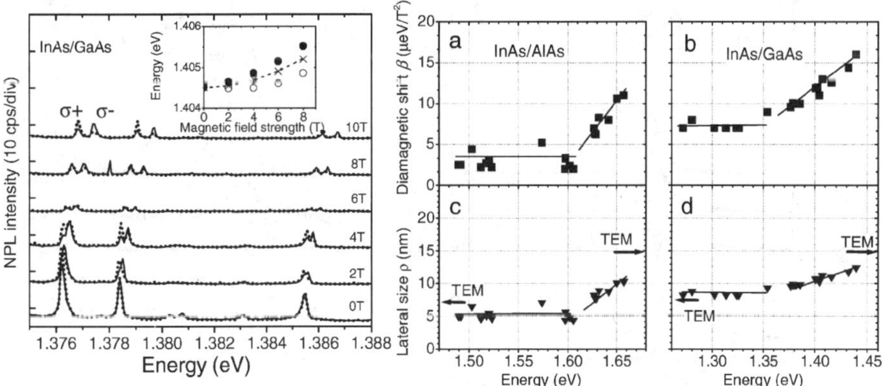

Fig. 4. σ- (solid) and σ+ (dotted) circularly polarized components of the magneto-NPL spectra of InAs/GaAs QDs. Insert shows dependence of the energy of σ- (solid circles) and σ+ (open circles) components and their average value (cross - experiment, dashed curve - quadratic approximation) versus magnetic field strength for one particular QD.

Fig. 5. Diamagnetic shift (a and b) and lateral size (c and d) of InAs/AlAs (a, c) and InAs/GaAs (b, d) QDs versus emission energy. Horizontal arrows in c and d show dot base measured from TEM data. The constant dot sizes in c and d are for the small dots of the bimodal distribution whereas the dots that vary in size are the larger ones in the distribution.

energy ranges of InAs/AlAs and InAs/GaAs QDs (see Fig. 5a-b). For low emission energy β has nearly constant values 4 and 7 μeV/T² for InAs/AlAs and InAs/GaAs QDs, respectively. For higher emission energies β gradually increases, reaching values 10 μeV/T² for InAs/AlAs QD and 15 μeV/T² for InAs/GaAs QDs. For InAs/AlAs QDs the increasing diamagnetic coefficient, which corresponds to an increase of the dot size, correlates well with decreasing Zeeman splitting (from ~1 to 0.5 meV at 10 T). For InAs/GaAs QDs the Zeeman splitting (0.2-0.7 meV at 10 T) depends only weakly on the energy. Using these measured values of the diamagnetic coefficient and the value of exciton effective mass 0.053,[9] we calculated the lateral size of InAs/GaAs and InAAs/AlAs QDs as was described in ref.[3] The calculated dot sizes are plotted in Fig 5c and d. They agree well with transmission microscopy measurements.[5]

Our observation of increasing QD size with increasing emission energy for the larger dots in the bimodal distribution is quite unexpected[10] and implies lower In content for larger dots, which compensates the usual quantum confinement effects. According to eight-band k·p calculations,[11] a change of InAs QD base dimension from 10 to 20 nm leads to a decrease of the confinement energy of 200 meV. Thus smaller dots, having ~100 meV lower emission energy, must have ~300 meV lower bandgap. Taking into account the change of the bandgap of InGaAs by ~100 meV per 10% of In[12] one can estimate the difference in In composition for the extreme sizes of the large dots of the bimodal distribution to be 30%.

Magnetic anomaly in single InP QD.

Fig. 6a and b presents circular polarized NPL spectra of two InP QDs measured at magnetic field B=0, 2, 4 … 10T. Our NPL scanning experiments have shown that bands A and

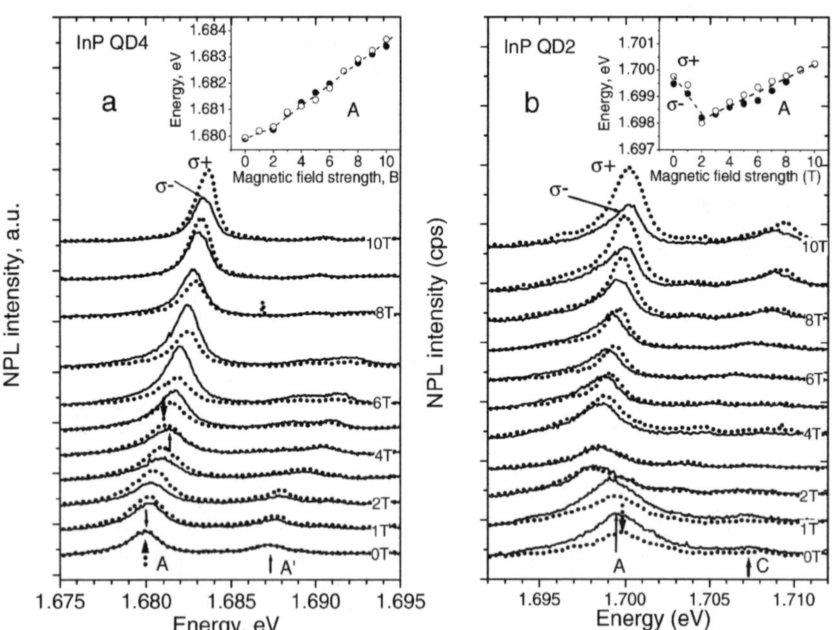

Fig.6 Magneto-NPL circularly polarized components (σ- - solid, σ+ - dotted) of normal (left) and anomaly (right) InP QDs. Vertical arrows indicate position of the emission manifold components (A, C) and neighboring dots (A'). Inserts show position of the energy of the main peak (A) versus magnetic field. Dashed lines are drawn for clarity.

A' in Fig. 6a are related to different QDs located 200 nm from each other whereas the bands A and C in Fig. 6b belong to one QD.

The A band in Fig. 6a (QD4) at zero field is not polarized. With increasing magnetic field it becomes circular polarized. The dominant emission of this band is σ^+-polarized for B=1-3 and 9-10T and σ^--polarized for B=5-7T. Zeeman splitting varies from +0.3 meV for 4T to -0.3 meV for 8T. The band shifts to higher energies with magnetic field. The shift has two linear regions with the slopes 0.15 and 0.43 meV/T for fields 0-2 and 2-10T, respectively. The total high energy shift at 10T is equal to 3.5 meV.

The A band in Fig. 6b (QD2) shows different magnetic field behavior. First, unlike QD4, it has strong circular polarization and Zeeman splitting at zero magnetic field. Both circular polarization and Zeeman splitting disappear at 3T. Thus the spectrum of QD2 at 3T becomes similar to the spectrum of QD4 at zero magnetic field (Fig.6a). Second, the band has a strong *low energy shift* (slope~0.8 meV/T) for magnetic fields 0-2T. For higher fields it has a high energy shift with a slope of 2.5 meV/T.

Our observations of the circular polarization and negative magnetic field shift are quite unusual and give strong evidence that the InP QD in Fig.6b has a strong internal magnetic field.

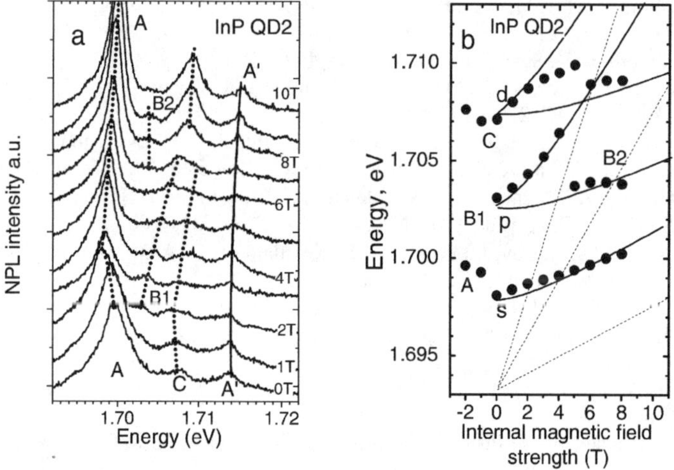

Fig. 7. Unpolarized magneto NPL spectra (a) and energy (b) of emission components of anomaly InP QD (see Fig.6b) versus "internal" magnetic field (solid circles) together with Darwn-Fock energy levels (solid curves). Dashed lines show energy of Landau levels.

The value of this field (B_0) can be estimated by the value corresponding to the compensation of the negative shift and the circular polarization (2-3T in Fig. 6b). At the present time we have no explanation of the observed effect and we did not find similar experimental observations in the literature. Our measurements of the magnetization using a SQUID magnetometer did not reveal any magnetic impurities in our samples, thus indicating that the origin of the built-in magnetic field must be some sort of electron-electron and electron-nuclear interaction in the QD.

As an initial step to explain our observation we plot in Fig.7b the energy of the manifold components of the emission of QD2 (see Fig. 7a) versus "internal" (B_i=B-B_0) magnetic field together with energy levels of Darwin-Fock Hamiltonian.[13] We obtained a good agreement between experimental and measured energy shifts in Fig. 7b and revealed a transition of electrons from the second to the first Landau level at B_i=4T.

CONCLUSION

In conclusion we used near-field optical scanning microscopy to study magneto-photoluminescence spectra of single InAs/AlAs, InAs/GaAs and InP/GaInP quantum dots. Using measurements of diamagnetic shift we found that increasing of the size of InAs QDs is accompanied with intermixing of the dot and matrix material. We observed strong circular polarization of the emission of InP QDs at zero magnetic fields indicating build-up fields as high as 2T.

AKNOWLEDGEMENT

This work was supported by the W. M. Keck Foundation.

REFERENCES.

[1] J. Levy, V. Nikitin, J. M. Kikkawa, D. D. Awschalom, N. Samarth, J. Appl. Phys. **79**, 6095 (1996).

[2] Y. Toda, S. Shinomori, K. Suzuki and Y. Arakawa, Phys. Rev. B, **58**, R10 147 (1998).

[3] A. M. Mintairov, T. H. Kosel, J. L. Merz, P. A. Blagnov, A. S. Vlasov, V. M. Ustinov, and R. E. Cook, Phys. Rev. Lett. **87**, 277401 (2001).

[4] G. Ortner, M. Bayer, A. Larionov, V. B. Timofeev, A. Forchel, Y. B. Lyanda-Geller, T. L. Reinecke, P. Hawrylak, S. Fafard and Z. Wasilewski, Phys. Rev. Lett. **90**, 086404 (2003).

[5] A. M. Mintairov, P. A. Blagnov, O. V. Kovalenkov, C. Li, J. L. Merz, S.Oktyabrsky, V. Tokranov, A. S. Vlasov, D. A. Vinokurov, Mat. Res. Soc. Proc. **722**, (2002).

[6] A. M. Mintairov, A. S. Vlasov, J. L. Merz, O. V. Kovalenkov, D. A. Vinokurov, MRS Symposium Proceedings, **618**, p. 207-212, (2000).

[7] D. Hessman, P. Castrillo, M.-E. Pistol, C. Pryor, and L. Samuelson, Appl. Phys. Lett. **69**, 749 (1996); G. Guttroff, M. Bayer, A. Forchel, D. V. Kazantsev, M. K. Zundel and K. Erbel, Phys. Stat. Sol. (a) **164,** 291 (1997);
P. G. Blome, M. Wenderoth, M.Hubner, R. G. Ulbrich, J. Porsche, and F. Scholz, Phys. Rev. B, **61**, 8382 (2000).

[8] D. Hessman, J. Persson, M.-E. Pistol, C. Pryor, and L. Samuelson, Phys. Rev. B **64**, 233308 (2001).

[9] M. Bayer, S. N. Walck, T. L. Reinecke, A. Forchel, Phys. Rev. B, **57**, 6584 (1998); Paskov : P. P. Paskov, P. O. Holtz, B. Monemar, J. M. Garcia, W. V. Schoenfeld, and P. M. Petroff, Phys. Rev. B, **62**, 7344 (2000); A. Polimeni, S. T. Stoddart, M. Henini, L. Eaves, P. C. Main, K. Uchida, R. K. Hayden, N. Miura, Physica E **2**, 662 (1998).

[10] S. Anders, C. S. Kim, B. Klein, Mark W. Keller, R. P. Mirin, A. G. Norman, Phys. Rev. B, **66** 125309 (2002).

[11] O. Stier, M. Grundmann, ,and D. Bimberg, Phys. Rev. B, **59**, 5688 (1999).

[12] Properties of Lattice-Matched and Strained Indium Gallium Arsenide, edited by P. Bhattacharya (INSPEC, London, 1993).

[13] L. Jacak, P. Hawrylak, A. Wojs "*Quantum Dots*", Springer, Berlin, 1998.

Self-Assembled QDs, Nanoparticles and Nanowires

Mat. Res. Soc. Symp. Proc. Vol. 789 © 2004 Materials Research Society N9.5

Dark exciton signatures in time-resolved photoluminescence of single quantum dots

Jason M. Smith[1], Paul A. Dalgarno[1], Richard J. Warburton[1], Brian D. Gerardot[2] and Pierre M. Petroff[2]
[1]School of Engineering and Physical Sciences, Heriot-Watt University, Edinburgh EH14 4AS, U.K.
[2] Materials Department and QUEST, University of California, Santa Barbara, California 93106, U.S.A.

ABSTRACT

Time-resolved photoluminescence of single charge tuneable quantum dots allows us to probe the differences in recombination dynamics between neutral and negatively charged excitons. We find that the luminescence decay from a neutral exciton contains a second lifetime component of several nanoseconds that is not present in the luminescence from singly or doubly charged excitons. We attribute the slowly decaying component to excitation cycles in which the initial exciton formed in the dot is dark, with angular momentum M = 2, and which subsequently scatters into the bright state with M = 1.

The nature of the scattering mechanism is revealed by the dependence of the lifetime on the electrical bias applied across the charge-tuneable device. That the lifetime changes by an order of magnitude within a short bias range implies that the dark-to-bright transmutation does not occur through a simple spin flip. Rather it appears to come about by the dot briefly entering a higher energy charging state which allows exchange of the existing electron with another from the n-type contact region. We model the lifetimes and relative intensities of the two decay components using a simple rate equation analysis.

INTRODUCTION

Dark excitons are electron-hole complexes which do not couple to the optical field and are thus impossible to probe directly using absorption or luminescence techniques. As they do not recombine radiatively, dark excitons can be much more long-lived than bright excitons, offering potential for use in memory storage or quantum information processing applications. Studies of dark excitons generally rely upon some form of symmetry breaking to allow access with an optical interaction.[1,2]

The lowest energy excited state of most semiconductor quantum dots is dark, with the spin of the excited electron aligned with the hole resulting in a total angular momentum of M=2. In InGaAs Stranski Krastanow quantum dots, which are the subject of this study, this state lies approximately 0.5 meV below the lowest bright state, with M=1, where the splitting is a result of the electron-hole exchange interaction. In an isolated dot, little communication is expected between the dark and bright states, as such a process requires a spin flip of one of the carriers that is not facilitated by acoustic phonon scattering.[3]

In this paper we report time-resolved photoluminescence (TRPL) measurements of single Stranski Krastanow quantum dots embedded in charge tuneable heterostructures.[4] We compare TRPL from the neutral exciton (e:h, labelled X^0) with that from the singly charged exciton (2e:h, labelled X^{1-}) and find evidence for an electrically controlled transition between the dark and bright states that involves a transition via a higher energy charging state and thereby an exchange of electrons with the contact.

PHYSICS OF CHARGE TUNING IN QUANTUM DOTS

First let us review the general form of the single dot PL spectrum vs bias, shown in figure 1. The main PL lines are labelled according to the initial state of the transition, whereby X^{n-} corresponds to a dot occupancy of a single hole and n+1 electrons.

We assume that the electron charging barrier between the dots and the n+ contact is transparent enough that relaxation to the lowest energy charging state occurs on a time scale faster than the radiative lifetime; ie. that the Fermi energy in the contact region extends into the dots themselves. Under this assumption we can model the bias dependent switching by considering the excitation energy of each charging state as a function of bias. In addition to the labelling convention used in figure 1, we denote a single hole with no electrons as h, and write

$$\begin{aligned} E_h &= E_0 + \Delta E_{CB} - \frac{(V_0 - V_g)}{7} \\ E_{X^0} &= E_0 - E_C^{eh} \pm E_{exch}^{eh} \\ E_{X^{1-}} &= E_0 - 2E_C^{eh} + E_C^{ee} + E_{exch}^{ee} - \Delta E_{CB} + \frac{(V_0 - V_g)}{7} \end{aligned} \tag{1}$$

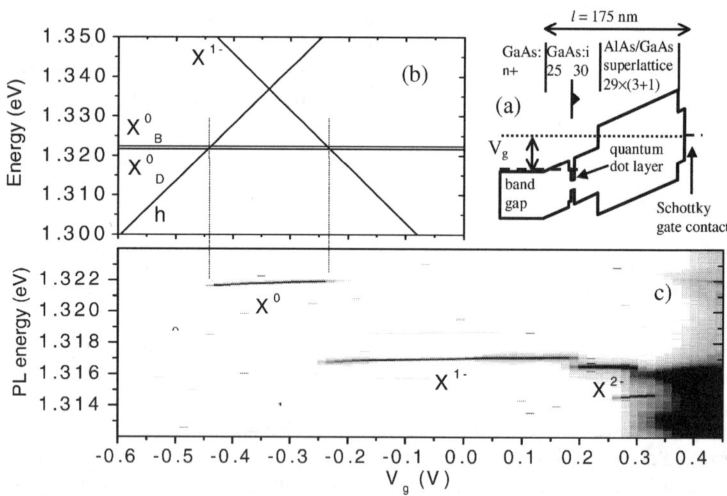

Figure 1. (a) Schematic band profile of the charge tuneable heterostructure; (b) Energies of first three charging states (equation 1) as functions of gate voltage; (c) Photoluminescence spectrum vs gate voltage showing the X^0, X^{1-}, and X^{2-} lines.

where E_0 is the energetic separation of the conduction and valence band 1s single particle levels, ΔE_{CB} is the flat-band electron ionisation energy of the dot, V_0 is the Schottky barrier height, and subscripts C and $exch$ refer to Coulomb and exchange terms respectively. The factor seven in the voltage-dependent term is a result of the lever arm effect given the layer thicknesses in the device.

Figure 1b shows the excitation energies from equation set (1) plotted against gate voltage, with parameters selected to fit the photoluminescence data in Figure 1c. The gate voltages at which switching between charging states occur are clearly identified by the crossing points in figure 1b. Note that the X^0 is split into its bright and dark components by twice the electron-hole exchange energy. Transitions directly between the two X^0 levels require a spin-flip and are therefore expected to occur on time scales much longer than the radiative lifetime of the bright state.[3] The result is that despite the dark X^0 state being of a lower energy than the bright state, a bright X^0 survives long enough upon capture to recombine before relaxing to the dark state, and we can observe strong X^0 luminescence even if the exchange splitting is much greater than the thermal energy kT.

The dynamics of electronic relaxation in these devices are such that we expect an equal probability for the capture into the dot of a bright or dark X^0 exciton. After photogeneration of an exciton in the wetting layer, the light electron effective mass leads to its rapid tunneling through the low potential barrier and into the n-type contact. The hole, with heavier effective mass, is far more likely to remain in the wetting layer long enough that it is captured into a dot. The capture of a hole by a dot then instigates capture of an electron, which may have either spin, from the contact region. The entire process from photogeneration to formation of the X^0 occurs on a time scale of ~100 ps.

EXPERIMENTAL

Photoluminescence (PL) was carried out using a confocal microscope with a diffraction limited spatial resolution of 600 nm at the emission wavelength of around 950 nm. Excitation is with a 830 nm pulsed diode laser. The luminescence is spectrally dispersed using a grating monochromator, and PL spectra recorded using a liquid nitrogen-cooled Si charge-coupled device (CCD) array. The spectral resolution of the complete PL setup is 0.1 meV. Time-resolved PL (TRPL) is carried out by time-correlated single photon counting (TCSPC), using a silicon avalanche photodiode at the second spectrometer output. Lifetimes are fitted to the recorded TCSPC data using iterative reconvolution of the instrumental response function. By this method, temporal resolution of around 100 ps is achievable for a data set with a high signal-to-noise ratio.

Measurements were made on several different dots from a single charge-tuneable hetero-structure, at a temperature of 4.2K. In all of the measurements described here, the excitation intensity was kept low enough that biexciton features are absent from the PL data, and no saturation effects were observed in the TRPL data. This corresponds to the weak excitation limit, of <1 photogenerated hole per dot, and enables the TRPL data to be fitted with simple exponential decay functions.

Raw TRPL data recorded at six gate voltage and spectral positions on Fig. 1c are shown in Fig. 2. At all three positions on the X^{1-} line, a single exponential decay with a lifetime around 0.6 ns is observed, characteristic of the radiative recombination of a bright exciton in these dots.[5] The X^{2-} luminescence, not shown, also reveals a mono-exponential decay of similar lifetime.

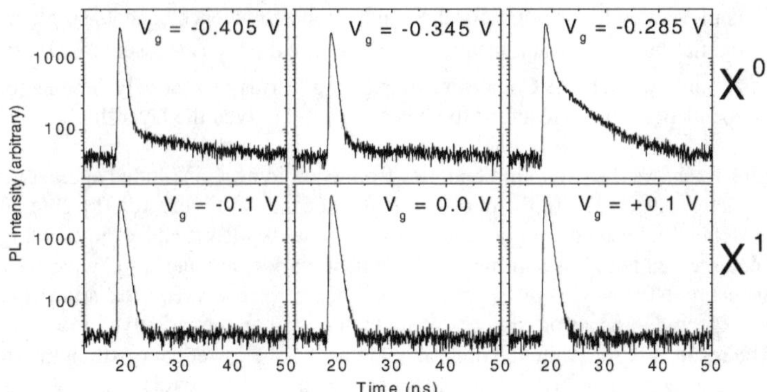

Figure 2. Time-resolved photoluminescence data taken from X^0 and X^{1-} at different gate voltages (see Fig. 1c). The three data sets from the X^0 line have a gate voltage dependent second lifetime component, whereas those from X^{1-} exhibit a fast, monotonic decay.

TRPL from the X^0 line shows, in addition to a component with a 0.6 ns decay, a second decay component which varies considerably in lifetime and strength across the voltage range.

Closer inspection of the dependence of the two decay lifetimes of X^0 on gate voltage reveals that whilst the shorter lifetime remains constant over most of the voltage range, the longer lifetime is longest (~15 ns) near the centre of the voltage range and decreases exponentially to converge with the shorter lifetime at the edges (Fig. 3a). The relative time-integrated intensity of each decay component is shown in Fig. 3b, which reveals that the longer lifetime component dominates at either end of the bias range whilst the shorter lifetime component dominates near the middle of the bias range.

In other single dots from the same sample we observe broadly similar behaviour, but with variation in two respects. Some show markedly different gradients in the portion of the plot where the second lifetime depends exponentially on gate voltage. In others the second lifetime component continues to lengthen and its intensity continues to decrease towards the middle of the voltage range until it is too weak to measure. We will discuss these variations in the context of the model developed in the following section.

A SIMPLE RATE EQUATION MODEL

Comparison between the lifetimes in Fig. 3 and the energy diagram in Fig. 1 suggests that the explanation for the strong bias dependence in the secondary lifetime is the proximity of higher energy states at either end of the X^0 bias range. Temporary occupation of either of these states can lead straightforwardly to the exchange of the electron in the dot with another electron from the contact region. For instance, at gate voltages slightly higher than that of the X^0 switch-on, the single hole charging state h is only a small energy above X^0. Little energy is required to remove the electron from the dot into the contact region of the device, after which it will be energetically favourable for another electron to tunnel in to restore the X^0 charging state. At the other end of the bias range, it is the X^{1-} charging state that is close to the X^0, and a small energy input will allow a second electron to tunnel into the dot briefly before one of the two is ejected.

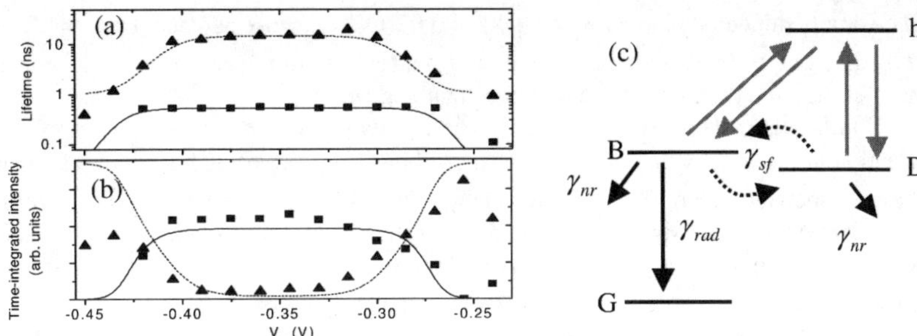

Figure 3. Gate voltage dependence of (a) the lifetimes and (b) the relative integrated intensity of each of the TRPL decay components comparing the recorded data (scatter) and the rate equation model (line) The fast component is indicated by squares / solid line, while the slow component is indicated by triangles / dashed line. (c) Four-level schematic diagram from which the rate equation model is constructed (see text).

Both of these processes provide a mechanism for an effective spin flip, as the electron present in the dot after the process has an equal probability of having either spin, regardless of the spin of the original electron.

A simple dynamic model of the system is shown in Fig. 3c. Levels B and D represent the bright and dark exciton states, while G represents the empty dot ground state. The arrows in gray represent interaction between the bright and dark exciton states and the higher state, here labelled h. The dotted arrows do not represent direct communication between B and D, but can be used in place of the gray arrows to simplify the rate equations, since relaxation from level h is fast enough that it remains empty. The resulting switching between the bright and dark exciton states is taken to occur at a rate γ_{sf} in either direction. Radiative recombination from the bright state occurs at a rate γ_{rad}, and a non-radiative loss mechanism with rate γ_{nr} is included for completeness.

The appropriate rate equations are

$$\frac{dn_B}{dt} = -n_B\left(\gamma_{sf} + \gamma_{nr} + \gamma_{rad}\right) + n_D\gamma_{sf}$$
$$\frac{dn_D}{dt} = -n_D\left(\gamma_{sf} + \gamma_{nr}\right) + n_B\gamma_{sf} \qquad (2)$$

These can be solved straightforwardly, revealing that n_B decays bi-exponentially with rates $\gamma_{1,2} = P \pm Q$ where $2P = \gamma_{rad} + 2\gamma_{nr} + 2\gamma_{sf}$ and $2Q = \sqrt{\gamma_{rad}^2 + 4\gamma_{sf}^2}$. The relative time-integrated intensities of the two PL decay components are then given by $I_1 = (\gamma_2 - \gamma_{rad} - \gamma_{nr})/Q\gamma_1$ and $I_2 = (-\gamma_1 + \gamma_{rad} + \gamma_{nr})/Q\gamma_2$.

To complete the model, we note that the lifetime appears to vary exponentially with voltage in the sloped portions of Fig. 3, suggesting that γ_{sf} varies exponentially with the energy ΔE of the higher energy state above X^0. We thus use a trial expression $\gamma_{sf} = \gamma_0 \exp(-\Delta E / E_{char})$ where γ_0 and E_{char} are fitting parameters. γ_{rad} and γ_{nr} are assumed to be independent of gate voltage.

Choosing parameters $\gamma_{rad} = 1.8 \text{ ns}^{-1}$, $\gamma_{nr} = 0.07 \text{ ns}^{-1}$, $\gamma_0 = 10 \text{ ns}^{-1}$ and $E_{char} = 1.2$ meV results in the line graphs in Fig 3, reproducing the main features of the gate voltage dependence of both the lifetimes and the intensities to an excellent degree.

Similar data sets have been recorded for different quantum dots, for which we observe variation in the fitted value of E_{char} between about 1 and 3 meV. This comes as something of a surprise since the rate of change of the charging energies for h and X^{1-} with gate voltage is dependent only on the sign of the charge of the state and on the device lever-arm factor, which does not vary significantly from dot to dot. Measurements on a single dot at slightly elevated temperatures (T<10K) reveal an apparent exponential increase in γ_0 at a rate of $d\log(\gamma_0)/dT \approx 0.5 \text{ K}^{-1}$, but, significantly, no apparent change in E_{char}.

The physical mechanism for excitation into the higher charging state is most likely the absorption of an acoustic phonon. This process will be stronger than that in an isolated quantum dot since momentum in the plane of the device can be conserved by the participation of the electron ejected into or captured from the contact region. Additionally, a portion of the energy required to excite the dot to charging state h (X^{1-}) can be obtained by the electron being ejected into an empty state just below (captured from an occupied state just above) the Fermi energy. We are currently performing the calculations to reveal whether these processes can fully explain the observed behaviour.

A final point of interest from Fig. 3 is the lack of voltage dependence of the longer lifetime in the centre of the bias range. Combined with the dramatic weakening of this component relative to the faster decay component, this suggests that we are measuring the non-radiative lifetime of the dark exciton. Some dots do not show such a flat region in the second lifetime, indicating that the non-radiative lifetime varies from dot to dot, although the longer lifetime component becomes too weak to measure when the lifetime climbs much above 20 ns due to the current configuration and sensitivity of our measurements. To our knowledge these are the first measurements of the non-radiative lifetime of a dark exciton in a quantum dot - a quantity that could have important implications for the use of dark excitons in practical applications.

ACKNOWLEDGEMENTS

This work was funded by the UK Engineering and Physical Sciences Research Council. J M S is supported by a Personal Fellowship from the Scottish Executive and the Royal Society of Edinburgh.

REFERENCES

1. M. Bayer, O. Stern, A. Kuther and A. Forchel, Phys. Rev. B **61**, 7273 (2000).
2. M. Bayer, G. Ortner, O. Stern et al, Phys. Rev. B **65**, 195315 (2002).
3. A. V. Khaetskii, and Y. V. Nazarov, Phys. Rev. B. **61**, 12639 (2000).
4. R. J. Warburton, C. Schäflein, D. Haft, F. Bickel, A. Lorke, K. Karrai, J. M. Garcia, W. Schoenfeld, and P. M. Petroff, Nature **405**, 926 (2000).
5. G. Wang, S. Fafard, D. Leonard, J. E. Bowers, J. L. Merz, and P. M. Petroff, Appl. Phys. Lett. **64**, 2815 (1994).

Mat. Res. Soc. Symp. Proc. Vol. 789 © 2004 Materials Research Society N9.8

Magnetic Characterization of $CoFe_2O_4$ Nanoparticles

G. Lawes[1], B. Naughton[2], D.R. Clark[2], A.P. Ramirez[3], R. Seshadri[2]
[1]Los Alamos National Laboratory, Los Alamos, NM 87544
[2]Materials Department, UC Santa Barbara, Santa Barbara, CA 93106
[3]Bell Laboratories, Lucent Technologies, 600 Mountain Ave, Murray Hill, NJ

We have synthesized $CoFe_2O_4$ nanoparticles with length scales ranging from 3.5 nm to 14.2 nm. We have characterized the magnetic properties of these samples using both DC and AC magnetization, and find some slightly anomalous behavior in two of the samples. We tentatively attribute these features to interactions between the magnetic nanoparticles.

There is a great deal of interest in understanding the physical basis for the magnetic properties of nanoparticles in order to facilitate their incorporation into a wide range of commercial applications. By studying the magnetic characteristics of $CoFe_2O_4$ nanoparticles using bulk measurement techniques, we are able to probe the properties of both the individual nanoparticles and interactions in these systems. In this report, we discuss our magnetic characterization of a series of $CoFe_2O_4$ nanoparticles grown using an aqueous co-precipitation technique. In addition to DC magnetization at fixed fields and temperatures, we also investigated the magnetic properties using AC susceptibility measurements. The long term goal of this research is to understand interparticle interactions in magnetic nanoparticles.

Bulk $CoFe_2O_4$ forms an inverse spinel structure, with the Co^{2+} located principally on the octahedral sites, while Fe^{3+} occupy both the octahedral and tetrahedral sites. The competition between these magnetic ions produces ferrimagnetic ordering below 798K[1] with a low temperature saturation magnetization of 93 emu/g. The magnetic properties of $CoFe_2O_4$ nanoparticles are sensitive to finite size effects. Particles smaller than 40 nm are single domain with significant magnetic anisotropy[2] which leads to superparamagnetic behaviour. Furthermore, the spins on the surface of the nanoparticles are subject to a different magnetic environment than the bulk spins[3].

The $CoFe_2O_4$ nanoparticles were synthesized using an aqueous co-precipitation technique[4]. The samples were prepared by combining cobalt and iron nitrates in water with the addition of NaOH in the presence of polyethelyne glycol (PEG-6000) as a capping agent. After centrifuging the particles, they were washed with water and dried in a vacuum oven at 60C. This synthesis technique has the advantage that relatively large quantities (>2g) can be prepared in a single batch. The $CoFe_2O_4$ nanoparticles were characterized via X-ray diffraction (XRD) using a commercial diffractometer (Philips X'PERT MPD). A representative spectrum is displayed in Figure 1; the nominal size of the nanoparticles is 14.2 nm. The XRD pattern shows several Bragg peaks which are indexed by a cubic spinel lattice. The sample is composed primarily of single crystal nanoparticles where the widths of the peaks are determined mainly by the finite size of the particles together with instrumental broadening. We used Debye-Scherrer formalism to estimate the size of the nanoparticles, and obtain mean nanoparticle sizes of 3.5, 5.6, 8.7, and 14.2 nm for the samples investigated in this study. TEM investigations of the particles show sizes consistent with the XRD patterns, and also suggest that there is negligible agglomeration in this system.

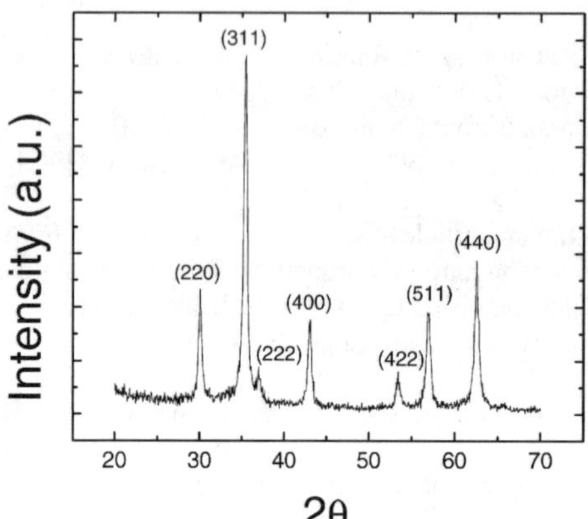

Figure 1. X-ray diffraction pattern for 14.2 nm $CoFe_2O_4$ nanoparticles.

We characterized the magnetic properties of the $CoFe_2O_4$ nanoparticles in a commercial SQUID magnetometer, using approximately 50 mg of sample for each measurement. In all cases, the mass was corrected to reflect the contributions of only the $CoFe_2O_4$ component by subtracting off the contribution from the capping agent, where the ratio of cobalt ferrite to PEG was determined by thermogravimetric analysis. The susceptibility of PEG used in the capping layer was measured separately, and contributed less than 0.02% of the total susceptibility at T=5K. Figure 2 plots the field-cooled (FC) magnetization as a function of temperature (for nominal particle sizes of 3.5, 5.6, 8.7, and 14.2 nm) and the zero-field cooled (ZFC) magnetization (for 3.5 nm) in an applied field of 1 kOe. All samples showed a divergence between the ZFC and FC magnetization; we plot only the data for the 3.5 nm nanoparticles for clarity.

The low temperature zero field cooled measurements show only a small moment, while the field cooled measurements show a much larger magnetization. This effect arises from superparamagnetic blocking associated with magnetocrystalline anisotropy in the single domain nanoparticles [2]. This blocking temperature can be expressed as $T_B = KV/k_B \log(\tau/\tau_0)$, where K is the magnetocrystalline anisotropy energy density, V is the particle volume, and τ and τ_0 are characteristic scales for the measurement and microsopic relaxation times respectively. For small $CoFe_2O_4$ nanoparticles (less than ~10 nm or so), this blocking temperature is typically below 300K, but there is significant variation [5-7]. The fact that the peak in the ZFC magnetization for our smallest nanoparticles (nominally 3.5 nm) is larger than 350K suggests that the magnetocrystalline anisotropy K is larger than expected for these samples, or that the samples are larger than what we find from

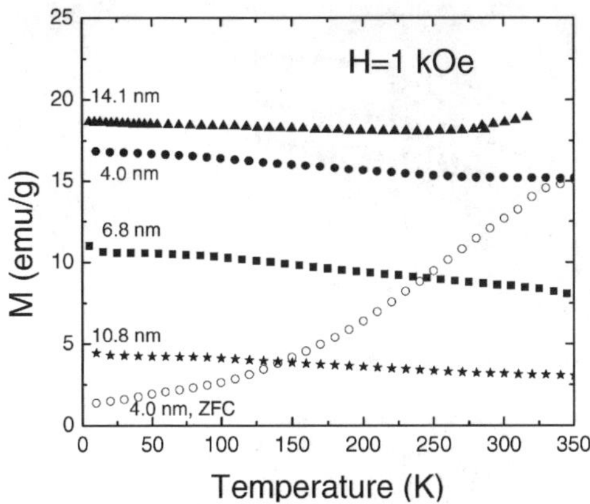

Figure 2. FC (solid symbols) and ZFC (open symbols) DC magnetization curves for $CoFe_2O_4$ nanoparticles, measured at H= 1 kOe.

XRD. Alternatively, there could be interparticle interactions present which modify Eq. (1).

Figure 2 also shows that there are significant differences between the magnitudes of the magnetization for the different nominal particle sizes. This is more clearly demonstrated in the hysteresis loops plotted at T=5K (and at T=300K for the 3.5 nm nanoparticles) in Figure 3. The saturation magnetization for most nanoparticles is suppressed below the bulk value of 93 emu/g, but is expected to be a monotonic function of particle size (neglecting any possible ferromagnetic contribution from the surface spins). The differences in saturation magnetization in Figure 3 show very strong and non-monotonic dependence on nominal particle size. There are several possible explanations for this behavior. It is possible that the size of these nanoparticles estimated by XRD does not accurately represent the magnetic size of the particles. This is unlikely, because the elevated values for the superparamagnetic blocking temperatures point to larger rather than smaller sizes for the nanoparticles. Alternatively, there could be additional interparticle interactions which can modify the magnetic behavior

We can analyze the hysteresis loops in Figure 3 to extract additional information about the magnetic properties of $CoFe_2O_4$ nanoparticles. One important parameter for characterizing the magnetic properties of nanoparticles is the squareness ratio of the hysteresis loop, defined as the ratio of the remanent magnetization to the saturation magnetization. For the 14.2 nm sample and 3.5 nm samples, the squareness ratios are M_r/M_s=0.66 and M_r/M_s =0.54 respectively, which are intermediate between the expected values for axial anisotropy (M_r/M_s =0.5) and cubic anisotropy (M_r/M_s =0.83). The values

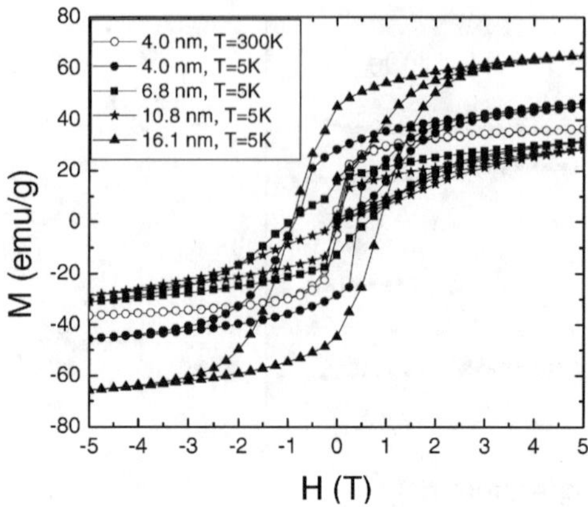

Figure 3. Hysteresis loops at T=5 K (solid symbols) and T=300 K (open symbols) for $CoFe_2O_4$ nanoparticles.

of M_r/M_s for the other two samples are significantly smaller (0.35 for the 8.7 nm nanoparticles, and 0.37 for the 5.6 nm particles). Similar values for M_r/M_s have been observed in 5 nm $CoFe_2O_4$ nanoparticles at T=5K [7] where the small value was attributed to interparticle interactions.

The final set of magnetic measurements we used to characterize the $CoFe_2O_4$ nanoparticles were AC susceptibility measurements. This technique measures the differential magnetization (dM/dH) at a fixed frequency and is sensitive only to the field induced changed in magnetization, and not the total magnetic moment like DC measurements. Since AC measurements probe the spin dynamics of the system (on timescales set by the signal frequency) they can give insight into superparamagnetic blocking. However, for the $CoFe_2O_4$ nanoparticles under investigation, this temperature region is inaccessible in our magnetometer.

Figure 4 plots the AC susceptibility, measured at a frequency of 100 Hz in a field of 100 Oe, for field-cooled samples of $CoFe_2O_4$ nanoparticles. The increase in χ' with increasing temperature is consistent with an increasing fraction of nanoparticles fluctuating freely as the blocking temperature is approached from below. Furthermore, there is no evidence of a peak in χ' below 350K, which is consistent with the separation between the ZFC and FC curves occurring at high temperatures in Figure 2.

One point of interest is the low temperature AC susceptibilities shown in the inset to Figure 4. The 8.7 nm and 5.6 nm nanoparticles clearly show a peak in χ' at approximately 5K, while the susceptibility for the other two samples remains flat down to the lowest temperatures. A peak has also been observed in the susceptibility of Pd

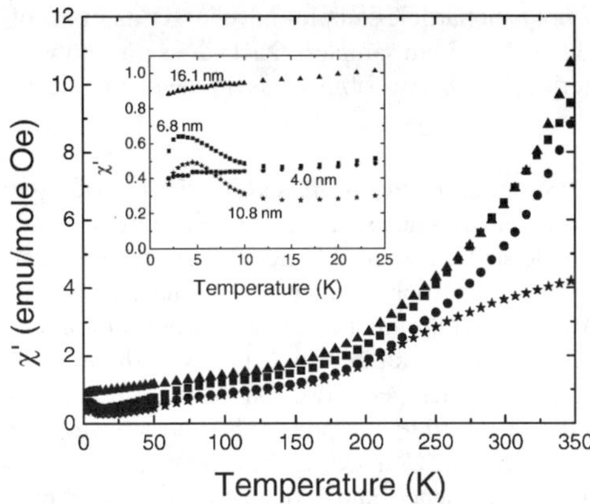

Figure 4. Real part of the AC magnetic susceptibility measured at 1 kHz in a field of 100 Oe. The symbols are labelled in the inset. INSET: Low temperature AC susceptibility.

nanoparticles and was attributed to interactions between spins on neighboring particles [8]. However, the Pd measurements were done on highly monodisperse samples, and the peak in susceptibility was observed at temperatures below 1K, so it is unclear whether the same mechanism is applicable for $CoFe_2O_4$ nanoparticles. In any case, it is extremely suggestive that the 5.6 nm and 8.7 nm samples, which show a reduced saturation moment and small M_r/M_s ratio (see Figure 3) also show a low temperature AC susceptibility peak, while the 3.5 nm and 14.2 nm samples, which have hysteresis loops more reminiscent of independent nanoparticles governed simply by magnetic anisotropy, show no such feature.

We have investigated the magnetic properties of $CoFe_2O_4$ nanoparticles, ranging in size from 3.5 to 14.2 nm, grown using a novel aqueous co-precipitation process which produces large quantities of sample. ZFC/FC DC magnetization measurements suggest that the magnetic properties are dominated by superparamagnetic blocking, although the blocking temperature is slightly higher than measured for other $CoFe_2O_4$ nanoparticles of comparable sizes. Magnetic hysteresis loops at T=5 K are consistent with non-interacting magnetic particles with cubic magnetocrystalline anisotropy for the largest particles, and uniaxial magnetic anisotropy for the smallest particles. On the other hand, the intermediate sized particles show evidence for interparticle interactions; the saturation magnetization and the squareness ratio M_r/M_s are both substantially smaller than expected. Furthermore, these intermediate sized samples show pronounced peaks in the low temperature AC susceptibility, near T=5K, peaks which are absent from the non-interacting 3.5 nm and 14.2 nm nanoparticles. However, it is impossible to exclude the possibility that the 5.6 and 8.7 nm nanoparticle samples have larger amounts of disordered $CoFe_2O_4$ than the 3.5 and 14.2 nm samples. XRD does not show a large background from diffuse scattering, but this possibility has not been completely eliminated.

These results lead to several questions about the origin and importance of interparticle interactions in systems of $CoFe_2O_4$ nanoparticles. 1) Under what circumstances are these interactions present? Particle size is presumably not a relevant parameter, since neither the largest nor the smallest nanoparticles evidenced any magnetic coupling. 2) Is the low temperature peak in AC susceptibility related to interparticle coupling, and if so, how? We have also measured the low temperature heat capacity of these systems to probe the thermodynamic features at low temperatures. These measurements show hints of intriguing behavior, but the data are hard to interpret. 3) Is it possible to tune the interparticle interactions during synthesis? Controlling these interactions (which affects saturation magnetization, magnetic remanence, and perhaps the coercivity) can be important for optimizing the materials properties for commercial applications. Answering these questions will involve careful measurements on well-characterized $CoFe_2O_4$ nanoparticle samples and will ultimately provide important insight into collective behavior in magnetic systems.

This research was supported by NSF-IGERT DGE-9987618 and NSF-GOALI DMR02-03785 (UCSB) and work at LANL was supported by Laboratory Directed Research and Development. We would also like to thank E.N. Butler for helpful discussions about X-ray diffraction.

REFERENCES

1 G.A. Sawatzky, F. van der Woude, and A.H. Morrish, *J. Appl. Phys.* **39**, 1204 (1968).
2 A.J. Rondinone, A.C.S. Samia, Z.J. Zhang, *Appl. Phys. Lett.* **76**, 3624 (2000).
3 M. Grigorova *et al.*, *J. Magnetism and Magnetic Materials* **183**, 163 (1998).
4 T. Sato, K. Haneda, M. Seki, and T. Iijima, Appl. Phys. A50, 13 (1990)
5 A.T. Ngo, P. Bonville, M.P. Pileni, *J. Appl. Phys.* 89, 3370 (2001).
6 N. Hanh, O.K. Quy, N.P. Thuy, L.D. Tung, and L. Spinu, *Physica* **B327**, 382 (2003).
7 V. Blaskov *et al.*, *J. Magnetism and Magnetic Materials* **162**, 331 (1996).
8 Y. Volokitin, J. Sinzig, L.J. de Jongh, G. Schmid, M.N. Vargaftik, and I.I. Moiseev, *Nature* **384**, 621 (1996).

Mat. Res. Soc. Symp. Proc. Vol. 789 © 2004 Materials Research Society N9.9

Atomic Organization In Magnetic Bimetallic Nanoparticles : An Experimental And Theoretical Approach

Marie-Claire Fromen, Samuel Dennler, Marie-José Casanove, Pierre Lecante, Joseph Morillo and Pascale Bayle–Guillemaud[1]
CEMES, CNRS, 29 rue J. Marvig, B.P.4347, 31055 Toulouse cedex, France
[1]CEA, DRFMC, 17 rue des Martyrs, 38054 Grenoble, France

ABSTRACT

Ultrafine bimetallic CoRh nanoparticles synthesized by a soft chemical route with compositions ranging from pure cobalt to pure rhodium are investigated using high-resolution and energy filtering transmission electron microscopy techniques as well as wide angle x-ray scattering. In parallel, they are simulated with the use of an n-body semi-empirical interaction model: quenched molecular dynamics and Monte-Carlo Metropolis simulated annealing are performed on these nanoparticles in order to find their most stable isomers as a function of composition and size. A progressive evolution from an original polytetrahedral structure to the face-centered cubic structure with increasing Rh content is observed in these particles. Strong tendency to Co surface segregation is both experimentally evidenced and confirmed by the simulations.

INTRODUCTION

Nanometer-sized metallic particles are well known to exhibit unique strong size-related magnetic effects [1]. Moreover, alloying with a 3d metal is an effective way to induce spin polarization in 4d metals and to obtain at the same time a large magnetic moment with a high anisotropy. Magnetic 3d-4d bimetallic nanoparticles then present a particular interest. Indeed, a strong enhancement of the saturation magnetization in ultrafine CoRh particles (in the 1.6 to 2 nm range) compared to the predicted value in the bulk alloy was recently demonstrated [2] and later confirmed by density functional theory calculations on much smaller particles [3]. This motivates a systematic study of the atomic organization in these CoRh alloyed particles, about 2 nm large, synthesized by co-decomposition of two organometallic precursors under dihydrogen in the presence of a polymer (the polyvinylpyrrolidone, or PVP polymer).

In the bulk, the CoRh alloy forms a continuous series of solid solutions and crystallizes in the hexagonal close-packed (hcp) structure below 50.5 at % Rh and in the face-centered cubic structure (fcc) for higher Rh contents. However, size reduction is likely to modify the phase diagram of this bimetallic alloy. Indeed, pure cobalt particles synthesized by the same chemical route, were seen to exhibit a very original polytetrahedral structure [4]. Besides, in such fine particles, the very high ratio of surface atoms against volume atoms considerably increases the surface effects. Preferential surface segregation of one of the chemical species, usually neglected in macroscopic samples, can here drastically modify the chemical distribution inside the particles. In order to investigate both the structural and chemical organization in these ultrafine materials, a wide range of experimental and modeling techniques have been implemented. The structural results have been obtained through high-resolution transmission electron microscopy (HRTEM) and wide-angle x-ray scattering (WAXS) techniques. The species distribution was examined by energy filtering transmission electron microscopy (EFTEM). However, chemical maps could be obtained only in bigger CoRh particles (4 to 7 nm large) stabilized by hexadecylamine (HDA), an organic ligand. As element-sensitive techniques reach their limits when investigating such small clusters, an n-body semi-empirical interaction model has been developed and applied to their simulation, in order to elucida their chemical organization.

EXPERIMENTAL DETAILS

TEM specimens for HRTEM and EFTEM investigations were prepared by slow evaporation of droplets of a solution colloid –methanol on carbon supported copper mesh. The HRTEM experiments were performed on a Philips CM 30/ST operated at 300 kV with a point resolution of 0.19 nm. EFTEM was carried out on a Jeol 3010 operated at 300 kV and fitted with a post-column filter (GATAN 200 model). Both high resolution and filtered images were recorded on CCD cameras in order to be processed. In particular, a three window procedure was used for chemical mapping: two filtered images are acquired before the edge, in order to extrapolate the background, and one is acquired after the edge in order to obtain information on the analyzed element. WAXS specimens consisted in small amounts of the fine powder (obtained after drying) sealed in glass capillaries. Measurements of the x-ray intensity scattered by the samples irradiated with graphite-monochromatized molybdenum K_α ($\lambda = 0.071069$ nm) radiation were performed using a dedicated two-axis diffractometer. Time for data collection was typically 20 hours at room temperature in the range $0° < \theta < 65°$. Data were reduced following the procedure already described in [4] in order to obtain the experimental radial distribution function (RDF). Analysis of this function provides the metal-metal bond length (position of the first peak), d_{mm}, and the order extent inside the particles (position of the last significant peak, also called coherence length hereafter).

RESULTS AND DISCUSSION

Three different compositions, namely $Co_{0.75}Rh_{0.25}$, $Co_{0.5}Rh_{0.5}$ and $Co_{0.25}Rh_{75}$, respectively named in the following, Co3Rh1, Co1Rh1 and Co1Rh3, have been investigated through both experimental and classical atomic scale simulation approaches.

<u>Experimental results</u>

TEM observations show that all the bimetallic particles synthesized in PVP present narrow size distributions (mean diameter close to 2 nm) and a very good dispersion in the polymer matrix. The fcc structure was identified by HRTEM in Co1Rh3 particles while no periodic lattice fringes were observed at low Rh content. The RDFs of the different specimens were processed and compared with the simulated RDFs of different models. The size of the models was chosen according to both the coherence length measured on the experimental RDF and the mean diameter measured by TEM. If some structural disorder occurs at the particle surface, the coherence length is slightly smaller than the mean diameter, otherwise they are quite comparable. A second feature to be adjusted is the mean metal-metal bond length, d_{mm}, which must be the same in the model and in the experimental RDF. Figure 1 reports the comparisons between the experimental RDFs in Co3Rh1 and Co1Rh3 particles and the RDF of

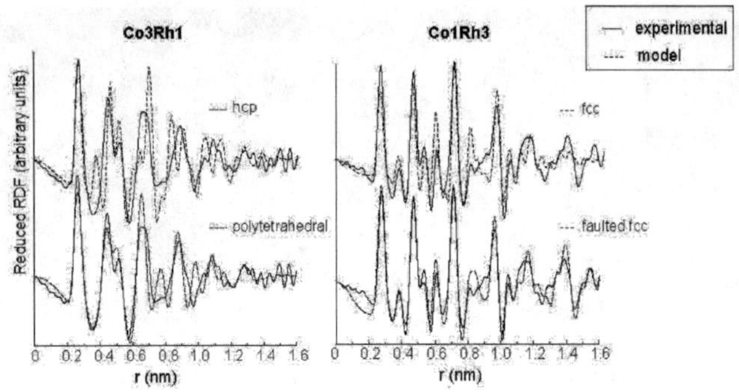

Figure 1. Comparisons between the experimental RDF of Co3Rh1, respectively Co1Rh3, ultrafine particles with the simulated RDF of models in the same structure as the bulk alloy (upper plots) and models presenting the best agreement (lower plots).

models with same size and d_{mm}. Clearly, the Co-rich particles do not crystallize in the expected hcp bulk structure, the best agreement being obtained for a polytetrahedral 105-atoms model (see [4]) with a d_{mm} equal to 0.263 nm. On the contrary and in agreement with the HRTEM results, Co1Rh3 particles crystallize in the fcc structure although with slight disorder. Indeed, the best fit was obtained with a model containing a mixture of perfect and faulted fcc particles, with d_{mm} equal to 0.269 nm. Investigating the various compositions showed two main features. First, increasing the Rh content leads to more compact atomic organization. Secondly, the d_{mm} bond lengths in the bimetallic particles are always quite higher than expected from the Vegard's law that applies very well in the bulk alloy. This last feature could be attributed either to the difference in the respective weights of Co and Rh in the scattering of x-rays, but also to a rhodium enriched core of the particle. Indeed, core atoms are more heavily linked than surface atoms and their contribution to the first metal-metal bond length should be higher. Previous investigations using extended x-ray absorption fine structure, [5], and more recent ones [6], evidence in these particles a strong tendency for cobalt to oxidize compared to Rh. Although not a definitive clue, this tendency is coherent with a rhodium core in the particles. New studies have been performed using EFTEM, but on bigger particles in order to obtain a significant signal. Figure 2 presents three filtered images of the same group of particles. The first one, which corresponds to the so-called zero-loss image (taken with the elastically transmitted beam), already presents an important contrast inside the particles: dark nuclei surrounded by greyer areas are clearly distinguished. This contrast is an amplitude contrast, which arises from different sources and in particular the thickness, the nature of the elements and the structural orientation. The Rh and Co chemical maps, respectively obtained by filtering around the Rh-M_{45} and the Co-L_{23} edges, clearly reveal the Rh-rich nature of the nuclei while cobalt atoms are spread all around these nuclei, although we cannot rule out the presence of Co in the particle core. This direct evidence for cobalt surface segregation has however been obtained in particles stabilized by a ligand instead of a polymer and this ligand could present a higher affinity with cobalt, so attracting this element towards the surface. Moreover they are larger than the previously studied ones.

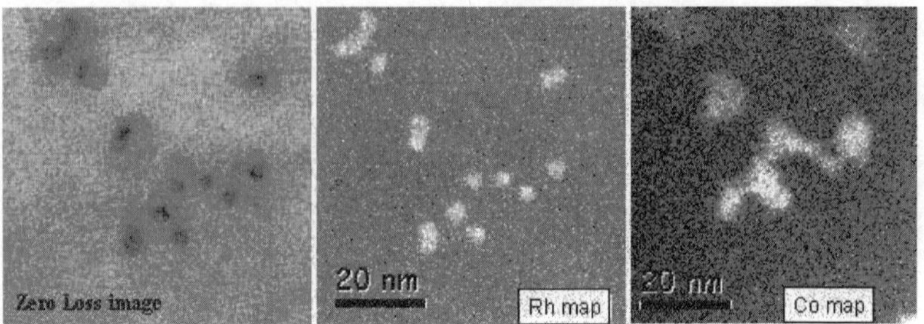

Figure 2. EFTEM images of Co3Rh1 particles synthesized in HDA. From left to right: zero-loss image, Rh map and Co map.

Atomic scale simulation

In order to ultimately resolve the chemical order and possible surface Co segregation on these 2 nm large CoRh particles, an n-body interaction model derived within the second-moment approximation of the tight binding model, well suited to describe the electronic structure in transition metals, has been developed [7]. A full description of this interaction model will soon be reported. Briefly, in this model the total energy for an atom *i* is given by:

$$E_i = -\sqrt{\sum_j \xi_{ij}^2 e^{-2q_{ij}(\frac{r_{ij}}{r_{0ij}}-1)}} + \frac{1}{2}\sum_j A_{ij} e^{-p_{ij}(\frac{r_{ij}}{r_{0ij}}-1)} \qquad (1)$$

where the r_{0ij} are fixed normalizing distances, the nearest-neighbor distances of the pure metals ($i=j$) and their mean value ($i \neq j$) and r_{ij} is the distance between atoms *i* and *j*. The first term, the n-body one, describes the electronic band energy while the second term, a pair one, characterizes the short-range core electron clouds repulsion. The 12 parameters ξ_{ij}, A_{ij}, p_{ij} and q_{ij} are fitted so that the potential suitably reproduces different characteristics of the studied material. The 2x4 parameters ($i=j$) describing the homo-atomic interactions were taken from [8]. The 4 remaining hetero-atomic parameters ($i \neq j$) were fitted, with the Merlin 3.0 optimization software [9], on ab initio DFT-GGA cohesion energies and unit-cell parameters results [10] of different chemically ordered alloys with various compositions (Co3Rh1, Co1Rh1 and Co1Rh3). This peculiar choice of the fitted data was made in order to ensure a good representation of chemical and structural behavior of the alloy as a function of its composition. Since surface effects play a crucial role in the studied nanoparticles simple low index surface excess energies have been calculated with the obtained model. As they compare very well with the ab-initio ones [10] the degree of confidence one can have on the model seems reasonable. Further tests including other relevant energies (dissolution, vacancy and antisite) are in progress.

Quenched molecular dynamics and simulated annealing Monte-Carlo Metropolis (MCM) runs with atomic exchanges and relaxations have been performed on a wide range of clusters with different sizes and compositions. In this paper, our aim is mainly to check the chemical distribution inside the particles, so that we will only discuss the results obtained by MCM. Figure 3 shows a comparison between the potential energies of three 420-atom Co3Rh1 clusters

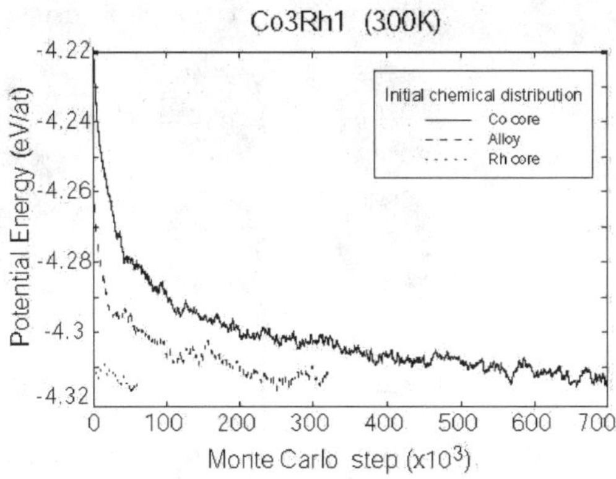

Figure 3. Potential energy vs MCM step (exchange + relaxation) for 420-atom Co3Rh1 particles with different initial chemical distributions. Note that the potential energy of the disordered and Co core particles converge towards that of the Rh core, which remains essentially unchanged.

with hcp structure. It shows that, during relaxation, the three clusters with initial Co core, Rh core and random chemical organizations reorganize in order to present the lower potential energy: that of the Rh-core particle. This result perfectly agrees with the experimental results obtained by EFTEM on the hcp particles with same composition. Figure 4 reports the final distributions of Rh and Co atoms in the particles with different compositions after relaxation. Clearly, a cobalt surface segregation is observed in all samples. This tendency is therefore obtained through both modeling and experimental approaches, which definitely reinforces the confidence that one can put on single evidences.
Segregation in an AB alloy is usually described as resulting from three competing different driving forces, namely: i/ the excess surface energies, ii/ the atomic sizes and iii/ the alloying effect. This last driving force is ineffective here to induce segregation as CoRh forms a continuous series of bulk solid solutions. Due to a slightly lower Co surface energy, cobalt surface segregation should be expected while the bigger size of Rh atoms should favor Rh surface segregation when they are in minority. The two first driving forces therefore compete in Co3Rh1 particles. Our results conclude to a predominance of the surface energy driving force. Interestingly some cobalt is always encountered at the very center of the Co-rich particles.

CONCLUSION

Structural investigation of ultrafine CoRh bimetallic particles has been successfully performed using a wide range of dedicated techniques. A structural evolution from a non-compact polytetrahedral structure in Co-rich particles towards the fcc structure in Rh-rich ones has been pointed out. Both experiments and atomic scale simulations provide clear evidences for cobalt surface segregation, highly favorable for magnetic moment exaltation.

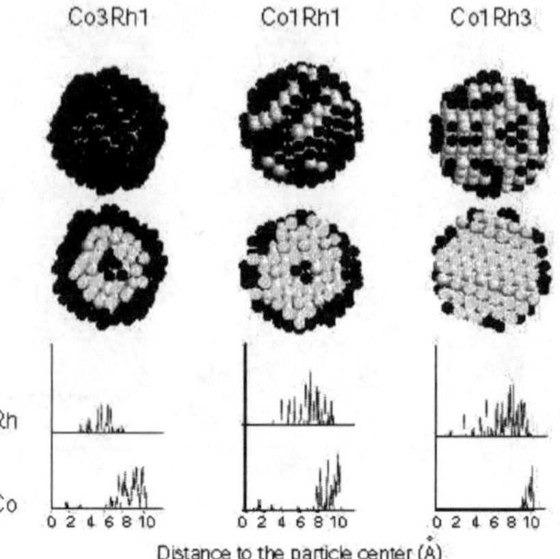

Figure 4. Atomic organization in 420-atom particles with different compositions obtained after Monte-Carlo Metropolis relaxation at 300K. Top: general view of the particles; middle: cross-sections of the particles, bottom: line profile as a function of the distance to the particle center. Black balls Co, grey balls Rh.

ACKNOWLEDGEMENTS

We gratefully acknowledge Drs D. Zitoun, C. Amiens and B. Chaudret who synthesized the CoRh particles and also for the always very fruitful discussions we had together. Most of the calculations were performed on the CALMIP/UPS-Toulouse regional computing center under project N° P0146.

REFERENCES

1. A.J. Cox, J.G. Louderback, and L.A. Bloomfield, *Phys. Rev. Lett.* **71**, 923 (1993)
2. D. Zitoun, M. Respaud, M.-C. Fromen, M.-J. Casanove, P. Lecante, C. Amiens and B. Chaudre *Phys. Rev. Lett.* **89**, 37203 (2002).
3. S. Dennler et al., *Surf. Sci.* **532-535** (2003) *; Eur. Phys J. D* **24**, 237 (2003).
4. F. Dassenoy, M.-J. Casanove, P. Lecante, M. Verelst, E. Snoeck, A. Mosset, T. Ould Ely, C. Amiens and B. Chaudret, *J. Chem. Phys.* **112**, 8137 (2000).
5. M.-J. Casanove, P. Lecante, M.-C. Fromen, M. Respaud, D. Zitoun, C. Pan, K. Philippot, C. Amiens, B. Chaudret, F. Dassenoy, *Mat. Res. Soc. Symp. Proc.* **704**, Nanoparticulate Materials 349 (2002)
6. M.-C. Fromen et al . *to be published*
7. C. Mottet, G. Treglia and B. Legrand, *Phys. Rev. B* **66**, 045413 (2002)
8. F. Cleri and V. Rosato, *Phys. Rev. B* **48**, 22 (1993)
9. http://nrt.cs.uoi.gr/merlin/
10. S. Dennler et al., accepted in *J. Phys. Cond. Mat.* (2003)

Mat. Res. Soc. Symp. Proc. Vol. 789 © 2004 Materials Research Society N9.12

Deposition of Functionalised Gold Nanoparticles by the Layer-by-Layer Electrostatic Technique

S. Paul[a,*], M. Palumbo,[a] M. C. Petty,[a] N. Cant[b] and S. D. Evans[b]

[a]School of Engineering and Centre for Molecular and Nanoscale Electronics, University of Durham, Durham DH1 3LE, UK
[b]Department of Physics and Astronomy, University of Leeds, Leeds LS2 9JT, UK

ABSTRACT

Intensive research is currently underway to exploit the intriguing optical and electronic behaviour of nano-sized particles. The basis of the unique properties of these particles is the smallness of their size; dimensions on the nanometre scale can result in interesting quantum mechanical phenomena, such as Coulomb blockade. There are currently a number of ways by which the nanoparticles can be deposited onto solid substrates. Here, we report on the use of the layer-by-layer electrostatic method, which has shown much promise in the context of deposition of thin films of certain organic materials. In this technique, layers of oppositely charged materials are generated by dipping an appropriate substrate into solutions of polyelectrolytes. For example, the polybases poly(ethyleneimine) (PEI), when adsorbed on a substrate, produce a positively charged surface. We have deposited carboxylic acid (-COOH) derivatised gold nanoparticles onto a PEI-coated silicon substrate and an amine funtionalised silicon substrate. The distribution of the gold nanoparticles was compared using atomic force microscopy.

INTRODUCTION

Nanoparticulate materials of semiconductors [1, 2] and metals [3, 4] are currently the focus of intense research. The physical properties of such small-scale structures can be tailored for particular applications. For example, their electronic and chemical behaviour can be changed simply by controlling the particle size. Nanoparticles also provide useful building blocks from which complex molecular architectures can be built. In the field of microelectronics, the 1-20 nm scale length that is typically associated with nanoparticles paves the way for achieving high device densities. Quantum mechanical effects, such as Coulomb blockade, may also be exploited in device structures such as the single electron transistor [5].

We have recently demonstrated the use of gold nano-particles in the fabrication of memory devices [2]. The encouraging aspect of this work was that the deposition of the gold nano-particles was carried out at room temperature, both by the Langmuir-Blodgett as well as via chemical routes. Attempts have been made in the past to utilise nano-crystallites of silicon to build memory devices (Tiwari *et. al* [6]). However, the nanoparticle deposition was carried out at high temperatures.

[*] Present address: Department of Ceramic and Materials Engineering, The State University of New Jersey 607 Taylor Road, Piscataway, NJ 08854-8065, USA

An alternative to silicon-based technologies (which generally requires operation at high temperatures) is to use organic materials. The devices generated from such materials can be inexpensive, manufactured on flexible and cheap substrates and scalable. For example, pentacene can be used as a substitute for amorphous silicon in making thin film transistors. Therefore, it is important to look for a technique which can help attach metal or semiconductor nano particles to these organic materials. In this report, we investigate the use of the polymer polyethyleimine (PEI) to deposit nano-particles to any surface. In addition, we investigate an amine terminated silicon surface to which gold nano-particles can be attached. In these two techniques, the distribution of the attached gold particles and the particle density have been studied.

EXPERIMENTAL DETAILS

The gold particles used in this work were prepared using a chemical route and details of the preparation can be found elsewhere [6]. The surface of Au particles (i.e the acid –COOH group) is negatively charged, allowing them to attach to a positively charged surface. Two different methods were used to prepare the positively charged surface.

In the first method, polyethyl(imine) (PEI) was used to charge the substrate layer. 1 mg of PEI was dissolved in 1 ml of tris(hydroxymethyl)aminometane, $C_4H_{11}NO_3$ (tris), buffer solution.. To this, a definite quantity of hydrochloric acid was added to bring the solution to a pH of 6.5. A number of substrates (namely Si with a thin oxide, glass, pentacene on glass) were soaked in this solution for 10 minutes. At the end of this time, the substrates were dried with a nitrogen gun.

The alternative approach used in our work was to functionalise the surface of the substrate. A 10 % silane solution (1 ml of 3-aminopropyltriethoxysilane in 9 ml toluene) was prepared in a small sample vial. The solution was kept in a nitrogen ambient. The silicon substrate was then placed in this solution for 1 hour (and in the nitrogen ambient). The substrate was subsequently washed in a toluene solution and then sonicated in toluene and dichloromethane solution for 2 minutes each. This process was repeated twice to remove all unwanted materials from the surface of the substrate. The funtionalised substrate was dried with nitrogen. Finally, the substrate was held under running deionised water for 2 minutes (to encourage the charging of the amino group) before being exposed to the nanoparticle solution.

The gold nanoparticles were deposited by dipping the substrate into a solution of acid (ie -COOH) derivatised Au-nanoparticles. The gold nano-particles (negatively charged) and the positively charged substrates surface mutually attract if the pH of the nanoparticle containing solution is adjusted correctly. Atomic Force Microscopy (AFM) was used to study the resulting surfaces.

RESULTS AND DISCUSSION

The schematic diagram of functionalised gold particles is shown in figure 1. These consist of a metallic gold core of an average diameter of 5 nm, as determined by transmission electron microscopy, figure 2. The particles are soluble in methanol and do not coagulate.

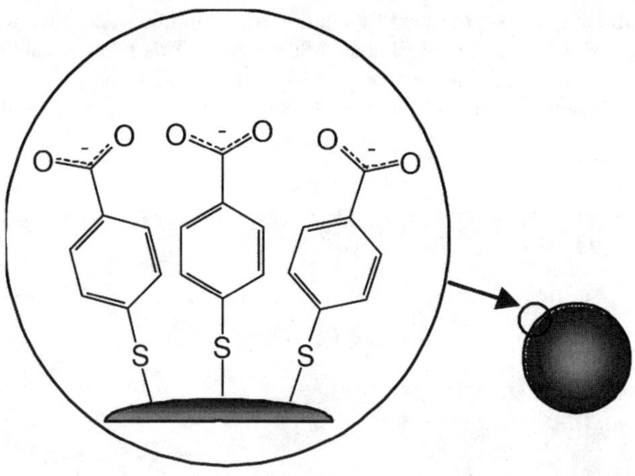

Figure 1. Schematic diagram of functionalised gold nano particles.

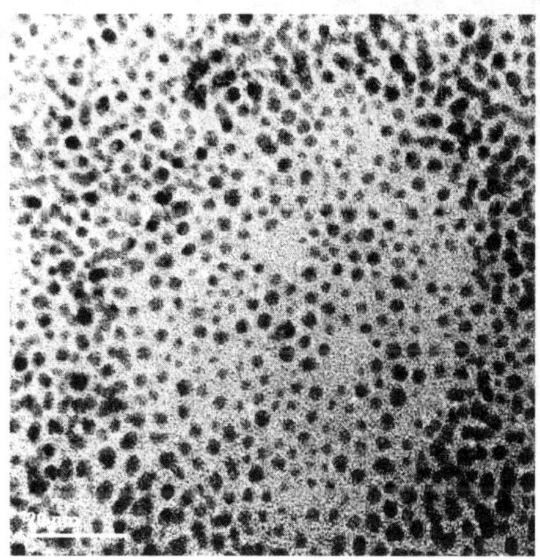

Figure 2. Transmission Electron Microscope Image of the gold nano-particles.

The silicon substrate was kept in the PEI solution for 10 minutes and dried with nitrogen Figure 3 shows an AFM image of the PEI coated Si substrate. This reveals that the Si substrate is fully covered with PEI and exhibits a degree of roughness. The PEI polymer is soft and its smearing effect can be seen in the AFM image. This image was taken in the contact mode of the AFM.

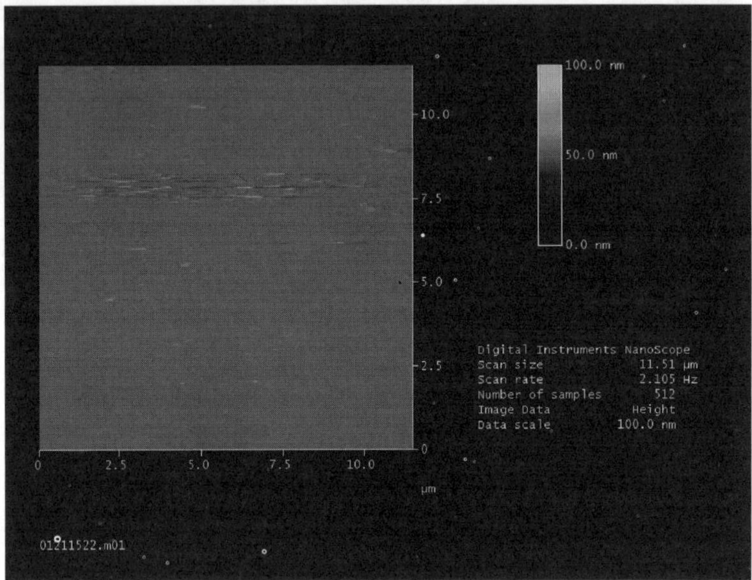

Figure 3. AFM images of polyethylimine coated Si substrate.

Figure 4 shows an AFM image of the gold nano-particles deposited on the amine terminated silicon substrate. It is evident from the image that the gold nano-particles are deposited densely on the substrate. In contrast, figure 5 shows gold deposited particles on the PEI covered silicon substrate. The image shows that the attachment of the gold nano-particles is not as dense and uniform as the amine terminated surface. This was found to be the case for all the samples we have examined. It has been suggested by Schmitt *et. al* [8] that PEI is in the form of coils on the surface of a substrate. This may lead to non-uniform distribution of surface charges and hence the observed distribution of the gold particles.

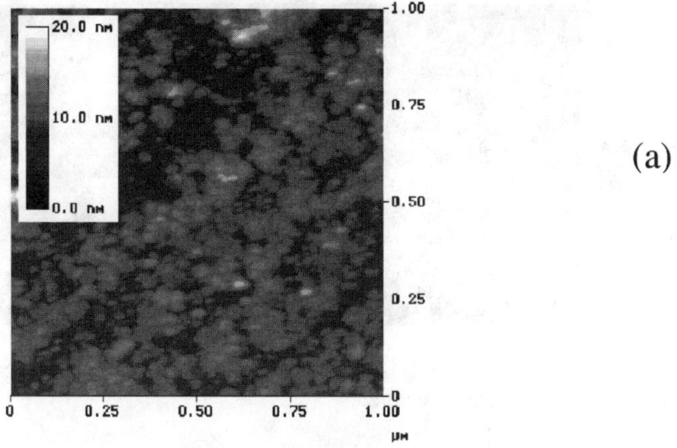

(a)

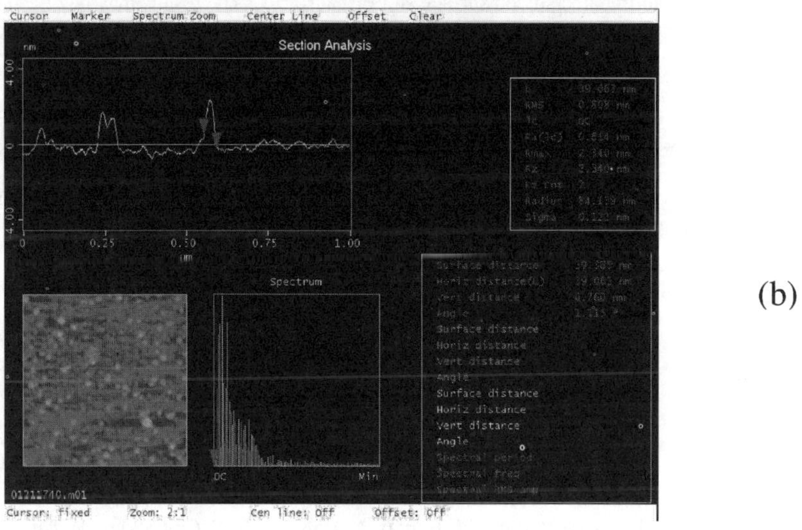

(b)

Figure 4. AFM image of the gold particles deposited on (a) amine terminated silicon surface (b) PEI covered silicon subtrate.

We have made attempts to increase the density of the nanoparticles by using alternate layers of positively (PEI) and negatively charged layers poly(ethylene-*alt*-maleic acid) (PMAE). Figure 5 shows the AFM images of gold particles deposited on single PEI layer, figure 5(a), and on a three layer multilayers architecture with the PEI on the top, figure 5(b).

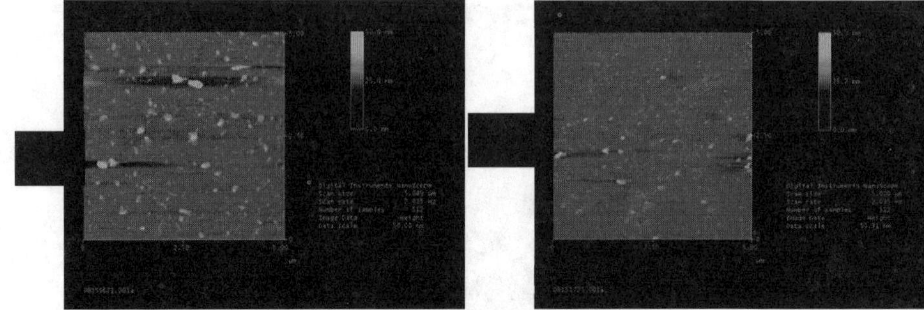

Figure 5. AFM images of gold particles on (a) single PEI layer and (b) a three layer PEI/PMAE/PEI structure..

In the latter case, the gold nano-particles seem to be less aggregated on the surface. This may b due to the straightening of the PEI layer by the PMAE.

In summary, we have investigated the attachment of acid derivatised gold nano particles to amine terminated silicon surface and PEI surfaces. The results may be relevant in the realisatio of memory devices using metallic or semiconductive nanoparticles. For example, the amine terminated surfaces (Figure 4 (a)) have been used to make organic and silicon hybrid memory devices [6].

ACKNOWLEDGEMENTS

This work was funded by the European Community under the Information Technologies Program (1998-2002): project FRACTURE IST-2000-26014.

REFERENCES

1. K. Grieve, P. Mulvaney and F. Grieser, *Current Opinion in Colloid & Interface Sci.* **5,** 168 (2000)
2. S. Paul, C. Pearson, A. Molloy, M. A. Cousins, M. Green, S. Kolliopoulou, P. Dimitraki P. Normad, D. Tsoukalas and M C Petty, *Nano Lett.* ,**3**, 533 (2003).
3. M. Brust and C. J. Kiely, *Colloids & Surfaces A*, **202** 175 (2002).
4. K. Yano, T. Ishii, T. Sano, K. Mine, K. Murai, T. Hashimoto, T. Kobayashi, T. Kure an K. Seki, *Proc. IEEE*, **87** 633 (1999).
5. S. Tiwari, F. Rana, H. Hanafi, A. Harstein, E. F. Crabbé and K. Chan *Appl. Phys. Lett.*, **68**, 1377 (1996).
6. S. Kolliopoulou, P. Dimitrakis, P. Normad, H. L. Zhang, N. Cant, S. D. Evans, S. Paul, C. Pearson, A. Molloy, M. C. Petty and D. Tsoukalas, *J. Appl. Phys.*, **94**, 5234 (2003).
7. H.-L. Zhang, S. D Evans, J. R Henderson, R.E Miles and T.-H. Shen, *Nanotechnology*, **13**, 439 (2002).
8. J. Schmitt, P. Mächtle, D. Eck, H. Möhwald and C.A Helm, *Langmuir*, **15**, 3256 (1999).

Metallic, Magnetic and Semiconductor Nanoparticles: Growth and Characterization

Mat. Res. Soc. Symp. Proc. Vol. 789 © 2004 Materials Research Society N10.1

Seeded and Non-Seeded Methods to Make Metallic Nanorods and Nanowires in Aqueous Solution

Tapan K. Sau and Catherine J. Murphy

Department of Chemistry and Biochemistry, University of South Carolina, Columbia, SC 29208.

ABSTRACT

Gold and silver nanorods/wires of different dimensions were obtained by wet-chemical seeded and non-seeded growth methods.

INTRODUCTION

Considerable attention has been attracted to prepare nanomterials confined in various dimensions to harvest their unique size- and shape-dependent physicochemical properties such as optical, electronic, electrical, magnetic, catalytic, etc. for their potential applications in nanodevices.[1] We have reported[2] the synthetic methods for the production various aspect ratio Au and Ag nanorods and nanowires. Here we describe in addition to so-called 'seed-mediated' process, 'non-seeded' processes to synthesize Au and Ag nanorods and nanowires. We explore the effects of variation of [Au^{3+}]/[seed] ratio and surfactant on the seeded Au nanorod preparation, and two new one-pot synthetic methods for the preparation of Au nanorod and Ag nanorod/wires.

EXPERIMENTAL

$HAuCl_4.3H_2O$, l-Ascorbic acid (AA), cetyltrimethylammonium bromide (CTAB), tri-sodium citrate, and silver nitrate were used as purchased (Aldrich).

A 10mL solution of Au seed particles was prepared by the reduction of 2.5 x 10^{-4} M $HAuCl_4.3H_2O$ by 4.0 x 10^{-4} M ascorbic acid containing 2.5 x 10^{-4} M tri-sodium citrate. The seed solution was used between one and two hour after its preparation.

Typically appropriate amount of such as 0.125 mL 0.01 M $HAuCl_4.3H_2O$, 0.020 mL 0.1 M ascorbic acid (AA), and finally seed particles were added one by one to a given quantity of 0.1 M CTAB solution. The reaction mixture was gently mixed (by inversion of test tube) after addition of each reactant. The reversal in the order of addition of ascorbic acid and seed particles did not give rise to any noticeable change. Addition of the yellow solution of $HAuCl_4.3H_2O$ to the CTAB solution led to a brown-yellow color. The brown-yellow color disappeared on the addition of ascorbic acid.

In the one-pot synthesis of Au nanorods, appropriate quantity of silver nitrate was added to a solution containing desired amount of CTAB, $HAuCl_4.3H_2O$, and AA. In a typical experiment, 0.030 mL silver nitrate (0.01 M) was added to a solution containing 4.8 mL CTAB (0.1 M), 0.200 mL $HAuCl_4.3H_2O$ (0.01 M), and 0.032 mL AA (0.1 M). The reaction mixture was gently mixed after the addition of silver nitrate. A good yield of short aspect ratio rods were obtained in the range roughly between 3.0×10^{-4} M and 6.0×10^{-4} M $HAuCl_4.3H_2O$, and 6.0×10^{-4} M and 8.0×10^{-4} M $AgNO_3$ for 0.1 M CTAB. One-pot synthesis of silver nanorods/wires involves boiling of aqueous $AgNO_3$ with trisodium citrate in the presence of small quantity of NaOH.[3]

Au nanorods were obtained as usual by centrifuging the sample solution twice and re-dispersing the desired solid residue in DI water. No separation by centrifugation is required for silver nanowires/rods. TEM grids were prepared by casting a drop of the solution and drying the drop in air. TEM images were obtained on a Hitachi H-8000 and JEOL electron microscope. UV-Vis-NIR spectra of the samples were obtained by a Varian Cary 500 Scan UV-Vis-NIR spectrophotometer.

RESULTS AND DISCUSSION.

Seeded Au Nanorods.

$[Au^{3+}]$/[seed] ratio and [CTAB] have important roles in determining the rod-length, width, and the yield. Here we explore these parameters one by one. Fig. 1 shows the Vis-NIR absorption characteristics of the samples containing rods of different average lengths. With the decrease in the number of seed particles longer rods are formed which is evident from the red shift of the peak position for longitudinal surface plasmon oscillations. Fig. 2 summarizes the effects of variation of $[Au^{3+}]$ and [seed] on the rod-length, width, and the average diameter of the coexisting spheres. The rod-length, width, and the diameter of the coexisting spheres increase with the decrease in [CTAB]. The effect is more pronounced when $[Au^{3+}]$ is more.

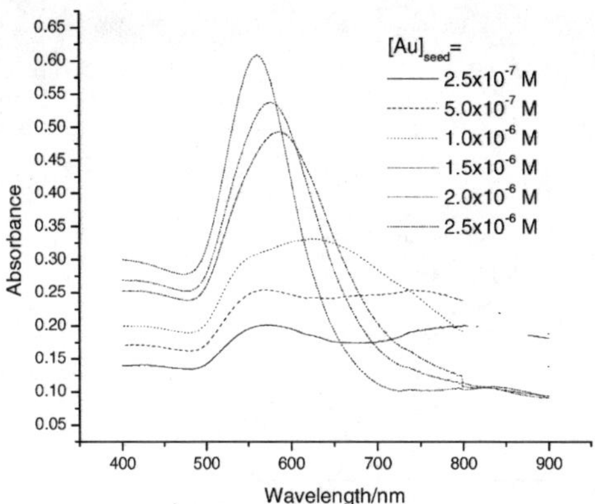

Fig.1. Vis-NIR absorption spectra of the nanorod containing solutions twelve hr after their preparation by seeded growth method. Conditions: $[Au^{3+}]=1.25x10^{-4}$ M; $[AA]=2.0x10^{-4}$ M; $[Au]_{seed}=2.5x10^{-6}$ M and $[CTAB]=8.0x10^{-3}$ M.

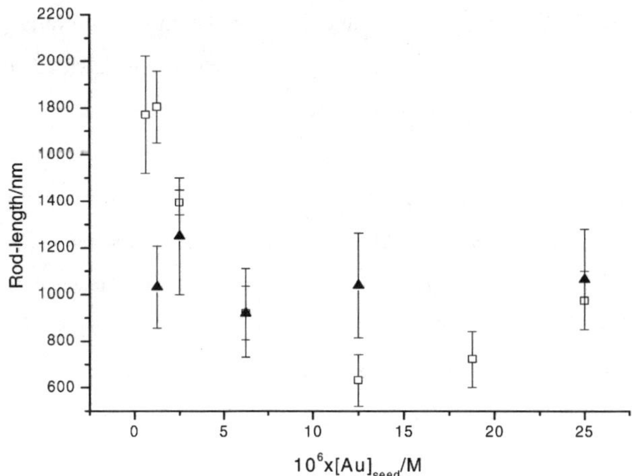

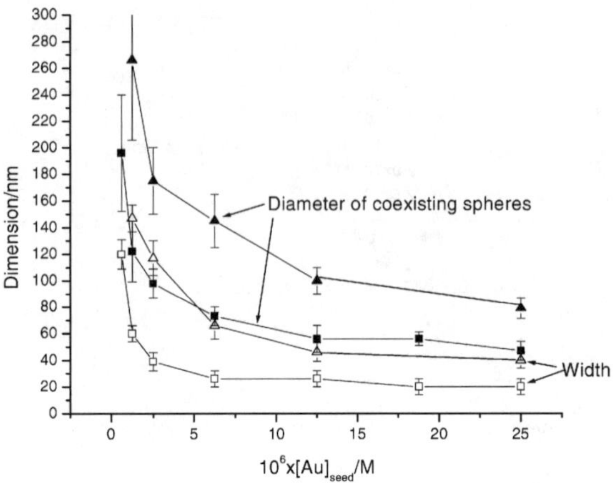

Fig. 2. Variation of rod and coexisting sphere dimensions with the variation of [Au^{3+}] and [seed]. Conditions: [Au^{3+}]=1.25×10^{-4} M; [AA]=2.0×10^{-4} M (squares) and [Au^{3+}]=5.0×10^{-4} M; [AA]=8.0×10^{-4} M (triangles); $[Au]_{seed}$=2.5×10^{-6} M and [CTAB]=0.1 M.

CTAB is not only an indispensable constituent in the rod formation, variation in its concentration has profound effect on it too. The rod-length and yield diminish with a decrease in [CTAB] for a given [Au^{3+}]/[seed] ratio, whereas the width of the rod and the diameter of the coexisting spheres increase slightly (Fig. 3). It is believed that both transverse and longitudinal growths occur simultaneously but with different rates due to the different interaction energy of CTAB molecules at different faces of the seeds/rods and the competition for the surface by Au^0 and CTAB molecules.[2c,4] The transverse growth could be more in the presence of smaller quantity of CTAB, since the lateral faces of the rod are less well protected under these conditions. Similarly the bi-layer formation by the CTAB is possibly disturbed with the falling concentration of CTAB thus decreasing the rod-length and yield.[2d,5]

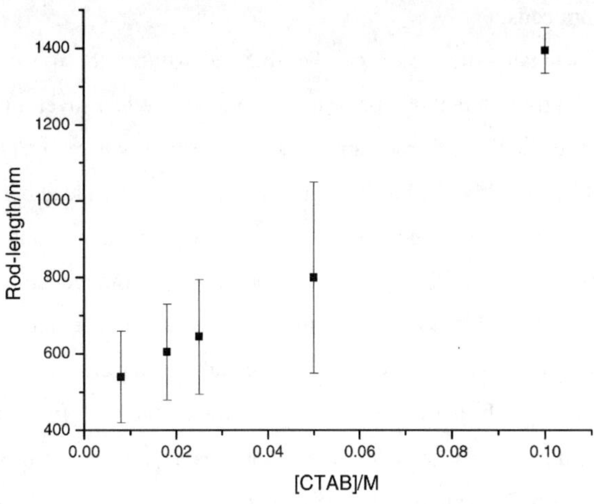

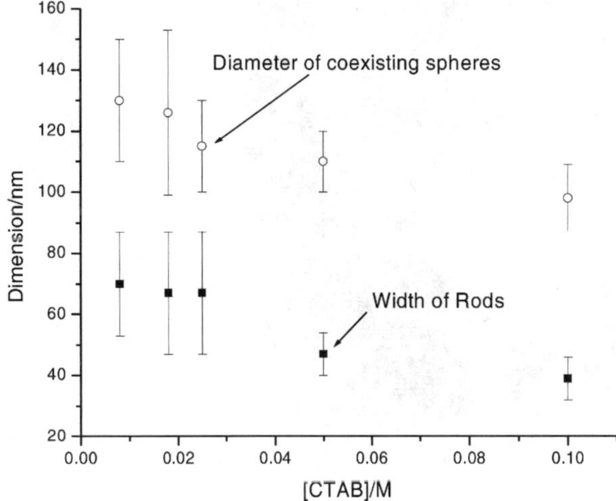

Fig.3. Variation of rod and coexisting sphere dimensions with the variation of [CTAB]. Conditions: Conditions: $[Au^{3+}]=1.25x10^{-4}$ M; $[Au]_{seed}=2.5x10^{-6}$ M; $[AA]=2.0x10^{-4}$ M and [CTAB]=0.1 M.

"Non-Seeded" Au nanorods.

Au nanorods of small aspect ratio can be obtained without the use of so-called seed particles following a slightly different one-pot synthetic protocol. When silver nitrate was used in place of Au seed particles, such short rods appeared over a small window of [Au^{3+}] and [Ag^{+}] range for a particular [CTAB] (Fig. 4). Interestingly when $AgNO_3$ was added before the addition of AA, there was practically no formation of particles. We believe that addition of silver nitrate into the AA containing solution results in the formation of Ag^0 by the reduction of Ag^+ by AA and thereby small silver particles formed in situ act as seeds for the development of gold particles. In fact, the formation of silver particles could be detected by naked eyes if a concentrated solution is used. In the absence of AA, silver nitrate may form AgBr in CTAB, which dissolves deep into CTAB micelles being unavailable for the further reaction. The formation of Au nanorods over a window of [Au^{3+}] and [CTAB] suggest again that a suitable

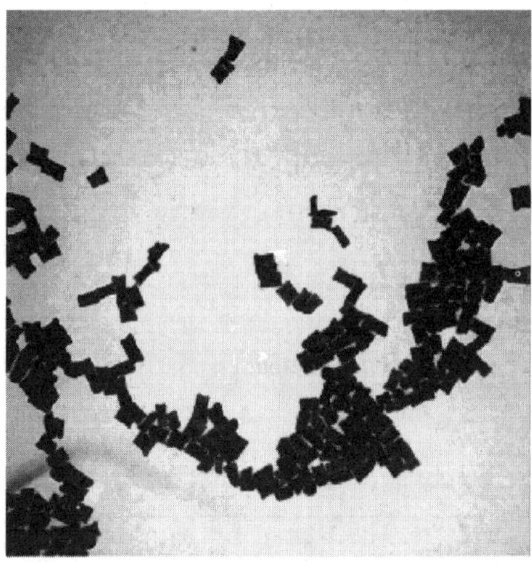

Fig.4. TEM image of Au nanorods obtained by non-seeded method. Conditions: [Au^{3+}]=4.0x10^{-4} M; [AA]=6.4x10^{-4} M; [CTAB]=0.1 M and 6.0x10^{-5} M. The scale bar= 100 nm.

supply rate of Au^0 to the seed particle and proper local environment are necessary for rod formation.

"Non-Seeded" Ag Nanowires/rods

An extreme example of a non-seeded approach to modify silver nanowires is given in our recent report.[3] In that preparation, silver nitrate, trisodium citrate, and sodium hydroxide boiled together with no preformed seeds at all, resulting in Ag nanowires that are about 30 nm in diameter and up to 10 micron long.

Kinetics of Rod/Wire Development

In an attempt to follow the evolution of nanorods/wires, kinetic studies were carried out. The time required to develop the rod completely depends on the $[Au^{3+}]$ and [seed] for a particular [CTAB]. Fig. 5 shows the TEM images taken during the initial period of Au nanorod development. Here underdeveloped rods could be found even up to about 15 min of their evolution. The rod length/width, as expected, increased with time. The TEM images in Fig. 6 show the evolution of silver wires from the initial small silver particles to small rods and finally to wires.

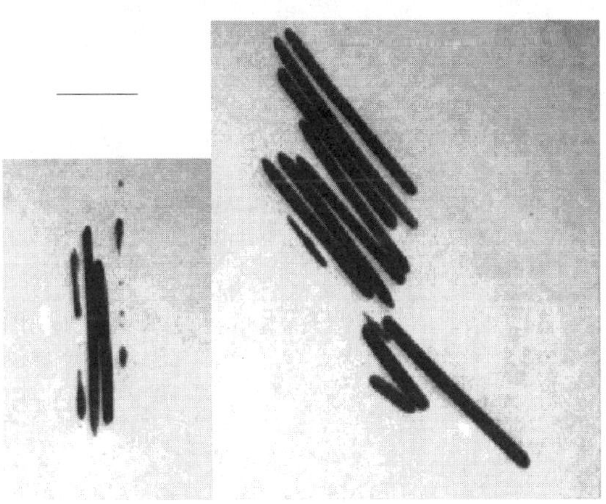

A

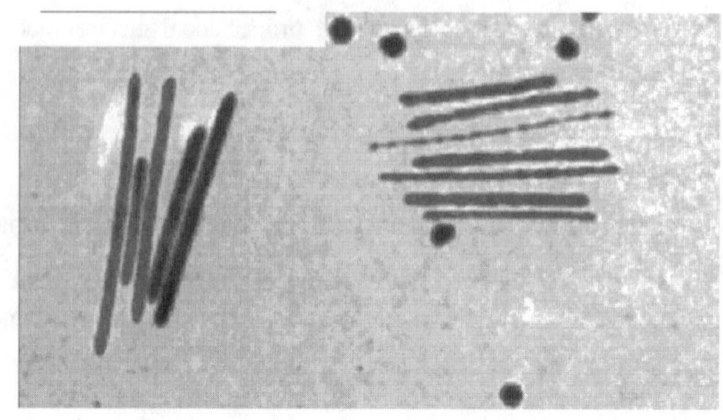

B

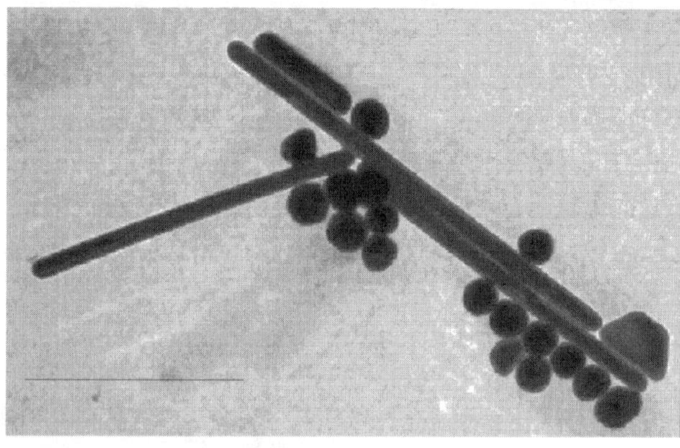

C

Fig 5. Representative TEM images at various stages of Au nanorod evolution by seeded growth method. Conditions: $[Au^{3+}]=1.25x10^{-4}$ M; $[Au_{seed}]=2.5x10^{-6}$ M ; $[AA]=2.0x10^{-4}$ M and [CTAB]=0.1 M. A=5 min; the scale bar equals to 100 nm; B=10 min; scale bar=500 nm; and C=45 min; scale bar=500 nm.

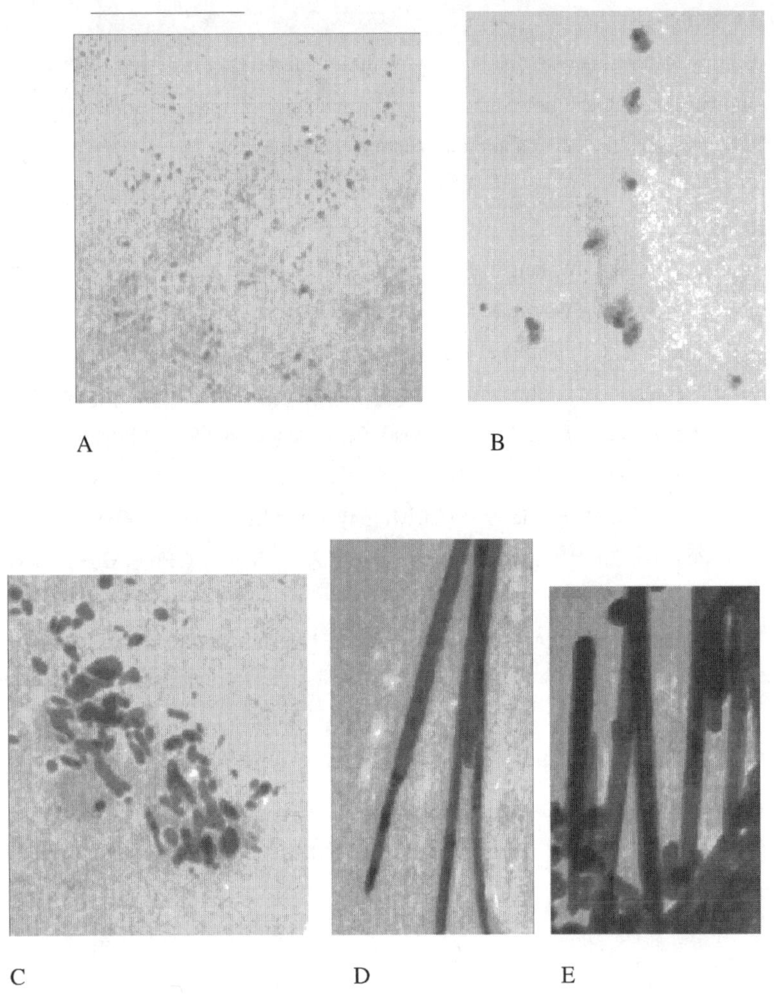

A B

C D E

Fig 6. Evolution of silver nanowire/rod. TEM pictures are taken at A: 10 min after the start of boiling of the reaction mixture, which was right before the addition of second batch of silver nitrate; B: 5 min; C: 20 min ; D: 30 min; and E: 60 min after the addition of second batch of silver nitrate. Conditions: see ref 3.The scale bar=500 nm.

CONCLUSION

Both seeded and non-seeded one-pot syntheses can produce gold and silver nanorods and silver nanowires, and in addition to the length and width of the rods/wires, their appearance and yield can be varied by the manipulation of reaction conditions.

REFERENCES

1. a) F. Favier, E.C. Walter, M.P. Zach, T. Benter and R.M. Penner, Science **293**, 2227 (2001). b) S. Iijima, Nature **354**, 56 (1991).
2. a) N.R. Jana, L. Gearheart and C.J. Murphy, Chem. Commun. 617 (2001); b) Adv. Mater. **13**, 1389 (2001). c) C. J. Johnson, E. Dujardin, S.A. Davis, C.J. Murphy and S. Mann, J. Mater Chem. **12**, 1765 (2002). d) J. Gao, C.M. Bender and C. J. Murphy, Langmuir **19**, 9065 (2003).
3. K.K. Caswell, C.M. Bender, and C.J. Murphy, Nano Lett. **3**, 667 (2003).
4. J.M. Petroski, Z.L.Wang, T. C. Green, and M.A. El-Sayed, J. Phys. Chem. B **102**, 3316 (1998).
5. B. Nikoobakht and M. A. El-Sayed, Langmuir, **17**, 6368 (2001).

Mat. Res. Soc. Symp. Proc. Vol. 789 © 2004 Materials Research Society N10.4

A Thermogravimetric Study of Alakanethiolate Monolayer-Capped Gold Nanoparticle Catalysts

Mathew M. Maye, Sandy Chen, Wai-Ben Chan, Lingyan Wang, Peter Njoki, I-Im. S. Lim, Jennifer Mitchell, Li Han, Jin Luo, and Chuan-Jian Zhong*

Department of Chemistry, State University of New York at Binghamton, Binghamton, NY 13902. *cjzhong@binghamton.edu

ABSTRACT

The application of molecularly-capped gold nanoparticles (1-5 nm) in catalysis (e.g., electrocatalytic oxidation of CO and methanol) requires a thorough understanding of the surface composition and structural properties. Gold nanoparticles consisting of metallic or alloy cores and organic encapsulating shells serve as an intriguing model system. One of the challenges for the catalytic application is the ability to manipulate the core and the shell properties in controllable ways. There is a need to understand the relative core-shell composition and the ability to remove the shell component under thermal treatment conditions. In this paper, we report results of a thermogravimetric analysis of the alkanethiolate monolayer-capped gold nanoparticles. This investigation is aimed at enhancing our understanding of the relative core-shell composition and thermal profiles.

INTRODUCTION

Catalysis plays a vital role in chemical processing, environmental protection and fuel cell technology. The use of gold nanoparticles as catalysts has attracted increasing interest recently [1-4], due to the pioneer work of Haruta [5] who demonstrated the high catalytic activities for CO and hydrocarbon oxidation at metal oxide supported gold nanoparticles of less than ~10 nm. We have recently shown that the preparation of nanoscale gold catalysts can be achieved using alkanethiolate monolayer-capped nanoparticles. These nanoparticles consist of metal or alloy nanocrystal cores with organic molecular wiring or linkage shells that aid to define the interparticle spatial property [6-8]. This approach to the nanoparticle preparation and processing is useful because it allows us to address many of the fundamental issues related to size, shape, aggregation, poisoning, and surface engineering of nanoparticles in catalysis (e.g., fuel cell catalysis).

To exploit the catalytic properties of the core-shell type nanoparticles, we need to establish effective means to remove the organic shell while making the nanocrystal core's morphological properties under control. The focus of this work is to assess the relative core-shell composition and thermal profile in thermal treatment of the core-shell gold nanoparticle catalysts using thermogravimetic analysis (TGA) technique. We studied Au nanoparticles of ~2 nm core sizes ($Au_{2\text{-}nm}$) capped with decanethiol (DT). The results provide information for us to assess thermal activation parameters.

EXPERIMENTAL

Chemicals. Major chemicals included decanethiol, (DT, 96%), hydrogen tetracholoroaurate ($HAuCl_4$, 99%), tetraoctylammonium bromide (TOABr, 99%), sodium borohydride ($NaBH_4$, 99%), toluene (Tl, 99.8%), hexane (Hx, 99.9%), ethanol (EtOH) and methanol (MeOH, 99.9%). All chemicals were purchased from Aldrich and used as received. Water was purified with a Millipore Milli-Q water system.

Synthesis. Gold nanoparticles of 2-nm core size ($Au_{2\text{-nm}}$) encapsulated with decanethiolate (DT) monolayer shell were synthesized by Schiffrin's two-phase protocol [9]. Briefly, $AuCl_4^-$ was first transferred from aqueous $HAuCl_4$ solution to Tl solution by phase transfer reagent TOABr. Thiols (e.g., DT) were added to the solution at a 2:1 mole ratio (DT/Au), and an excess of aqueous $NaBH_4$ was slowly added for the reaction. The produced DT-encapsulated Au nanoparticles were subjected to solvent removal via rotary evaporation, and cleaned using ethanol 5 times. DT-capped gold nanoparticles with a core size of ~5-nm ($Au_{5\text{-nm}}$) were also studied, which were produced by a thermally-activated processing route developed in our lab [10].

Instrumentation. Thermogravimetric analysis (TGA) was performed on a Perkin-Elmer Pyris 1-TGA. Typical samples weighed ~4 mg, and were heated in a platinum dish. Samples were heated in either N_2 or O_2 from 55-500 ^{0}C at rates ranging from 5-20 ^{0}C/min.

Transmission electron microscopy (TEM) was performed on Hitachi H-7000 Electron Microscope (100 kV). The nanoparticle samples dissolved in hexane or toluene solution were drop cast onto a carbon-coated copper grid sample holder followed by natural evaporation at room temperature.

Electrochemical measurements were performed using an EG&G Model 273A potentiostat. A three-electrode cell was employed, with a platinum coil as the auxiliary electrode and a standard calomel electrode (SCE) as the reference electrode.

RESULTS AND DISCUSSION

Alkanethiolate monolayer-capped gold nanoparticles with 2-6 nm core diameters have been extensively characterized. Figure 1 shows a representative TEM micrograph for the DT-capped gold nanoparticles synthesized by the Schiffrin's two-phase protocol [9]. A size analysis of the nanoparticles in the TEM micrograph yielded 2.0 ± 0.5 nm for the diameter of the nanoparticle cores. The well-isolated individual nanocrystal character and the average interparticle edge-to-edge distance are consistent with the presence of the DT capping shells on the gold nanocrystals.

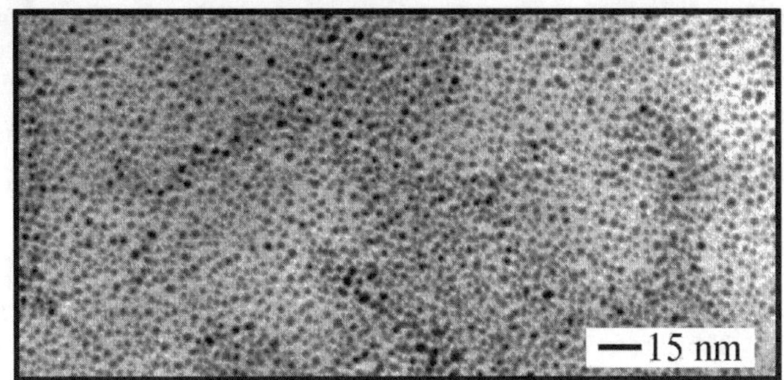

Figure 1: TEM micrograph of decanethiolate-capped gold nanoparticles (2.0 ± 0.5 nm).

Figure 2 shows a representative TGA curve for DT-$Au_{2\text{-}nm}$. The experiment was run from 55 to 500 ^{0}C at 10 ^{0}C/min rate in oxygen atmosphere.

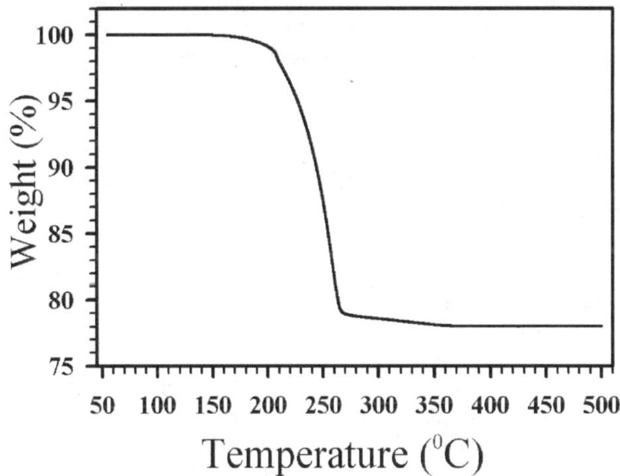

Figure 2: TGA profiles for DT-$Au_{2\text{-}nm}$ heated from 55^{0}C 500 ^{0}C at 10 ^{0}C/min in oxygen.

The TGA curve clearly shows that the removal of the organic shell starts at ~180 ^{o}C, and is complete at ~350 ^{o}C. The assessment for the complete removal is consistent with our XPS data for the nanoparticles after the thermal treatment, which showed the absence of detectable sulfur species [8]. In the thermal profile, the additional mass change (~2%) at 270-350 ^{o}C is currently under investigation. Possible sources could include residual carbon and traces of TOABr in the nanoparticle sample [13].

Table 1: A Comparison of theoretical and experimental weight percentage of organics for $DT\text{-}Au_{2\text{-}nm}$ nanoparticles

Nanoparticle	Size (nm)	**n**	**m**	Organic (wt.) % Theoretical	Organic (wt.) % Experimental
$DT\text{-}Au_{2\text{-}nm}$	2.0 ± 0.5	293	87	21	22

n= number of Au atoms per nanoparticle, **m**= number of capping thiolates per nanoparticle

The mass change percentage determined from the TGA curve is 22%. In Table 1, a comparison is made between the theoretical and experimental weight percentages for $DT\text{-}Au_{2\text{-}nm}$ nanoparticles. The experimental value is in close agreement with the theoretical value (21%). The theoretical value is obtained from model calculations. The model calculation can be based a simple spherical model for the core-shell Au nanoparticles, or a truncoctahedron model [13]. In the simple spherical model, we calculate the number of gold atoms and the number of capping thiolates using the average particle diameter and the average surface coverage of thiolates. In the truncoctahedron model, which is a more realistic description of the Au nanocrystal cores capped by alkanethiolate shells [9, 13-15], we calculate the number of gold atoms and the number of capping thiolates using available literature data [13-15]. By analyzing the existing literature data, we obtained the number of Au atoms (*n*) and for the number thiolates (*m*) for the nanoparticle in the general formula, $(Au)_n(RS)_m$. For $DT\text{-}Au_{2\text{-}nm}$ of 2-nm core diameter, the calculated n and m are listed in Table 1. These values allowed us to calculate the organic weight percentage based on the following equation.

$$Organic(\%) = \frac{(m)(MW_{RS})}{[(m)(MW_{RS}) + (n)(MW_{Au})]} \times 100$$

Our experiment has shown that extensive cleaning of the nanoparticles using ethanol is important in obtaining the expected weight percentage. A series of TGA measurements using different particle core sizes and different chain lengths for the capping alkanethiolates is in progress for assessing the core-shell composition.

We have also used *in-situ* AFM technique to probe the thermal activation of the $DT\text{-}Au_{2\text{-}nm}$ nanoparticles assembled on planar substrates using molecular linkers [8]. The data, together with XPS and FTIR data, has evidenced the removal of the organic shells/linkers with controllable local aggregation of the nanoparticle cores. The data are comparable with those after removal of organic shells/linkers using electrochemical method [16].

The electrocatalytic data for the nanoparticle catalysts activated under anodic potential polarization (electrochemical activation) and 250-300 ^{0}C treatment (thermal activation) in methanol oxidation reaction are compared. The results are consistent with our recent report [8] on $NDT\text{-}Au_{2\text{-}nm}$/glassy carbon electrode after electrochemical and thermal activation treatments for the electro-oxidation of methanol in 0.5 M KOH electrolyte. The nanoparticle assembly followed a one-step "exchange-crosslinking-precipitation" method recently developed in our laboratory [11]. The thickness of each thin film (~5 layers) was controlled by immersion time [11b]. The results are for the 2-nm gold nanoparticle catalysts that are either thermally activated (by heating at 250 ^{0}C for 30 minutes in oxygen) or electrochemically activated (by potential polarization to +800 mV vs. SCE). For the electrooxidation of methanol in alkaline medium, the thermally-activated catalysts showed a higher current density (~2x) than its electrochemically-activated counterpart. Also, the oxidation potential is slightly less positive than that for the electrochemically-activated film. The electrocatalytic oxidation was characterized by a large anodic wave at +250~270 mV [6,7]. Our recent XPS results have shown that the thermally activated DT-$Au_{2\text{-nm}}$ catalysts have undetectable levels of sulfur upon activation, with only a small percentage of oxygen [8]. In comparison with the thermally-activated catalysts, the electrochemically-activated catalysts showed the presence of a higher levels of oxygen (~7 %) and sulfur (~0.7 %) [16].

CONCLUSION

The experimental results have demonstrated that the organic shell weight percentage for the alkanethiolate monolayer-capped gold nanoparticles can be determined by TGA measurements. The determined organic weight percentage is in close agreement with theoretically-calculated percentage based on a truncoctahedron model for the 2-nm sized nanoparticle core and a dense packing model for the alkanethiolates. Our experiments have also shown that the electrocatalytic activity of the thermally-activated gold nanoparticle catalysts towards methanol oxidation is higher than that obtained with electrochemically-activated gold nanoparticle catalysts. Our on-going work is further addressing issues related to the correlation of the catalytic activity with the relative core-shell compositions and structures.

ACKNOWLEDEMENT

Financial support of this work from the National Science Foundation (CHE 0316322) and Petroleum Research Fund administered by the American Chemical Society (40253-AC5M) is gratefully acknowledged. M.M.M. thanks the Department of Defense (Army Research Office) for support via a National Defense Science & Engineering Graduate Fellowship.

REFERENCES

1. (a) M. M. Maye, J. Luo, L. Han, N. N. Kariuki, C. J. Zhong, *Gold Bulletin*, 36, 75 (2003).
2. (b) G.C. Bond, D. T. Thompson, *Gold Bulletin*, 33, 41 (2000).
3. (a) A. Sanchez, S. Abbet, U. Heiz, W.-D. Schneider, H. Hakkinen, R.N. Barnett, U. Landman, *J. Phys. Chem. A,* 103, 9573 (1999). (b) J.W. Yoo, D. Hathcock, M.A. El-Sayed, J. Phys. Chem. A, 106, 2049 (2002). (c) W.T. Wallace, R.L. Whetten, *J. Am. Chem. Soc.,* 124, 7499 (2002).
4. G. Schmid, S. Emde, V. Maihack, W. Meyer-Zaika, S. Peschel, *J. Mol. Catal. A–Chem.,* 107, 95 (1996)
5. M. Haruta, Catal. Today, 36, 153 (1997). (b) M. Haruta, M. Date, *Appl. Catal. A–Gen.*, 222, 427 (2001). (c) M. Valden, X. Lai, D.W. Goodman, *Science*, 281, 1647 (1998).

6. M.M. Maye, Y.B. Lou, C.J. Zhong, *Langmuir,* 16, 7520 (2000). (b) Y. B. Lou, M.M. Maye, L. Han, J. Luo, C.J. Zhong, *Chem. Commun.*, 473, (2001). (c) J. Luo, Y. Lou, M.M. Maye, C.J. Zhong, M. Hepel, *Electrochem. Commun.,* 3, 172, (2001).
7. C.J. Zhong, M.M. Maye, *Adv. Mater.*, 13, 1507 (2001). (b) J. Luo, M.M. Maye, Y.B. Lou, L. Han, M. Hepel, C.J. Zhong, *Catal. Today*, 77, 127-138 (2002).
8. J. Luo, V. W. Jones, M. M. Maye, L. Han, N. N. Kariuki, C. J. Zhong, *J. Amer. Chem. Soc.*, 124, 13988, (2002).
9. M. Brust, M. Walker, D. Bethell, D.J. Schiffrin, R. Whyman, *J. Chem. Soc., Chem. Comm.*, 801. (1994)
10. (a) M.M. Maye, W.X. Zheng, F.L. Leibowitz, N.K. Ly, C.J. Zhong, *Langmuir,* 16, 490 (2000). (b) M.M. Maye, C.J. Zhong, *J. Mater. Chem.*,10, 1895 (2000).
11. (a) F.L. Leibowitz, W.X. Zheng, M.M. Maye, C.J. Zhong, *Anal. Chem.,*71, 5076, (1999). (b) L. Han, M.M. Maye, F.L. Leibowitz, N.K. Ly, C.J. Zhong, *J. Mater. Chem.,11*, 1258 (2001).
12. C.A. Waters, A.J. Mills, K.A. Johnson, D.J. Schiffrin, Chem. Commun., 4, 540 (2003).M.J.
13. Hostetler, J.E. Wingate, C.J. Zhong, J.E. Harris, R.W. Vachet, M.R. Clark, J.D. Londono, S.J. Green, J.J. Stokes, G.D. Wignall, G.L. Glish, M.D. Porter, N.D. Evans, R.W. Murray, *Langmuir*, 14, 17, (1998).
14. R.H. Terrill, T.A. Postlethwaite, C.-H. Chen, C.-D. Poon, A. Terzis, A. Chen, J.E. Hutchison, M.R. Clark, G. Wignall, J.D. Londono, R. Superfine, M. Falvo, C.S. Johnson, E.T. Samulski, R.W. Murray, *J. Am. Chem. Soc.* 117, 12537 (1995).
15. R.L. Whetten, J.T. Khoury, M.M. Alvarez, S. Murthy, I. Vezmar, Z.L. Wang, P.W. Stephens, C.L. Cleveland, W.D. Luedtke, U. Landman, *Adv. Mater.* 8,428, (1996).
16. M.M. Maye, J. Luo, Y. Lin, M. H. Engelhard, M. Hepel, C. J. Zhong, *Langmuir*, 19, 125 (2003).

Poster Session III

Multi-color Luminescence from Surface Oxidized Silicon Nanoparticles

K. Sato[*,***], K. Hirakuri[*], M. Iwase[**], T. Izumi[***] and H. Morisaki[****]
[*]Department of Electronic and Computer Engineering, Tokyo Denki University, Ishizaka, Hatoyama, Hikigun, Saitama, 350-0394, JAPAN
[**]Department of Materials Science, Tokai University, 1117 Kitakaname, Hiratsuka, Kanagawa, 259-1292, JAPAN
[***]Department of Electronics, Tokai University, 1117 Kitakaname, Hiratsuka, Kanagawa, 259-1292, JAPAN
[****]Department of Electronic Engineering, The University of Electro-Communications, 1-5-1 Chofugaoka, Chofu, Tokyo, 182-8585, JAPAN

ABSTRACT

We fabricated the multi-color electroluminescent (EL) device using hydrofluoric (HF) acid solution treated and oxidized silicon (Si) nanoparticles. Strong red luminescence was obtained from the HF treated Si nanoparticles based EL device under a low forward bias of 4.0 V. On the other hand, green and blue luminescence, which could be seen with naked eye under room illumination, was observed for the oxidized Si nanoparitcles based EL device at the forward bias below 9.5 V, because of reduction of size due to oxidation onto the Si nanoparticle surface. Furthermore, the red/green/blue lights showed good stability for aging of a long period of time in air by the formation of oxidized layer on the surface. These results indicate that the EL devices developed in this study can realize as application to future flat panel display.

INTRODUCTION

Silicon (Si) nanoparticles are useful candidate materials for development of new flat panel displays including electroluminescent (EL) display and field emission display, because it exhibits high efficient photoluminescence [1] and injection-type EL [1] at room temperature. Recently, some research groups have been developed a multi-color EL devices using porous silicon and Si nanoparticles as an application to Si-based display device [2-4]. The EL devices show red/green/blue luminescence in N_2 atmosphere [2]. However, they have been demonstrated low efficiency and unstable luminescence in air. The low efficiency and the poor stability for their luminescence is caused by the re-formation of high-density Si dangling bonds, which is a non-radiative recombination center, onto the Si nanoparticle/SiO_2 layer interface, because the Si nanoparticle surface has most unstable passivation in air. Moreover, the EL devices require the high voltage above 10.0 V, because of formation of thick oxidized (SiO_2) layer on the surface [3,4].

From above feature of the multi-color EL devices, it is necessary to reduce the Si dangling bonds and the SiO_2 layer on the surface. They can be improved by the hydrofluoric (HF) acid solution and the oxidation techniques. The HF treatment leads to decrease of Si dangling bonds by hydrogen termination. And, the oxidation treatment leads to formation of SiO_2 layer as stable passivation, and reduction of size, which is directly related to luminescent color.

In this paper, we fabricate the EL devices using HF treated and surface oxidized Si nanoparticles (HF treated and oxidized EL devices) to realize the Si-based EL display, which has high brightness, good stability, low voltage and full color. Moreover, we report surface compositional of Si nanoparticles, and luminescent and electrical properties from the EL devices.

EXPERIMENTAL DETAILS

The Si nanoparticle based multi-color EL devices were fabricated by manufacturing process as shown in Fig. 1. First, the formation of Si nanoparticle, which was used as luminous layer of EL device, was carried out. Substrate was used p-type Si (100) with resistivity of 2-9 Ωcm. The Si nanoparticles were formed in the SiO_2 layer with thickness of 3 μm by co-sputtering of Si and SiO_2 targets and subsequently annealing at 1100 °C for 60 min (Fig. 1(a)) [5]. The Si nanoparticles with approximately 2.5 nm average sizes were confirmed by using high resolution-transmission electron microscope (HRTEM) measurement. The Si nanoparticles, then, were treated in HF acid solution for 60 min to remove thick SiO_2 layer on the surface after annealing (Fig. 1(b)). The film thickness of the HF treated sample was reduced to approximately 1 μm by the etching of SiO_2 layer. And, the Si nanoparticles were uniformly dispersed on surface of the sample. After the HF treatment, the Si nanoparticles were carried out the thermal oxidation to vary the size, which is closely related to the luminescent color (Fig. 1(c)). The thermal oxidation was carried out to some HF treated samples. The HF treated Si nanoparticles were oxidized by annealing at 600 °C and 900 °C in oxygen gas atmosphere for 30 sec. Next, the formation of top and bottom electrodes to the luminous layer was carried out (Fig. 1(d)). The indium tin oxide (ITO) electrode with scale of 2×2 mm and thickness of 200 nm was deposited by radio frequency (RF) sputtering technique on the surface of luminous layer. Ohmic contact with thickness of 70 nm onto back side of the Si substrate was formed by the evaporation of aluminum (Al) film. The ITO and the Al electrodes, then, were respectively connected with lead wire in silver (Ag) paste to apply the voltage between two electrodes of the EL device.

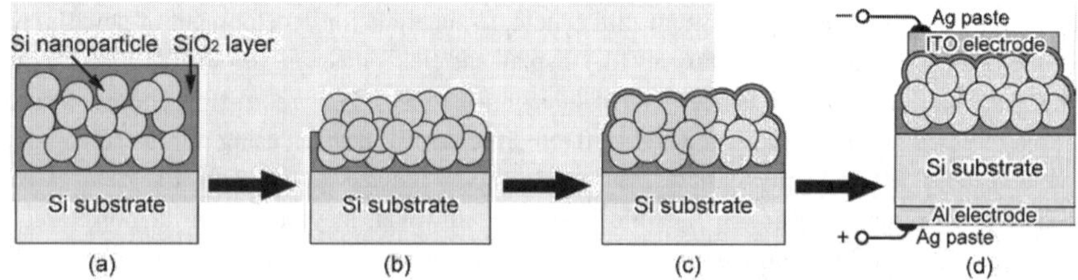

***Figure 1**. Schematic illustration of manufacturing process for EL device.*

The surface compositional, the luminescent and electrical properties of the EL device were characterized by using FT-IR, EL, current-voltage (*I-V*) measurements. The oxidation on Si nanoparticles surface was confirmed by using the FT-IR spectroscopy. The luminescence spectra were detected by using a photonic multichannel analyzer (Hamamatsu M5098) and were measured by applying the voltage up to ±20.0 V at room temperature. The *I-V* characteristics of EL device were measured by applying the voltage up to ±20.0 V at room temperature. The forward bias condition corresponds to the case in which a negative voltage is applied to the ITO electrode with respect to the Al electrode.

RESULTS AND DISCUSSION

Figure 2 shows the *I-V* characteristics of the HF treated and the oxidized EL devices in linear scale. These EL devices showed a rectifying behavior. The threshold forward voltage of current for each EL device was increased with an increase of the oxidation temperature. The HF treated EL device and the oxidized EL devices at 600 °C and 900 °C was at threshold voltage of 4.0 V, 9.0 V and 9.5 V, respectively. The light emission from these EL devices was observed by applying the forward bias above this threshold voltage.

Figure 3 shows the luminescence spectra for the EL devices used in Fig. 1. The HF treated EL device exhibited the strong and stable red luminescence with a peak of 670 nm in air [1]. On the other hand, the luminescent color from the oxidized EL device strongly depended on the oxidation temperature. The oxidized EL devices at 600 °C and 900 °C emitted green light at 560 nm and blue light at 410 nm, respectively. The green/blue luminescence could be strongly and stably seen with naked eye in air.

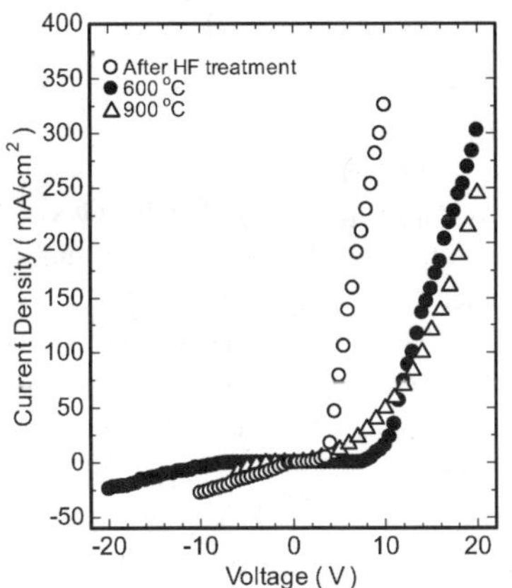

***Figure 2.** I-V characteristics of the HF treated and the oxidized EL devices.*

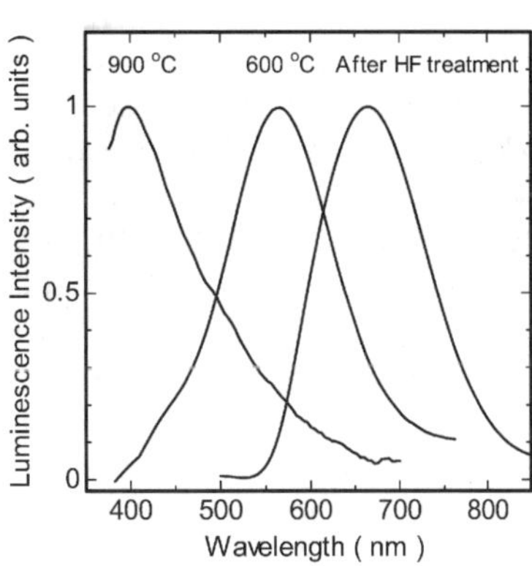

***Figure 3.** Luminescene spectra of the HF treated and the oxidized EL devices.*

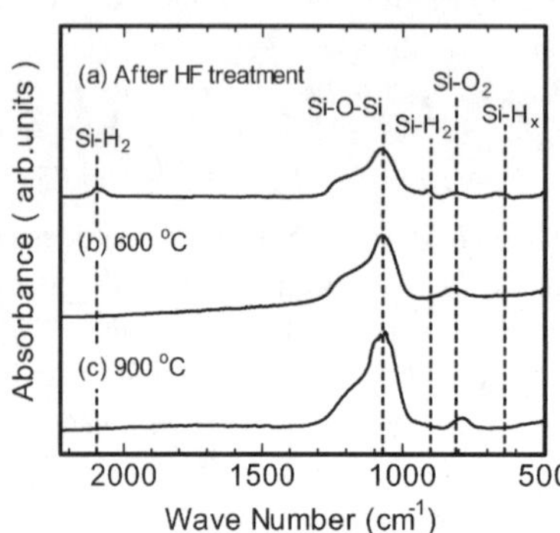

Figure 4. *FT-IR spectra of the luminous layer in the HF treated and the oxidized EL devices.*

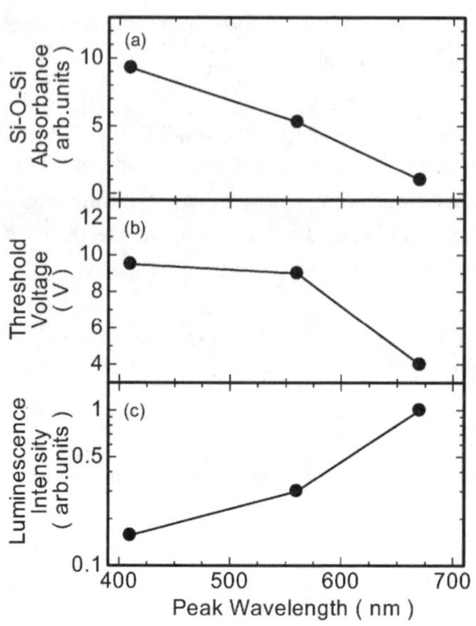

Figure 5. *Correlation between (a) IR absorption intensity of Si-O-Si (1070 cm^{-1}) (b) threshold voltage and (c) luminescence intensity for red, green and blue light emitted EL devices.*

Figure 4 shows the FT-IR spectra for the luminous layer in EL devices used in Fig. 2. The IR absorption peaks around 640, 810, 900, 1070 and 2100 cm^{-1} were observed for the HF treated Si nanoparticles as shown in Fig. 4(a). The peaks around 640, 900 and 2100 cm^{-1} were attributable to Si-H_x wagging mode, Si-H_2 scissors mode and Si-H_2 stretching mode, respectively [6]. And the peaks around 810 and 1070 cm^{-1} were corresponded to Si-O_2 and Si-O-Si stretching modes, respectively [6]. Therefore, the Si nanoparticle in the HF treated EL device is considered to terminate the Si atoms on the surface with hydrogen and oxygen atoms. On the other hand, the oxidized Si nanoparticle at 600 °C and 900 °C showed only the peaks of Si-O_2 and Si-O-Si stretching modes, and the peaks of Si-H_x wagging mode, Si-H_2 scissors and stretching modes were not found as shown in Figs. 4(b) and (c). These FT-IR results revealed that almost Si atoms on the Si nanoparticle surface in the oxidized EL device are terminated with oxygen atoms.

Figure 5 shows the correlation between (a) IR absorption intensity of Si-O-Si (1070 cm^{-1}) (b) threshold voltage and (c) luminescence intensity for red, green and blue light emitted EL devices. The blue-shift of peak wavelength from red light to blue light was generated with the increase in the Si-O-Si intensity. The Si-O-Si intensity for the EL device showed green/blue luminescence was about 5 and 10 times higher than that of the red luminescence, respectively, as shown in Fig. 5(a). Moreover, the increase of Si-O-Si intensity also results in an increase of the threshold voltage (Fig. 5(b)). The threshold voltage from red light to green light rapidly increased to about 2.3 times, while the threshold voltage from green light to blue light slightly increased to about 1.1 times. On the other hand, the increase of Si-O-Si intensity led to a decrease of the luminescence intensity (Fig. 5(c)). The intensity of green/blue luminescence decreased to about 1/3 and 1/5 lower than that of red luminescence.

From the above results, it was clarified that the Si nanoparticles based EL device emits strong (External quantum efficiency of ~ 0.35 %) and good stable (continuous operation above 10 days) red/green/blue light at low voltage can be achieved by relatively simple HF and oxidation techniques. The high efficiency and the good stable red luminescence at low forward bias of 4.0V was observed for the HF treated EL device. The HF treatment makes resistance of the EL device decrease by etching thick SiO_2 layer in the annealed film and the resultant realization of effective carrier (electron and hole) injection into the Si nanoparticles by applying the low forward bias of 4.0 V. Moreover, it is confirmed from electron spin resonance (ESR) measurement that the HF treatment also makes Si dangling bonds act as non-radiative recombination center at Si nanoparticle/SiO_2 layer interface compensate with hydrogen atoms [7]. Because of removal of non-radiative recombination center, the injected carrier into Si nanoparticles effectively flow to radiative recombination center and emits strong and stable red light in air. From these effects of HF treatment, the red light emitted EL device realized high brightness, good stability and low voltage. On the other hand, the oxidized EL device showed the green/blue luminescence. The thermal oxidation technique used in this study can freely control the size, which is closely related to the luminescent color, by the replacement of Si atoms existing most outside region of Si nanoparticle and oxygen atoms. In fact, it is confirmed from X-ray diffraction (XRD) measurement that the size is rapidly reduced with increasing the oxidation temperature up to 900 °C, because the increase of oxidation temperature leads to the increase in oxidation quantity onto the Si nanoparticle surface [8]. The oxidized EL devices at 600 °C and 900 °C had the Si nanoparticle with average size of 2.2 nm and 1.9 nm, respectively. Therefore, the green/blue luminescence is caused by the reduction of size under the thermal oxidation process. However, the thermal oxidation technique results in the increase of threshold voltage and the decrease of luminescence intensity with the blue shift of luminescent color. The increase of threshold voltage is due to the existence of thicker SiO_2 layer on the Si nanoparticles surface by the increase in oxidation temperature, because the thickness of SiO_2 layer thickens to supply much oxygen onto the surface under the high oxidation temperature. The EL device, which has thicker SiO_2 layer, can be injected carrier into Si nanoparticles through the SiO_2 layer by applying higher voltage. Therefore, the green/blue light emitted EL devices required the forward bias of 9.0 V and 9.5 V, which is higher than the red light emitted one, to inject carrier into the Si nanoparticles. Moreover, the decrease of luminescence intensity is the influence of the re-formation of Si dangling bonds onto the Si nanoparticle/SiO_2 layer interface under the growth process of SiO_2 layer. In our experiment, the increase of Si dangling bonds with increasing the oxidation temperature up to 900 °C is confirmed by using ESR measurement, because of intense desorption of hydrogen atoms, which terminates Si dangling bonds after HF treatment, at temperature above 200 °C [8]. Therefore, the injected carrier into Si nanoparticles flow to both radiative recombination center and non-radiative recombination center, and thereby the green/blue luminescence can be sufficiently seen with naked eye though the intensity of their luminescence weakens further than that of red luminescence. Furthermore, the good stability of their luminescence in air is advantageous even in the thermal oxidation technique to form SiO_2 layer, which is stable passivation. Consequently, the red/green/blue light emitted EL devices using Si nanoparticle fabricated in this study is useful for the development of Si-based new flat panel display.

CONCLUSIONS

We have studied surface compositional of Si nanoparticles, and the luminescent and electrical properties from the HF treated and the oxidized EL devices. The HF treated and the oxidized EL devices showed the strong and the stable red/green/blue luminescence in air under the forward bias below 9.5 V. The realization of multi-color luminescence with strong intensity and good stability is due to the reduction of size by the oxidation, and the existence of Si nanoparticle with low-density Si dangling bonds and SiO_2 layer on the surface. Consequently, the EL device developed in this study can be sufficiently expected for realization to future flat panel display.

ACKNOWLEDGMENTS

This work was supported by Research and Education Grant-in-Aid of the Illuminating Engineering Institute of Japan.

REFERENCES

1. K. Sato, T. Izumi, M. Iwase, Y. Show, S. Nozaki and H. Morisaki, *Mater. Res. Soc. Proc.* **638**, F14.30.1 (2001).
2. H. Mizuno and N. Koshida, *Mater. Res. Soc. Proc.* **536**, 179 (1999).
3. Mingxiang Wang, Kunji Chen, Lei He, Wei Li, Jun Xu and Xinfan Huang, Appl. Phys. Lett. **73**, 105 (1998).
4. D. Muller, P. Knápek, J. Fauré, B. Prevot, J. J. Grob, B. Hönerlage and I. Pelant, *Nucl. Instr. And Meth. In Phys. Res.* **B148**, 997 (1999).
5. K. Sato, T. Izumi, M. Iwase, Y. Show, H. Morisaki, T. Yaguchi and T. Kamino, *Appl. Surf. Sci.* **216**, 376 (2003).
6. W. Theiβ, *Surf. Sci. Rep.* **29**, 91 (1997).
7. K. Sato, Y. Sugiyama, T. Izumi, M. Iwase, Y. Show, S. Nozaki and H. Morisaki, *Electrochem. Soc. Proc.* **99-22**, 240 (1999).
8. K. Sato, K. Hirakuri, M. Iwase, T. Izumi and H. Morisaki, *National Conference of Nano Science and Technology*, May, 30, 2003 [in Japanese].

Mat. Res. Soc. Symp. Proc. Vol. 789 © 2004 Materials Research Society N11.2

Exciton photoluminescence and energy transfer in nanocrystalline Si/ Si dioxide superlattice structures

V. Yu. Timoshenko (a), O. A. Shalygina (a), M. G. Lisachenko (a), P. K. Kashkarov (a), D. Kovalev (b), J. Heitmann (c), M. Zacharias (c), B. V. Kamenev (d), L. Tsybeskov (d)

(a) Moscow State M.V. Lomonosov University, Physics Department, 119992 Moscow, Russia

(b) Munich Technical University, Physics Department E16, 85747 Garching, Germany

(c) Max-Planck-Institute of Microstructure Physics, Weinberg 2, 06120 Halle, Germany

(d) Electrical and Computer Engineering Department, New Jersey Institute of Technology, University Heights, Newark NJ 07102

ABSTRACT

Photoluminescence (PL) of nanocrystalline Si (nc-Si) assemblies formed by thermal crystallization of amorphous Si/SiO_2 and SiO/SiO_2 superlattices (SLs) has been investigated at different temperatures and excitation conditions. The low temperature resonant PL spectroscopy reveals phonon-assisted excitonic recombination. At room temperature the samples formed from a-SiO/SiO_2 SLs possess relatively high PL quantum yield (~ 1%). The PL transients have non-exponential decay, which indicates the exciton energy transfer in nc-Si ensembles. The excitonic energy of Er-doped nc-Si SL structures can be almost completely transferred to Er ions incorporated in SiO_2 matrix that results in a strong emission line at 0.81 eV.

INTRODUCTION

Nanocrystalline Si (nc-Si) structures fabricated by controlled crystallization of amorphous Si/SiO_2 [1] and SiO/SiO_2 [2] superlattices (SLs) are attractive for future electronic and optical applications because of their well defined structure and full compatibility with the standard Si technology. The structures formed from a-Si/SiO_2 SLs have a well-defined order in the direction perpendicular to the layer, but are disordered laterally due to the grain boundaries [1]. The electrical transport in such structures at low temperatures has been shown to be governed by resonant hole tunneling via quantized valence band states in nc-Si [3]. The nc-Si assemblies formed from a-SiO/SiO_2 SLs consist of 3D-arranged size-controlled Si nanocrystals, which exhibit rather strong room temperature photoluminescence (PL) [2]. The Er doping of nc-Si SL structures results in a material which can emit efficiently at 1.5 μm [4]. In the present work, the PL properties of the both types of nc-Si structures are investigated under different temperatures and excitation conditions. We demonstrate that the excitonic energy transfer influences the photoexcited electron-hole recombination in nc-Si ensembles.

EXPERIMENTAL DETAILS

Amorphous SiO/SiO_2 SLs were prepared by alternating reactive evaporation of SiO powder in vacuum or in oxygen atmosphere on c-Si substrate (see for details [2]). The thickness of SiO layers d_{SiO} was varied from 2 to 6 nm, while the SiO_2 layer thickness d_{SiO2} was 2-4 nm. The number of SiO/SiO_2 periods was typically 30-70. The samples were subjected to a conventional furnace annealing at 1100°C in N_2-atmosphere for 1 hour. Some part of the samples were Er-implanted at 300 keV with dose of $2 \cdot 10^{15}$ cm^{-2}. The Er-implanted samples were subjected to a thermal annealing at 900°C for 64 min in order to anneal the implantation defects. The mean nanocrystal size $d_{nc\text{-}Si}$ was found to be close to d_{SiO} and remains constant after Er implantation and following annealing. The inset of Fig. 1 shows a transmission electron microscopy (TEM) cross-sectional image of a nc-Si/SiO_2 SL structure.

Samples based on a-Si/SiO_2 SLs were prepared on c-Si substrates by magnetron sputtering followed by controlled crystallization. Finally they consist of 10-20 alternating layers of nc-Si ($d_{nc\text{-}Si}= d_{Si} = 4$ nm) and a-SiO_2 (d_{SiO2}=1-2 nm) (see for details [1, 3]).

A HeCd laser (E_{exc} =2.8 eV) or a N_2-laser (E_{exc} =3.7 eV, pulse duration 10 ns) were used for the PL excitation. The resonant PL excitation was done by a Ti-sapphire laser (E_{exc}=1.4 – 1.7 eV) pumped by an Ar^+-laser. The intensities of both CW and pulsed lasers were chosen to be kept in a linear regime of the PL excitation. The PL in the spectral region from 0.7 eV to 2.5 eV was detected by using a spectrometer equipped with a CCD array or InGaAs photodiode. The PL spectra were corrected for spectral response of the measurement system. Transient PL experiments were carried out using a mechanical modulation of the exciting light. The PL transients in the spectral range from 1.1 to 2.2 eV were measured by a S1-photomultimpler with a preamplifier (time response 10 μs).

RESULTS AND DISCUSSIONS

Figure 1 shows PL spectra of different nc-Si/SiO_2 structures at room temperature. The PL intensity of the samples formed from a-SiO/SiO_2 SLs is much higher than that for the a-Si/SiO_2 based.

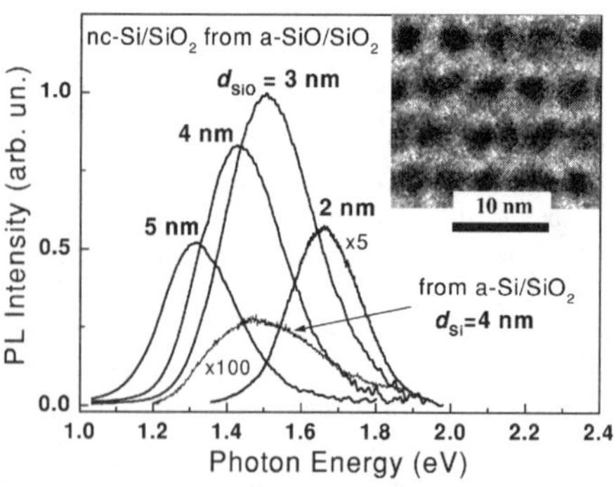

Figure 1. Room temperature PL spectra of nc-Si structures prepared from a-SiO/SiO_2 (d_{SiO} = 5, 4, 3 and 2 nm) and a-Si/SiO_2 (d_{Si} = 4 nm) SLs. E_{exc} =2.8 eV. Inset: TEM cross-sectional image of an a-SiO-based structure with d_{SiO} = 3 nm.

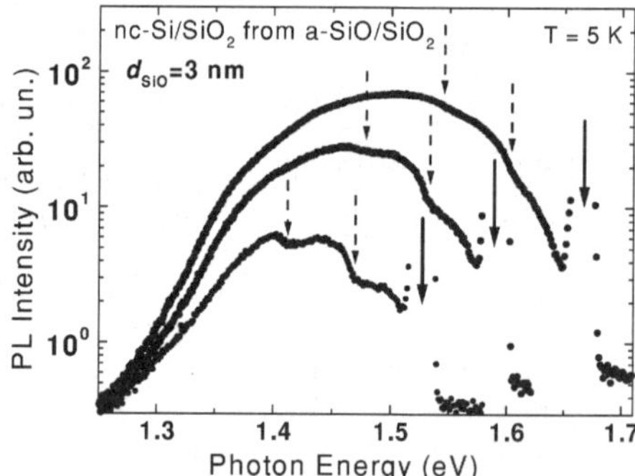

Figure 2. Resonantly excited PL spectra of a sample formed from a-SiO/SiO_2 SL. The excitation energies and momentum conserving TO and 2TO phonon replicas are indicated by solid and dashed arrows, respectively.

The maximal PL quantum yield (~1%) is observed for the former samples with d_{SiO} = 3-4 nm. The estimated yield gives probably the lowest limit because the exciting light is not completely absorbed in the investigated nc-Si structures due to their small thickness (< 200 nm). The rather high PL yield of the samples formed from a-SiO/SiO_2 SLs can be explained by a good passivation of non-radiative recombination defects and the optimal spacing of size-controlled Si nanocrystals. These structural properties result in a suppression of non-radiative losses of photoexcited carriers. Indeed, the PL yield of our nc-Si SLs is almost independent of the temperature.

Figure 2 presents the resonantly excited PL spectra of a sample produced from an a-SiO-based SL. The momentum conserving TO and 2TO phonon replicas at about 56 and 112 meV below the excitation line, respectively, become more pronounced when the excitation photon energy falls into the PL band. The similar phonon related features were also observed in the resonant PL spectra of nc-Si structures formed from a-Si/SiO_2 SLs [5]. Note, that the phonon replicas are not very sharp in our nc-Si structures. Furthermore, the ratio between the 2TO- and TO-related PL features as well as between the TO-related and no-phonon ones is about 2-3, what is smaller than that for porous Si [6]. This fact indicates an additional breaking of the momentum conservation rule for the exciton radiative transition in nc-Si/SiO_2 structures.

Figure 3 shows typical PL transients of nc-Si SLs at different temperatures. The PL transients $I_{PL}(t)$ can be well fitted by a stretch exponential function, defined as

$$I_{PL}(t) = I_0 \exp\left\{-\left(t/\tau_0 \right)^{\beta}\right\}, \qquad (1)$$

where τ_0 is the mean lifetime and β is the dispersion parameter. The stretched exponential decay with $\beta < 1$ corresponds to the existence of a distribution of PL lifetimes, which can be due to a diffusive motion of photoexcited carriers in nc-Si ensemble [7]. Also, the non-exponential decay can be caused by an energy transfer from nc-Si, which does not imply the charge motion. For instance, it can be realized via the Förster mechanism of dipole-dipole interaction [8].

Figure 4 depicts the temperature dependence of τ_0 and β obtained from the stretched exponential fits of the PL transients presented partially in Fig. 3.

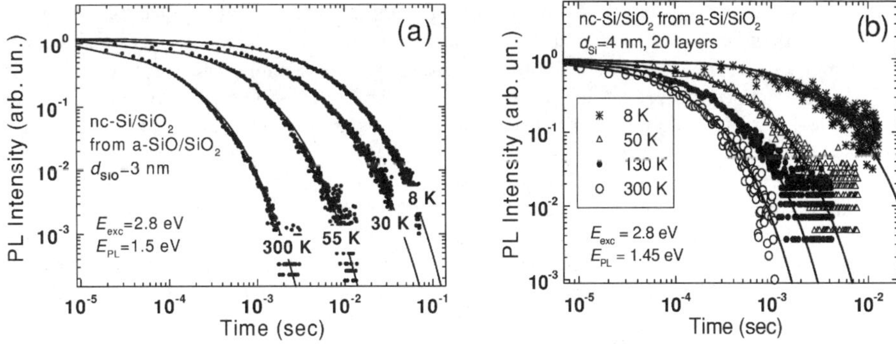

Figure 3. PL transients of nc-Si structures at different temperatures. The samples prepared from a-SiO/SiO_2 (a) and a-Si/SiO_2 (b) SLs. The lines are fits according to Eq. (1).

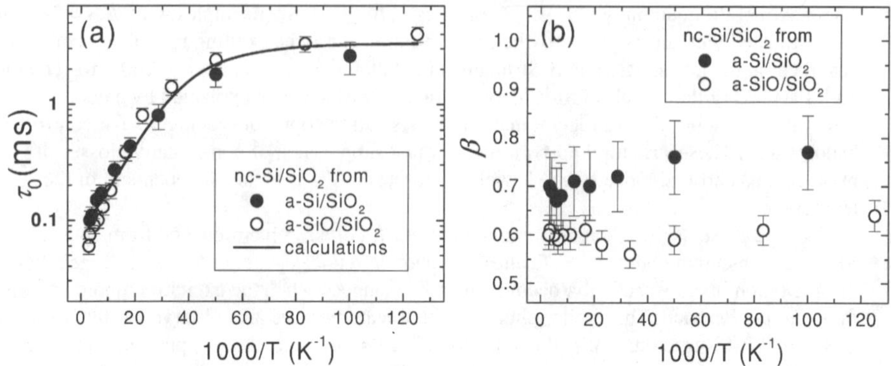

Figure 4. Temperature dependence of τ_0 (a) and β (b) obtained from the PL transient fitting as in Fig.3. The curve through the values of τ_0 is calculated according to Eq. (2).

The temperature dependence of τ_0 for our samples is well described by a simple model of the radiative singlet-triplet excitonic recombination [9]. The exciton lifetime can be written as

$$\tau_0 = \left(3 + \exp(-\Delta / kT)\right) / \left(3\tau_T^{-1} + \tau_S^{-1} \exp(-\Delta / kT)\right) , \qquad (2)$$

where Δ is the exciton exchange splitting energy, τ_T and τ_S are the triplet and singlet lifetimes, respectively. The fit of the experimental data according to Eq.(2) gives the following parameters: Δ=9 meV, τ_T =3 ms and τ_S=20 μs, which are close to the results for Si nanocrystals distributed randomly in SiO_2 [10]. Note, that the obtained value of Δ is ~2 times larger, while τ_T, and τ_S are 2-3 times smaller than the corresponding parameters for the porous Si PL at the same detection energy [6]. This fact can be explained by different boundary conditions for excitons in nc-Si/SiO_2 and in porous Si. In particular, the shorter τ_S for nc-Si/SiO_2 means faster rate of the radiative recombination, which correlates with the observed enhancement of the no-phonon transition in the resonantly excited PL. Also, the lifetimes can be shorted by the energy transfer.

As one can see from Fig. 4 (b), at room temperature β is equal to 0.7 and 0.6 for the samples made from a-Si- and a-SiO-based SLs, respectively. While β increases slightly with temperature decrease, at any temperature it is smaller for the a-SiO-based structure. This fact evidences a larger lifetime distribution in the a-SiO-based SLs. Since this type of nc-Si structures is characterized by the small dispersion of $d_{\text{nc-Si}}$ [2], the lifetime distribution should be caused by the transfer of charge or/and energy of photoexcited carriers through nc-Si arrays. This process differs from that in porous Si, where the thermally activated dependence of β was observed and attributed to the trap-controlled hopping mechanism [7].

We believe that the non-exponential PL decay in nc-Si/SiO_2 SL structures originates from the energy transfer to neighboring Si nanocrystals, i.e. the exciton migration. The small size dispersion of Si nanocrystals as well as the thin SiO_2 barriers between them (see inset of Fig. 1) favor the exciton migration in the nc-Si ensemble at least for the in-plane direction. In general, the migration is possible through SiO_2 barriers even without tunneling because it can be realized via the long-range Coulomb interactions, e.g. via the Förster mechanism.

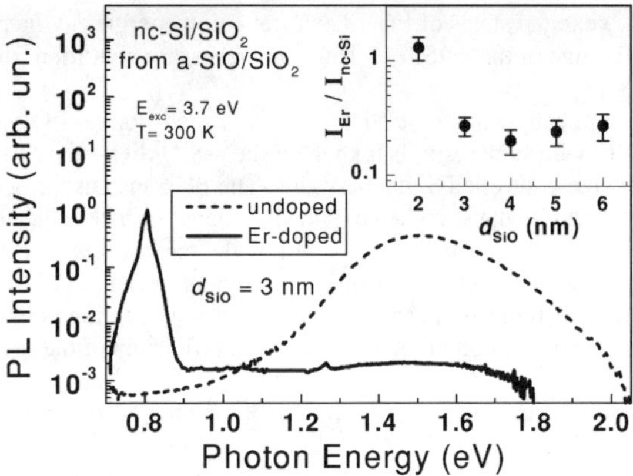

Figure 5. Room temperature PL spectra of undoped and Er-doped nc-Si structures formed from a-SiO/SiO_2 SL. Inset: ratio between the total yields of the exciton PL in the undoped samples and Er^{3+} emission in the Er-doped structures *versus* d_{SiO}.

The possibility of energy transfer through the nc-Si/SiO_2 SL structure can be employed to excite optically active centers like rare earth ions embedded in SiO_2. Figure 5 shows typical room temperature PL spectra of undoped and Er-doped nc-Si/SiO_2 structures formed from a-SiO/SiO_2 SLs. A strong quenching of the intrinsic excitonic PL at ~1.5 eV and an appearance of the Er^{3+} emission at 0.8 eV are observed for the Er-doped sample. The similar effects take place for all Er-implanted samples formed from a-SiO/SiO_2 SLs. We note, that the Er-related PL of the reference sample (a homogeneous 200 nm thick a-SiO_2 layer implanted with the same Er dose) was 3-4 orders of magnitude weaker. Therefore the Er^{3+} ions in the nc-Si/SiO_2 structures are sensitized by nc-Si.

The quantitative analysis of the efficiency of the energy transfer from nc-Si to Er^{3+} is possible when the ratio between the spectrally integrated intensities of the Er^{3+} PL in Er-doped structures I_{Er} (0.75-0.9 eV) and the exciton PL in the undoped reference samples $I_{nc\text{-}Si}$ (1.1-2.1 eV) is considered. The inset of Fig. 5 shows that the ratio I_{Er}/I_{nc} is about 0.2-0.3 for the samples with d_{SiO} = 3-6 nm and it reaches 1.3±0.3 for d_{SiO} = 2 nm. The last value implies that the energy transfer from nc-Si to Er^{3+} competes successfully with non-radiative recombination in the nc-Si structure and even initially "dark" nanocrystals into them can contribute to the Er^{3+} sensitization. It means that the energy transfer event occurs very fast, i.e. on a submicrosecond time scale.

It should be noted, that Si nanocrystals in the samples produced from a-SiO/SiO_2 SLs are spatially closely assembled and thicknesses of SiO_2 barriers between them are typically 1-2 nm and 2-4 nm for the in-layer and perpendicular to layer directions, respectively (see inset in Fig. 1). Since the optically active Er^{3+} ions are mainly located in the SiO_2 matrix, they are maximum 2 nm away from Si nanocrystals. For such short distances the coupling between excitons in nc-Si and Er^{3+} should be very efficient, considering both the Förster transfer mechanism [8] or higher order multipole interactions [11]. The energy transfer between Si nanocrystals can further increase the efficiency of the Er^{3+} sensitization. The observed dependence of $I_{Er}/I_{nc\text{-}Si}$ on d_{SiO} evidences the coupling between the excitonic states in nc-Si and upper (2nd, 3rd, and 4th) excited states of Er^{3+}. Their transition energies to the ground state of Er^{3+} lie within the spectral range of the nc-Si PL. Thus, the necessary condition for the Förster-Dexter transfer is satisfied [8,11].

In conclusion, the optically excited nc-Si structures formed from a-SiO/SiO_2 and a-Si/SiO_2 SLs show excitonic PL, whose intensity is higher for the a-SiO/SiO_2-based samples due to the better arrangement of size-controlled Si nanocrystals. The PL transients of both types of the structures are characterized by the stretched exponential decay, which indicates the energy transfer among nc-Si. The energy transfer is favored by the narrow size distribution of Si nanocrystals and small distances between them. The efficient excitation of Er^{3+} in the Er-doped nc-Si/SiO_2 SL structures is found and explained by the strong coupling between the excitons in nc-Si and Er^{3+} ions in surrounding SiO_2 matrix. The high efficiency of the excitonic PL as well as the possibility of its almost complete conversion into the Er^{3+} emission provide perspectives for the optoelectronic applications of the size-controlled nc-Si/SiO_2 SL structures.

ACKNOWLEDGMENTS

The authors would like to thank M. Schmidt and G. Lenk for the assistance in the sample preparation. The research described in this publication was supported partially by Award No. RE2-2369 of the U.S. Civilian Research and Development Foundation (CRDF). J.H. and M.Z. acknowledge the support of the Volkswagen Stiftung.

REFERENCES

1. L. Tsybeskov, K. D. Hirschman, S. P. Duttagupta, M. Zacharias, P. M. Fauchet, J. P. McCaffrey, and D. J. Lockwood, *Appl. Phys. Lett.* **72**, 43 (1998).
2. M. Zacharias, J. Heitmann, R. Shcholz, U. Kahler, M.Schmidt, J. Bläsing, *Appl. Phys. Lett.* **80**, 661 (2002).
3. L. Tsybeskov, G. F. Grom, R. Krishnan, L. Montes, P. M. Fauchet, D. Kovalev, J. Diener, V. Timoshenko, F. Koch, J. P. McCaffrey, J.-M. Baribeau, G. I. Sproule, D. J. Lockwood, Y. M. Niquet, C. Delerue, and G. Allan, *Europhys. Lett.* **55**, 552 (2001).
4. M. Schmidt, J. Heitmann, R. Scholz, M. Zacharias, *J. Non-Cryst. Sol.* **299-302**, 678 (2002).
5. G. F. Grom, D. J. Lockwood, J. P. McCaffrey, H. Labbe, P. M. Fauchet, B. White, J. Dienner, D. Kovalev, F. Koch, and L. Tsybeskov, *Nature* 407, 358-361 (2000).
6. D. Kovalev, H. Heckler, G. Polisski, F. Koch, *Phys. Stat. Sol. (b)* **215**, 871 (1999).
7. L. Pavesi and H. E. Roman in *Microcrystalline and Nanocrystalline Semiconductors*, edited by R. W. Collins, Ch. Ch. Tsai, M. Hiroshe, F. Koch, and L. Brus (Mat. Res. Soc. Symp. Proc. **358**, Pittsburgh, PA, 1995) pp. 545-554.
8. Th. Förster, *Annal. Phys.* **6**, 55 (1948).
9. P.D.J. Calcott, K.J. Nash, L.T. Canham, M.J. Kane, and D. Brumhead, *J. Phys.: Condens. Matter* **5**, L91 (1993).
10. S. Takeoka, M. Fujii, S. Hayashi, *Phys. Rev.* **B 62**, 16820 (2000).
11. D. L Dexter, *J. Chem. Phys.* **21**, 836 (1953).

Mat. Res. Soc. Symp. Proc. Vol. 789 © 2004 Materials Research Society N11.6

Photocurrent in a hybrid system of 1-thioglycerol and HgTe quantum dots

Hyunsuk Kim, Kyoungah Cho, Byungdon Min, Jong Soo Lee, Man Young Sung, Sung Hyun Kim[1] and Sangsig Kim

Department of Electrical Engineering, Korea University, Seoul 136-701, Korea

[1]Department of Chemical and Biological Engineering, Korea University, Seoul 136-701, Korea

ABSTRACT

Photocurrent mechanism in a hybrid system of 1-thioglycerol and HgTe quantum dots(QDs) was studied for the first time in the intra-red (IR) range. 1-thioglycerol-capped HgTe QDs were prepared using colloidal method in aqueous solution; the synthesis and size of the HgTe QDs were examined by x-ray diffraction, Raman scattering, and high-resolution transmission electron microscopy. Absorption and photoluminescence spectra of the capped HgTe QDs revealed the strong excitonic peaks in the range from 900 to 1100nm, because of their widened band gap due to the shrinkage of their sizes to about 3 nm. The wavelength dependence of the photocurrent for the hybred system of the 1-thioglycerol and HgTe QDs was very close to that of the absorption spectrum, indicating that charge carriers photoexcited in the HgTe QDs give direct contribution to the photocurrent in the medium of 1-thioglycerol. In this hybrid system, the photo-excited electrons in the HgTe QDs are strongly confined, but the photo-excited holes act as free carriers. Hence, in the photocurrent mechanism of the this hybrid system, only holes among electron-hole pairs created by incident photons in the HgTe QDs are transferred to 1-thioglycerol surrounding HgTe QDs and contribute photocurrent flowing in the medium of 1-thioglycerol.

INTRODUCTION

Semiconductor quantum dots exhibit unique physical properties with respect to bulk regimes. Their advantages include the tunable effective bandgap, the enhanced quantum efficiency due to discrete energy density of states, and the isotropic absorption efficiency independent of polarization of incoming light. Because of these advantages, semiconductor quantum dots (also called nanocrystals or nanoparticles) are very prospective materials for a variety of optoelectronic devices including light-emitting diodes (LEDs)[1-2] and photodetectors [3-8].

For HgTe QDs, the quantum confinement of charge carriers in these QDs due to the shrinkage of their sizes leads to their effective bandgap in the range from ultra-far infrared to near infrared wavelength [9]; note that bulk HgTe has been known as semi-metal material with the bandgap of near zero eV at room temperature. In this study, room-temperature photocurrent of a hybrid system of 1-thioglycerol-capped HgTe QDs in the visible and infrared wavelength range is investigated and a simple model is suggested for the description of charge transport mechanism in this hybrid system.

EXPERIMENTS

1-thioglycerol-capped HgTe QDs were synthesized in aqueous solution by the colloidal method [9]. The solution containing the QDs was dropped and dried on silicon substrates, for preparation of films. The synthesized HgTe QDs were characterized by x-ray diffraction(XRD), Raman scattering, and high-resolution transmission electron microscope(HRTEM) and their optical properties were examined with absorption and photoluminescence spectroscopies. The films were contacted laterally with a separation of ~ 1mm to the electrodes for photocurrent measurement. The light source for wavelength-dependent photocurrent measurement was a 450-W xenon lamp dispersed by a monochromator and a SR830 lock-in amplifier was utilized to measure the photocurrent between the electrodes. All the measurements were performed at room temperature.

RESULTS

Figure 1 shows XRD pattern(inset is HRTEM image) (a), and Raman scattering spectrum (b) HgTe QDs prepared by colloidal method; for comparison, XRD pattern and Raman scattering spectrum of HgTe microcrystallites with an average size of 44 μm are given in this figure. The XRD pattern of the synthesized HgTe QDs matches well that of the HgTe microcrystallites with the exception of the broadening of the XRD peaks for the HgTe QDs. The broadening originates from the nanometer-scaled diminution of the crystallite sizes. The HRTEM image(the inset of Fig. 1(a)) confirms that the synthesized HgTe QDs have high crystalline qualities and the average size of HgTe particles is about 5nm in diameter. In the Raman scattering spectrum of the HgTe QDs (Fig. 1(b)), the characteristic vibration bands peaking at 119 and 139 cm^{-1} correspond to the TO and LO phonon lines, respectively. These peak positions match well the corresponding values for the HgTe microcrystallites. (121 and 145 cm^{-1}, respectively)

Figure 2 shows PL and absorption spectra, and photocurrent taken for the colloidally synthesized HgTe QDs capped by 1-thioglycerol. In the PL spectrum of the HgTe QDs in aqueous solution (on the right of Fig. 2(a)), a broad emission band peaks at 1050 nm, ranging from 800 to 1300 nm; this band is attributed to the recombination of excitons. In the absorption spectrum of the HgTe QDs in aqueous solution (on the left of Fig. 2(a)), a strong excitonic absorption band peaks at ~900 nm. The energy difference between the peak positions of the emission and absorption bands may be caused by the Frank-Condon shift. The photocurrent of

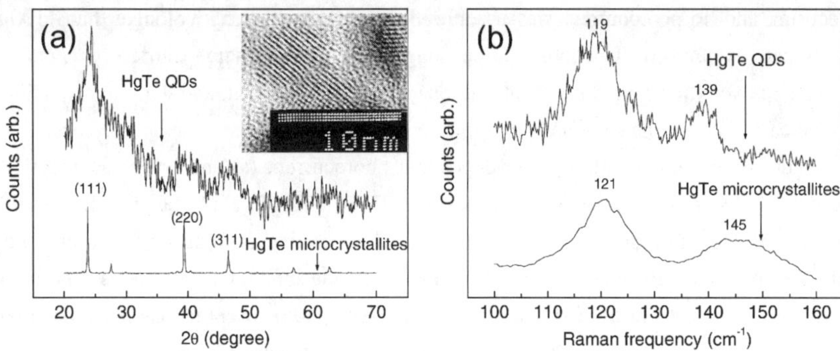

Figure 1. XRD patterns (a) (the inset is HRTEM image) and Raman scattering spectra of the HgTe QDs and microcrystallines.

the hybrid system of the 1-thioglycerol-capped HgTe QDs is measured with chopped light of wavelength selected in the range from 1100 nm to 650 nm (Fig. 2(b)). 1-thioglycerol acts not only as capping material of the HgTe QDs but also a medium in which the HgTe QDs are scattered and isolated, so that this capping material provides the channels for the photocurrent; it is suggested here that the medium consists of the necked capping 1-thioglycerol. The lineshape of the photocurrent as a function of wavelength is very close to that of the absorption

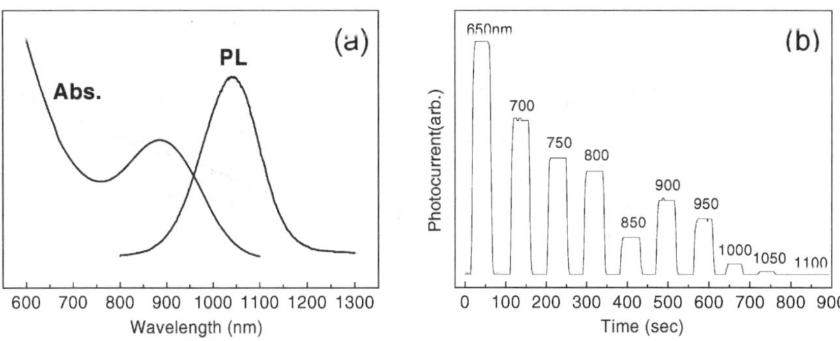

Figure 2. PL and absorption spectra (a) of the HgTe QDs in aqueous solution, and the photocurrent (b) of a hybrid system of 1-thioglycerol capped HgTe QDs with chopped light of wavelengths selected in the range from 1100nm to 650 nm.

spectrum, and no photocurrent was observed for light of wavelengths longer than 1100 nm (or, below-gap excitation). This observation suggests that free charge carriers generated by the above-gap excitation give direct contributions to the photocurrent, and that the excitation of free charge carriers from the sub-bandgap trap states doesn't contribute to any photocurrent.

Figure 3 shows excitation-power-dependent photocurrent (a) and I-V characteristics (b) for the hybrid system of the 1-thioglycerol-capped HgTe QDs. In Fig. 3(a), the photocurrent as a function of excitation power of the continuous 514.5nm line from an Ar ion laser is compared with the integrated PL intensity shown in the inset; the integrated PL intensity is obtained by integrating PL spectrum from 900 nm to 1200 nm. The photocurrent increases exponentially with the enhanced excitation power, while the integrated PL intensity represents a linear increase. The exponential increase of the photocurrent might imply the existence of a threshold barrier for the photocurrent channel, and the linear increase of the integrated PL intensity indicates that electron-hole pairs are generated in linear proportional to the number of incident photons; with raised excitation power, the photocurrent efficiency increases exponentially, and the radiative efficiency is independent. Figure 3(b) shows I-V characteristics of the hybrid system of the 1-thioglycerol-capped HgTe QDs in the dark and under illumination. A distinct photocurrent difference of about 100 nA for dark and illumination states is observed at 3 V and for the 514.5-nm wavelength excitation (100mW), but the dark current is significantly large; the photocurrent difference increases exponentially with the raised excitation power, nevertheless, the photocurrent is substantially lower than the dark current for excitation power weaker than

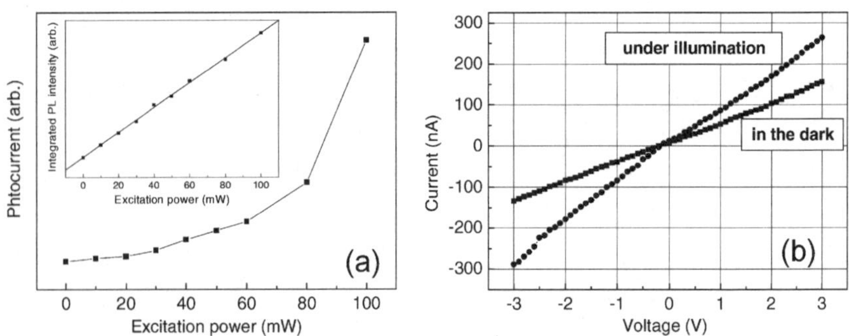

Figure 3. Photocurrent (a) as a function of excitation power of the continuous 514.5nm line from an Ar ion laser (the inset represents the integrated PL intensity), and I-V curves (b) of the hybrid system of the 1-thioglycerol-capped HgTe QDs in the dark and under illumination (100mW, 514.5nm).

1mW, corresponding to about 3mW/cm^2 in radiation. The origin of such high dark current is supposed to be wetted 1-thioglcerol. A piece of evidence for this is that dark current drops remarkably in vacuum. The medium of 1-thioglcerol is highly hydroscopic at ambient atmosphere, so charged ions dissolved in water (absorbed by 1-thioglcerol) may form the dark current with applied voltage.

Energy diagram of a 1-thioglycerol-capped HgTe QD and illustration describing photocurrent mechanism in the hybrid system of 1-thioglycerol-capped HgTe QDs are shown in Fig. 4. The energy diagram is drawn in Fig. 4 on the basis of the relative energy positions of the conduction band and valence band for the 1-thioglycerol and bulk HgTe with respect to vacuum, and the PL peak position of the HgTe QDs; the averaged effective bandgap of the HgTe QDs is estimated to be ~1.18eV from the PL peak position of 1050nm. In the energy diagram, the topmost of the valence band for the HgTe QDs is lower by 0.33eV in energy than that for 1-thioglycerol, so photogenerated holes may not be confined in the HgTe QDs. In other words, for the HgTe QDs, the subbands for holes in the valence band do not vary with decrease of particle size, because of the lower energy of the valence band of 1-thioglycerol than that of HgTe bulk. In contrast, the subbands for electrons in the conduction band is raised in energy as the particle size is smaller, because electrons are confined in the potential well of the conduction band edge of HgTe with the potential barriers of the conduction band of 1-thioglycerol. The widening of the effective bandgap with the shrinkage of the particle size is due to the blue shift of the subbands for the electrons in the conduction band. In this hybrid system, the photo-excited electrons in the HgTe QDs are strongly confined, but the photo-excited holes act as free carriers. Hence, in the photocurrent mechanism of this hybrid system, only holes among electron-hole pairs created by incident photons in the HgTe QDs are transferred to 1-thioglycerol surrounding the HgTe QDs and they contribute to the photocurrent flowing in the medium of 1- thioglycerol.

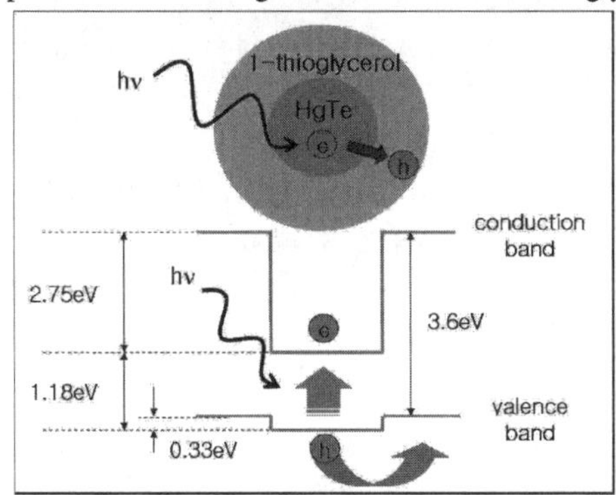

Figure 4. Energy diagram of a 1-thioglycerol-capped HgTe QDs

CONCLUSIONS

The photocurrent mechanism for a hybrid system of 1-thioglycerol-capped HgTe QDs was studied for the first time in the visible-infrared wavelength range. Absorption and photoluminescence spectra taken for the capped HgTe QDs reveal strong excitonic peaks in the wavelength range from 800 to 1150 nm, because of their widened effective band gap due to the shrinkage of their sizes to about 5 nm. The wavelength dependence of the photocurrent for the hybrid system was very close to that of the absorption spectrum. Energy diagram of this hybrid system illustrates that photogenerated electrons in the HgTe QDs are strongly confined, but that photogenerated holes act as free carriers. Hence, in the photocurrent mechanism of this hybrid system, the photogenerated holes in the HgTe QDs contribute to the photocurrent flowing the media of the capping 1-thioglycerol.

ACKNOLEDGEMENTS

This research (paper) was performed for the Carbon Dioxide Reduction & Sequestration Center, one of the 21st Century Frontier R&D Programs funded by the Ministry of Science and Technology of Korea(M102KP010001-02K1601-01310) and National R&D Project for Nano Science and Technology

REFERENCES

1. V. L. Colvin, M. C. Schlamp, and A. P. Alivisatos, Nature **370**, 354 (1994).
2. Mingyuan Gao, Constanze Lesser, Stefan Kirstein, Helmuth Mohwald, Andrey L. Rogach, and Horst Weller, J. Appl. Phys. **87**, 2297 (2000).
3. D. S. Ginger, and N. C. Greenham, J. Appl. Phys. **87**, 1361 (2000).
4. C. A. Leatherdale, C. R. Kagan, N. Y. Morgan, S. A. Empedocles, M. A. Kastner, and M. G. Bawendi, Phys. Rev. **B62**, 2669 (2000).
5. Nicole Y. Morgan, C. A. Leatherdale, M. Drndic, Mirna V. Jarosz, Marc A. Kastner, and Moungi Bawendi, Phys. Rev. **B66**, 075339 (2002).
6. R. A. M. Hikmet, D. V. Talapin, and H. Weller, J. Appl. Phys. **93**, 3509 (2003)
7. J. Nanda, K. S. Narayan, Beena Annie Kuruvilla, G. L. Murthy, and D. D. Sarma, Appl. Phys. Lett. **72**, 1335 (1998).
8. K. S. Narayan, A. G. Manoj, J. Nanda, and D. D. Sarma, Appl. Phys. Lett. **74**, 871 (1999).
9. Andrey Rogach, Stephen Kershaw, Mike Burt, Mike Harrison, Andreas Kornowski, Alexander Eychmuller, and Horst Weller, Adv. Mater. **11**, 552 (1999).

Mat. Res. Soc. Symp. Proc. Vol. 789 © 2004 Materials Research Society N11.8

VISIBLE LIGHT EMISSION FROM ERBIUM DOPED YTTRIA STABILIZED ZIRCONIA

Michael Cross and Walter Varhue
Materials Science Program, Dept of Electrical and Computer Eng
University of Vermont, Burlington VT 05405, U.S.A

ABSTRACT

One of the major shortcomings of silicon (Si) as a semiconductor material is its inability to yield efficient light emission. There has been a continued interest in adding rare earth ion impurities such as erbium (Er) to the Si lattice to act as light emitting centers. The low band gap of Si however has complicated this practice by quenching and absorbing this possible emission. Increasing the band gap of the host has been successfully tried in the case of gallium nitride (GaN) [1] and Si-rich oxide (SRO) [2] alloys. A similar approach has been tried here, where Er oxide (ErO_x) nanocrystals have been formed in a yttria stabilized zirconia (YSZ) host deposited on a Si (100) substrate. The YSZ is deposited as a heteroepitaxial, insulating layer on the Si substrate by a reactive sputtering technique. The Er is also incorporated by a sputtering process from a metallic target and its placement in the YSZ host can be easily controlled. The device structure formed is a simple metal contact/insulator/phosphor sandwich. The device has been found to emit visible green light at low bias voltages. The advantage of this material is that it is much more structured than SiO_2 which can theoretically lead to higher emission intensity.

INTRODUCTION

Numerous methods have been developed to create an Er doped LED, however, none these have utilized YSZ. Successful light emitting devices have been created in poly-Si [3], SRO [2-5], Si nanocrystals (SNC) [6,7], SiO_2 [8], Si nitride (SiN) [9], and most recently in GaN [1,10]. All of these systems have similar basic characteristics. The primary goal is to produce a device which is compatible with current silicon processing techniques in order to be integrated into business as usual (BAU) fabrication. Secondly, the material must have a wide band gap so as not to attenuate the emitted radiation through band to band absorption. In addition, the light emitting material must have very low defect levels. Defects lead to non-radiative recombination and reduce or eliminate emission. Finally, the device must be able to operate efficiently at room temperature. Yttria stabilized zirconia has been considered as a possible replacement for SiO_2 as the gate oxide in MOS devices. Published results suggest ease of integration with current Si manufacturing techniques. Having a single crystal structure, YSZ can support the electron acceleration required for impact excitation. High quality crystalline Er doped YSZ films suggest that recombination defects will also be minimized.

In order for the device to emit light, the Er must form ErO_x complexes. This process is carried out during the anneal step. Multiple annealing conditions were investigated in order to optimize the device performance. Annealing has two main benefits: it repairs defects in the bulk film and promotes the formation of the ErO_x nanocrystals. Here, the Er atoms become surrounded by six oxygen atoms forming ErO_x complexes.

Device operation is governed by hot electron impact excitation. The collision between the accelerated electrons and ErO_x nanoparticles excites an electron within the 4f shell [11]. As that electron returns to a lower energy state, a photon may be emitted. There are multiple possible transitions in Er, one of which is approximately 2.3eV, which is in the visible region of the spectrum. By using alternating layers of YSZ and YSZ doped with Er, the light that is produced is in theory optically polarized and parallel with the direction of the substrate surface. Further, the radiation from the Er doped layers will also in theory travel preferentially in a direction parallel with the Er doped layers. This will occur either as a result of the higher index of refraction of the doped oxide material or the reflection from the metal rich Er doped layer.

EXPERIMENT

The fabrication of these thin film devices was done via Electron Cyclotron Resonance Enhanced Reactive Sputtering (ECR-ERS). The device itself consists of a super lattice of YSZ and YSZ containing ErO_x nano crystallites. The main device parameters varied were YSZ layer thickness, Er layer thickness, and number of Er layers.

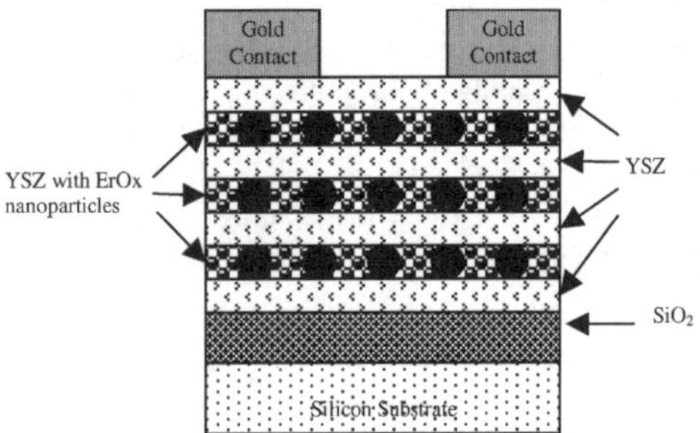

Figure 1: YSZ/Er device structure.

# Er	Thickness Er (Angstrom)	Thickness YSZ (Angstrom)
1	5	30
4	10	45
9	20	80
13	540	90
20		95

Table 1: Device material variables.

In order to create the alternating structure, a shutter was constructed and attached to the Er source. By mechanically rotating the shutter at a fixed rate, the desired stack structure was created. The wafers were *ex-situ* cleaned in a bath of acetone, rinsed in methanol, rinsed in ultra-pure de-ionized water and then dipped in an ultrasonically agitated 5% hydrofluoric acid solution to remove the native SiO_2 layer. To re-grow a very thin layer of SiO_2, the pieces were finally dipped in a solution of ultra-pure de-ionized water: ammonium hydroxide: hydrogen peroxide (5:1:1). The YSZ layers were deposited with RF power of 300 W and a pressure of 5 mTorr. The Er was deposited with RF power of 50 W and an Ar pressure of 5 mTorr. Microwave power supplied to the ECR source was 150 W. The substrate was heated to approximately 800°C during the entire deposition. Gold contacts were thermally evaporated onto the substrate using a shadow mask.

For high-vacuum and Ar-forming gas anneals, a R.D. Webb Red Turbo vacuum furnace was employed. The samples were annealed at 1050°C for a 2 hour duration. The vacuum level during hi-vac anneals was approximately 10^{-6} Torr and the vacuum level during Ar anneals was approximately 10^1 Torr with an Ar flow of 1400 SCCM. The rapid thermal anneal samples were heated at 1100°C for 30 seconds in a H_2/N_2 forming gas. Air anneals were performed in a tube furnace at 1050°C for 2 hours at standard pressure.

RESULTS AND DISCUSSION

Multiple characterization techniques were used to characterize the thin film stack. A Phillips MRD X-ray diffractometer was utilized to determine film quality. In order to determine the effect of the Er layer thickness on crystal quality, multiple samples were prepared and analyzed. As can be seen in Figure 2, as the Er layer thickness increases, the overall quality (as measured by full-width at half maximum of the 2theta-omega X-ray peak) of the film degrades.

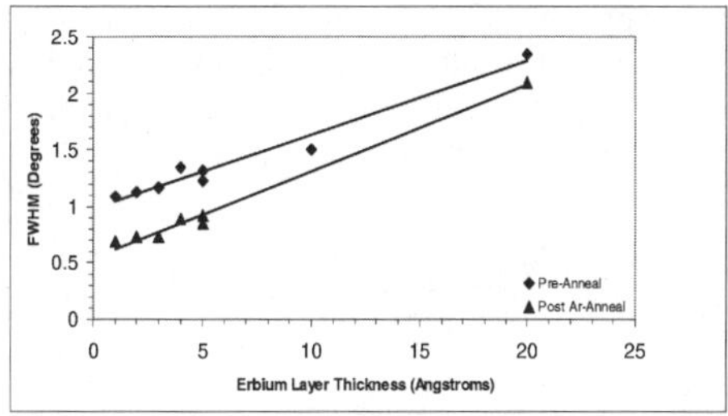

Figure 2: Film response to varying Er layer thickness.

The fact that the crystallinity of the YSZ film is only mildly affected by the intentional introduction of defects (Er atoms) is astonishing. This suggests that the YSZ host is a very stable environment for the ErO_x complexes. The film quality was also measured after the anneal process. There are two general trends that can be seen in Figure 3. First, as the overall stack thickness increases, so does the quality. Second, the anneal at high-vacuum or with Ar forming gas provided the best improvement in crystalline quality.

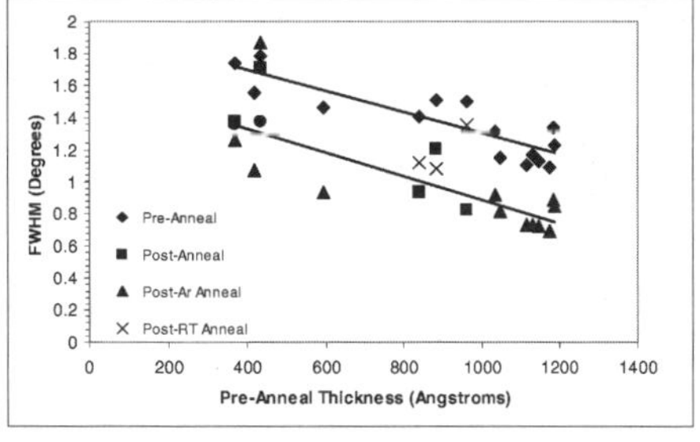

Figure 3: Film quality improvement after anneal.

Thin film thickness and index of refraction were measured with the Rudolph Research AutoEL-III ellipsometer. As can be seen in Figure 4, an anneal in high-vacuum, Ar forming gas or air all lead to an increase in film thickness. The majority of this change can be attributed to the growth of a SiO_2 layer between the Si substrate and the first YSZ layer. As a result of the densification of the YSZ film, the thickness for the rapid thermal anneal remained the same or decreased slightly.

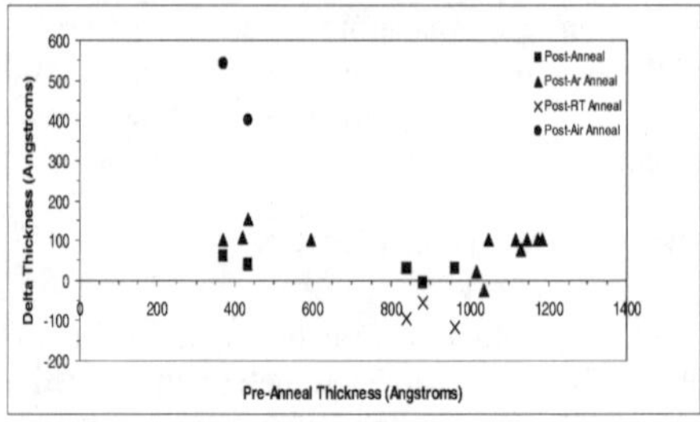

Figure 4: Film thickness change due to anneal.

In order to determine the composition of the thin film, secondary ion mass spectrometry (SIMS) was utilized. The alternating structure is well defined. As expected, there is a maximum in Er concentration where there is a minimum of YSZ (Y and Zr in this plot). This particular sample has nine layers of Er with a targeted thickness of 5 Å.

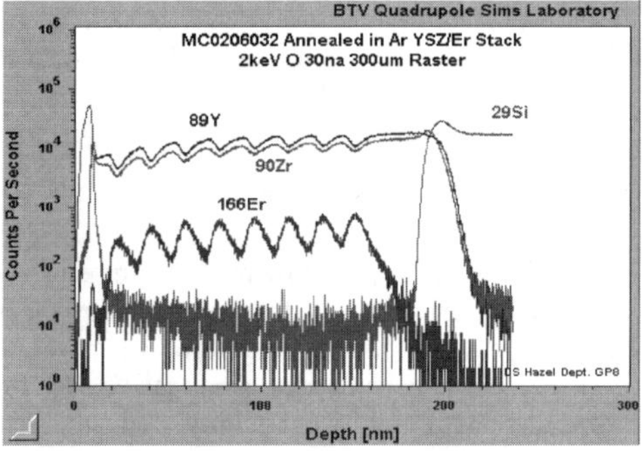

Figure 2: SIMS plot of YSZ/Er structure.

A Hewlett-Packard 4145A semiconductor parametric analyzer was used to determine the electrical and optical characteristics of the films.

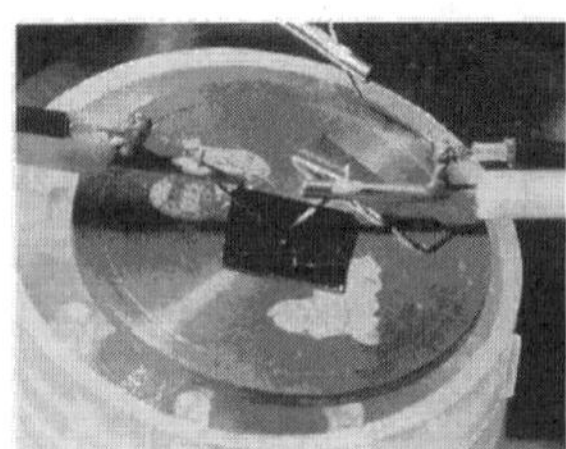
Figure 3: Electrical/optical analysis set-up.

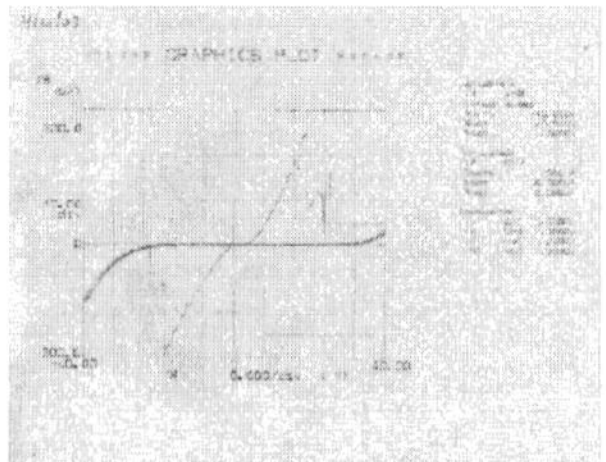
Figure 4: I-V indicating oxide break-down and light emission.

Spikes in the I-V curve indicate points where breakdown occurs and light is emitted. It is hypothesized that there is one spike per layer, until a high enough potential is established in order to penetrate to the bottom Er doped layer. At this point, a conducting path through the top layers is formed and continuous emission is observed. The voltage scale in this plot is -40V to +40V. The current scale is from -200 mA to +200 mA. When biased to -60V, the LED begins to emit a continuous green light. From this point, the intensity increases as the voltage is increased. After breakdown to the final Er layer has occurred, the voltage can be decreased while still maintaining visible light emission. Visible emission was observed at -28V, but for photographic considerations the device was biased at -80V on the right contact while the left contact was grounded. Both photographs in Figure 8 were taken with 1600 speed film with an exposure time of 5 min.

Figure 5: Photograph of room temperature emission, actual dot size is approximately 1mm.

It should be noted that the color on the cathode contact was green to the eye but over exposure led to a blue-ish color on film. The orange color of the anode contact is an interesting phenomenon which will be further investigated in future experiments. As the contacts were opaque metallic gold, the actual emission intensity would be greater if a transparent (ITO) contact was utilized.

SUMMARY

Continuous room temperature Er doped YSZ green light emission has been observed. Film analysis shows alternating layers of high quality YSZ and YSZ doped with Er. Intensity and reliability appear to be a function of metal contact rather than the Er superlattice. Transparent contacts will be investigated to improve device performance. Future work will involve spectral analysis and investigation of device operation via plasmon phenomena.

The authors would like to thank Paul Kozlowski of IBM Yorktown for the rapid thermal anneals and Mark Lavoie and Scott Hazel from IBM Microelectronics for their SIMS work

REFERENCES

1. M. Garter, R. Birkhahn, A.J. Steckl, and J. Scofield, MRS Internet J, Nitride Semiconductor, Res 4S1, G11.3 (1999).

2. T. Whitaker, Compound Semiconductors, **8**, no. 11 (2002).
3. T. Chen, A. Agarwal, A. Thilderkvist, j. Michel, L. Kimerling, Journal of Electronic materials, **29**, no. 7, 973-978 (2000)
4. M. Castagna, S. Coffa, M. Monaco, L.Caristia, A. Messina, R. Mangano, C. Bongiorno, Physica E, **16**, 547-553 (2003)
5. F. Gourbilleau, P. Choppinet, C. Dufour, M. Levalois, R. Madelon, C. Sada, G. Battaglin, R. Rizk, Physica E, **16**, 341-346 (2003)
6. D. Pacifici, A. Irrera, G. Franzo, M. Miritello, F. Iacona, F. Priolo, Physica E, **16**, 331-340 (2003)
7. J. De La Torre, A. Souifi, A. Poncet, C. Busseret, M. Lemiti, G. Bremond, G. Guillot, O. Gonzalez, B. Garrido, J. Morante, C. Bonafos, Physica E, **16**, 326-330 (2003)
8. S. Wang, H. Amekura, A. Eckau, R. Carius, Ch. Buchal, Nuclear Instruments and Methods in Physics Research B, **148**, 481-485 (1999)
9. M. Bell, V. Anjos, Physica E, **16**, 137-138 (2003)
10. J.Zavada, M. Thaik, U. Hommerich, J. MacKenzie, C. Abernathy, F. Ren, H. Shen, J. Pamulapati, h. Jiang, J. Lin, R. Wilson, Symposium G: GaN and Related Alloys, G11.1
11. T. Chen, A. Agarwal, A. Thilderkvist, j. Michel, L. Kimerling, Journal of Electronic materials, **29**, no. 7, 973-978 (2000)

Mat. Res. Soc. Symp. Proc. Vol. 789 © 2004 Materials Research Society N11.13.1

Optical Properties of CdSe Nanoparticle Assemblies

F. Rafael Leon[1,2], Natalia Zaitseva[1], Daniele Gerion[1], Thomas Huser[1], Denise Krol[1,2]
[1]Lawrence Livermore National Laboratory,
7000 East Avenue
Livermore, CA 94550, U.S.A.
[2]University of California at Davis
3001 EU III, One Shields Avenue
Davis, CA 95616, U.S.A.

ABSTRACT

We report on three-dimensional fluorescence imaging of micron-size faceted crystals precipitated from solutions of CdSe nanocrystals. Such crystals have previously been suggested to be superlattices of CdSe quantum dots [1,2]. Possible applications for these materials include their use in optical and optoelectronic devices. The micron-size crystals were grown by slow evaporation from toluene solutions of CdSe nanocrystals in the range of 3-6 nm, produced by traditional wet-chemistry techniques. By using a confocal microscope with laser illumination, three-dimensional raster-scanning and synchronized hyper-spectral detection, we have generated spatial profiles of the fluorescence emission intensity and spectrum. The fluorescence data of the micro-crystals were compared with spectra of individual nanocrystals obtained from the same solution. The results do not support the assertion that these microcrystals consist of CdSe superlattices.

INTRODUCTION

Assemblies of CdSe nanocrystals (quantum dots) have potential applications in new optical, electronic and optoelectronic devices. The multipole interaction between neighboring dots suggests that they may prove useful in computational and data storage devices, and their tunable bandgap could allow for the construction of photodiodes and light-emitting diodes in the visible range [1]. The fluorescence emission wavelengths of the quantum dots are tunable by varying their size.

Because the optical properties of the CdSe semiconductor nanoparticles strongly depend on their size, surface structure, and interactions with their environment, it is important to assess how they behave when closely-packed into micrometer-scale assemblies. It has recently been suggested [2] that microcrystals grown from solutions of CdSe nanocrystals consist of crystalline superlattices of CdSe quantum dots. For individual quantum dots to condense into an ordered lattice they must be similar enough to bond together into a regular pattern. For this reason, it was speculated that an ordered superlattice of nanocrystals would consist of similar nanocrystals and thus be more homogeneous than the precursor solution, which typically contains a distribution of nanocrystals of different sizes.

In the present study we have used scanning confocal fluorescence microscopy to characterize micron-size crystals grown from solutions of CdSe nanocrystals. The fluorescence spectra of these crystals are compared to those of individual nanocrystals obtained from the same solution in order to determine whether the fluorescence of the crystals is consistent with an ordered superlattice of CdSe nanocrystals, and if so, how uniform the nanocrystal size distribution is inside the micro-crystal.

EXPERIMENTAL DETAILS

Sample preparation

Core CdSe nanocrystals were prepared by standard air-free techniques, using TOPO (trioctylphosphine oxide) as a solvent-surfactant and a mixture of $Cd(CH_3)_2$ and Se as a high temperature precursor [1,3,4]. Please note that the quantum dots used for this study consisted of just the TOPO-terminated CdSe core particle, not the much brighter CdSe/CdS system that is often used in biological applications [5]. The nanocrystals were washed by MeOH three times, dried under nitrogen and dissolved in toluene. The solutions contained quantum dots with a size range of 3-6 nm, as determined by TEM. Although it is possible to prepare samples which are more monodisperse, the wide size distribution did not prevent the subsequent precipitation of microcrystals from the solution. Initial diluted solutions were left to concentrate by slow evaporation of toluene until they reached saturation to form microscopic faceted crystals. At this point, a drop of solution was placed onto a microscope slide and covered by a Petri dish. Faceted microscopic crystals grew in the drop by slow evaporation of toluene. After the solution dried out completely, crystals were observed under the microscope.

Fluorescence experiments

For the fluorescence experiments both micro-crystals and individual quantum dots were measured. For the measurements of the micro-crystals the crystals were simply placed onto a glass microscope slide.

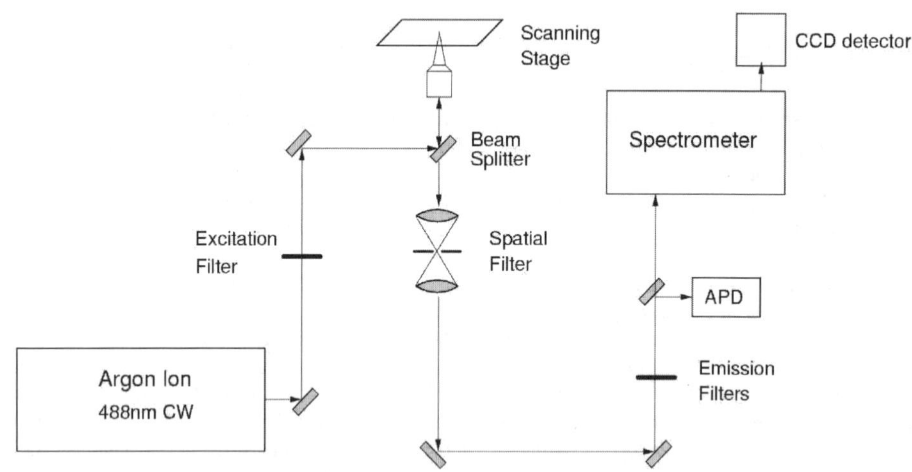

Figure 1: Schematic diagram of the confocal microscope setup used to obtain fluorescence spectra from single microcrystals and individual CdSe quantum dots.

Figure 1 shows a schematic diagram of the confocal set-up used in our experiments. A collimated Ar+ 488nm excitation laser was filtered using a 488NB3 excitation filter (Omega Optical) and focused through an Olympus 100x 0.90NA air objective to a focal volume on the order of one cubic micron. The microcrystals were then raster-scanned through this focus using a Polytec PI P-500 series piezoelectric stage. Typical scan ranges of 100 □m by 100 μm and step sizes of 1 μm were used. The emitted fluorescence was epi-detected using the same microscope objective. After passing through a spatial filter, this light was chromatically filtered using a 498 AELP long-pass filter (Omega Optical) and BG500 glass filter (Schott) and then sent into an Oriel MS257 spectrometer equipped with a Roper Spec-10:100B CCD multichannel detector. The spectrometer grating had a groove spacing of 300 lines/mm. Synchronized spectral acquisition was used to acquire spectra every micron during the scanning in order to develop full spatial fluorescence profiles of the crystals. The acquisition time was 50ms per pixel, and a 100X100 micron scan with 10,000 pixels had a total acquisition time of 500 seconds.

For the single quantum dot experiments the precursor solutions of nanocrystals were diluted and spin-coated onto glass coverslips in order to spatially isolate single nanocrystals. The coverslips were then excited using the same laser with 8 μW of power. The confocal microscope used was also similar, although a Zeiss 1.4NA 100x oil objective was used. Fluorescence intensity detection was performed using an EG&G SPCM avalanche photodiode. The acquisition time was 1 ms per 1-micron square pixel. Once located, spectra were taken of single isolated nanocrystals, and these spectra were then compared to the spectra taken of the nanocrystal assemblies.

RESULTS AND DISCUSSION

Fluorescence spectra were obtained from several select places on each microcrystal. These locations are indicated in figure 2a, and the corresponding spectra are displayed in figure 2b. Notice that there is negligible variation in spectral profile or intensity from one place to another, even for the spectrum taken from point 2, i.e. outside the microcrystal. The differences in intensity are due to slight variations in focusing and corresponding confocal collection efficiency. The profile of each spectrum is nearly the same, with a FWHM of approximately 70nm.

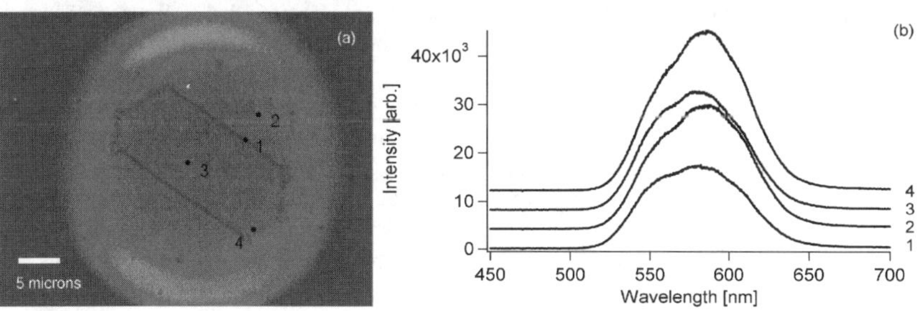

Figure 2. Fluorescence spectra (b) taken from the points on the microcrystal indicated in the white light image (a). The spectra are vertically offset for clarity.

The only notable difference in these spectra is a slight shift in the overall spectral mean, which is due to slight changes in the peak emission wavelength. The spectrum covers the range from about 520 nm to 650 nm with a slight tail to redder wavelengths.

Because of their broad spectral distribution, it was assumed that these spectra resemble ensemble measurements of many quantum dots. In order to confirm this, spectra were taken of single CdSe quantum dots isolated from the same solution from which the microcrystals were grown. By diluting the solutions and dispersing the dots onto a glass coverslip, the nanometer-sized dots can be spatially separated to a point, where the average distance between adjacent dots is greater than the diameter of the diffraction-limited microscope focus. Subsequently, individual dots can be isolated for spectroscopy. A 10X10-micron scan of several dots is displayed in figure 3a. Some of the nanocrystals luminesce more brightly than others. The brightest spots in the image are the result of clusters of several quantum dots. Emission spectra of these clusters have multiple peaks and indicate the presence of multiple dots. The spectra of several single dots and a representative spectrum of an assembled microcrystal are shown in figure 3b. The spectra of single CdSe quantum dots have a characteristic spectral width on the order of 20 nm. The difference in emission peak wavelengths between the different dots is due to the different sizes of the dots. Note that the spectra of the individual dots fall under the spectral envelope of the micro-crystal fluorescence, suggesting that the fluorescence of the crystal does not originate from a narrow size distribution of quantum dots.

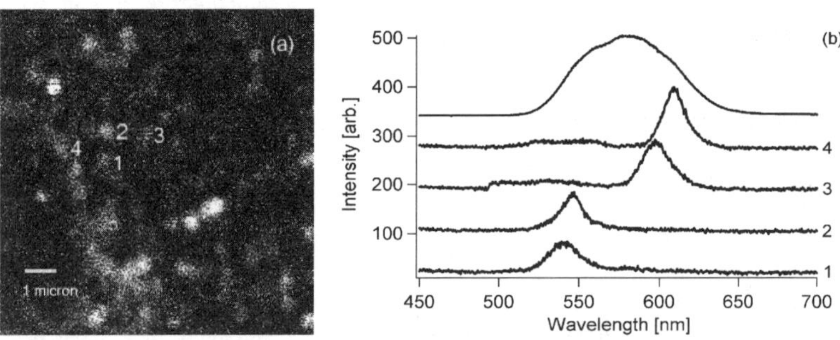

Figure 3. (a) A scanned fluorescence micrograph showing isolated nanocrystals and (b) spectra of four single dots (curves 1-4, taken from positions 1-4 in figure (a)) together with a representative spectrum of an assembled microcrystal (top curve).

As shown in figure 2b, the emitted fluorescence spectra and intensities obtained from the microcrystal sample vary little with respect to position across the microcrystal surface. In order to verify this fact, a fluorescence intensity scan was performed on several microcrystals. The results of a typical scan are shown in figure 4a. The measured intensity variation has such low contrast that the microcrystal is not discernible from the surrounding background. A white outline was added to the image to indicate the position of the microcrystal. This position was determined by correlating the scan with white-light images. The observed uniform intensity indicates that the microcrystals were covered and surrounded by fluorescent quantum dots. In order to separate the microcrystals from the unassembled fluorescent contaminants, microcrystals bound to the glass surface were rinsed with pure toluene. Similar scans were performed on washed crystals, and the results of such a scan are shown in figure 4b. The crystal is in the top right of the picture, and fluorescence is only observed from the meniscus region that forms during the drying process and from several remaining spots of powder on its surface. As the toluene washed over the microcrystal, it dissolved the nanocrystal powder. As the toluene dried, a small number of solvated nanocrystals precipitated within the meniscus at the edge of the microcrystal. The microcrystal is 2 microns thick and its top surface is therefore slightly above the plane of the microscope slide. Nanocrystals also collected in small drops on the surface of the microcrystal and the surrounding glass slide. The collections of nanocrystals on the microscope slide in the bottom left of the figure are slightly out of focus, and therefore less bright.

The fact that the washed microcrystals were found to not fluoresce led us to believe that they were not ordered arrays of nanocrystals. One possible explanation is that the microcrystals consist of monoclinic Selenium. Further experiments using TEM and XRD are under way to precisely determine the nature and structure of these crystals.

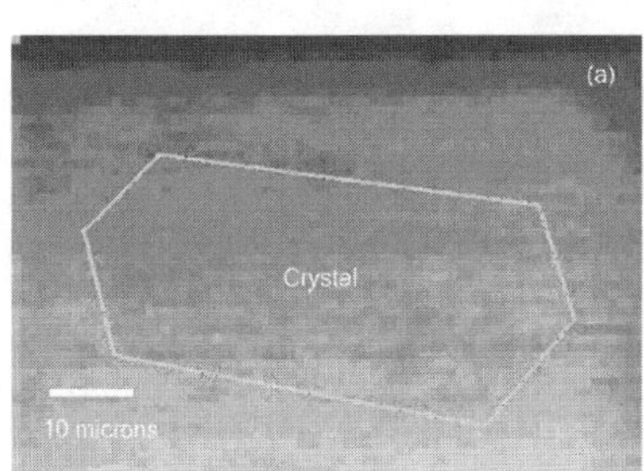

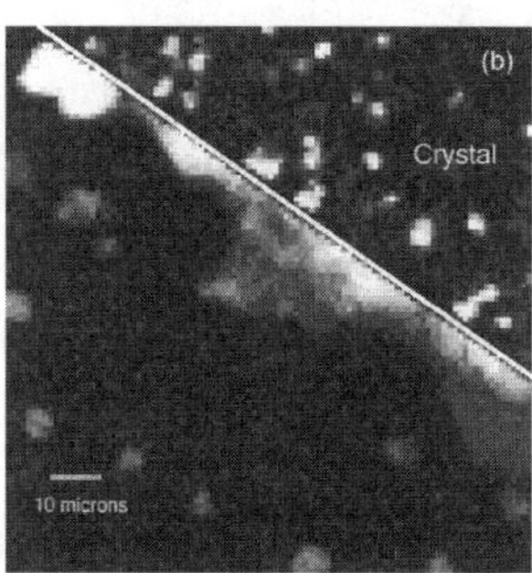

Figure 4. Fluorescence intensity scans of (a) unwashed and (b) washed microcrystals. The white outlines were drawn in to indicate the outline of the crystals as determined from white-light microscopy.

ACKNOWLEDGMENTS

This work was performed under the auspices of the U.S. Department of Energy, National Nuclear Security Administration by the University of California, Lawrence Livermore National Laboratory under contract no. W-7405-Eng-48. The authors would also like to thank the SEGRF student fellowship program at LLNL for supporting this work.

REFERENCES

1. C. B. Murray, C. R. Kagan, and M. G. Bawendi, Annual Review of Materials Science, **30**, 545-610 (2000).
2. A. L. Rogach, D. V. Talapin, E. V. Shevchenko, A. Kornowski, M. Haase, H. Weller, Adv. Func. Mat., **12** (10), 653-664 (2002).
3. L. Qu and X. Peng, J. Am. Chem. Soc., **124** (9), 2049-2055 (2002).

4. Peng, X. G., Schlamp, M. C., Kadavanich, A. V., Alivisatos, A. P., J. Am. Chem. Soc., **119**, 7019-7029 (1997).
5. X. Michalet, F. Pinaud, T. D. Lacoste, M. Dahan, M. P. Bruchez, A. P. Alivisatos, S. Weiss, Single Molecules, **2** (4), 261-276 (2001).

Mat. Res. Soc. Symp. Proc. Vol. 789 © 2004 Materials Research Society N11.17

Optical Interactions and Photoluminescence Properties of Wide-Bandgap Nanocrystallites

Leah Bergman, Xiang-Bai Chen, Jesse Huso, Althea Walker, John L. Morrison, Heather Hoeck, and Margaret K. Penner.
Department of Physics, University of Idaho, Moscow, ID 83844-0903

Andrew P. Purdy
US Naval Research Laboratory, Chemistry Division, Washington DC 20375

Abstract
The UV-photoluminescence (PL) properties of GaN and ZnO nanocrystallites and nanocrystallite ensembles were studied utilizing micro-photoluminescence. We address the origin of the light emissions of the nanocrystallite as to whether it is due a bandgap or excitonic recombination process. The other topic presented here focuses on the interaction of the laser with a collective of crystallites; we address the phenomena of intensity saturation at a high density of laser excitations as well as the impact of the vacuum state on the PL characteristics. Our analysis indicates that the PL of both GaN and ZnO nanocrysallites is excitonic-like and very similar to the behavior of the free exciton in bulk materials. Additionally, we attribute the intensity saturation of GaN and ZnO to the laser heating and heat trapping which takes place in the enclosure of the nanocrystallite ensemble. In vacuum the PL energy was found to exhibit a strong PL energy redshift relative to the PL in air. We attribute the observed shift to a thermal effect and analyze it in terms of the conditions enabling a convective cooling in the ensemble: the mean free path of air in atmospheric pressure and in vacuum relative to the interparticle separation inside the ensemble.

Introduction
GaN, due to its wide band gap ~ 3.5 eV and the ability to grow superb quality material, is one of the most promising materials for applications in blue light emitting diodes as well as lasers, high-speed field effect transistors, and high temperature and high frequency electronic devices [1]. Another highly promising wide bandgap material is ZnO of bandgap ~ 3.3 eV. ZnO has attracted interest as a blue light emitting material [2-3], a substrate for GaN-based devices [4], and a transparent conductor in solar cells [5-6]. Recent efforts have expanded investigation into the field of GaN and ZnO nanostructures including nanowires, nanorods, and various submicron powders. These structures can potentially be utilized for applications that involve high density, ultra-lightweight, and cost effective optical devices [7-11]. For such applications knowledge concerning the optical properties of light emission such as the underlying recombination mechanism of the PL and the PL efficiency, as well as the factors affecting these properties, is crucial to the realization of optical nanodevices.

Our previous results concerning the optical properties of GaN nanocrystallites showed that when the crystallites were grouped in a relatively large and semi-compact ensemble (of size ~ 2-30 μm diameter and average porosity ~ 200 nm), the PL peak position exhibits a significant redshift accompanied by a line broadening; however, no major redshift was observed for smaller ensembles (~ 300-600 nm diameter and of similar porosity) [8]. Our results were explained in terms of the thermal effects, due to laser heating and heat trapping inside the porous enclosure of the large ensemble, which affect the photoluminescence properties. In contrast, for a small size ensemble the heat can radiate out to the environment and no significant PL peak position was observed [8]. In this paper we present a further analyses concerning the underlying optical recombination process in GaN and ZnO nanocrystallite ensembles and the factors affecting it. We found that the PL concurs with excitonic emission accompanied by intensity saturation. The intensity saturation was found to be due to efficiency quenching originating from thermal effects taking place in the enclosure of the ensemble. Moreover, in order to gain insight into the topic of heat transfer dynamics in the ensemble, we investigated the PL emission energies of samples placed in air as well as in vacuum. The basis for that analysis lies in the fact that for an ensemble with interparticle separation at the submicron scale, a convective cooling by air is expected which would depend on the extent of the vacuum state. We found that the PL emission in vacuum is significantly redshifted relative to that in air, indicative of excess heating occurring in the vacuum. This phenomenon is discussed and analyzed in terms of convective cooling and the values of the mean free path of air relative to the interparticle separation distance.

Experimental

Our experiments utilized the CW-SHG LEXEL laser with a wavelength of 244 nm (5.08 eV), and a JY-Horiba Raman/PL system consisting of a high-resolution T-64000 triple monochromator and a UV microscope capable of focusing a spot size of ~ 600 nm in diameter. Additionally, the system has a CCD camera and a TV screen, enabling the viewing of the laser spot as well as the area sampled. The GaN powder was synthesized via an ammonothermal process, the method details of which are published elsewhere [7], and the ZnO was obtained from Sigma-Aldridge.

Results and Discussion

Although there are still ongoing investigations, the origins of the light emissions in GaN and ZnO thin films and bulk material are, for most part, well understood. The consensus is that the main PL line at room temperature in GaN is due to the free A-exciton [12] and the UV-PL in ZnO is also of excitonic nature [13]. In this section of the paper we present analysis concerning the underlying recombination mechanisms of the PL as well as the effect of laser heating on the PL efficiency of the nanocrystallites. It has been established that the luminescence intensity, I, can be expressed as [14-20]:

$$I = \eta I_0^{\alpha} \qquad (1)$$

In this relation I_0 is the power of the exciting laser radiation, η is the emission efficiency, and the exponent α represents the radiative recombination mechanism. For excitonic recombination $1 < \alpha < 2$; for bandgap emission $\alpha \approx 2$; and α is less then 1 when an impurity is involved in the transitions as well as for donor-acceptor transitions [15-20].

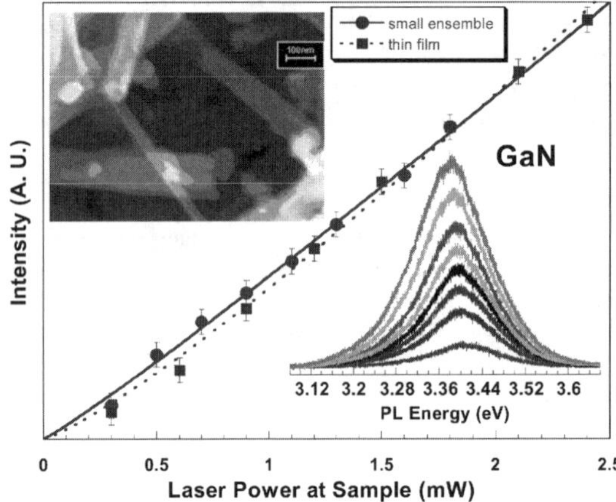

Figure 1. The behavior of the PL intensity as a function of laser power for small ensemble of GaN nanorods and for GaN thin film. The insets show the SEM (100 nm scale) of GaN nanorods and its PL spectra as a function of laser power.

Figures 1 and 2 present the PL intensity as a function of the laser power for a GaN and ZnO small ensemble of nanocrystallites as well as for bulk materials. Using Equation 1 to fit the data, we found α=1.2 and α=1.1 for the GaN film and the GaN nanocrystallites respectively, implying free-exciton type transition. Similarly, we found α=1.5 and α=1.1 for ZnO bulk and ZnO nanocrystallites that indicates the PL to be of exciton nature. In contrast, the light emission from the larger ensembles is more convoluted, as is depicted in Figures 3 and 4: at high laser power the PL intensity for both materials exhibits a strong saturation. Furthermore, as can be seen in the insets to Figures 3 and 4, the PL peak positions exhibit a significant redshift (which for the small ensembles is not as prominent). Additionally, the experiments on the large ensembles were done in two routes: first via increasing the laser power in consecutive steps (the full dots in Figures 3 and 4), and then decreasing the power in similar power

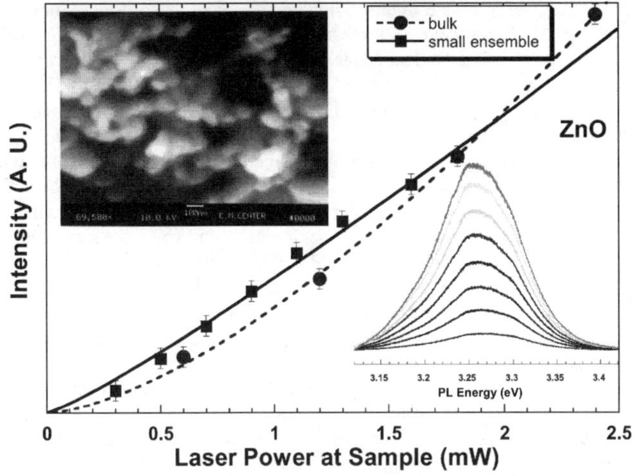

Figure 2. The behavior of the PL intensity as a function of laser power for small ensemble of ZnO nanocrystallites and for ZnO bulk. The insets show the SEM (100 nm scale) of ZnO powder and its PL spectra as a function of laser power.

intervals (open dots). As can be seen in Figure 3 the GaN powder exhibits intensity saturation which was recoverable: i.e., the intensity has the same value for a given laser power independent of the route. In contrast, the ZnO ensemble has anomaly saturation and irrecoverable intensity. Our previous work concluded that in an ensemble of GaN nanoparticles laser heating and heat trapping in the porous region of the ensemble is taking place and as a result the photoluminescence properties, PL peak position, and linewidth are being modified via the thermal effect [8].

Redshift photoluminescence may be the result of several mechanisms, specifically bandgap renormalization (BGR) [21] as well as radiative or exciton transfer dynamics in confined structures [22]. BGR is expected to take place when a high dopant concentration is present in a semiconductor, and as a result the bandgap energy is lowered via the exchange energy mechanism. However, a competing mechanism known as band filling is also expected to occur when a high dopant density is present; the conduction-band lower states are being filled and as a result the PL is blueshifted [21]. Thus the resultant

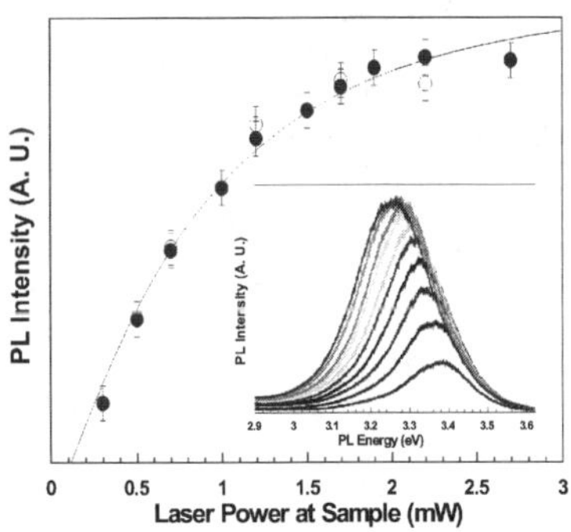

Figure 3. The behavior of the PL intensity as a function of laser power for a large ensemble of GaN nanorods and its spectra.

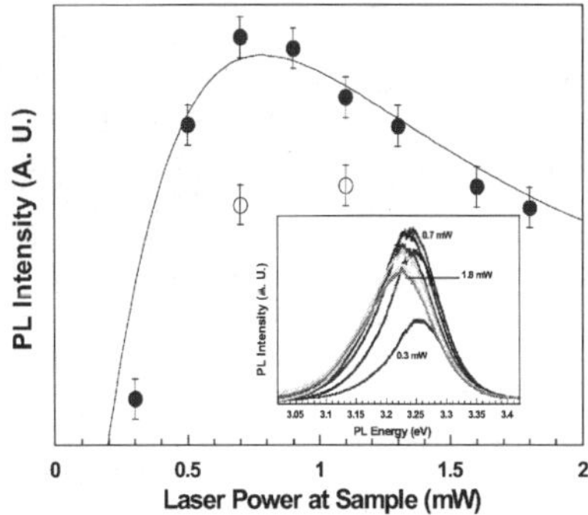

Figure 4. The behavior of the PL intensity as a function of laser power for a large ensemble of ZnO powder and its spectra.

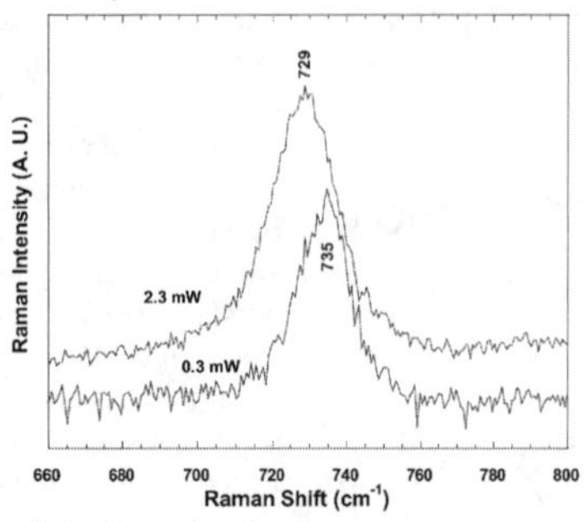

Figure 5. The Raman spectra of the A1(LO) mode of the GaN nanorods for two laser powers.

PL of a heavily doped semiconductor is determined by the extent of the two opposing mechanisms. Recent works on Si doped GaN films addressed the issue of the PL energy in the doped films in terms of the BGR and filling mechanisms [23-24]; in these works the room temperature PL exhibited only 10-20 meV redshift when the dopant concentration was increased by ~ two orders of magnitude or more. If our observed redshift were due to the activation of free carriers via laser excitation, then the small as well as the large ensembles should exhibit the same energy shift, since BGR and band filling are intrinsic properties of a crystallite. The other possible cause of the redshift PL, the transfer mechanisms, can occur when distances of a few nanometers are in play; however, as is depicted in the SEM in Figure 1, our GaN nanorod ensemble has a much larger scale. Again, if the PL transfer mechanism in a confinement were the origin of the observed redshift, one would expect it to be independent of the size of the ensemble, which is not the case. Our experimental results can alternatively be understood in terms of a thermal effect in which heat is being retained in the porous media and as a result a thermal PL redshift is taking place. Previously, it has been formulated that the temperature dependence PL is of the form [25]:

$$E(T) = E(0) - \frac{2\alpha}{\exp(\Theta/T) - 1} \tag{2}$$

This relation represents the modification of a band gap of a semiconductor due to the electron-phonon interaction, In Equation 2, α is a measure of the strength of the interaction, and $\Theta = E_p / k_B$ where E_p is the average phonon energy. We used Equation 2 with the parameters we found previously for GaN films [26]. Our calculations indicate that in the large ensembles the temperature can be as high as ~ 600 K. Our preliminary studies on the phonon dynamics of the GaN nanorods support our PL analysis. Figure 5 depicts the Raman spectra of the A1(LO) mode of a large ensemble of GaN nanorods (~ 10 μm) at two different laser powers. At a relatively low power ~ 0.3 mW the Raman peak is at 735±1 cm^{-1} which is the expected value for the A1(LO) mode in GaN films and bulk [27]. However, for elevated laser power ~ 2.3 mW the peak position shifted to 729±1 cm^{-1}. No shift was observed for a small ensemble nor for GaN film. Previous studies on temperature dependence of Raman of free standing GaN and epitaxial film have indicted that at near room temperature the A1(LO) is at ~ 736 cm^{-1}, and for temperature of ~ 660 K is at ~ 730 cm^{-1} [28]. In light of the reported findings, the temperature inferred from our observed Raman peak shift thus concurs with the temperature predicted via the PL analysis (~ 600K). In light of the above discussion concerning the PL red shift we surmise that the laser heats the particles via absorption, and upon cooling to equilibrium temperature the heat is being transferred into the voids. For a large ensemble due to the convoluted geometry heat is being trapped inside the voids, while for a small ensemble the heat can percolate out to the environment.

An additional impact of this thermal effect is the intensity saturation of the GaN ensemble at elevated laser power. In contrast to GaN, the PL of ZnO exhibits anomalous and unrecovered intensity (~ 25% losses) characteristics as can be seen in Figure 4. The anomalous behavior of the photoluminescence of the ZnO ensemble can be hypothesized to originate from oxygen vacancies which are being formed via the heat and which act as nonradiative centers for the UV-luminescence. Previous work linked the presence of oxygen vacancies in ZnO to the variations in the luminescence [29]. The ZnO powders used in this research consist of almost round particles of average size of 200 nm and voids of similar size. Redshift due to the transfer mechanism among the particles in this size regime is not expected; confinement effects are prominent only in sizes of a few nm. More notably, if a confinement effect were present it would be independent of the ensemble size, and that is not the case as was demonstrated by our

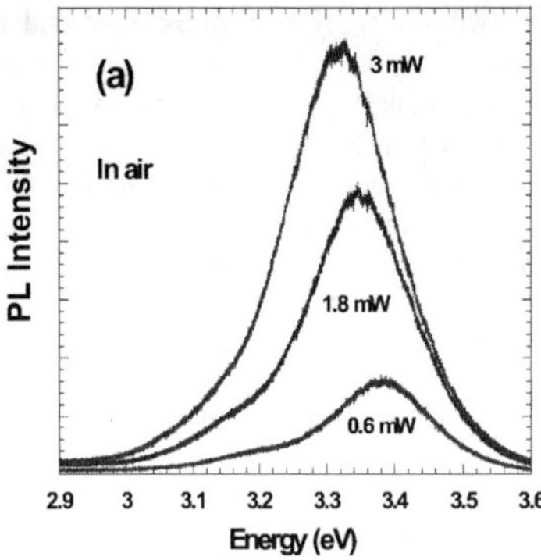

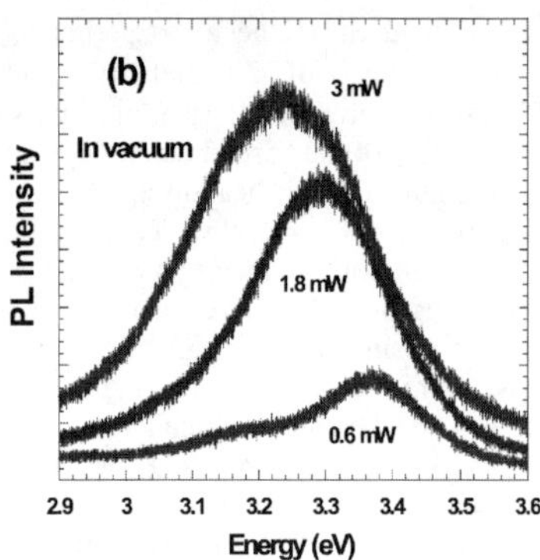

Figure 6. The behavior of the PL intensity as a function of laser power for ensemble of GaN nanorods: spectra acquired (a) in air, and (b) in vacuum.

results. We are currently addressing the topic of the UV and the visible green PL of ZnO powders, and are looking into the issue of whether laser damage induces excess oxygen vacancies or other deep complexes [2, 30] which might explain our results concerning the abnormal PL intensity saturation.

Direct measurements of the temperature and heat transfer mechanisms due to laser heating in an ensemble of nanoparticles is a challenging undertaking that will be addressed in a future study; however, in order to gain further insight into this issue we investigated the PL emission energies of ensembles of the GaN nanopowders placed in air as well as in vacuum. The basis for that analysis lies in the fact that for an ensemble with interparticle separation at the submicron scale, a convective cooling by air is expected which would depend on the extent of the vacuum state. Figure 6 (a, b) shows the PL spectrum of the GaN nanorods as a function of laser power in air and in vacuum. Our data indicate that the PL properties of the ensemble in vacuum are more susceptible to the laser power than those of ensembles in air. For laser power of 3 mW, the PL energy in vacuum redshifts by ~ 80 meV relative to that in air accompanied by a significant thermal spectral line broadening. Our finding can be explained in terms of the efficiency of convection in the two environments and can be approximated from the mean free path λ for air, [31-32]:

$$\lambda \approx 5\times10^{-3}/P, \tag{3}$$

where P is the pressure in units of Torr, and the mean free path, λ, is in cm. Our optical system has a vacuum cell with a pressure of 10^{-4} Torr. For this vacuum state, $\lambda \cong 50$ cm, a value that is much larger than the dimension of the average interparticle separation distance of ~ 300 nm (see Figure 1). Under such conditions, convective cooling becomes inefficient, and the ensemble retains more of the heat that causes the PL redshift. In contrast, the mean free path for air at atmospheric pressure is ~ 65 nm which is less than the interparticle separation distance, and partial cooling via convection in a percolated fashion can take place, so that the PL is less affected by the laser heating. In a simplified model, the overall heat transfer dynamics of the ensemble in thermal equilibrium occur by the heating of the particles via the laser and the cooling via convection, radiation, and thermal conduction. Convection and thermal cooling need a transfer media, while radiation process is always present. The bulk bandgap value of GaN is ~3.5 eV and the laser excitation energy in our experiments is 5.08 eV; thus, heating of the crystallites occurs principally via the absorption which is significant at that laser energy (absorption length ~ 50 nm) [33]. Therefore the laser heating occurs via the principal light beam which due to the voids in the ensemble can penetrate a certain distance, as well as via light scattering in the enclosure.

Since some of the particles are in contact with each other, thermal conductivity can take place from the upper layer of the ensemble to the lower layers and to the substrate. We investigated the impact of the substrate used in the air experiments using two types of substrates: diamond and glass of conduction coefficients of 2300 and 1.0 $W/m \cdot K$ respectively [32]. We found no significant difference in the PL results between the two, and we speculate that due to the short absorption length and the compact geometry of the ensemble, the laser-particle interaction occurs mainly at the top few microns of the large ensemble and that heat dissipation via the substrate-conduction in our ensemble configuration is not as significant as convective cooling. To verify this hypothesis we measured the laser-power transmission after exiting the ensemble (and compensating for the absorption in the substrate) and found that only a negligible fraction of the laser light is transmitted through the ensemble.

In conclusion, we investigated the optical properties of wide bandgap nanocrystallite ensembles. Our analyses indicate that when dealing with sub-micron crystallites, the properties of the ensemble as a whole have to be taken into account along with the underlying properties of the individual crystallites. We found that the PL for GaN and ZnO concurs with the exciton model; moreover, intensity saturation was observed and was attributed to result from laser heating and heat trapping in the enclosure of the relatively large ensemble. Additionally, the PL of the ZnO ensembles exhibits an anomalous saturation behavior that was attributed to the laser heating damage. Lastly, PL in a vacuum state exhibits thermal shift relative to that in air, indicative of the importance of the convective cooling process in the porous media.

Acknowledgments

Leah Bergman gratefully acknowledges NSF CAREER DMR-0238845, NSF-EPS-0132626. Andrew Purdy gratefully acknowledges the Office of Naval Research.

References

1. (a) Semiconductors and semimetals Vol. 57, edited by J. I. Pankove and T. D. Moustakas (Academic, San Diego, 1999). (b) H. Morkoc, S. Strite, G. B. Gao, M. E. Lin , B. Sverdlov, and M. Burns, J. Appl. Phys. 76, 1363 (1994).
2. A. Van Dijken, E.A. Meulenkamp, D. Vanmakelbergh, and A. Meijerink, J. of Luminesc. **90**, 123 (2000).
3. D.C. Look, Mater. Sci. Engin. B **80**, 383 (2001).
4. J.E. Nause, III-Vs Review **12** (4), 28-31 (1999).
5. T. Minami, MRS Bulletin **25** (8), 38 (2000).
6. A. Nuruddin and J.R. Abelson, Thin Solid Films **394**, 49 (2001).
7. A. P. Purdy, Chem. Matter. 11, 1648 (1999).
8. a) L. Bergman, X. B. Chen, A. Purdy, Appl. Phys. Lett. 83, 764 (2003). b) L. Bergman, et. al. MRS Proceedings, V. **776**, Q1.1 (spring 2003)
9. W. -Q. Han, and A. Zettl, Appl. Phys. Lett. **81**, 5051 (2002).
10. M. W. Lee, H. Z. Twu, C. -C. Chen, and C. -H. Chen, Appl. Phys. Lett. **79**, 3693 (2001).
11. X. Duan and C. M. Lieber, J. Am. Chem. Soc. **122**, 188 (2000).
12. B. Monemar, Phys. Rev. B **10**, 676 (1974).
13. L. Wang and N.C. Giles, J. Appl. Phys. **94**, 973 (2003).
14. Shirong Jin, Yanlan Zheng, and Aizhen Li, J. Appl. Phys. **82**, 3870 (1997).
15. T. Taguchi, J. Shirafuji, and Y. Inuishi, Phys. Status Solidi B **68**, 727 (1975).
16. D. E. Cooper, J. Bajaj, and P. R. Newmann, J. Cryst. Growth **86**, 544 (1988).
17. Z. C. Feng, A. Mascarenhas, and W. J. Choyke, J. Lumin. **35**, 329 (1986).
18. Q. Kim and D. W. Langer, Phys. Status Solidi B **122**, 263 (1984).
19. T. Schmidt, K. Lischka, and W. Zulehner, Phys. Rev. B **45**, 8989 (1992).
20. J. E. Fouquet and A. E. Siegman, Appl. Phys. Lett. **46**, 280 (1984).
21. J. Wagner, Phys. Rev. **B29**, 2002 (1984).
22. Th. Forster, "Excitation Transfer" in ***Comparative Effects of Radiation***, PP.300 (Wiley and Sons, New York 1960).
23. M. Yoshikawa, M. Kunzer, J. Wagner, H. Obloh, P. Schlotter, R. Schmidt, N. Herres, and U. Kaufmann, J. Appl. Phys. **86**, 4400 (1999).
24. I-H Lee, J.J. Lee, P. Kung, F.J. Sanchez, and M. Razeghi, Appl. Phys. Lett. **74**, 102 (1999).
25. L. Vina, S. Logothetidis, and M. Cardona, Phys. Rev. **B30**, 1979 (1984).
26. L. Bergman, M. Dutta, M.A. Stroscio, S.M. Komirenko, R. J. Nemanich, C.J. Eiting, D.J.H. Lambert, H.K. Kwon, and R. D. Dupuis, Appl. Phys. Lett. **76**, 1969 (2000).
27. Leah Bergman, Mitra Dutta, and Robert J. Nemanich. "Raman Analysis of Wide Band Gap Nitrides; Film, Crystals, and Superlatices", In Raman Scattering in Materials Science Science p. 273 (Editors: R. Merlin and W.H. Weber, Springer Verlag **2000**).
28. M.S. Liu, L.A. Bursill, S. Prawer, K.W. Nugent, Y.Z. Tong, and G.Y. Zhang, Appl. Phys. Lett. **74**, 3125 (1999).
29. H. Zhou, H. Alves, D.M. Hofmann, W. Kriegseis, B.K. Meyer, G. Kaczmarczyk, and A. Hoffmann, Appl. Phys. Lett. **80**, 210 (2002).
30. K. Vanheusden, W.L. Warren, C.H. Seager, D.R. Tallant, J.A. Vigt, and B.E. Gnade, J. Appl. Phys. **79**, 7983, (1996).
31. Alexander Roth, Vacuum technology (North-Holland, New York, 1976), p.37
32. CRC Handbook of Chemistry and Physics, edited by David R. Lide (CRC Press, 78th ed., 1997-1998).
33. J.F. Muth, J.H. Lee, I.K. Shmagin, R.M. Kolbas, H.C. Casey, B.P. Keller, U.K. Mishra, and S.P. DenBaars, Appl. Phys. Lett. **71**, 2572 (1997).

Mat. Res. Soc. Symp. Proc. Vol. 789 © 2004 Materials Research Society N11.18

Formation and Properties of Silicon/Silicide/Oxide Nanochains

Hideo Kohno[1], Yutaka Ohno[1], Satoshi Ichikawa[2], Tomoki Akita[2], Koji Tanaka[2], and Seiji Takeda[1]
[1]Department of Physics, Graduate School of Science, Osaka University, 1-16 Machikaneyama, Toyonaka, Osaka 560-0043, JAPAN
[2]National Institute of Advanced Industrial Science and Technology (AIST Kansai), 1-8-31 Midorigaoka, Ikeda, Osaka 563-8577, JAPAN

ABSTRACT

Silicon/silicide/oxide nanochains, which have an alternate arrangement of silicon/copper silicide composite nanoparticles and oxide nanowires, were fabricated using silicon/oxide nanochains as templates. Photoluminescence was observed at room temperature and considered to be due to recombination of excitons in oxide. Electrostatic potential in the material was also investigated by electron holography. No visible potential bending at the silicon/silicide interface was detected.

INTRODUCTION

Self-organized formation of composite nanomaterials with heterointerfaces and their characterization have been attracted much attention in recent years [1-3]. This class of nanostructures sometimes exhibits rich functionalities, and can be a candidate for building blocks of future devices of bottom-up basis. On the one hand, electron holography has been applied to studying Si *p-n* junction [4], an AlGaN/InGaN/AlGaN diode [5] and Au/TiO_2 catalysts [6], and is also considered to be an optimal tool for investigating electrostatic potential and charge distribution in nanomaterials with self-formed heterointerfaces at nanometer-scale.

In this paper, we report that more complex hetero-nanostructures can be fabricated by using templates through a self-organized process. Silicon/oxide nanochains [7-14] which have an alternate arrangement of silicon nanocrystallites and oxide wires in one-dimension forming a chain-like structure were used as templates to fabricate silicon/silicide/oxide composite nanomaterials [15]. In addition to the luminescence property, electrostatic potential across a heterointerface was investigated by electron holography.

EXPERIMENTAL DETAILS

The templates, silicon/oxide nanochains [7-14], were grown via the well-know vapor-liquid-solid (VLS) process [16, 17]. To make catalytic particles for the VLS growth of the silicon/oxide nanochains, gold about 10 nm thick was deposited on a silicon substrate (*n*-type, $10^{14}/cm^3$) and heated to 500 °C for 30 min in an evacuated silica container. Then, the silicon substrate was sealed in a new container and heated to 1200 °C for 2 h for the silicon/oxide nanochain growth. For fabricating silicon/silicide/oxide nanochains [15], the silicon/oxide nanochains were sealed with a small piece of copper in an evacuated silica container and heated to 700 °C for 10 min to infuse copper into silicon/oxide nanochains. Conventional transmission electron microscopy (TEM) observations were performed on a 200kV-TEM, JEOL JEM-2010. PL was measured on a TEM, JEOL JEM-2000EX which was equipped with optical measurement system [18, 19] at room temperature. The 514.5nm emission of an Ar ion laser was used for excitation. Electron holograms were recorded on a 300kV-TEM, JEOL JEM3000F with a field emission gun at AIST-Kansai. The fringe spacing of the hologram was about 0.25 nm. Phase images were reconstructed using the JEOL HIPS software.

RESULTS AND DISCUSSION

Structure and formation mechanism

A TEM image of the templates, silicon/oxide nanochains, is shown in figure 1a. Silicon nanocrystallites, which were nearly spherical, were covered with and connected by amorphous silicon oxide forming chain-like structures. After the silicon/oxide nanochains were heated with copper, the silicon/silicide/oxide nanochains were formed as shown in figure 1b, where the copper silicide was hemispherical and appeared in dark contrast. The result of energy-dispersive x-ray (EDX) analysis is shown in figure 2. No copper was detected when the electron probe was off the silicide region. Electron diffraction and EDX analysis showed that the copper silicide was η”-Cu_3Si [20].

The formation mechanism of the silicon/silicide/oxide nanochains is considered to be simple infusion of copper into silicon/oxide nanochains, followed by silicidation in knots. Due to the infusion of copper, the dimensions of nanochains increased through silicidation: the diameter of nanocrystallites from 13.6 ± 1.2 to 14.6 ± 2.2 nm, and the thickness of surface oxide from2.2 ± 0.2 to 3.9 ± 0.6 nm. Using these values, it can be concluded that about half of the silicon atoms in a silicon crystallite must have been ejected and oxidized on the surface contributing to the increase in the thickness of the surface oxide layer.

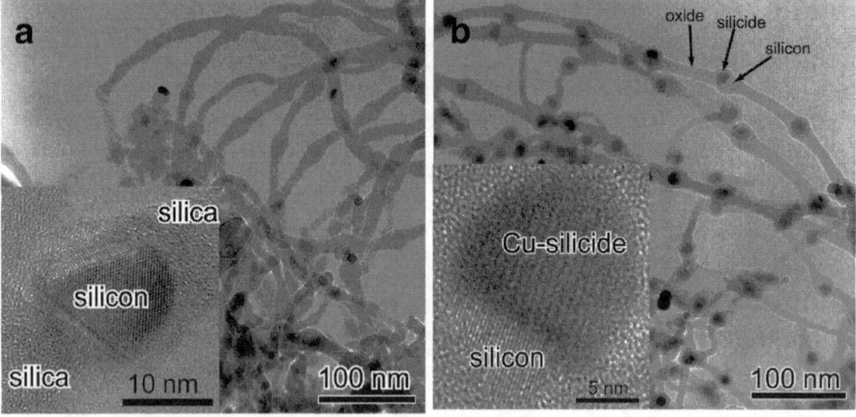

Figure 1. TEM images of (a) silicon/oxide nanochains and (b) silicon/silicide/oxide nanochains.

Photoluminescence

The PL of 2.2 eV (~560 nm) was observed from silicon/silicide/oxide nanochains at room temperature as shown in figure 3. The photon energy is much larger than that predicted by quantum confinement effect in a silicon nanocrystallite: since the silicon nanocrystallites in the nanochains were over 10 nm in diameter, no significant quantum confinement effect can be

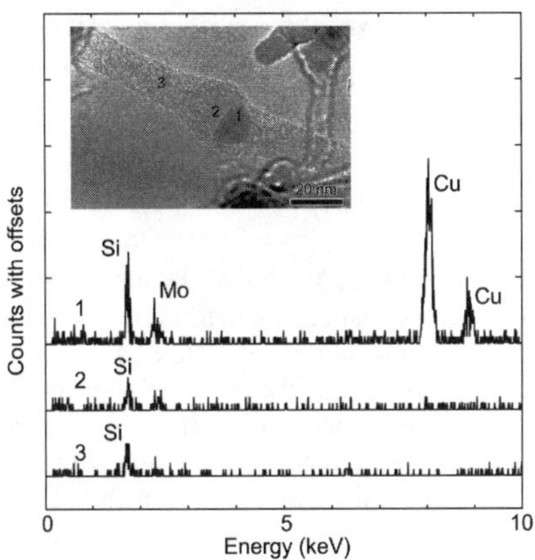

Figure 2. EDX spectra from the positions 1, 2, and 3 indicated in the inset (TEM image). The Mo peak was due to a Mo supporting grid.

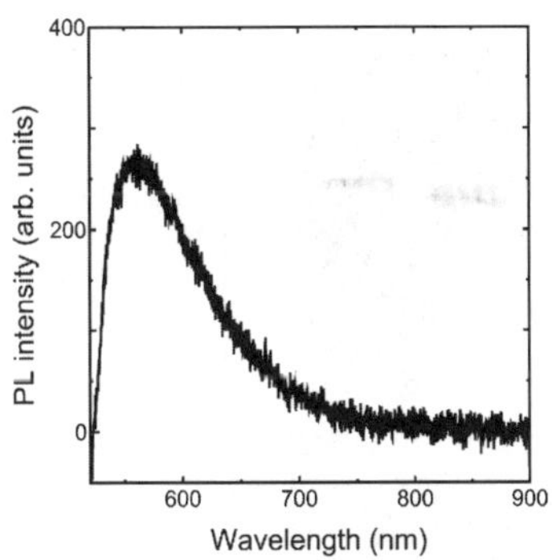

Figure 3. PL spectrum of silicon/silicide/oxide nanochains at room temperature excited by the 514.5nm emission of an Ar ion laser.

expected. We observed similar luminescence from fully oxidized silicon/oxide nanochains. Accordingly, the PL was very likely to originate from amorphous oxide which covers and connects nanocrystallites. Tentatively, the PL was attributed to the recombination of self-trapped excitons in oxide [21, 22].

Electrostatic potential

Electrostatic potential across a silicon/copper silicide nanointerface was investigated by electron holography. As is well known, phase shift of an exit electron wave, $\Delta\phi$, which can be measured by electron holography is proportional to the electrostatic potential, V_0, and the thickness, t, of a material,

$$\Delta\phi = C_{\mathrm{E}} V_0 t, \tag{1}$$

where C_{E} is a constant. Accordingly, given that the thickness is estimated by other methods such as conventional TEM, the potential can be determined. In figure 4, we show a phase image of a silicon/silicide/oxide nanochain reconstructed from a hologram, and its line profile. The silicon/silicide interface is abrupt, thus the observation is almost along the interface. The phase image clearly revealed the difference in potential and a line profile (figure 4b) was examined in detail. Since the reported mean inner potential of silicon and silicon dioxide is almost the same [23], the profile on silicon and oxide was fitted by the same curve. We also assumed that the nanocrystallite and surface oxide was spherical. As shown in figure 4b, no distinct deviation of the experimental curve from the fitting around the silicon/silicide interface, namely no potential bending, was observed. Accordingly, no strong electrostatic field was formed at the interface.

Values of mean inner potential of silicon and copper silicide were determined from the fitting curves as 6.8 and 14.0 V, respectively. These values are much smaller than those

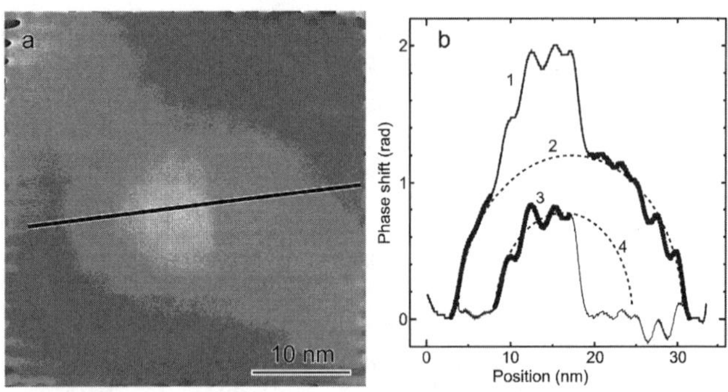

Figure 4. (a) Phase image of a silicon/silicide/oxide nanochain and (b) its line profile. Curve 1: experimental data, curve 2: fitting to silicon and oxide region, curve 3: curve 1 – curve 2, and curve 4: fitting to the curve 3.

calculated to be 9.8 and 17.7 V for silicon and copper silicide, respectively. Furthermore, reported values for silicon ranges from 9.26 to 12.1 V [23], which are not consistent with our experimental value. We tentatively attribute the difference to charging of the material due to incident electrons.

CONCLUSIONS

By infusing metal into template nanostructures, silicon/oxide nanochains, we have fabricated more complex self-organized nanostructures, silicon/silicide/oxide nanochains, where silicon/copper silicide composite nanoparticles were covered with and connected by amorphous silicon oxide. PL of 2.2 eV was observed from nanochains at room temperature and attributed to the recombination of excitons in oxide. Electrostatic potential across the silicon/silicide interface was investigated by electron holography and no visible charge transfer was detected at the interface.

ACKNOWREDGEMENTS

This work was supported in part by a Grant-in-Aid for Young Scientists (No. 14702019) from the Ministry of Education, Science, Sports and Culture.

REFERENCES

1. Y. Wu, R. Fan and P. Yang, *Nano Lett.* **2**, 83 (2002).
2. J. Hu, M. Ouyang, P. Yang and C. M. Lieber, *Nature* **399**, 48 (1999).
3. M. S. Gudiksen, L. J. Lauhon, J. Wang, D. C. Smith and C. M. Lieber, *Nature* **415**, 617 (2002).
4. S. Frabboni, G. Matteucci, G. Pozzi and M. Vanzi, *Phys. Rev. Lett.* **55**, 2196 (1985).
5. M. R. McCartney, F. A. Ponce, J. Cai and D. P. Bour, *Appl. Phys. Lett.* **76**, 3055 (2000).
6. S. Ichikawa, T. Akita, M. Okumura, M. Haruta, K. Tanaka and M. Kohyama, *J. Electron. Microsc.* **52**, 21 (2003).
7. H. Kohno and S, Takeda, *Appl. Phys. Lett.* **73**, 3144 (1998).
8. H. Kohno, S. Takeda and T. Iwasaki, *Mater. Res. Soc. Symp. Proc.* **571**, 293 (2000).
9. H. Kohno, S. Takeda and K. Tanaka, *J. Electron Microsc.* **49**, 275 (2000).
10. H. Kohno and S. Takeda, *J Crystal Growth* **216**, 185 (2000).
11. H. Kohno, T. Iwasaki, and S. Takeda, *Solid Sate Commun.* **116**, 591 (2000).
12. H. Kohno, K. Tanaka and S. Takeda, *Mater. Res. Symp. Proc.* **638**, F13.3.1 (2001).
13. H. Kohno, T. Iwasaki, Y. Mita, M. Kobayashi, S. Endo and S. Takeda, *Phisica B* **308-310**, 1097 (2001).
14. H. Kohno, T. Iwasaki and S. Takeda, *Mater. Sci. Eng. B* **96**, 76 (2002).
15. H. Kohno and S. Takeda, *Appl. Phys. Lett.* **83**, 1202 (2003).
16. R. S. Wagner and W.C. Ellis, *Appl. Phys. Lett.* **4**, 89 (1964).
17. N. Ozaki, Y. Ohno, and S. Takeda, *Appl. Phys. Lett.* **73**, 3700 (1998).
18. Y. Ohno and S. Takeda, *Rev. Sci. Instr.* **66**, 4866 (1995).
19. Y. Ohno and S. Takeda, *J. Electron Microsc.* **51**, 281 (2002).

20. J. K. Solberg, *Acta Cryst. A* **34**, 684 (1978),
21. C. Itoh, T. Suzuki and N. Itoh, Phys. Rev. B 41, 3794 (1990).
22. K. Tanimura, C. Itoh and N. Itoh, J. Phys. C 21, 1869 (1998).
23. M. Gajdardziska-Josifovska amd A. H. Carim, "Applications of electron holography", *Introduction to electron holography*, ed. E. Völkl, L. F. Allard and D. C. Joy (Kluwer, 1999) pp267-293.

Mat. Res. Soc. Symp. Proc. Vol. 789 © 2004 Materials Research Society N11.19

Preparation of Metal Oxide Nanowires by Hydrothermal Synthesis in Supercritical Water

Yukiya Hakuta, Hiromichi Hayashi and Kunio Arai
Supercritical Fluid Research Center
National Institute of Advanced Industrial Science and Technology (AIST)
Nigatake 4-2-1, Miyagino-ku, Sendai 983-8551, JAPAN

ABSTRACT

Three kinds of single metal oxide (MnO_2, ZnO, and AlO(OH)) and one complex metal oxide ($K_2O{\cdot}6TiO_2$) having nano structure of wire, rod and ribbon were rapidly synthesized by hydrothermal synthesis in supercritical water. Aqueous $Mn(NO_3)_2$, $Zn(NO)_2$, $Al(NO_3)_3$ solutions and mixtures of TiO_2 sols and KOH solutions were used as starting materials, respectively. Syntheses of these nanostructured materials were performed by a flow type apparatus. We investigated the relationship between reaction parameters (temperature, pH and reaction time) and morphologies of the products. Reaction temperatures were 350, 400, and 420 °C. Reaction time is in the range of 1.8 - 116 s. The product was characterized by X-ray diffraction (XRD) and transmission electron microscopy (TEM). The results show that particle morphologies strongly depend on pH for MnO_2 and ZnO. MnO_2 nanowires with 10 nm in diameter and ZnO nanorods with 50 nm in diameter were obtained from acidic metal salt solutions. In the case of AlO(OH), temperature and time were key parameters for crystal growth. In the case of K_2O6TiO_2, larger fibrous particles with 50 nm in diameter were obtained at higher reaction temperature.

INTRODUCTION

One-dimension materials such as rod, wire and ribbon are of interest in wide fields of semiconductor, electric, optical devices due to these specific physical and chemical properties that are different form bulk materials. Formation of carbon nanotube or some metal nanowires (e.g. Ag, Si) has shown tremendous success [1]. Recently production of nanorod or nanowire of metal oxides has been reported by many researchers. Two methods are establishing to produce nanostructured metal oxide. One is anodic alumina template (AAT) techniques. Production of nanorods or nanowires of NiO[2], SiO_2[3], TiO_2[4], ZrO_2[5], In_2O_3[6-8], $LiCoO_2$[9], $LiNiO_2$[10] was reported by AAT methods. The other is epitaxial growth method by which several metal oxide nanowires such as TiO_2[11] and ZnO[12-14] were successfully synthesized. Besides these methods, hydrothermal synthesis is one of the promising methods for nano structure materials, too and ZnO [15] and $ZnWO_4$ [16] nanorods were synthesized by hydrothermal routes. Considering from the practical mass production, these methods have some problems such as requirement of complex multi-step treatments and use of expensive raw materials. It is desirable to develop simple and practical method for the production of nanostructure materials.

One of methods for producing metal oxide nanoparticles is hydrothermal synthesis in supercritical water using a flow type apparatus [17]. Immediate formation of metal oxide nano particles was achieved by feeding aqueous metal salt solution into supercritical water in a flow reactor. The specific features of this method as reviewed by Adschiri [18] are (i) nano particle formation, (ii) controllability of redox reaction, (iii) variety of morphology of particle obtained. These features could concern with the variation of thermodynamic and transport properties of water near the critical point.

In this paper, we demonstrate the production of nano-structured materials of three single metal oxides and a complex oxide by hydrothermal synthesis in supercritical water. We investigated the effects of experimental parameters including temperature, time, and pH, on particle morphologies and discuss the predominant parameters.

EXPERIMENTAL

Materials

Manganese nitrate ($Mn(NO_3)_2$: WAKO Chemicals Co. Ltd), zinc nitrate ($Zn(NO_3)_2$, Wako chemicals Co. Ltd, and aluminum nitrate ($Al(NO_3)_3 9H_2O$: WAKO Chemicals Co. Ltd) were used as raw materials. For K_2O6TiO_2 production, titanium oxide sols (STS01: Ishihara Sangyo Kaisya Ltd.) and aqueous potassium hydroxide (KOH: WAKO Chemicals Co. Ltd) solution were used as starting materials. The starting solutions were prepared by dissolving these metal salts into distilled water to be designated concentration.

Method

Figure 1 shows a schematic diagram of experimental apparatus. The metal salt solution was fed by a high pressure pump at 4 cm^3/min of flow rate. For the rapid heating distilled water was fed by pump at 11.0 cm^3/min of flow rate and then preheated by an external electric furnace. Two streams were mixed at a mixing point. The reactant was heated to the reaction temperature, and then hydrothermal reaction occurred. The reactant was introduced to a reactor that was maintained at the reaction temperature by an external furnace and then quenched by cooling at the end of the reactor. The reaction time was adjusted by changing reactor volumes. The cooled reactant was decompressed through a back-pressure regulator and recovered in the reservoir. Almost solid-products were captured by an in-line filter (pore size 500 nm) in front of the back-pressure regulator. Reaction pressure was controlled to 30 MPa by the back-pressure regulator.

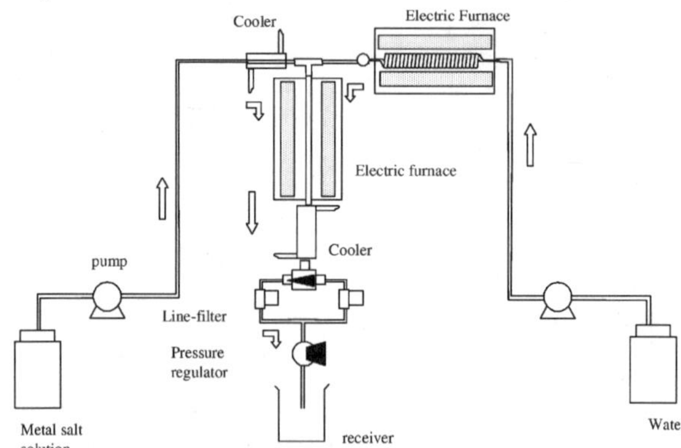

Figure 1. Schematic diagram of experimental apparatus for producing nano-structured materials via hydrothermal synthesis in supercritical water.

Analyses

The solid-products captured in the in-line filter were dried in an oven at 60 °C for 24 hours. Concentrations of the remained metal ions in the reservoir were determined by ICP technique to evaluate conversions of metal ions. Crystal structure of products was determined by X-ray diffraction (XRD: RIGAKU RINT 2000). Particle morphologies and size were measured by transmission electron microscopy (TEM: FEI, TECNAI G20).

RESULTS AND DISCUSSION

Production of manganese dioxide nanowire

Nanostructured manganese dioxides are expected to use as highly effective electrode materials or stable catalyst for a supercritical water oxidation process. We investigated the effects of reaction temperature, pH, and oxidant (H_2O_2) on products in hydrothermal reaction of 0.025M $Mn(NO_3)_2$ solutions. Table 1 shows experimental conditions and results for manganese oxide production. As shown in table 1, crystal structure of products strongly depended on both reaction temperature and pH. At 350 °C, product was Mn_2O_3. With increasing reaction temperature to 400 °C, the products changed from Mn_2O_3 to MnO_2. The manganese ion (Mn^{2+}) in a starting solution was oxidized by HNO_3 or O_2 that was formed by thermal decomposition of HNO_3. The results revealed that the oxidation of Mn^{2+} proceeds at higher temperature. According to TEM, morphology of Mn_2O_3 was spheres with 50 nm in diameter and that of MnO_2 was nanowire with 10 nm in diameter.

Next, the influence of pH of starting solution on products was investigated at 400 °C. With increasing pH, the product changed from MnO_2 through Mn_2O_3 to Mn_3O_4, and the Mn^{2+} conversion increased. Figure 2 shows TEM images of products under various pH. Crystal growth of fibrous MnO_2 (Mn4) in the presence of HNO_3 was observed by TEM (Figure 2(a)). The morphology of Mn_2O_3(Mn6) and Mn_3O_4 (Mn7) was sphere and rhombic, respectively, as shown in Figure 2(b) and (c). From these results, both oxidation of Mn(II) and dissolving manganese ions are important to grow fibrous MnO_2 particles. So, we conducted an experiment to examine the effect of oxidant. Figure 2(d) showsMnO_2 particles obtained in the presence of peroxide. Larger MnO_2 wires (Mn8) were obtained as expected.

Table 1. Experimental conditions and results for producing manganese oxide

Sample	Temp. °C	Additive and concentration / M	pH	Conversion of Mn(II)	Product	Morphologies
Mn1	350	None	4.81	0.87	Mn_2O_3	Sphere
Mn2	380	None	4.81	0.84	Mn_2O_3, βMnO_2	Sphere, Wire
Mn3	400	None	4.81	0.80	βMnO_2	Wire
Mn4	400	NHO_3 / 0.05	2.05	0.83	βMnO_2	Wire
Mn5	400	KOH / 0.05	8.73	0.95	βMnO_2, Mn_2O_3	Wire, Sphere
Mn6	400	KOH / 0.075	11.58	0.95	Mn_2O_3	Sphere
Mn7	400	KOH / 0.1	11.98	0.99	Mn_3O_4	Rhombic
Mn8	400	H_2O_2 /0.01	4.57	0.83	βMnO_2	Wire

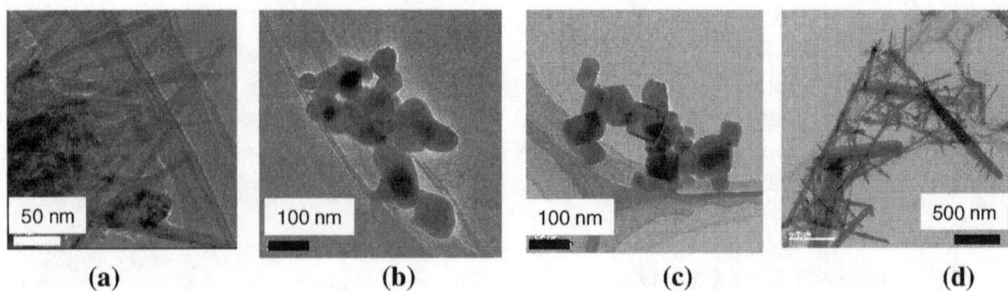

(a) (b) (c) (d)

Figure 2 TEM images of manganese oxide particles with various crystal structures.
(a) MnO_2 nanowires (Mn4), (b) Mn_2O_3 nanoparticles (Mn6), (c) Mn_3O_4 rhombic plate (Mn7), and (d) Larger MnO_2 wires obtained in the presence of H_2O_2 (Mn8).

Production of Zinc Oxide nanorod

Nanostructured zinc oxide can be used as sensitive gas sensor or phosphor materials. In the case of manganese oxide system, the pH was one of main factors on both particle morphologies and crystal structure. So we investigated the effect of pH on ZnO particle formation using 0.02M zinc nitrate solutions with various pH adjusted by addition of KOH. Reaction temperature and time were 400 °C and 1.8 s, respectively. Table 2 shows experimental conditions and results for ZnO formation. Conversion of zinc ions increased with increasing pH. According to XRD analyses, particles obtained in all experiment were ZnO (Zincite).

Figure 3 shows TEM images of ZnO particles obtained from zinc nitrate solution with various pH. In the case of acidic condition, ZnO nanorods were obtained (Zn1, Zn2). With increasing pH, particle morphologies changed from rod to sphere. The morphology of ZnO particles concerned with the conversion of zinc ion as descried in Table 2. The conversions of zinc ion in acidic condition (Zn1, Zn2) were lower than those in basic condition (Zn3, Zn4). These dissolving zinc ions in the reaction field seemed to contribute to the crystal growth of ZnO rod, which was analogous to the case of MnO_2 particles. On the other hand, in the basic condition, hydrothermal reaction and crystal growth was finished immediately (> 90% of conversion), and so spherical particle formed.

Table 2 Experimental conditions and results for producing ZnO particles

Sample	Temp. °C	KOH conc. M	pH	Conversion of Zn(II)	Morphologies
Zn1	400	0	5.46	0.57	Rod
Zn2	400	0.02	6.55	0.74	Rod, Sphere
Zn3	400	0.04	10.53	0.91	Sphere
Zn4	400	0.06	12.45	0.92	Sphere

Production of AlO(OH) nano-ribbon

The nanostructured AlO(OH) particles are expected to use translucent conductive materials or catalysts having high specific surface areas. Production of AlO(OH) particles by this method has been studied[19][20]. The morphologies of particle depended on the initial

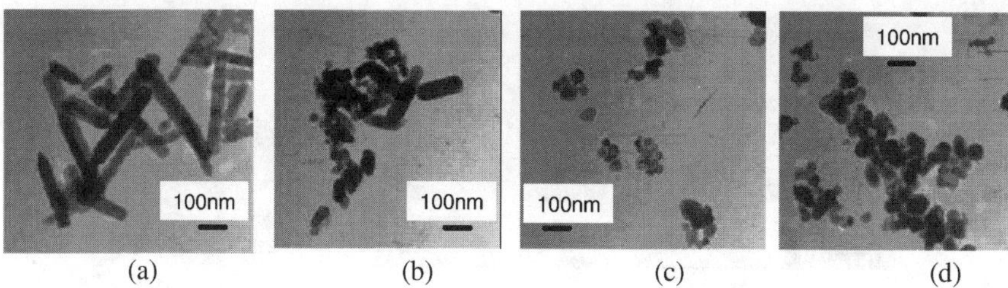

(a) (b) (c) (d)

Figure 3 TEM images of zinc oxide particles obtained form various pH of (a) 5.46, (b)6.55, (c) 10.53 and (d) 12.45.

aluminum nitrate concentration and reaction temperature [19]. In addition, the pH of starting solution is a key parameter to determine the direction of the crystal growth [20]. In this study, we investigated the effect of reaction time on particle growth at 350 °C and 400 °C using 0.05M $Al(NO_3)$ solutions (pH = 1.90).

Products obtained in all experiments were identified as AlO(OH) by means of XRD analyses. Figures 4(a)(b) shows TEM images of AlO(OH) obtained at 350°C and 30MPa for various reaction times. Particles of 38 s of reaction times were rhombic and those of 116 s were hexagonal plate. On the other hand, AlO(OH) particles did not grow with increasing reaction times in supercritical water. In the case of MnO_2 and ZnO, the particles grew by inclusion of the dissolving metal ions. In this AlO(OH) system, particle growth was observed even if conversion of aluminum ions achieved to 96 % at 38 s of reaction times as described in Table 3. We think that these crystal growth can be cause by Ostwald ripening due to high solvent power of subcritical water.

Table 3. Experimental conditions and results for producing AlO(OH) particles

Sample	Temp. °C	Reaction time Sec	Conversion of Al(III)	Morphologies
Al1	350	38	0.96	Rhombic
Al2	350	116	0.99	Ribon
Al3	400	21	0.99	Rhombic
Al4	400	63	0.99	Rhombic

Prodcution of $K_2O{\cdot}6TiO_2$ nanowire

In above section, we showed the availability of preparation of single metal oxide nanowires for single metal oxide. Finally, we would like to introduce the case of complex metal oxide. The production of K_2O6TiO_2 was demonstrated as an example, which was used as photocatalytic materials [21].

From the results, crystal structure strongly depended on the reaction temperature and KOH concentration. Single phase of K_2O6TiO_2 was obtained at the reaction temperature at 400 °C or higher and it is required that KOH concentration in higher than 0.4M. Figure 4(c) shows TEM image of KTO particles at reaction temperature of 400°C. Crystalline nano wires with 10 nm in diameter were obtained. With increasing reaction temperature from 400°C to 420°C, crystal growth proceeded and larger wires with 50 nm in diameter were obtained.

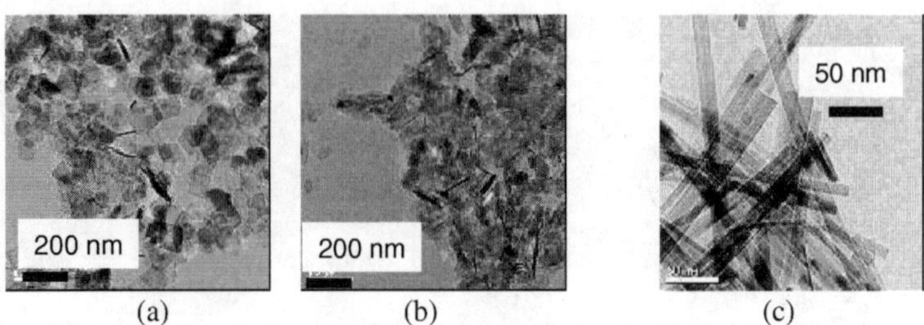

(a) (b) (c)

Figure 4. TEM image of AlOOH particles and K_2O6TiO_2 nanowires.
AlOOH particles were obtained at 350°C and 30MPa for various reaction times of (a) 38s and (b) 116 s. (c) K_2O6TiO_2 nanowire obtained at 400°C and 30MPa for 2.1 s.

CONCLUSION

We proposed a simple and practical method for producing nanostructured materials via hydrothermal synthesis in supercritical water. We revealed that there were key reaction parameters to obtain nanostructured metal oxides. MnO_2 nanowires and ZnO nanorods were obtained from acidic solution under supercritical condition. AlO(OH) nano ribbons were obtained by crystal growth in subcritical water having high solvent power. Larger K_2O6TiO_2 wires were obtained at higher reaction temperature of 420°C. These materials are expected to apply for effective electrode for the battery, phosphor, and catalyst.

REFERENCE

[1] J. Hu, T. W. Odom, C. M. Lieber, *Acc. Chem. Res.* **32**, 435 (1999).
[2] Y. Lin, T. Xie, B. Cheng, B. Geng, L. Zhang, *Chem. Phys. Lett.* **380**, 521 (2003).
[3] Z. Pan, S. Dai, D. B. Beach, D.H.Lowndes, *Nano Letters* **3**, 1279(2003).
[4] S.J. Limmer, T.L. Hubler, G. Cao G, *J. Sol-Gel Sci. Tech.* **26**, 577(2003).
[5] H. Xu D. H. Qin, Z. Yang, H. L. Li, *Mater. Chem. Phys.* **80**, 524 (2003).
[6] M.J. Zheng, L.D. Zhang, G.H. Li, X. Y. Zhang, X.F. Wang, *Appl. Phys. Lett.* **79**, 839(2001).
[7] M. Zheng, L. Zhang, X. Zhang, J. Zhang, G. Li, *Chem. Phys. Lett.* **334**, 298(2001).
[8] H. Yang, Q. Shi, B. Tian, Q. Lu, F. Gao, S. Xie, J. Fan C. Yu,; B. Tu,D. Zhao, *J. Am. Chem. Soc.* **125**, 4724(2003).
[9] Y. Zhou, C. Shen, H. Li, *Solid State Ionics.* **146**, 81(2002).
[10] Y. Zhou, J. Huang, C. Shen, H. Li, *Mater. Sci. Eng. A* **335**, 260(2002).
[11] S. K. Pradhan, P. J. Reucroft, F. Yang, A. Dozier, *J. Cryst. Growth* **256**, 83(2003).
[12] S. Y. Li, C. Y. Lee, T. Y. Tseng, *J. Cryst. Growth* **247**, 357(2003).
[13] K. S. Kim, H. W. Kim, *Physica B* **328**, 368(2003).
[14] K. Ogata K, K. Maejima, S. Fujita, *J. Cryst. Growth* **248**, 25(2003).
[15] X. M. Sun, X. Chen, Z. X. Deng, Y. D. Li, *Mater. Chem. Phys.* **78**, 99(2003).
[16] S. J. Chen, J. H. Zhou, X. T. Chen, J. Li, L. H. Li, J. M. Hong, Z. Xue, X. Z. You, *Chem. Phys. Lett.* **375**, 185(2003).
[17] Y. Hakuta, T. Haganuma, K. Sue, T. Adschiri, K. Arai K, *Mater. Res. Bull.* **38**, 1257(2003).
[18] T. Adschiri, Y. Hakuta, K.Arai, *Ind. Chem. Eng. Res.* **39**, 4901(2000).
[19] T. Adschiri, K. Kanazawa, K. Arai, *J. Am. Ceram. Soc.* **75**, 1019(1992).
[20] Y. Hakuta, T. Adschiri, H. Hirakoso, K. Arai, *Fluid Phase Equilibiria* **158**, 733(1999).

Mat. Res. Soc. Symp. Proc. Vol. 789 © 2004 Materials Research Society N11.21

Superparamagnetic Iron Oxide Nanoarticles for Biomedical Applications : a focus on PVA as a coating

M.Chastellain, A. Petri and H. Hofmann
Powder Technology Laboratory, Swiss Federal Institute of Technology (EPFL), Switzerland

ABSTRACT

Nanoscaled particles showing a superparamagnetic behavior have been intensively studied these past years for biomedical applications and water-based ferrofluids turned out to be promising candidates for various in vivo as well as in vitro applications. Nevertheless, the lack of well-defined particles remains an important problem. One of the major challenges is still the large-scale synthesis of particles with a narrow size distribution. In this work iron oxide nanoparticles are obtained by classical co-precipitation in a water-based medium and are subsequently coated with polyvinyl alcohol. The thus obtained ferrofluids are studied and a focus is made on their colloidal stability.

INTRODUCTION

The recent development of a large variety of ferrofluids has led to a range of new biomedical and diagnostic applications [1-9]. Various applications such as magnetic separation techniques or magnetic resonance imaging (MRI) are relatively well established on the market, whereas drug delivery and targeting are still in a development stage. An important application in the field of drug targeting is related to the healthcare of the elderly, in particular at diseases of the musculo-skeletal system. Many of these diseases are characterized by severe inflammation, disability and pain [10]. If a better control of inflammation can be obtained, dosages can be reduced, side effects on other tissues can be eliminated, and immediate action can be taken. The application of static external magnetic fields in order to target the drug-loaded particles to an area of inflammation e.g. a joint may provide an important component of such treatments. Although a large interest exists in the field of radiology and oncology from the medical side, more fundamental research is still needed to show the feasibility of the concept.

A major drawback for a lot of applications remains the lack of well-defined and well-characterized particles. Growing attention is paid to iron oxide nanoparticles embedded in a polymer matrix. The matrix fulfils several demands : on the one hand it acts as a stabilizer, on the other hand it determines the physicochemical properties of the material and allows further surface functionalization. In this study ferrofluids based on polyvinylic alcohol (PVA) coated iron oxide particles have been prepared by alkaline co-precipitation of ferric and ferrous chlorides in aqueous solution. Different techniques have been applied in order to obtain extensive information about the iron oxide core composition, structure and size distribution. Photon correlation spectroscopy was used for PVA coating characterization.

EXPERIMENTAL DETAILS

All chemicals were analytical reagent grade and were used without further purification. Ultra-pure de-ionized water (Seralpur delta UV/UF setting, 0.055 µS/cm) was used in all synthesis steps. D-9527 Sigma cellulose membrane dialysis tubing with a molecular weight cut off at 12'000 was used for dialysis. The obtained ferrofluids were stored at 4 °C.

Superparamagnetic PVA coated nanoparticles were prepared by alkaline co-precipitation of ferric and ferrous chlorides in aqueous solution [11]. Solutions of $FeCl_3{\cdot}6H_2O$ and $FeCl_2{\cdot}4H_2O$ were mixed and precipitated with a six time smaller volume of concentrated ammonia while stirring vigorously. The black precipitate, which immediately formed was washed several times with ultra-pure water until the pH decreased from about 10 to 7. The sample was centrifuged for 5 minutes at 5'000 g. The supernatant was discarded. The solid was collected and refluxed in a mixture of 0.8M nitric acid and 0.21M aqueous $Fe(NO_3)_3{\cdot}9H_2O$ for 1 hour. During this step the formation of nitric oxide could be observed. The system was allowed to cool down to room temperature, the remaining liquid was discarded, and 100 ml of ultra-pure water was added to the slurry, which immediately dispersed. The obtained black suspension was dialysed for 2 days against 0.01M nitric acid. The suspension was then centrifuged for 15 minutes at 30'000 g. The supernatant was collected and mixed with different solutions of polyvinylic alcohol.

Light scattering measurements for particle size investigations were carried out on a photon correlation spectrometer (PCS) from Brookhaven equipped with a BI-9000AT digital autocorrelator. Measurements were performed using a 661 nm laser illumination and the detector position was set to 90°. The CONTIN method was used for data processing. All suspensions were diluted in 0.01 M nitric acid prior to measurement and showed a pH of 2.3. The water used for dispersion medium preparation was always filtered on 20 nm ceramic filters. The iron oxide concentration was always 0.03 mg/ml. Viscosity and refractive index of pure water were used for as dilution media characteristics. The theoretical refractive index of magnetite was used to calculate the number weighted distribution from the raw intensity weighted values.

DISCUSSION

X-ray diffraction, transmission electron microscopy, atomic force microscopy and magnetic characterization were carried out on bare iron oxide particles [12]. The different data agreed to show crystalline almost spherical particles with an average diameter of 9±1 nm. Bare iron oxide particles are usually stable under acidic conditions (pH around 2) but flocculation arises at physiological pH (around 7). For biomedical applications a coating is thus required.

The result of photon correlation spectroscopy for bare particles is presented below (see figure 1). A more detailed description of these results (quantitative approach) is given in reference [12]. Bare particles present an average hydrodynamic diameter of 13 nm. The shoulder is thought to be due to the presence of agglomerates as described elsewhere [13]. The latter might by reversible dynamic structures evolving with time.

The diameter measured with PCS (average of 13 nm) was bigger than what was found for dried particles (XRD and TEM). The difference was attributed to an electrolyte layer associated to the particles. Such a structure moving with particles is known to sometimes be as thick as 3 nm in the case of nanoscale iron oxide particles [14].

When adding PVA, the size is found to increase (see figure 1a and 1b). The PVA conformation in water is complex [15] and its interactions with the iron oxide nanoparticles are not known. They are nevertheless thought to arise through hydrogen bridging [16] leading to a PVA-based hydrogel structure, which should form around the particles. The PCS hydrodynamic diameter measured for both bare and coated particles is summarized below (see Table 1). A PVA over iron oxide mass ratio of 0.72 leads to almost no noticeable size increase (Fig. 1a). When increasing the polymer amount the average diameter is shifted towards bigger values (Fig. 1a). Figure 1b shows that a mass ratio of 1.44 (average of 21 nm) gives similar results to what is obtained with a ratio of 7.2 and 14.4 (average of 19 nm and 22 nm respectively). The critical value of 1.44 is thus found to correspond to a trend change, in this case a peak shift limit. Above this critical concentration, the average diameter increase arises due to a broadening of the distribution tail on the large size side. This observation is confirmed by zeta potential measurements, which show that a ratio of 1.44 corresponds to the surface adsorption saturation. It is to be noted here that after adding PVA, the size distributions seem to deviate from the classical log-normal shape appropriate to describe bare particles. This tendency is attributed to the formation of an increasing number of agglomerates made of a few iron oxide nanoparticles embedded in a PVA-hydrogel structure.

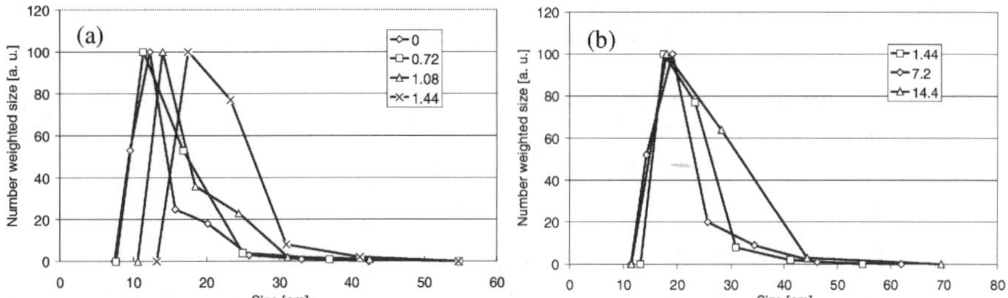

Figure 1. PCS number weighted hydrodynamic diameter distribution for bare and PVA coated particles suspended in 0.01 M nitric acid (pH of 2.3). (a) Adding PVA (from a PVA/iron oxide mass ratio of 0 up to 1.44) leads to a peak shift towards higher sizes. (b) For higher polymer contents (from a PVA/iron oxide mass ratio of 1.44 up to 14.4) the size increase is rather due to a size distribution broadening.

Table 1. Summary of iron oxide particles PCS hydrodynamic average diameters : the statistical error is always given.

PVA/iron oxide mass ratio	Average diameter [nm]
0	13±10
0.7	14±9
1.1	17±7
1.4	21±4
7.2	19±6
14.4	22±6

CONCLUSION

All data show primary crystalline iron oxide particles with an average diameter of about 9 nm and a lognormal size distribution ranging between 5 and 20 nm [12]. The PCS hydrodynamic diameter of bare particles is measured to be 3 nm thicker. The difference agrees well with the assumption of an electrolyte layer moving along with the particles described elsewhere. Adding PVA leads to a size increase. Up to an iron oxide over PVA mass ratio of 1.44, the main peak shifts towards higher values, which is attributed to the formation of a coating-like structure. Above this limit corresponding to the surface saturation with polymer, the distribution broadens probably because of the formation of agglomerates.

PVA coated particles were studied in vitro and showed suitable biocompatibility. The ferrofluids were then successfully used intraarticularly and periarticularly in short term experiments in the carpal and stifle joint of experimental sheeps [17].

Further experiments have been carried out with amino-modified PVA and particles could successfully be modified so that to be taken up by tumor cells according to an active, saturable mechanism [18].

ACKNOWLEDGMENTS

Part of this work was funded by the European MAGNANOMED project (EC contract GRD5-CT2000-00375).

REFERENCES

1. Roger, J., Pons, J. N., Massart, R., Halbreich, A., Bacri, J. C. Eur. Phys. J. 5 (1999) 321
2. Lübbe, A. S., Bergemann, C., Brock, J., McClure D. G. J. Magn. and Magn. Mater. 194 (1999) 149
3. Taylor, J. I., Hurst, C. D., Davies, M. J., Sachsinger, N., Bruce, I. J. J. Chromatogr. A 890 (2000) 159
4. Jung, C. W., Jacobs, P. J. Magn. Reson. Imaging, 13 (1995) 661
5. Jung, C. W, Rogers, J. M., Groman, E. V. J. Magn. and Magn. Mater. 194 (1999) 210
6. Šafarik, I., Šafarikova, M., J. Chromatog. B 772 (1999) 33
7. Goodwin, S., Peterson, C., Hoh, C., Bittner, C. J. Magn. and Magn. Mater. 194 (1999) 132
8. Bergemann, C., Müller-Schulte, D., Oster, J., Brassard, L., Lübbe, A. S. J. Magn. and Magn. Mater. 194 (1999) 45
9. Sheng, R., Flores, G. A., Liu, J. J. Magn. and Magn. Mater. 194 (1999) 167
10. http://www.niams.nih.gov, National Institute of Arthritis and Muskulosceletal and Skin Diseases
11. Van Ewijk, G. A., Vroege, G. J., Philipse, A. P. J. Magn. and Magn. Mater. 201 (1999) 31
12. Submitted paper, M. Chastellain, A. Petri and H. Hofmann, J. Coll. and Interface Sc.
13. Massart, R., Cabuil, V. J. Chim. Physique 84 (1987) 967
14. J.-C. Bacri, R. Perzynski, D. Salin, V. Cabuil, R. Massart, J. Magn. and Magn. Mater. 62 (1986) 36
15. Li, H., Zhang, W., Xu, W., Zhang, X. Macromolecules 33 (2000) 465
16. Jolivet, J. -P., De la Solution à L'oxyde, CNRS Editions, Paris, 1994
17. Paper to be submitted in collaboration with B. von Rechenberg (MSRU, Tierspital, Zürich, Switzerland)
18. Paper to be submitted in collaboration with L. Juillerat (CHUV, Lausanne, Switzerland)

Mat. Res. Soc. Symp. Proc. Vol. 789 © 2004 Materials Research Society N11.23

ZnS nanoparticles synthesis and characterization

Yvonne Axmann[1], Alke Petri[1] and Heinrich Hofmann[1]
[1]Powder Technology Laboratory, Materials Science and Engineering Department, Swiss Federal Institute of Technology, EPFL, CH-1015 Lausanne, Switzerland

ABSTRACT

$ZnS:Mn^{2+}$ nanoparticles were synthesized via a wet chemical method with L-cysteine as the stabilizing agent. The obtained aqueous dispersions show an orange luminescence, which is typical for the $^4T_1 \rightarrow {}^6A_1$ transition within the Mn^{2+} d-orbitals. The fluorescence quantum yield has been determined with quinine sulphate as a dye reference. It can be increased after formation of a SiO_2 shell around the particles by a factor of three. The particle size was determined with transmission electron microscopy (TEM), X-ray diffraction and PCS measurements.

INTRODUCTION

Fluorescence is a widely used tool in biological and medical analysis. Conventional dye molecules impose stringent requirements on the optical system used to make these measurements. Their narrow excitation spectrum makes simultaneous excitation difficult in most cases, and their broad emission spectrum with a long red tail introduces detection problems [1]. Furthermore, for multicolour experiments it is not possible to excite a whole group of molecules with only one laser, which would be the ideal case. Another disadvantage of these organic dyes is the so-called photobleanching, i.e. the fluorescence degradation during UV exposure.
These problems could be overcome with inorganic semiconductor nanocrystals. Here the excitation over the band gap leads to a broad excitation spectrum and the fluorescence is emitted in a narrow and symmetric spectrum either by recombination of the electron and the whole over the band gap (e.g. CdSe, CdS) or by d-orbital transitions for doped semiconductors like $ZnS:Mn^{2+}$. In the first case different colours for multicolour experiments can be achieved by varying the particle sizes since the band gap energy increases with decreasing particle size, in the second case by variation of the dopand. On this article we present Mn^{2+} doped ZnS as a model system for doped semiconductor nanoparticles.

EXPERIMENTAL DETAILS

All chemicals were analytical reagent grade and were used without further purification. Ultra-pure deionised water (Seralpur delta UV/UF setting, 0.055 µS/cm) was used in all synthesis steps. D-9527 Sigma cellulose membrane dialysis tubing with a molecular weight cut off at 12'000 was used in all dialysis steps.

Fluorescent, L-cysteine coated $ZnS:Mn^{2+}$ nanoparticles were prepared by precipitation of the sulphide from aqueous solution in the presence of the surfactant. To do this, solutions of $ZnSO_4{\cdot}7H_2O$ (1M, in 0.01M HCl), L-cysteine (0.25M) and $MnCl_2{\cdot}4H_2O$ (0.01M, 0.1-0.35 mol%) were mixed and the $ZnS:Mn^{2+}$ precipitated with a $Na_2S{\cdot}xH_2O$ (pH 13.5) (1M, S^{2-}/Zn^{2+} ratio of 2.0) solution in a N_2 atmosphere without stirring. Due to the basic shock with the Na_2S

injection $Zn(OH)_2$ precipitates first. The formed white precipitate slowly redisolves under formation of $ZnS:Mn^{2+}$ nanoparticles during incubation under N_2 for 30 min and a 60°C thermal treatment for one hour. The obtained dispersions looked transparent and were purified by dialysis against either NaOH (0.01M) or ultra pure deionised water. The powders were isolated by freeze-drying.

For SiO_2 coated particles 3-(mercaptopropyl)methoxysilane (1 mM) was added to the dispersion and agitated for one night to partially replace the cysteine. After addition of sodium silicate ($SiO_2 \cdot NaOH$) the dispersions were dialysed against 0.01M NaOH after 5 days of agitation under N_2.

DISCUSSION

Fluorescence characteristics

It is well established that when the particle size of a semiconductor becomes comparable to the de Broglie wavelength of an electron or hole, three dimensional quantum confinement, imposed by the dimensions of the nanocrystal, occurs. The shift of the absorption edge of the semiconductor to higher energy provides experimental evidence to such quantum confinement and corresponds to an increase of the energy gap between the valence band and the conduction band of the semiconductor. This effect was used to observe particle growth during synthesis. To do so samples of the dispersion have been taken after 30, 60, 90, and 120 minutes respectively and examined for their absorption properties. In figure 1 these spectra are super imposed so that the red shift of the absorption edge due to particle growth can be observed.

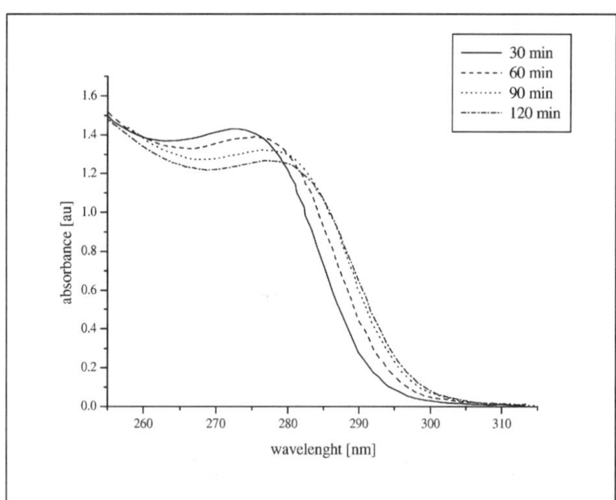

Figure 1. Absorption spectra recorded at different times during synthesis illustrating particle growth.

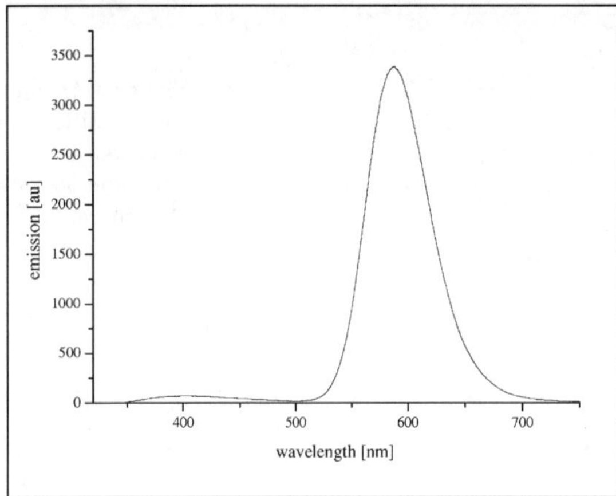

Figure 2: Fluorescence emission spectrum of Mn^{2+} doped ZnS nanocrystals

Figure 3 shows a typical photoluminescence spectrum of ZnS:Mn nanoparticles. The sample was excited at 310 nm and shows an orange luminescence with a maximum at 585 nm, which is characteristic for the $^4T_1 \rightarrow {}^6A_1$ transition of Mn^{2+} ions in a crystalline ZnS-matrix [2]. To determine the position of Mn^{2+} in the nanocrystals ESR measurements have been reported in literature. Generally, two distinct signals can be distinguished from ESR spectra of ZnS:Mn: a hyperfine structured spectrum and a broad background [3, 4, 5]. The broad line has been assigned to the Mn^{2+} ions occupying the Zn^{2+} sites in ZnS [3, 4, 5]. The broadening is attributed to the Mn-Mn dipolar interactions in the aggregated nanoparticles [4]. The hyperfine structured signal is associated with isolated Mn^{2+} ions near the surface of a nanoparticle (Mn^{2+} ions at the near-surface sites) [4, 5]. Magnetic measurements with our own particles in corporation with the group of Palacio at the university in Zaragossa showed so far an homogenous distribution of the Mn^{2+} in the ZnS particles, but conclusions of the exact position can not yet be made. Furthermore we observe a blue emission at about 400 nm. As pure band gap emission in our nanoparticles should be in the ultra violet, we suggest in analogy with Murphy et al., that the emission is due to shallow electron traps acting as recombination centres for photo generated charge carriers [6].

The most limiting factor for the luminescence intensity of nanoparticles is the radiationless recombination of the electron and the hole at the particle surface. Unsaturated valences at the surface act as recombination centres and so their saturation leads to an increase in luminescence intensity [7]. To saturate these "dangling bonds" the particles are coated with silica. This coating has many advantages, it is chemically inert, optically transparent [8], the shell has a sufficient lattice match with the ZnS and its refractive index is smaller than the one of ZnS so that the excitation light is diffracted into the particle core. To synthesize the shell some of the cysteine surfactant molecules are replaced by the mercaptosilane to give a point of attack to the silicate afterwards. The thus coated particles show fluorescence intensities, which are 3 times higher than the intensity obtained without the silica coating. In figure 3 this is shown in terms of

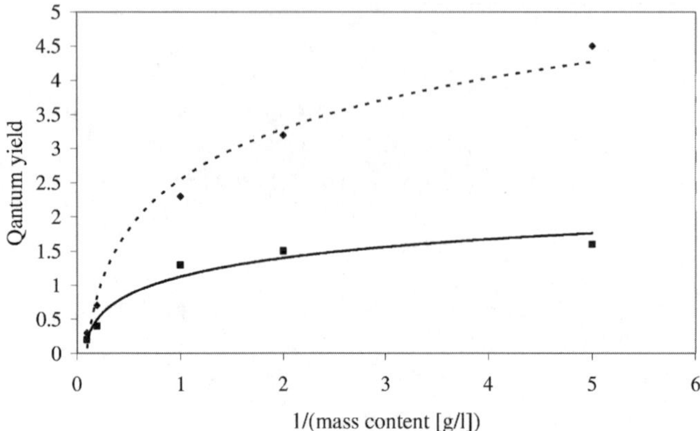

Figure 3: Quantum yiel data for different particl contents.
Solid line: ZnS:Mn
Dashed line: ZnS:Mn/Si

quantum yields. This is the only possibility to compare results obtained with different spectrometers. To calculate the quantum yield we compared the absorption and emission of a dye with known quantum yield with the absorption and emission of the particles with the following equation:

$$\Phi_{NP} = \frac{E_{NP} \cdot Abs_{dye} \cdot \eta_{NP}}{E_{dye} \cdot Abs_{NP} \cdot \eta_{dye}} \cdot \Phi_{dye}$$

where Φ is the quantum yield, E is the area under the emission peak of the dye or the nanoparticles, Abs the respective absorptions and η the viscosity of the solvent.

It can be seen in figure 3 that the quantum yield is also highly dependent on the concentration. The quantum yield increases with decreasing particle concentration; this phenomenon is called concentration quenching. There are two different types of quenching possible, dynamic or static quenching. In the case of dynamic quenching, the quencher must diffuse to the fluorophore during the lifetime of the excited state. Upon the collision the fluorophore returns to the ground state without emission of a photon. In the case of static quenching a complex is formed between the fluorophore and the quencher, and this complex is nonfluorescent. In either event the fluorophore must be in contact [9]. Since measurements at different temperatures showed that the quantum yields increase with decreasing temperature we assume a dynamic quenching process. Here higher temperatures facilitate the diffusion of the particles and therefore the quenching, were as for a static quenching process higher temperatures destabilize the formed nonfluorescent complex. So we think that the particles return to their ground state without emission of a photon but by collision of two particles.

In figure 3 the quantum yields are plotted against the inverse of the mass content. It can be seen, that they converge at an upper limit, quantum yield at infinite dilution, which means the quantum yield of a single particle.

Particle size measurements

Particle size distribution measurements for particles in the size range smaller than 10 nm are rather difficult to carry out. Many methods that work well between 10 and 100 nm get to their detection limit in this size range and dirt and dust present in the sample can lead to false results. To get a good first idea of the particle size transition electron microscopy (TEM) and high resolution TEM (HRTEM) are usually applied. The micrograph in figure 4 shows such a TEM image with a HRTEM micrograph of a single particle in the right upper corner. For sample preparation a drop of dispersion is dried on a copper grid. During the drying process particle agglomeration can occur and makes it often difficult to distinguish between an agglomerate of two or three particles and a single particle. In the particle size distribution shown in figure 4, 500 particles of one sample have been counted and their diameter measured. When it was difficult to decide if it was a single particle or an agglomerate it was counted as a single particle which leads to the second maximum at about 18 nm. The number distribution which was obtained by counting was transformed into a volume distribution with $V=4/3\pi r^3$.

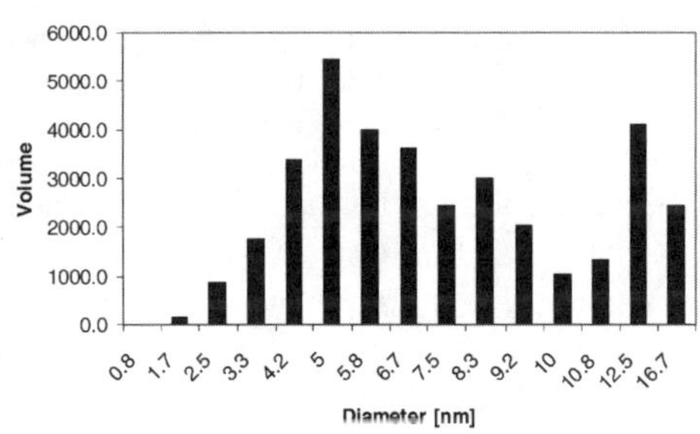

Figure 4: TEM micrograph and corresponding particle size distribution for 500 ZnS :Mn nanoparticles

Another possibility to information about the particle size is to use Scherer's equation and to calculate the crystallite size with the help of the peak broadening for the X-ray diffraction spectra.

$$d = 57.3 \cdot \frac{k \cdot \lambda}{\beta} \cdot \cos \Theta$$

where k is a geometric factor taken to be 1, λ is the x – ray wavelength (for Cu K_α radiation, λ = 1.541 Å), Θ is the diffraction angle, and β is the half width of the diffraction peak at 2 Θ [10]. The mean crystal size of Mn^{2+} - doped ZnS nanoparticles were calculated to be 3.6 ± 0.4 nm. This is smaller as the estimated size from the nanocrystals in the TEM images but it has to be taken into account, that with the X-ray diffraction only the crystalline part of the particles can be detected. So the 3.5 ± 0.4 nm correspond to the size of the crystalline regions within the particles and not to the effective particle size. So the results for Scherer's equation can be seen as a lower limit of the particle size.

PCS measurements give a size of 5.0 ± 0.5 nm which corresponds to the maximum of the distribution obtained for the TEM micrographs. Prior to the measurements the dispersions have been centrifuged to remove dust from the samples. For PCS measurements the particle diameter enters with d^6 in the intensity signal so dust particles give a strong signal so that the particles cannot be detected anymore. Of course it cannot be excluded that we also loose particle agglomerates during the centrifugation process. The program that was used for data requisition does not carry out a Fourier transformation so that only a medium value for the particle diameter is obtained and not a size distribution.

So the lower limit for the particle size has to be set to about 4 nm, a maximum value of about 5 nm is obtained from the TEM Micrographs, which is consistent with the result obtained from PCS measurements. At higher sizes the TEM distribution shows a second maximum, which is probably due to agglomerates.

CONCLUSIONS

Manganese doped ZnS nanoparticles have been synthesized in aqueous dispersions. Their absorption band is shifted to wavelength smaller than 340 nm, which is the bulk value for ZnS, indicating a quantum size effect which proves the presence of nanoparticles. The intensity of their orange emission at 585 nm is concentration dependent due to a dynamic fluorescence quenching process. The particle size has been measured with different methods to be at about 5 ± 0.5 nm.

ACKNOWLEDGEMENTS

Financial support from the Commission for Technology and Innovation (TOP NANO 21 No: 4679.1) is gratefully acknowledged.

REFERENCES

1. M. Bruchez Jr., M. Maronne, P. Gin, S. Weiss and A. P. Alivisatos, *Science* **281**, 2013 (1998)
2. H.-E. Gumlich, *J. Lumin.* **23**, 73 (1981)
3. L. Levy, N. Feltin, D. Ingert, M.P. Pileni, *J. Phys. Chem. B*, **101**, 9153 (1997)
4. T.A. Kennedy, E.R. Glaser, P.B. Kelin, R.N. Bhargava, *Phys. Rev. B*, **52**, R14356 (1995)
5. G. Counio, S. Esnouf, T. Gacoin, J.P. Boilot, *J. Phys. Chem.*, **100**, 20021 (1996)
6. K. Sooklal, B. S. Cullum, S. M. Angel and C. J. Murphy, *J. Phys. Chem.* **100**, 4551 (1996)
7. A. P. Alivisatos, *J. Phys. Chem.* **100,** 13226 (1996)
8. K. P. Velikov, *Langmuir* **17,** 4779 (2001)
9. J. R. Lakowicz, *Principles of Fluorescence Spectroscopy*, Edition (Kluwer Academic/Plenum Publishers, New York, 1983) p. 257-260
10. S. W. Lu, B. I. Lee, Z. L. Wang, W. Tong, B. K. Wagner, W. Park and C. J. Summers, *J. Lumin.* **92**, 73 (2001)

Mat. Res. Soc. Symp. Proc. Vol. 789 © 2004 Materials Research Society N11.25

Platinum nanoparticles growth by means of pulsed laser ablation

R. Dolbec, E. Irissou, F. Rosei, D. Guay, M. Chaker and M.A. El Khakani*
INRS-Énergie, Matériaux et Télécommunications, 1650 Boul. Lionel-Boulet, C.P. 1020, Varennes, Québec, J3X 1S2, Canada.
*Corresponding author: elkhakan@inrs-emt.uquebec.ca

ABSTRACT

Platinum nanoparticles were deposited onto Highly Oriented Pyrolitic Graphite (HOPG) substrate by laser ablating a Pt target at room temperature into a vacuum chamber. By varying the helium background pressure (from 10^{-5} to 0.5 Torr) and the target-to-substrate distance (from 3 to 6 cm), we were able to explore a large range of kinetic energies (i.e., from ~4 to ~130 eV/atom) of the Pt ablated neutrals species impinging on the HOPG substrates. Thus, the effect of the kinetic energy on the size and the surface density of Pt nanoparticles has been investigated *ex-situ* by means of scanning tunneling microscopy (STM), transmission electron microscopy (TEM) and X-ray diffraction (XRD). The pulsed laser deposition technique is shown to produce Pt nanoparticles (of which diameter in the 1 - 4 nm range) with a relatively narrow size distribution. While the size of the PLD Pt nanoparticles is shown to be mainly influenced by the number of laser pulses, their shape is found to be more sensitive to kinetic energy of the Pt ablated species.

INTRODUCTION

The pulsed laser deposition (PLD) is a powerful and versatile method that has been used to deposit a variety of nanostructured materials under a very wide range of deposition conditions [1-3]. The extremely high instantaneous deposition rate of PLD together with the highly energetic ablated species are among the unique features that distinguish PLD from other conventional deposition methods [4,5]. In the early stages of growth, the supersaturated flux of the energetic ablated species leads to a high density of nucleation sites on the substrate and consequently to the formation of nano-islands or nanoparticles of the material being deposited. As the deposition process continues, the PLD leads to the formation of thin films (up to few 100s nm-thick) which are generally found to be nanostructured [2-7]. Nevertheless, the nanostructural characteristics of PLD thin films have been shown to be greatly influenced by the deposition conditions (e.g. background gas and pressure, laser fluence, target-to-substrate distance, substrate temperature, etc…). As an illustration, we have demonstrated that the grain size and nanoporosity of PLD nanostructured SnO_2 thin films can be monitored, to some extent, by controlling the substrate deposition temperature and the oxygen background pressure in the deposition chamber [6,7]. On the other hand, it has been shown that the nanocrystalline structure of PLD gold thin films can be correlated to the kinetic energy (K_E) of the ablated species [8, 9].

Despite the numerous studies reported on the PLD growth of thin films, much less effort has been dedicated so far to the study of the early stages of PLD growth that leads to the formation of nanoparticles. Of particular interest is to study the effect of the PLD deposition parameters on the nanoparticle characteristics. In this paper, we report a study of the growth of PLD platinum nanoparticles. It is shown that the size, the surface density and the shape of the Pt nanoparticles can be monitored by controlling the PLD deposition conditions.

EXPERIMENTAL

Platinum nanoparticles were deposited by ablating a pure polycrystalline Pt target using a pulsed KrF excimer laser (wavelength = 248 nm; pulse duration = 15 ns; repetition rate = 20 Hz) in a high-vacuum chamber (residual pressure ~10^{-6} Torr). The laser beam (on-target laser fluence of ~ 4 J/cm^2) was focused at an incident angle of 45° onto the Pt target (99.99% purity). To perform deposition under reproducible ablation conditions, the target was continuously moved across the laser beam via a dual rotation and translation motion of the target holder. The substrates consisted of freshly cleaved HOPG samples mounted on a holder positioned parallel to the target at various perpendicular distances in the 3-6 cm range. For TEM and XRD analysis purposes, the Pt nanoparticles were also deposited onto carbon-coated TEM microgrids and Si(100) substrates, respectively. The deposited Pt thickness (t) was controlled by varying the number of laser pulses between 5 and 1000. The Pt nanoparticles were grown at room temperature under three distinct deposition conditions where the target-to-substrate distance (D_{ts}) and the helium background pressure were selected in order to obtain K_E values of 4 eV/at., 45 eV/at. and 130 eV/at., as determined by time of flight emission spectroscopy [10].

The Pt nanoparticles deposited onto HOPG substrates were characterized *ex-situ* using a commercial scanning tunneling microscope (STM; NanoScope III, Digital Instruments) operated at room temperature in ambient air. All STM images were acquired using a tungsten etched tip at a scanning rate of 4 Hz, with typical tunneling current and bias voltage of about 1 nA and 1 V, respectively. The samples were also investigated by TEM using a Hitachi H-9000NAR microscope at an acceleration voltage of 300 kV. The crystalline structure of the PLD Pt thin films deposited on Si(100) substrates was investigated by XRD using a Bruker-AXS D8 ADVANCE X-ray diffractometer (Cu K_α radiation), at grazing angle of 1°.

RESULTS AND DISCUSSION

Figures 1a and 1b present typical STM images of Pt nanoparticles deposited, at the same K_E = 4 eV/at., with two different laser pulse numbers (i.e., 50 and 500, respectively). Both images clearly show that PLD Pt nanoparticles, which are randomly distributed on the HOPG surface, present a relatively round shape regardless of their size.

The samples deposited with 50 laser pulses (Fig. 1a) consist of discontinuous films composed of isolated Pt nanoparticles. From the corresponding size distribution histogram (see Fig. 1d), the mean diameter (d_m) of the deposited nanoparticles is found to be of 1.4 ± 0.35 nm. (d_m is determined as the average value of the FWHM of cross-sectional scans of more than 100 nanoparticles, as illustrated in Fig. 1c.) By counting the nanoparticles observed on various STM images, it has been possible to evaluate their surface density (N), which was found to be of ~5 × 10^{11} cm^{-2} in the case of 50 pulses (as in Fig. 1a). As the number of laser pulses is increased from 50 to 500, not only the diameter of the nanoparticles is found to double to a d_m value of 2.8 ± 0.5 nm, but more particularly the surface coverage drastically increased (N was found to reach a value of ~9 × 10^{12} cm^{-2}). As it can be clearly seen in Fig. 1b, the Pt nanoparticles are piled up to form a densely packed continuous film (of which thickness t is of ~ 1.5 nm) with no evidence of any particle agglomeration and/or neck formation. When the laser pulse number is increased beyond 500 (STM images not shown), the Pt nanoparticles that form the continuous films increase in size. On the other hand, the scan-profile presented in figure 1c shows that the height of the Pt nanoparticles is of about 0.6 nm and their diameter is of ~1.4 nm. This suggests that the nanoparticles deposited with 50 laser pluses onto HOPG are hemispherical in shape. Furthermore, while the nanoparticles grow in size (i.e., by increasing the number of laser pulses)

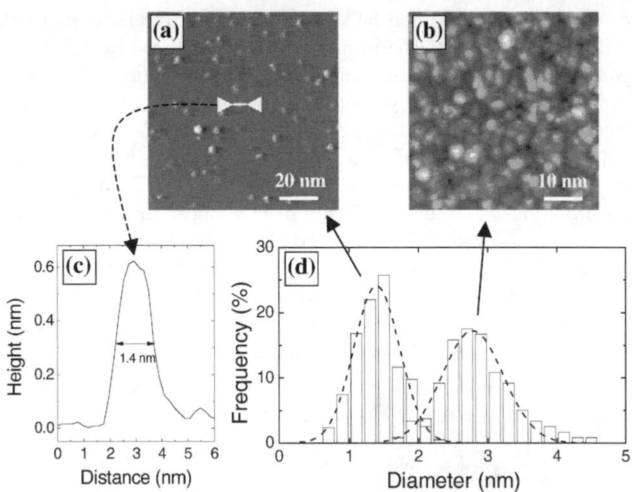

Figure 1. STM images of PLD Pt nanoparticles grown on HOPG substrates, under $K_E = 4$ eV/at, with **(a)** 50 and **(b)** 500 laser pulses. **(c)** Height profile of a Pt nanoparticle taken along the white line indicated by two white arrows in Fig. 1a. **(d)** Size distribution histograms of Pt nanoparticles shown in Figs. 1a and 1b. The error on the mean diameter (d_m) reported in the text corresponds to the FWHM of the Gaussian fit (dashed lines) of the experimental data of Fig. 1d.

their shape remains hemispherical for $K_E = 4$ eV/at. However, their shape is found to be sensitive to the kinetic energy (K_E) of the Pt species involved in the deposition process. Indeed, as K_E is increased to 130 eV/at. the shape of the Pt nanoparticles changes and become more flattened, as illustrated by the cross-sectional profiles of Fig. 2.

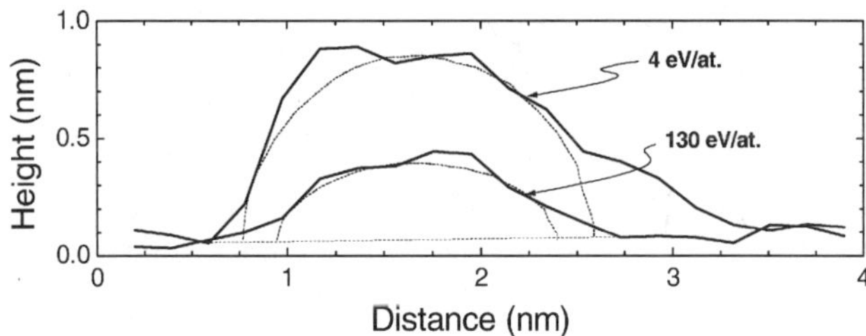

Figure 2. Cross-sectional profiles of Pt nanoparticles grown with 5 laser pulses at the two extreme K_E energies of 4 and 130 eV/at.

Finally, it is worth noting that the size dispersion obtained in this study is of about 20% (Fig. 1d). Such relatively narrow size dispersion is a promising feature for future investigation of the size-dependent properties of the PLD Pt nanoparticles. Indeed, it is among the narrowest size distributions reported so far for Pt nanoparticles synthesized by various deposition methods, including deposition-precipitation (23%), impregnation (21%) and photochemical deposition (25%) [11].

Figure 3 shows both TEM and STM micrographs of Pt nanoparticles deposited with 50 laser pulses and a K_E = 45 eV/at. The Pt nanoparticles examined by TEM were deposited onto TEM microgrids while those intended for STM characterizations were grown onto HOPG substrates. The TEM image (Fig. 3a) shows that Pt nanoparticles are randomly distributed on the substrate surface with a mean diameter d_m of 2.0 ± 0.4 nm and a surface density N of ~5 × 10^{12} cm^{-2}. On the HOPG substrate (Fig. 3b), the Pt nanoparticles present a d_m value of 2.5 ± 0.6 nm, which is rather consistent with the value determined above from TEM analysis. However, as it is clearly apparent from Fig.3, the surface density of Pt nanoparticles was found to be about twice higher on HOPG (i.e., N ~ 1 × 10^{13} cm^{-2}) than on the carbon coated microgrids. Such a discrepancy between the two N values is probably due to the different nature of the substrates involved (i.e., amorphous carbon versus atomically smooth HOPG). Indeed, different surface corrugations and potentials are known to affect the surface mobility of the landing Pt species, modifying thereby their nucleation and growth kinetics.

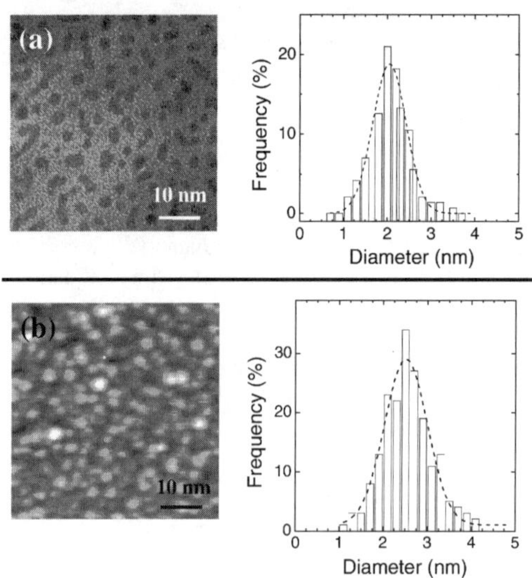

Figure 3. Typical (**a**) TEM and (**b**) STM micrographs of Pt nanoparticles deposited with 50 laser pulses at K_E = 45 eV/at. on both (**a**) carbon-coated TEM microgrids and (**b**) HOPG substrates. The corresponding size distributions for both micrographs are presented on the right-hand panel.

The dependence of d_m on the number of laser pulses for the three K_E conditions investigated here (namely, 4, 45 and 130 eV/at.) is reported in figure 4. For the three kinetic energies, d_m is found to vary in the same manner with the number of laser pulses (i.e., continuous increase with a tendency to saturate at higher pulse numbers). It is worth noting that, for all the K_E energies investigated here, the PLD grown Pt nanoparticles are found to be limited to a minimum d_m value of 1.2 ± 0.25 nm, at the lowest pulse numbers. As the number of laser pulses is increased, the nanoparticles grow in size to reach d_m values of about 3.2 nm with a slight tendency of smaller d_m for higher K_E values. On the other hand, our preliminary results [12] showed that the surface density N for samples deposited under high K_E was found to be higher

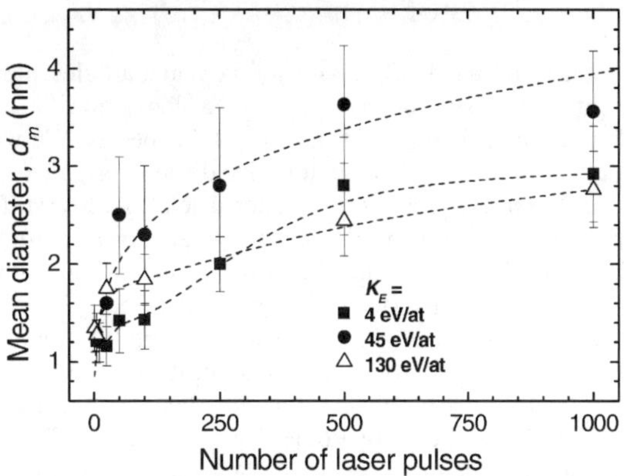

Figure 4. Variation of d_m as a function of the number of laser pulses for the three investigated kinetic energies: K_E = 4, 45 and 130 eV /at.

than that of samples deposited under low K_E. This result suggests that highly energetic ablated species creates more surface defects at the surface of the HOPG substrate, increasing thereby the density of nucleation centers for the diffusing Pt species. Such a high nucleation density is also consistent with the observed tendency to obtain relatively smaller d_m values at high K_E.

Finally, figure 5 displays a typical XRD spectrum of Pt thin film (~10 nm-thick) deposited onto Si(100) substrate at K_E = 45 eV/at. Although the intensity of the diffracted beam is rather weak due to the reduced thickness of the film, the obtained spectrum is a typical signature of polycrystalline Pt structure. By using the Scherrer formula [$d = \lambda / B \cos(\theta)$, where λ is the x-ray wavelength, 2θ is the XRD peak position and B is the peak FWHM], the mean crystallite size (d)

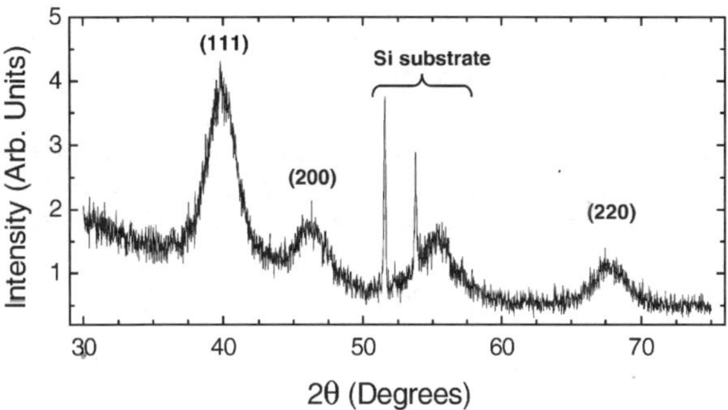

Figure 5. Typical XRD spectrum of PLD Pt thin films (~10 nm-thick) deposited on Si(100) substrate with K_E = 45 eV/at.

was estimated to ~3.8 nm. This value is in very good agreement with the d_m value of 3.6 ± 0.6 nm obtained from STM analysis (images not shown), confirming thereby the discrete character of the Pt nanoparticles observed by STM even when a continuous film is formed.

CONCLUSION

The early stages of the PLD growth of Pt nanoparticles have been investigated as a function of laser pulse number (the equivalent of a deposition time in continuous deposition methods) for different kinetic energies of the ablated Pt species. We were thus able to show that the PLD Pt nanoparticles present relatively low size dispersions, a feature that is highly desirable for the study of size-dependent properties of nanoparticles. At low surface coverage and low K_E, isolated Pt nanoparticles with a hemispherical shape are formed. As the laser pulse number is increased, the Pt nanoparticles grow in size with an increasing surface density. The trade-off between a high nucleation density (which is expected to be enhanced at high K_E values) and nanoparticle size growth seems to lead to a certain upper-limit value of d_m value of about 3.2 nm, regardless of the K_E. Moreover, K_E is found to influence much more the shape of the Pt nanoparticles rather than their size. Finally, the obtained agreement between the crystallite size, as determined from XRD, and the nanoparticle size revealed by STM is a clear indication that the Pt thin films consist of a densely packed arrangement of discrete Pt nanoparticles (or Pt nanograins).

ACKNOWLEDGMENTS

This work was financially supported by the Natural Science and Engineering Research Council (NSERC) of Canada. R. Dolbec is also grateful to the FQRNT (Fonds de recherche sur la nature et les technologies) for the fellowship. The authors would also like to thank Dr. A.M. Serventi for TEM observations.

REFERENCES

1. D.H. Lowndes, D.B. Geohegan, A.A. Puretzky, D.P. Norton and C.M. Rouleau, Science **273**, 898 (1996).
2. A.M. Serventi, M.A. El Khakani, R.G. Saint-Jacques and D.G. Rickerby, J. Mater. Res. **16**, 2336 (2001).
3. N. Braidy, M.A. El Khakani and G.A. Botton, Chem. Phys. Lett. **354**, 88 (2002).
4. P.O. Jubert, O. Fruchart and C. Meyer, Surf. Sci. **522**, 8 (2003).
5. D.B. Chrisey and G.K. Hubler, *Pulsed Laser Deposition of Thin Films*, John Wiley & Sons, New-York NY, 1994.
6. R. Dolbec, M.A. El Khakani, A.M. Serventi and R.G. Saint-Jacques, Sens. Actuators B **93**, 566 (2003).
7. R. Dolbec, M.A. El Khakani, A.M. Serventi, M. Trudeau and R.G. Saint-Jacques, Thin Solid Films **419**, 230 (2002).
8. E. Irissou, B. Le Drogoff, M. Chaker and D. Guay, Appl. Phys. Lett. **80**, 1716 (2002).
9. E. Irissou, B. Le Drogoff, M. Chaker and D. Guay, J. Appl. Phys. **94**, 4796 (2003).
10. E. Irissou, R. Dolbec, B. Le Drogoff, M. Chaker, M.A. El Khakani and D. Guay, to be published.
11. G.R. Bamwenda, S. Tsubota, T. Nakamura and M. Haruta, Catal. Lett. **44** 83 (1997).
12. R. Dolbec, E. Irissou, M. Chaker, D. Guay, F. Rosei and M.A. El Khakani, to be published.

Mat. Res. Soc. Symp. Proc. Vol. 789 © 2004 Materials Research Society N11.26

GaP Nanostructures : Nanowires, Nanobelts, Nanocables, and Nanocapsules

Hee Won Seo, Seung Yong Bae, Jeunghee Park*
Department of Chemistry, Korea University, Jochiwon 339-700, South Korea
E-mail : Parkjh@korea.ac.kr

ABSTRACT

Various GaP nanostructures such as nanowires, nanobelts, nanocables, and nanocapsules were synthesized by sublimation of ball-milled powders. They have a single-crystalline zinc blende structure with [111] growth direction. The morphology and structure were controlled by reactant gas, growth time, flow rate, and growth temperature. The size, morphology and properties of the nanostructures were examined by scanning electron microscopy, transmission electron microscopy, electron energy-loss spectroscopy (EELS), electron diffraction, energy dispersive x-ray spectroscopy, powder x-ray diffraction, and Raman spectroscopy using a 514.5 nm argon ion laser. The photoluminescence was carried out using the 458 nm line of an argon ion laser as the excitation source. The GaP nanowires are straight, cylindrical, and smooth in surface, with mean diameter of 40 nm and length up to 300 mm. The nitrogen-doped nanobelts and nanowires were synthesized by ammonia ambient gas. EELS data reveals that the nitrogen doping occurs mainly in the surface region. The PL spectrum shows the typical isoelectronic bound exciton peaks in the range of 2.11~2.25 eV, suggesting a concentration of (10^{18} cm^{-3} nitrogen atoms. We also synthesized two types of GaP nanocables; GaP nanowire sheathed with the amorphous silicon oxide layers and with the graphite layers. The core-shell diameter is under 30 nm and the outerlayer can be removed by acid treatment to produce the 10 nm diameter GaP nanowires. The GaP encapsulated with BCN nanotubes were synthesized under the ammonia flow using the ball-milled carbon-containing boron oxide powders. The number of BCN layers is typically 10~20.

INTRODUCTION

Since the discovery of carbon nanotubes (CNTs), one-dimensional (1D) nanostructures have attracted much attention as well-defined building blocks to fabricate nanoscale electronic and optoelectronic devices.[1-6] Recently, the modulation of the radial composition in nanowires has demonstrated their potential in nanodevice applications with diverse functions.[2,6] Gallium phosphide (GaP), a wide band gap semiconductor (E_g=2.24 eV at 300 K), is a potential material for optical and high-temperature electronic devices. The GaP nanowires and nanobelts were synthesized by various methods including laser ablation, carbon nanotube-confined reaction, and sublimation.[7-11] Lieber group synthesized by a laser ablation of gold-containing GaP target using 1064 nm.[7] S. T. Lee and coworkers synthesized by irradiating 248 nm laser on the gallium oxide (Ga_2O_3)/GaP mixture target.[8] They also prepared using the reaction of carbon nanotubes with Ga/Ga_2O_3 mixture in a phosphorus vapor atmosphere.[9] Our group synthesized the GaP nanowires and nanobelts via a sublimation of ball-milled GaP powders under argon or NH_3 atmosphere.[10-12] However, the coaxial nanocable structure of GaP has not been reported.

In this letter, we report a direct large-scale synthesis of the various GaP nanowires and N-doped GaP nanobelts via a sublimation of ball-milled GaP powders. Also, we report noble synthesis routes for the core/shell nanostructures of GaP/SiO_x, GaP/C, and GaP/BCN using thermal chemical vapor deposition (CVD) methods. The synthesis has been achieved by modifying the method that was used for the GaP nanowires.[10]

EXPERIMENTAL DETAILS

Ball-milled GaP (99.99%, Aldrich) powders were placed in a quartz boat located inside a quartz tube reactor. Silicon substrates were coated with 0.001M $HAuCl_4 \cdot 3H_2O$ (98+%, Sigma) or 0.01M $NiCl_2 \cdot 6H_2O$ (99.99%, Aldrich) ethanol solutions. The substrate was transferred at a distance of about 10 cm away from the boat. For the GaP/SiO_x nanocables, a piece of the Si substrate was placed underneath the ball-milled GaP powders. Argon flowed with a rate of 500 standard cubic centimeters per minute (sccm). The temperature of the source was set at 1000-1200 °C for 10-30 min and that of the substrate was approximately 800-850 °C. For the GaP/C nanocables, CH_4 was flowed with a rate of 20 sccm for 5-20 min, following the sublimation of the ball-milled GsP powders. In order to produce the GaP/BCN nanocables, B_2O_3 (99.9%, Aldrich) was ball-milled and a mixture of the ball-milled GaP and B_2O_3 powders was placed in the source boat. The temperature of the GaP source was set at 1000℃ and that of the substrate was approximately 950℃. The growth time was 10 min.

The size, structure, composition of the product were examined by scanning electron microscopy (SEM, Hitachi S-4300), high-resolution transmission electron microscopy (TEM, Jeol JEM-2010, 200 kV), electron diffraction (ED), energy dispersive X-ray spectroscopy (EDS), electron energy-loss spectrometer (EELS, GATAN GIF-2000) attached to TEM (Philips CM200), powder X-ray diffraction (XRD, Philips X'PERT MPD), and Raman spectroscopy (Renishaw) using a 514.5 nm argon ion laser. The PL measurement was conducted using the 458 nm line of an argon ion laser as a excitation wavelength. The laser power was about 1 kW/cm^2. The wavelength was precisely calibrated using the atomic lines from an argon calibration lamp.

DISCUSSION

Greenish yellow wool-like product was deposited on the substrates. Fig. 1(a) shows the SEM image of the high-density nanowires homogeneously grown on a large area of the substrate. The nanowires are nearly aligned with a length up to 300 μm. EDS analysis identifies Ga and P with an atomic ratio of nearly 1:1. The TEM image shows the general morphology of the GaP nanowires which are straight, cylindrical and smooth in surface (Fig. 1(b)). There are negligible amorphous outerlayers and no catalytic particles exist at the tip. The diameters are 20–60 nm

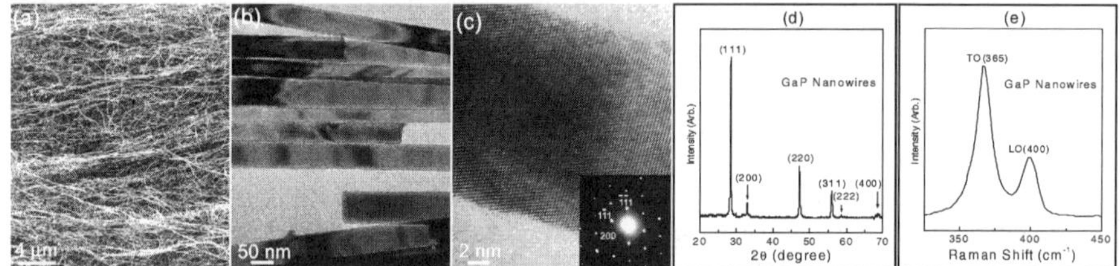

Figure 1. (a) SEM micrograph of high-density GaP nanowires grown on a large area of Ni-deposited alumina substrates via the sublimation of the ball-milled GaP powders. (b) Magnification image showing all the nanowires to be straight and cylindrical. (c) HRTEM image of one GaP nanowire with its corresponding SAED pattern. (d) XRD patterns and (e) Raman spectra of GaP nanowires. The excitation laser of Raman spectroscopy is a 514.5 nm argon ion laser.

with an average value of 40 nm. A high-resolution TEM (HRTEM) image reveals that the nanowires consist of nearly perfect GaP single crystals (Fig. 1(c)). The [111] direction of the cubic unit cell is aligned to the wire axis. The (111) fringes perpendicular to the wire axis are separated by about 3.1 Å, which is consistent with that of bulk crystalline GaP (a = 5.4506 Å; JCPDS Card No. 32-0397). The selected-area ED (SAED) patterns further confirm that the nanowire is single-crystalline with a [111] growth direction (inset). The XRD pattern of the straight GaP nanowires detached from the substrate is shown in Fig. 1(a) along with that of the microsized GaP crystals (GaP microcrystals) prepared by grinding GaP pieces. The peaks can be indexed to zinc blende structured GaP crystals. No other crystalline forms were detected. The position of each peak is the same for both samples, indicating no change of lattice constant for the nanowires. The sharpness of peaks is due to the high crystallinity of the nanowires. The room-temperature Raman spectra of GaP nanowires and GaP microcrystals are displayed in Fig. 1(e). In the spectrum of the GaP microcrystals, the first order phonon frequencies of the transverse optical (TO) and longitudinal optical (LO) modes are at 365 and 405 cm^{-1}, respectively, which is consistent with the reported values, 365 and 402 cm^{-1}. The corresponding peaks of the GaP nanowires appear at around 365 and 400 cm^{-1}, showing a downshift of the LO mode by about 5 cm^{-1}. The TO and LO peaks are somewhat broader than those of the microcrystals. The quantum confinement effect would be negligible for the nanowires having a diameter of 20–60 nm. Since the nanowires possess an extremely large surface area, surface defects would play a role in shifting and broadening the Raman peaks.

Figure 2(a) shows a SEM image for the high-density nanostructures grown on the substrate. The width is 200–500 nm with an average value of 300 nm. The nanobelts occupy more than 90% of the products. The nanobelt is tilted to show the rectangular cross section (Fig. 2(b)). The thickness is usually 1/10 of the width. Figure 2(c) and its enlarged scale (inset) display the more significant absorption of the *N–K* edges for the edge part compared to the center part of the nanobelt, indicating that the surface region is mainly doped with the N atoms. Figure 2(d) shows the PL spectrum of the nanobelts measured at 8 K. A series of the PL peaks appear in the energy range 2.11–2.25 eV. The separation and intensity of these peaks are nearly identical with those of well-known isoelectronic bound exciton peaks of the N-doped GaP crystal.[13-14] The only difference is that the peaks redshift from those of the bulk by about 70 meV. We could tentatively assign the peaks as follows. The peaks at 2.246 and 2.241 eV would be, respectively, *A* and *B* lines originated from the recombination of isoelectronic excitons bound to isolated nitrogen atoms. The peak at 2.237 eV would be *C* line, associated with the recombination of excitons bound to single neutral donors.

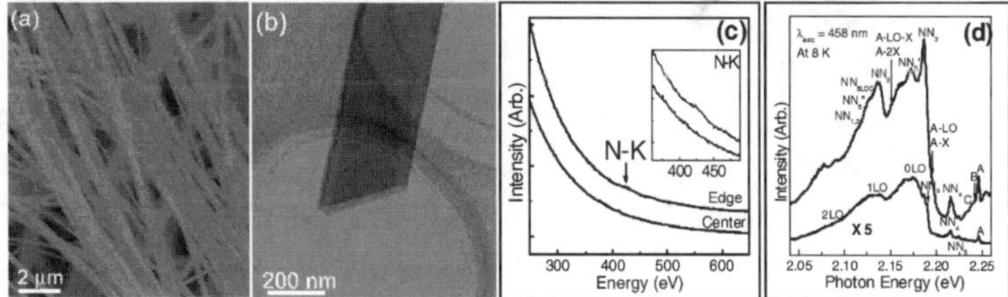

Figure 2. (a) SEM micrograph of high-density GaP nanobelts. (b) The tilted nanobelt shows the thickness that is 1/10 of the width. (c) EELS data showing that the N-K edges are more significant at the edges of the nanobelt. (d) PL spectra of the undoped and N-doped GaP nanobelts measured at 8K. The excitation wavelength is the 458 nm from an argon ion laser.

The peaks at 2.112, 2.187, 2.215, 2.224, and 2.232 eV would be labeled as $NN_{1,2}$, NN_3 , NN_4 , NN_5 , and NN_6 , respectively, come from the recombination of isoelectronic excitons bound to the N atom pairs. The phonon replicas and local mode of NN_3 and NN_5 are labeled NN_3\`, $NN_5$\`, NN_5\`\`, and NN_{5LOC} . The appearance of both *A* and *NN* peaks suggests that the concentration of the doped N atoms would be about 10^{18} cm^{-3}. The peaks at the energies below 2.20 eV; NN_3, NN_3\`, $NN_5$\`, etc., are superimposed on three broad bands that were seen from the PL of the undoped GaP nanowires.[10] The band centers are 2.08, 2.13, and 2.175, separated by about 48 meV which is close to the energy of the LO mode. Therefore, the bands can be assigned to the zero-phonon donor-acceptor pair (DAP) (0*LO*) emission and its replicas (1*LO* and 2*LO*). The DAP bands also redshift from those of the bulk by about 40 meV. The isoelectronic bound exciton peaks may reflect the global conduction band according to the Hopfield–Thomas– Lynch model.[15] The redshift indicates that the energy difference between the global conduction band and the topmost valence band would reduce due to the N doping. This reduction can be explained if we can make an assumption that the surface potential formed in the nanobelt affects significantly the band structure. We also suggest that the DAP band would be influenced by the surface potential that probably reaches to the whole thickness of the nanobelts. TEM image shows the nanocable morphology of this product (Figure 3(a)). The layers coat the straight nanowires along the entire length. The nanocables are rotated along the wire axis, showing that the diameter is not much changed. The cylindrical nanowires are thus uniformly coated with the layers. The outer diameter of the nanocables is 40-60 nm with an average value of 50 nm. The thickness of the outer layers is about 10 nm. A magnified TEM image reveals the cable structure that the layers coat uniformly the nanowires (Figure 3(b)). The thickness of outer layers is less than 10 nm. We were able to control the thickness by the C deposition time. Atomic-resolved TEM image reveals that the GaP core is nearly single crystalline with a few defects and the outer layers are almost amorphous. The EDX analysis identifies Ga and P with atomic ratio of nearly 1:1 (Figure 3(c)-top). The significant atomic (at.)% of Si and O components is detected. The EDX analysis identifies 1:1 ratio of Ga:P and a large amount of C component (Figure 3(c)-bottom). The Cu component originated from the TEM grid. Further analysis reveals that the amorphous outer layers are composed of pure C. Also, the amorphous SiO_x and C shells can be also easily eliminated by the acid treatment. Figure 3(d) shows that the SiO_x shell is completely dissolved in a 0.001 M HF and 4M HCl solutions.

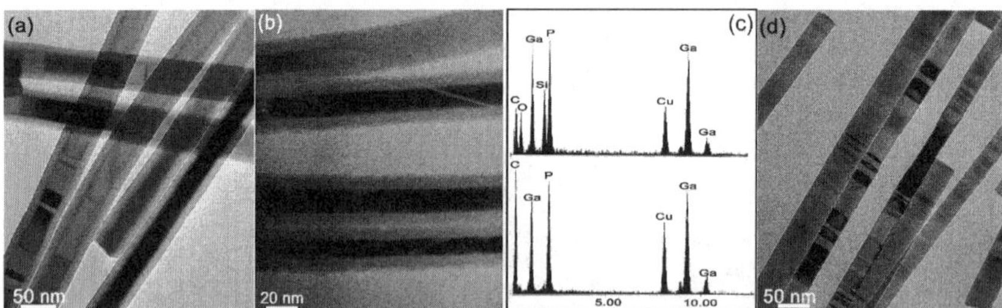

Figure 3. (a) TEM image reveals the nanocable structure. (b) An enlarged image reveals that the nanowires are sheathed homogeneously with the C shell. (c) EDX data shows Ga, P, and Si components, top (for (a)), Ga, P, and C components, bottom(for (b)). (d) Dissolving the shell via the acid treatment produces the pure GaP nanowires.

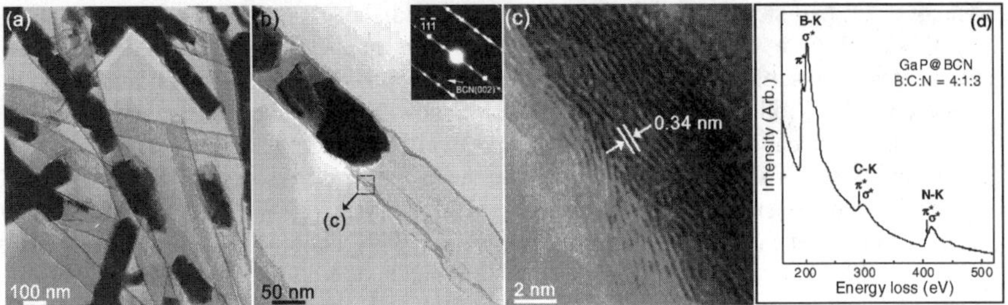

Figure 4. (a) TEM image for the general morphology of GaP/BCN nanocables that the GaP nanorods are encapsulated with the BCN nanotubes. (b) A magnified view for one GaP/BCN nanocable. Its corresponding SAED pattern confirms the [111] growth direction of GaP nanorods (inset). The arrow indicates the obvious [002] direction of the BCN nanotubes. (c) Atomic-resolved image shows the crystalline BCN nanotubes. (d) EELS data of the nanotubes show the K-shell ionization edges of B (B-K), C (C-K), and N (N-K), at 188 eV, 280 eV, and 400 eV, respectively. The atomic ratio B/C and N/C is 4 and 3, respectively.

Figure 4(a) shows a TEM image for the GaP core/BCN shell nanostructures that the GaP nanorods partially fill the BCN mutiwalled nanotubes. The diameter of the BCN nanotubes is about 100 nm and the thickness is about 5 nm. The number of BCN sheets is less than 20. The length of the GaP nanorods is 200-500 nm. The individual GaP nanorod core consists of nearly single crystal with the [111] growth direction (Figure 4(b)). The SAED pattern further proves the [111] growth direction (inset). The HRTEM image reveals the detailed feature for individual nanocables. The fringes of the crystalline graphite-like layers are separated by 0.33-0.34 nm, which is consistent with that of the BCN nanotubes (Figure 4(c)). The corresponding SAED pattern also identifies the [002] direction of the graphite-like layers as indicated by the arrows (inset of Fig. 4(b)). The EELS data (fig. 4(d)) was obtained for the hollow nanotube part, showing three distinct absorption features corresponding to the known K-shell ionization edges for boron (B-K), carbon (C-K), and nitrogen (N-K), respectively. A detailed inspection of the near-edge fine structure confirms the sp^2 hybridization state for B, C, and N, distinguished by the defined π^* features of the graphitic layers. The relative atomic ratios B/C and N/C are 4 and 3, respectively. The N/C ratio varies in the range 3-4, depending on the nanocables. The result suggests that the nanorubes are mostly composed of $B_4CN_{3\text{-}4}$.

CONCLUSIONS

In summary, we synthesized the N-doped GaP nanobelts that the width is 200–500 nm with an average value of 300 nm and the thickness is about 1/10 of the width. They were grown directly via the sublimation of ball-milled GaP powders under NH_3 flow. The nanobelts consist of single crystalline zinc blende structured GaP crystal whose [111] direction is parallel to the belt axis. The EELS data reveal that the N doping occurs mainly in the surface region of the nanobelts. The PL spectrum exhibits the typical isoelectronic bound exciton peaks in the range of 2.11–2.25 eV. The concentration of the doped N atoms is estimated to be about 10^{18} cm^{-3}. The redshift of the peaks from those of the bulk may be explained by the formation of surface potential. The N doping in the surface region would determine the PL properties of the nanobelts. Also, the GaP/SiO_x, GaP/C, and GaP/BCN core/shell nanocables were synthesized successfully via one- or two-step routes of CVD. We modified the experimental conditions used for the synthesis of pure GaP nanowires. The GaP/SiO_x nanocables were produced when the Si substrates were sublimated with the ball-milled GaP powders. The C outer layers were produced via the CVD of following the growth of GaP nanowires. The average diameter of the GaP/SiO_x and GaP/C nanocables is 50 and 30 nm, respectively. The length is up to 500 μm. The amorphous SiO_x and C shells sheath the GaP nanowires with uniform thickness of about 10 nm. The GaP nanowire core consists of the zinc blende structured crystals grown with the [111] direction. The GaP nanorods encapsulated with the multiwalled BCN nanotubes were synthesized using the CVD of the ball-milled GaP/B_2O_3/C powders under NH_3 flow. The average diameter of the GaP/BCN nanocables is 100 nm. The length of the GaP nanorods is 200-500 nm and the growth direction is [111]. The thickness of the BCN nanotubes is about 5 nm, corresponding to less than 20 BCN sheets. The EELS estimates the atomic ratio of B: C: N as about 4:1:3-4. Electron beam irradiation and chemical etching can easily eliminate the amorphous SiO_x and C outer layers. We showed the production of the hollow BCN nanotubes by dissolving the GaP nanorods by the acid.

ACKNOWLEDGMENTS

SEM and X-ray diffraction analysis were performed at Korea Basic Research Institute in Seoul.

REFERENCES

1. J. Hu, T. W. Odom, C. M. Lieber, *Acc. Chem. Res.* ***32***, 435 (1999).
2. X. Duan, Y. Huang, Y. Cui, J. Wang, C. M. Lieber, *Nature (London)* ***409***, 66 (2001).
3. M. S. Gudiksen, L. J. Lauhon, J. Wang, D. C. Smith, C. M. Lieber, *Nature (London)* ***415***, 617(2002).
4. L. J. Lauhon, M. S. Gudiksen, D. Wang, C. M. Lieber, *Nature (London)* ***420***, 57 (2002).
5. X. Duan, Y. Huang, R. Agarwal, C. M. Lieber, *Nature (London)* ***421***, 241 (2003).
6. H. J. Choi, J. C. Johnson, R. He, S. –K. Lee, F. Kim, Pauzauskie, P.; Goldberger, J.; Saykelly, R. J.; Yang, P. *J. Phys. Chem. B* ***107***, 8721 (2003).
7. X. Duan, C. M. Lieber, *Adv. Mater.* ***12***, 298 (2000).
8. C. Tang, S. Fan, M. Lamy de la Chapelle, H. Dang, P. Li, *Adv. Mater.* ***12***, 1346 (2000).
9. W. S. Shi, Y. F. Zheng, N. Wang, C. S. Lee, S. T. Lee, *J. Vac. Sci. Technol.* B ***19***, 1115(2001).
10. H. W. Seo, S. Y. Bae, J. Park, H. Yang, S. Kim, *Chem. Commun.* 2564 (2002).
11. H. W. Seo, S. Y. Bae, J. Park, H. Yang, M. Kang, S. Kim, *Appl. Phys. Lett.* ***82***, 3752 (2003).
12. H. W. Seo, S. Y. Bae, J. Park, M. Kang, S. Kim, *Chem. Phys. Lett.* ***378***, 420 (2003).
13. J. I. Pankove, *Optical processes in Semiconductors* , Dover, New York (1971)
14. V. K. Bazhenov and V. I. Fistul', Sov. Phys. Semicond. ***18***, 843 (1984).
15. A. S. Nasibov, N. N. Mel'nik, I. V. Ponomarev, S. V. Romanko, S. B. Topchii, A. N. Obraztsov, M.Yu Bashtanov, A. A. Krasnovskii, *Quantum Electron.* ***28***, 40 (1998).
16. F. Tuinstra, J. L. Koenig, *J. Chem. Phys.* ***53***, 1126 (1970).
17. Y. K. Yap, M.Yoshimura, Y. Mori, T. Sasaki, *Appl. Phys. Lett.* ***80***, 2559 (2002).
18. C. Y. Zhi, X. D. Bai, E. G. Wang, *Appl. Phys. Lett.* ***80***, 3590 (2002).

Controlled Structure of Gallium Oxide and Indium Oxide Nanowires

Hye Jin Chun, Seung Yong Bae[1] and Jeunghee Park[1]
Dept. of Chemistry, Korea University, Seoul 136-701, South Korea
[1]Dept. of Material Chemistry, Korea University, Jochiwon 339-700, South Korea

ABSTRACT

Gallium oxide (Ga_2O_3) and indium oxide (In_2O_3) nanostructures were synthesized by chemical vapor deposition (CVD). Ga_2O_3 nanowires were synthesized using Ga/Ga_2O_3 mixture and O_2. The diameter of the nanowires is 30-80 nm with an average value of 50 nm. They are consisted of single-crystalline monoclinic crystal. While the nanowires grown without catalyst exhibit a significant planar defect, the nanowires grown with nickel catalytic nanoparticles are almost defect-free. The growth direction of the nanowires grown without the catalyst is uniformly [010]. In contrast, the nanowires grown with the catalyst have random growth direction. X-ray diffraction, Raman spectroscopy, and photoluminescence are well correlated with the structural characteristics of the nanowires. The result provides an evidence for the catalyst effect in controlling the structure of nanowires. In_2O_3 nanostructures were also synthesized in a controlled manner by selecting the catalyst. The reactants were In and In/In_2O_3 mixture. The nanowires were produced using catalytic Au nanoparticles and Ga. But the unique bifurcated-structure nanobelts were instead grown without Ga. The nanowires have uniform [100] growth direction with rectangular cross-section. We converted the In_2O_3 nanowires to In_2O_3-Ga_2O_3 nanostructures.

INTRODUCTION

Since the discovery of carbon nanotubes [1], the pseudo-one-dimensional semiconductor nanostructuers are currently the subject of intense research because of the potential for nanoscale electronic and optoelectronic applications [2]. They are expected to exhibit unique physical properties, distinctive from those of the bulk materials. Monoclinic gallium oxide (β-Ga_2O_3) is an important wide band gap ($E_g \approx 4.9$ eV at 300 K) material due to good chemical and thermal stability. It has a variety of applications including transparent conducting oxide [3], optical emitter for UV [4], and gas sensors [5]. The synthesis and characterization of β-Ga_2O_3 nanostructures have been lately progressed. Various methods, e.g., arc discharge [6], laser ablation [7], thermal oxidation [8], carbothermal reduction [9], were developed by a number of research groups. Indium oxide (In_2O_3) is another important wide band gap ($E_g \approx 3.6$ eV at 300 K) transparent semiconductor. It has been widely used in optoelectronic devices as solar cells [10], flat panel display materials [11], and gas sensors [12]. In_2O_3 nanostructures were prepared by physical evaporation [13], sol-gel template method [14], an electrodeposition [15], and triblock copolymer template method [16]. However, despite the tremendous amount of syntheses, the control of the structure has not been much achieved yet. In this paper, we report a large-scale synthesis of β-Ga_2O_3 nanowires, In_2O_3 nanowires, and In_2O_3 nanobelts. The In_2O_3-Ga_2O_3 hetero-nanostructures were grown on the side of pre-grown In_2O_3 nanowires. The structure and morphology of the nanostructures could be controlled by the use of catalyst.

EXPERIMENTAL DETAIL

For the synthesis of β-Ga_2O_3 nanowires, Ga (99.999%, Aldrich) and Ga_2O_3 (99.999%, Aldrich) with a molar ratio of 4:1 were placed in a quartz boat located inside a quartz tube reactor. The alumina substrates were coated with 0.01 M ethanol solution of $NiCl_2 \cdot 6H_2O$ (99.99%, Aldrich), leading the deposition of Ni catalytic nanoparticles. The uncoated substrates or coated substrates were placed on the quartz boat. The temperature of the reactor was set at 1000-1200 °C. Ar flowed constantly with a rate of 500 sccm. The growth time was 30 min-1 h. When the uncoated substrates were used, the O_2/Ar flow (2-10/500 sccm) was maintained during the growth in order to maximize the yield of nanowires. For In_2O_3 nanowires and nanobelts, a mixture of In (99.99%, Aldrich)/In_2O_3 (99.99%, Aldrich)/Ga and In (99.999%, Aldrich) powders were placed in a quartz boat located inside a quartz tube reactor, respectively. Alumina substrates were coated with 10^{-4} M $HAuCl_4 \cdot 3H_2O$ (98+%, Sigma) ethanol solution, leading to the deposition of Au nanoparticles. The substrate was placed on the boat and the temperature was set at 900-1100 °C. Ar flowed constantly with a rate of 300-500 sccm. The growth time was 1 h. The size and structure of the product were examined using SEM (Hitachi S-4300), TEM (Jeol JEM-2010, 200 kV), ED, energy dispersive X-ray spectroscopy (EDS), powder XRD (Philips X'PERT MPD), and Raman spectroscopy (Renishaw 1000) using a 514.5 nm argon ion laser. PL was conducted at room temperature with the 325 nm line from a helium-cadmium (He-Cd) laser. The laser power was about 1 kW/cm^2.

DISCUSSION

Fig 1(a) shows SEM image for the high-density β-Ga_2O_3 nanowires grown without Ni catalytic nanoparticles. Let us denote the product as NWs-1. The length of nanowires is 10-20 μm. Parts b and c of Fig. 1 show the TEM and HRTEM images of NWs-1. All of the NWs-1 are grown uniformly with the [010] growth direction. The end of the NWs-1 is not flat, but rather sharp (Fig. 1(c)). The selected-area ED (SAED) pattern shows the [010] growth direction. The atomic-resolved image shows the (200) planes separated by 5.94 Å, which is consistent with that

Figure 1. (a) SEM, (b) TEM, and (c) HRTEM images for NWs-1, (d) SEM, (e) TEM, and (f) HRTEM images for NWs-2

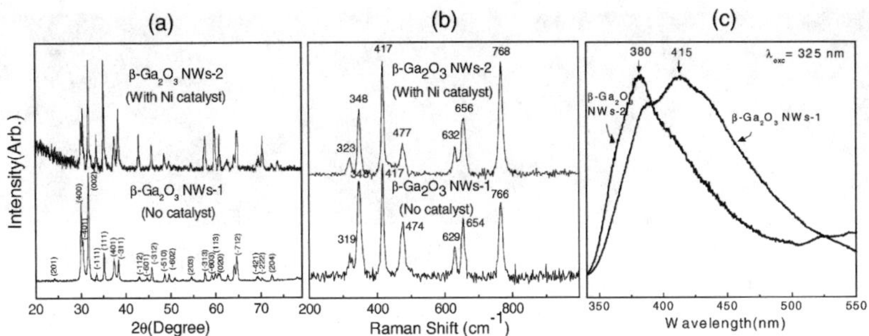

Figure 2. (a) XRD pattern, (b) Raman spectrum, and (c) PL spectrum of NWs-1 and NWs-2. The excitation wavelength is the 325 nm line of He-Cd laser.

of the monoclinic structure. A significant planar defect exists along the growth direction. Fig. 1(d) shows SEM image of β-Ga_2O_3 nanowires grown using the Ni catalyst. Let us denote these nanowires grown with the Ni catalyst as NWs-2. The length of nanowires is up to 500 μm, which is much longer than NWs-1. Parts e and f of Fig. 1 show the TEM and HRTEM images of NWs-2, respectively. The diameter of both NWs-1 and NWs-2 is 30-80 nm with an average value of 50 nm. The NWs-2 is longer and more flexible compared to NWs-1. Fig. 1(f) shows a HRTEM image and its SAED pattern for a nanowire of NWs-2 sample. The growth direction is tilted to the [$\bar{2}0\bar{2}$] direction with an angle of 15 degrees. All NWs-2 we observed have random growth direction. Atomic-resolved image reveals well-separated (111) planes with a distance of 2.51 Å. In contrast to the NWs-1, the NWs-2 show defect-free crystalline perfection. These results show that the structure of the nanowires can be controlled by the use of catalyst. Without the Ni catalyst the nanowires are growth with an uniform growth direction of [010]. The stacking faults exist along the growth direction. However, the nanowires grown with the Ni catalyst have a variety of growth directions and nearly defect-free crystalline structure. The NWs-2 are longer and more flexible compared to NWs-1.

The XRD pattern of NWs-1 and NWs-2 are displayed in Fig. 2(a), which is consistent with the reported one (JCPDS Card No. 41-1103). The XRD analysis verifies the highly crystalline nature of β-Ga_2O_3 crystals. Raman scattering spectra of NWs-1 and NWs-2 are displayed in Fig. 2(b). The peak position and relative intensity are consistent with those of the reported one [17]. It is noticeable that the peaks of NWs-1 show a red shift from that of the NWs-2 by an average value of 4 cm^{-1}. The red shift of NWs-1 is probably related with their

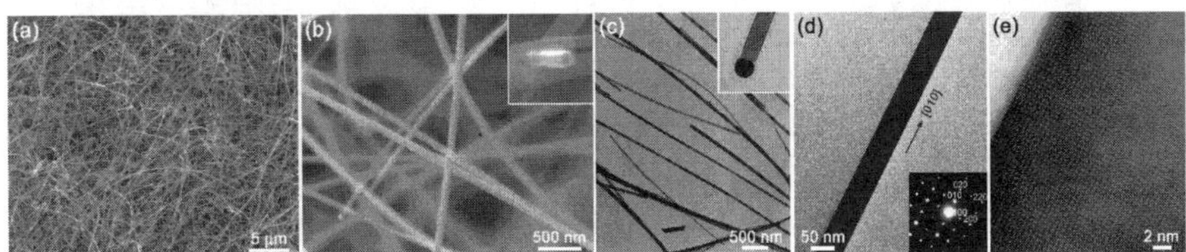

Figure 3. (a) SEM micrograph of the In_2O_3 nanowires. (b) A magnified SEM image. Inset shows a rectangular cross section. (c) TEM image showing a general morphology of the nanowires, (d) HRTEM image for a In_2O_3 nanowire and (e) its atomic-resolved image.

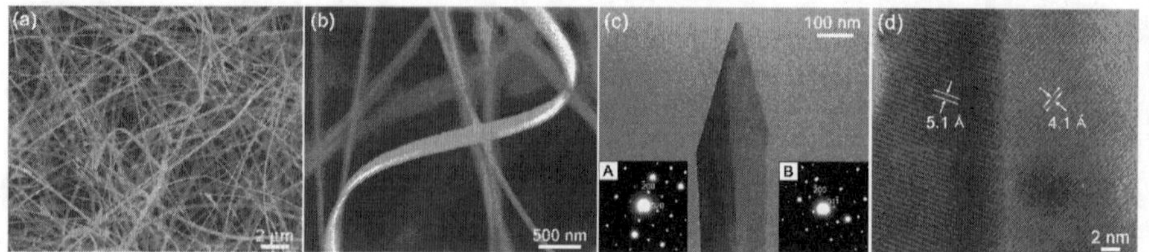

Figure 4. (a) SEM micrograph of the In_2O_3 nanobelts. (b) A magnified SEM image. (c) TEM image of a In_2O_3 nanobelt and (e) its atomic-resolved image.

defective crystalline structure [18]. Fig. 2(c) shows PL spectrum obtained from two β-Ga_2O_3 nanowire samples. Although the photon energy is below the band gap, a strong blue emission is observed. The NWs-1 exhibits a broad band centered at 415 nm. For NWs-2, the band center blueshifts to 380 nm and the bandwidth is narrower than that of NWs-1. The NWs-1 probably possesses the more atomic vacancies due to the planar defect, compared to the NWs-2. The energy of donor and acceptor formed by the atomic vacancies would distribute in the broader range, resulting in a broader and red shifted emission [19,20].

We've controlled the structure of In_2O_3 nanostructures by using Ga as co-catalyst. Fig. 3(a) shows SEM image of high-density In_2O_3 nanowires grown using Ga. A magnified SEM image shows that the nanowires have uniform diameter and smooth surface (Fig. 3(a) and (b)). Fig. 3(c), (d), and (e) correspond to the TEM and HRTEM images of the nanowires. The diameter of nanowires is distributed in the narrow range 30-50 nm with an average value of 40 nm (Fig. 3(c)). The tip of the nanowires sometimes attaches to the Au nanoparticles, as shown in the inset. The growth direction of In_2O_3 nanowires is [010] (Fig. 3(d)). Atomic-resolved image reveals that the distance between neighboring (200) fringes is about 0.5 nm (Fig. 3(e)).

Fig. 4(a) and (b) show the SEM image of the belt-shaped In_2O_3 nanostructures grown without using Ga. Fig. 4(c) corresponds to a TEM image of the bicrystalline nanobelt having a sharp tip. The angle of side edges is about 40 degrees. The ED pattern for the left side is taken using the [001] zone axis, revealing the $[\bar{3}\bar{1}0]$ growth direction (inset "A"). The right side has the $[\bar{6}1\bar{1}]$ growth direction, confirmed by the ED pattern taken using [011] zone axis (inset "B"). Fig. 4(d) corresponds to its atomic-resolved image. The measurement of XRD pattern and Raman spectrum confirms that the In_2O_3 nanowires and nanobelts have highly crystalline cubic structure.

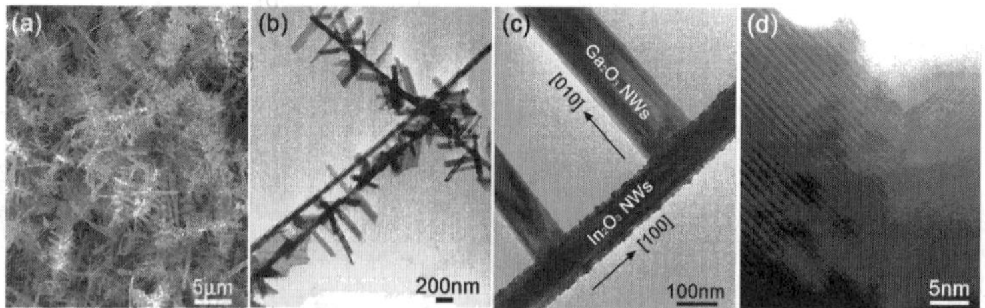

Figure 5. (a) SEM and (b) TEM images showing a general morphology of In_2O_3-Ga_2O_3 heterostructures, (c) HRTEM and (d) its atomic-resolved images for the interface between In_2O_3 nanowire and β-Ga_2O_3 nanorod.

Fig. 5 shows the In_2O_3-Ga_2O_3 hetero-nanostructures that the β-Ga_2O_3 nanorods are grown on the side of pre-grown In_2O_3 nanowires. SEM and TEM image shows pine-tree morphology, that the β-Ga_2O_3 nanorod array decorates fully the side of the β-In_2O_3 nanowires (Fig. 5(a) and (b)). Fig. 5(c) shows that the β-Ga_2O_3 nanorods are vertically grown on side of a In_2O_3 nanowire. The β-Ga_2O_3 nanorods have uniform growth direction of [010]. Fig. 5(d) corresponds to the atomic-resolved image for the interface region between β-Ga_2O_3 and In_2O_3 nanowires, showing defect-free β-Ga_2O_3 and In_2O_3 crystals.

CONCLUSION

We synthesized high-density β-Ga_2O_3 nanowires, In_2O_3 nanowires, In_2O_3 nanobelts, and β-Ga_2O_3-In_2O_3 nanostructures via thermal CVD. The β-Ga_2O_3 nanowires were synthesized using Ga/Ga_2O_3 mixture and O_2. The diameter of the nanowires is 30-80 nm with an average value of 50 nm. They are consisted of single-crystalline monoclinic crystal. While the nanowires grown without catalyst (called as NWs-1) exhibit a significant planar defect, the nanowires grown with nickel catalytic nanoparticles (called as NWs-2) are almost defect-free. The growth direction of the NWs-1 is uniformly [010]. In contrast, the NWs-2 have random growth direction. The XRD patterns confirm the highly crystalline nature of β-Ga_2O_3 for both NWs-1 and NWs-2. The NWs-1 shows a red shift of Raman peaks relative to the NWs-2 would be related with their significant planar defects. The PL exhibits a strong blue emission centered at 415 and 380 nm, respectively, for NWs-1 and NWs-2. The bandwidth is narrower for NWs-2, compared to that of NWs-1. The peak shift and broadening suggest that the emission would originate from the defect of the nanowires. The result provides an evidence for the catalyst effect in controlling the structure of nanowires. The In_2O_3 nanostructures were also synthesized in a controlled manner by selecting the catalyst. The reactants were In and In/In_2O_3 mixture. The nanowires were produced using catalytic Au nanoparticle and Ga. But the unique bifurcated-structure nanobelts were grown without Ga. The nanowires have uniform [100] growth direction with a rectangular cross-section. The nanobelts exhibit intrinsically bicrystalline structure that the nanobelts can be split into two single-crystalline nanobelts. The bicrystalline nanobelts usually have a width in the range of 300~500 nm and the sharp tip with a triangle shape. The (010) planes of the bicrystals are tilted toward the twin boundary with an angle of 13~20 degree. This may produce a large strain at the twin boundary, resulting in the separation of the nanobelts. The XRD pattern and Raman spectrum confirm the highly crystalline of In_2O_3 nanostructures. We also synthesized the Ga_2O_3-In_2O_3 nanostructures that the β-Ga_2O_3 nanorods are vertically grown on side of the In_2O_3 nanowire. The β-Ga_2O_3 nanorods have uniformly the [010] growth direction.

ACKNOWLEDGMENTS

This work was supported by Korea Science and Engineering Foundation (Project No.: R14-2003-033-0100-0). SEM and X-ray diffraction analysis were performed at Basic Science Research Center in Seoul.

REFERENCES

1. S. Iijima, *Nature* **354**, 56 (1991).
2. J. Hu, T. W. Odom and C. M. Lieber, *Acc. Chem. Res.* **32**, 435 (1999).
3. M. Passlack, E. F. Schubert, W. S. Hobson, M. Hong, N. Moriya, S. N. G. Chu, K.

Konstadinidis, J. P. Mannaerts, M. L. Schnoes and G. J. Zydzik, *J. Appl. Phys.* **77**, 686 (1995).
4. L. Binet and D. Gourier, *J. Phys. Chem. Solids* **59**, 1241 (1998).
5. M. Ogita, N. Saika, Y. Nakanishi and Y. Hatanaka, *Appl. Surf. Sci.* **142**, 188 (1999).
6. W. Q. Han, P. Kohler-Redlich, F. Ernst and M. Rűhle, *Solid State Commun.* **115**, 527 (2000).
7. J. Q. Hu, Q. Li, X. M. Meng, C. S. Lee and S. T. Lee, *J. Phys. Chem. B* **106**, 9536 (2002).
8. H. Z. Zhang, Y. C. Kong, Y. Z. Wang, X. Du, Z. G. Bai, J. J. Wang, D. P. Yu, Y. Ding, Q. L. Hang and S. Q. Feng, *Solid State Commun.* **109**, 677 (1999).
9. X. C. Wu, W. H. Song, W. D. Huang, M. H. Pu, B. Zhao, Y. P. Sun and J. J. Du, *Chem. Phys. Lett.* **328**, 5 (2000).
10. K. Sreenivas, T. S. Rao, A. Mansingh and S. Chandra, *J. Appl. Phys.* **57**, 384 (1985).
11. X. Li, M. W. Wanlass, T. A. Gessert, K. A. Emery and T. J. Coutts, *Appl. Phys. Lett.* **54**, 2674 (1989).
12. Y. Shigesato, S. Takaki and T. Haranoh, *J. Appl. Phys.* **71**, 3356 (1992).
13. Z. W. Pan, Z. R. Dai and Z. L. Wang, *Science* **291**, 1947. (2001).
14. B. Cheng and E. T. Samulski, *J. Mater. Chem.* **11**, 2901 (2001).
15. M. J. Zheng, L. D. Zhang, G. H. Li, X. Y. Zhang and X. F. Wang, *Appl. Phys. Lett.* **79**, 839 (2001).
16. H. Yang, Q. Shi, B. Tian, Q. Lu, F. Gao, S. Xie, J. Fan, C. Yu, B. Tu and D. Zhao, *J. Am. Chem. Soc.* **125**, 4724 (2003).
17. D. Dohy, G. Lucazeau and A. Revcolevschi, *J. Solid State Chem.* **45**, 180 (1982).
18. Y. H. Gao, Y. Bando, T. Sato, Y. F. Zhang and X. Q. Gao, *Appl. Phys. Lett.* **81**, 2267 (2002).
19. T. Harwig and F. Kellendonk, *J. Solid Sate Chem.* **24**, 255 (1978).
20. V. I. Vasil'tsiv, Ya. M. Zakharko and Ya. I. Prim, *Ukr. Fiz. Zh.* **33**, 1320 (1988).

Mat. Res. Soc. Symp. Proc. Vol. 789 © 2004 Materials Research Society N11.30

Control of Morphology and Growth Direction of Gallium Nitride Nanostructures

Seung Yong Bae, Hee Won Seo, Jeunghee Park,
Department of Chemistry, Korea University, Jochiwon 339-700 Korea

ABSTRACT

Various shaped single-crystalline gallium nitride (GaN) nanostructures were produced by chemical vapor deposition method in the temperature range of 900 – 1200 °C. Scanning electron microscopy, transmission electron microscopy, electron diffraction, x-ray diffraction, electron energy loss spectroscopy, Raman spectroscopy, and photoluminescence were used to investigate the structural and optical properties of the GaN nanostructures. We controlled the GaN nanostructures by the catalyst and temperature. The cylindrical and triangular shaped nanowires were synthesized using iron and gold nanoparticles as catalysts, respectively, in the temperature range of 900 - 1000 °C. We synthesized the nanobelts, nanosaws, and porous nanowires using gallium source/ boron oxide mixture. When the temperature of source was 1100 °C, the nanobelts having a triangle tip were grown. At the temperature higher up to 1200 °C the nanosaws and porous nanowires were formed with a large scale. The cylindrical nanowires have random growth direction, while the triangular nanowires have uniform growth direction [010]. The growth direction of the nanobelts is perpendicular to the [010]. Interestingly, the nanosaws and porous nanowires exhibit the same growth direction [011]. The shift of Raman, XRD, and PL bands from those of bulk was correlated with the strains of the GaN nanostructures.

INTODUCTION

One-dimensional nanostructures such as wires (or rods), belts, and tubes are currently the subject of intense research because of a pure scientific standpoint and their technological applications in electronics and optoelectronic nanodevices, etc.[1-4] These properties and applications are usually highly dependent on the size and the shape of nanostructures. In fact, the shape of nanostructures has significant influence on the physical properties that are important in many practical applications, such as light-emitting diodes and scanning-microscopy probes.[5,6] Therefore the development of synthesis whose size and shape are easily controlled is of great interest in this area. Among the various nanostructured materials, wurtzite structured gallium nitride (GaN) is widely accepted as a potential material for the optoelectronic application, due to its direct large energy bandgap (E_g=3.4 eV at 300 K), high thermal stability, and strong resistance to radiation.[7-9] Here we report that various 1-dimensional nanostructures of GaN were synthesized in a controlled manner via chemical vapor deposition at the temperature ranging from 900 to 1200 °C.[10-14] The cylindrical and triangular shaped nanowires were synthesized using iron (Fe) and gold (Au) nanoparticles in the temperature range of 900 - 1000 °C. When the gallium (Ga) source/ boron oxide (B_2O_3) mixture was used to reactant, we synthesized the nanobelts, nanosaws, and porous nanowires. The results indicate that the reaction temperature and the catalyst play an important role in controlling the structure. The morphology and optical properties of the GaN nanostructures were investigated through scanning electron microscopy (SEM), transmission electron microscopy (TEM), selected-area electron diffraction (SAED), X-ray powder diffraction (XRD), Raman spectroscopy and Photoluminescence (PL).

EXPERIMENTAL DETAILS

For the syntheses of cylindrical and triangular nanowires,[10,11] Ga metal (99.999 %, Aldrich) and GaN powder (99.99+ %, Aldrich) mixture was placed inside of a quartz tube reactor. NH_3 gas was introduced into the quartz tube and temperature of the substrate was approximately 900 - 1000 °C. When the Fe nanoparticles deposited on the silicon substrate were used, the cylindrical nanowires were synthesized. For the use of Au nanoparticles on the alumina substrate, triangular nanowires were synthesized. In addition to these nanostructures, we synthesized the nanobelts, nanosaws, and porous nanowires using Ga source/ B_2O_3 mixture in the temperature range of 1100 – 1200 °C.[12-14] When the temperature of sources was 1100 °C, the nanobelts were synthesized. At the temperature of 1200 °C, the nanosaws were synthesized using pure reactants ($Ga/Ga_2O_3/B_2O_3$) and porous nanowires were obtained using of ball-milled reactants. The morphology and structure of the products were examined by SEM (Hitachi S-4300), TEM (Jeol JEM-2010), SAED, and XRD (Philips X'PERT MPD). Raman scattering spectra were taken using a 514.5 nm argon ion laser. The PL of product was measured at room temperature. The excitation source was 325 nm line from a helium-cadmium (He-Cd) laser.

DISCUSSION

Figure 1(a) and (b) show SEM images for the high-density nanowires synthesized homogeneously over a large area of substrate using Fe and Au catalyst, respectively. They have very clean surface without any particles. Figure 1(c) shows a high-resolution TEM (HRTEM) image for a cylindrical GaN nanowires having a mean value of 25 nm. The inset is the corresponding SAED pattern, which is consistent with that of wurtzite GaN. The [0-21] direction is parallel to the long axis. It is observed that the single-crystalline cylindrical nanowires are grown with their own specific growth direction, but not with the identical direction. Figure 1(d) shows HRTEM image for the triangular nanowires and its SAED pattern, showing the single-crystalline wurtzite structured GaN crystal and the [010] growth direction. Upper inset proves undoubtedly the triangle cross-section of the nanowires. Therefore, We obtained a possible control of the morphology and growth direction using the catalyst. The cylindrical nanowires were grown with random growth direction when the Fe catalytic nanoparticles were used.[10] With Au nanoparticles, the triangular nanowires having the [010] growth direction were grown.[11,15]

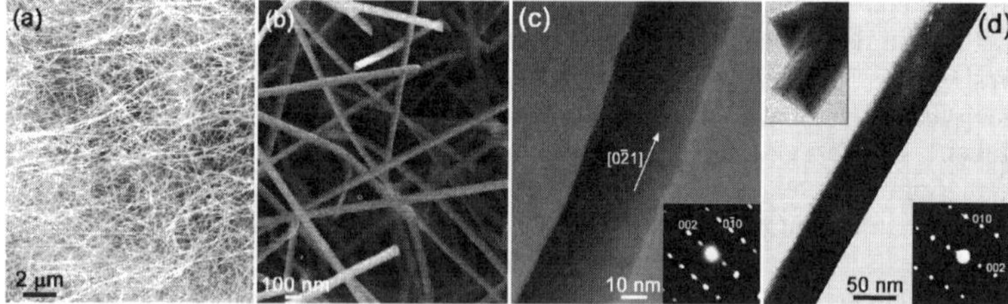

Figure 1. SEM images show (a) the cylindrical shaped nanowires grown on the substrate using Fe catalyst, (b) the triangular shaped nanowires grown using Au catalyst. (c) HRTEM images and SAED pattern of the cylindrical shaped nanowires whose growth direction is [0-21]. (d) HRTEM images and SAED pattern of the triangular nanowires with the [010] growth direction. The upper inset is magnification of the tip part of the triangular nanowires.

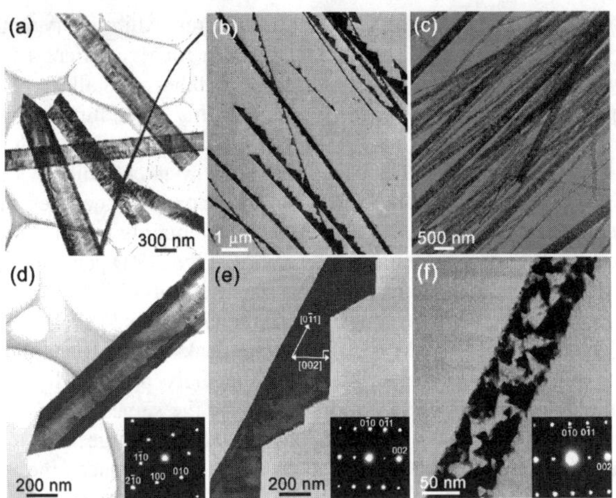

Figure 2. TEM images of the (a) nanobelts, (b) nanosaws, and (c) porous nanowires. (d)-(f) HRTEM images of the (d) nanobelt, (e) nanosaw, and (f) porous nanowire. The insets are the corresponding SAED pattern of individual GaN nanostructure.

Figure 2 shows TEM images for the various shapes of GaN nanostructures. Figure 2(a)-(c) display a typical morphology of the nanobelts, nanosaws, and porous nanowires, respectively. The TEM images of nanobelts and nanosaws reveal that they have not a round cross section. From the images of many nanobelts and nanosaws, the thickness of the belt plane can be estimated to be approximately 1/10 of the width. The porous nanowires are straight and cylindrical. They contain many pores inside whose size is 5-20 nm. Figure 2(d)-(f) exhibit HRTEM images and SAED patterns of individual nanostructure. They are all single crystalline. All of the nanobelts, we observed, have a triangle tip and uniformly [2-10] growth direction. The growth directions of nanosaws and porous nanowires are the same as [0-11].

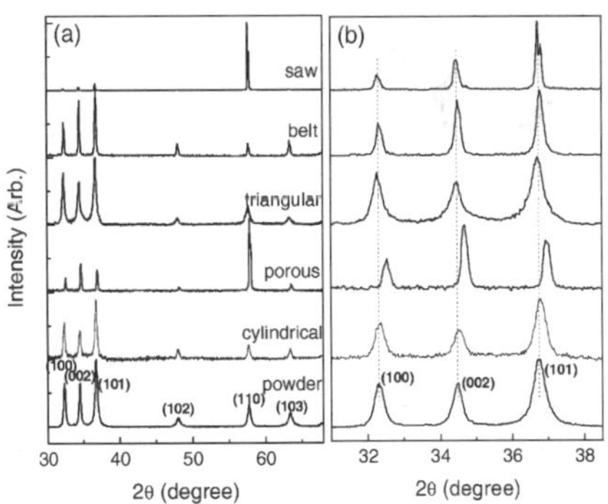

Figure 3. (a) A full-range XRD pattern taken from the GaN nanostructures. (b) The (100), (002), and (101) peaks are magnified.

The XRD pattern taken from the GaN nanostructures exhibits a typical one of GaN wurtzite structure. Figure 3(a) shows a full-range XRD pattern for five different GaN nanostructures and commercially available GaN powder (Aldrich). No other crystalline phases are detected. The XRD data provide another evidence for the highly crystalline wurtzite crystal of GaN nanostructures. It is noteworthy that the intensity of (110) peak from the nanosaws and porous nanowires is strongest and the other peaks are much weaker, which is different from the XRD pattern of GaN powder. Since the nanosaws and porous nanowires are lying with random orientation on the substrate, the XRD would take place mainly from the large (110) lattice planes parallel to the axis. Therefore, the relative intensity of the XRD peaks can support powerfully the growth direction of the nanostructures. The position of (100), (002), and (101) peaks are compared in figure 3(b). The lattice constants of the GaN powders agree with the reported values of bulk GaN crystals, a=3.19 and c=5.18 Å.[16] The peaks of the triangular nanowires, nanobelts, and nanosaws shift negligibly from those of the GaN powder. It indicates that the lattice constants of the GaN nanostructures would be nearly the same as those of the bulk. In contrast, the peaks of the cylindrical and porous nanowires show a large shift to the higher diffraction angle, suggesting the change of lattice constants. For instance, the shift of peaks for the cylindrical nanowires is $\Delta(2\theta)=0.07°$, 0.15°, and 0.10°, respectively, for (100), (002), and (101) peaks.[10] The peak shift toward a larger angle indicates that the separations of neighboring lattice planes along the wire axis are shorter than those of bulk GaN. We suggest that that they would undergo biaxial compressive stresses in the inward radial direction and the induced tensile uniaxial stresses in the growth direction. The largest shift of (002) peak suggests that the (001) and (100) planes would be under the strongest compressive and tensile stresses, respectively.

Raman scattering spectra of the triangular nanowires, cylindrical nanowires, and powder are displayed in figure 4. The first-order phonon frequencies of A_1(TO), E_2(high), and A_1(LO) modes were measured to be 533, 570, and 738 cm^{-1}, for the GaN bulk, respectively.[17,18] The GaN powder shows the peaks at the same position, 533, 569, and 730 cm^{-1}, which can be assigned to A_1(TO), E_2(high), and A_1(LO), respectively. The lower frequency shift as significant as 7 cm^{-1} was observed from the cylindrical nanowires. It is known that the compressive or tensile strains play an important role in the shift of E_2(high) peak.[10,19] The E_2(high) phonon band toward a lower frequency can be correlated with the reduction of the lattice constant. The Raman shift of the cylindrical nanowires provides another evidence of the tensile strains. No shift of the

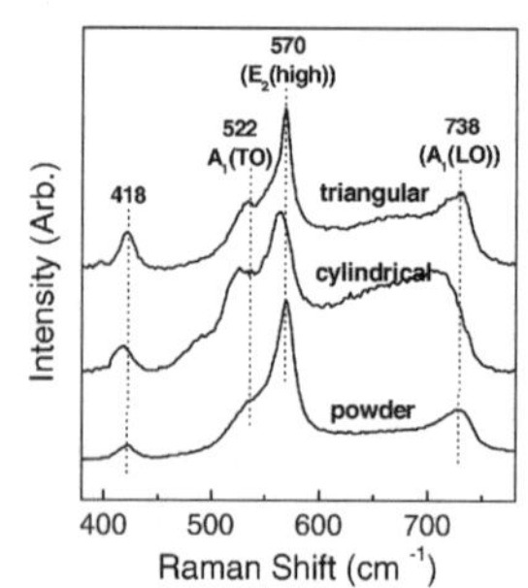

Figure 4. Raman scattering spectrum of the triangular nanowires, cylindrical nanowires, and powder. The excitation source is 514.5 nm argon ion laser.

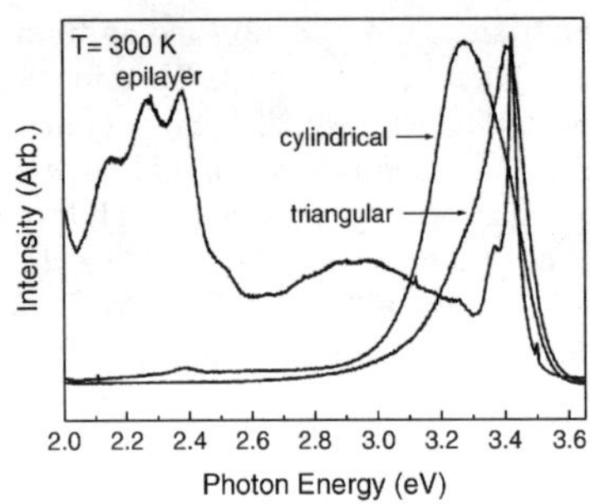

Figure 5. PL spectrum of the triangular nanowires, cylindrical nanowires and epilayer at 300 K. The excitation wavelength is the 325 nm of He-Cd laser.

E_2(high) peak implies that triangular nanowires would experience no strains.[11] Therefore the XRD and Raman data suggest that triangular GaN nanowires would be stress-free materials. The bandgap energy of triangular GaN nanowires is probably the same as that of the GaN bulk, which is distinct from the cylindrical GaN nanowires.

Figure 5 shows the PL spectra obtained for the triangular nanowires, cylindrical nanowires, and GaN epilayer at room temperature (300 K). The GaN epilayer was grown with a thickness of 3 μm on a sapphire substrate. The epilayer exhibits a typical yellow band in the range 2.0-2.6 eV, but the cylindrical nanowires and triangular nanowires show almost no yellow band. PL spectrum of epilayer shows a strong narrow peak at 3.42 eV, which is ascribed to the bound exciton. The cylindrical nanowires have the PL in the broader energy range with a peak (3.28 eV) at the lower energy than epilayer. The broad PL band is plausibly due to a superposition of the bound exciton emission from the differently strained domains. The much less thermal quenching compared to that of the epilayer could be correlated with a large binding energy, providing another piece of evidence for an existence of the strains inside the cylindrical nanowires. We suggest that the lower energy shift would be a result of the reduction of band gap due to strain of cylindrical nanowires. In contrast, the peak of triangular nanowires (3.41 eV) was very similar with those of epilayer, confirming the stress-free properties of the triangular nanowires compared with cylindrical nanowires grown with random direction.

CONCLUSION

In summary, we synthesized various shaped single-crystalline GaN nanostructures by thermal CVD using the catalyst-assisted reaction of Ga sources with NH_3 in the temperature range of 900-1200 °C. We controlled the GaN nanostructures by the catalyst and temperature. The cylindrical and triangular shaped nanowires were synthesized using Fe and Au nanoparticles in the temperature range of 900-1000 °C. When the Ga source/boron oxide mixture were used to reactant, the nanobelts, nanosaws, and porous nanowires were synthesized above 1100 °C. The cylindrical nanowires have random growth direction and show a large shift of XRD and Raman peaks. Also, we can observe the strong and broad blue band in the PL measurement at 300 K. These results suggest that cylindrical nanowires would undergo biaxial compressive stresses in the inward radial direction and the induced tensile uniaxial stresses in the growth direction. The

GaN nanowires having a triangle cross-section and uniform [010] growth direction show no shift of XRD, Raman, and the bound exciton peak. Therefore triangular nanowires having the band gap energy of the bulk would be free from the stress. In the case of the other GaN nanostructures, porous nanowires have [011] growth direction and large shift of XRD and Raman peaks. The cylindrical shaped porous nanowires experience a high level of strains. In contrast, the nanosaws and nanobelts are under no strain. We suggest that the cylindrical shape would play a major role in inducing the compressive strains. The shift of XRD, Raman, and PL from those bulk was correlated well the strains of the GaN nanostructures.

ACKNOWLEDGEMENT

This work was supported by Korea Science and Engineering Foundation (Project No.: R14-2003-033-0100-0). SEM and X-ray diffraction analysis were performed at Basic Science Research Center in Seoul.

REFERENCES

1. S. Iijima, *Nature* **354**, 56 (1991).
2. J. Hu, T. W. Odom, and C. M. Lieber, *Acc. Chem. Res.* **32**, 435 (1999).
3. M. H. Huang, S. Mao, H. Feick, H. Yan, Y. Wu, H. Kind, E. Weber, R. Russo, and P. Yang, *Science* **292**, 1897 (2001).
4. Y. Xia, P. Yang, Y. Sun, Y. Wu, B. Mayers, B. Gates, Y. Yin, F. Kim, and H. Yan, *Adv. Mater.* **15**, 353 (2003).
5. E. N. Bogachek, A. G. Scherbakov, and U. Landman, *Phys. Rev. B* **56**, 1065 (1997).
6. H. Zeng, S. Michalski, R. D. Kirby, D. J. Sellmyer, L. Menon, and S. Bandyopadhyay, *J. Phys.: Condens. Matter* **14**, 715 (2002).
7. S. N. Mohammad and H. Morkoç, *Prog. Quantum Electron.* **20**, 361 (1996).
8. F. A. Ponce and D. P. Bour, *Nature* **386**, 351 (1997).
9. S. Nakamura, *Science* **281**, 956 (1998).
10. H. W. Seo, S. Y. Bae, J. Park, H. Yang, G. S. Park, and S. Kim, *J. Chem. Phys.* **116**, 9492 (2002).
11. S. Y. Bae, H. W. Seo, J. Park, H. Yang, H. Kim, and S. Kim, *Appl. Phys. Lett.* **82,** 4564 (2003).
12. S. Y. Bae, H. W. Seo, J. Park, H. Yang, J. C. Park, and S. Y. Lee, *Appl. Phys. Lett.* **81**, 126 (2002).
13. S. Y. Bae, H. W. Seo, J. Park, and H. Yang, *Chem. Phys. Lett.* **373**, 620 (2003).
14. S. Y. Bae, H. W. Seo, J. Park, and H Yang, *Chem. Phys. Lett.* **376**, 445 (2003).
15. T. Kuykendall, P. Pauzauskie, S. Lee, Y. Zhang, J. Goldberger, and P. Yang, *Nano Lett.* **3**, 1063 (2003).
16. D. A. Neumayer and J. G. Ekerdt, *Chem. Mater.* **8**, 9 (1996).
17. T. Azuhata, T. Sota, K. Suzuki, and S. Nakamura, *J. Phys.: Condens. Matter* **7**, L129 (1995).
18. V. Yu. Davydov, Yu. E. Kitaev, I. N. Goncharuk, A. N. Smirnov, J. Graul, O. Semchinova, D. Uffmann, M. B. Smirnov, A. P. Mirgorodsky, and R. A. Evarestov, *Phys. Rev. B* **58**, 12899 (1998).
19. M. Klose, N. Wieser, G. C. Rohr, R. Dassow, F. Scholz, and J. Off, *J. Crystal Growth* **189/190**, 634 (1998).

Electrical, Electronic Properties and Devices I

Mat. Res. Soc. Symp. Proc. Vol. 789 © 2004 Materials Research Society N12.5

Monocrystalline InP Nanotubes

Erik. P.A.M. Bakkers[1], Louis F. Feiner[1], Marcel. A. Verheijen[1], Jorden A. van Dam[2], Silvano De Franceschi[2], and Leo Kouwenhoven[2]

[1]*Philips Research Laboratories, Prof. Holstlaan 4, 5656 AA Eindhoven, The Netherlands*

[2]Department of Nanoscience, Delft University of Technology, P.O.Box 5046, 2000 GA Delft, The Netherlands

ABSTRACT

Indium phosphide (InP) nanowires and nanotubes have been synthesized via the vapor-liquid-solid (VLS) growth mechanism. The wires as well as the tubes are crystalline and have the (bulk) zinc blende structure. Compared to the nanowires the nantubes are formed at higher temperatures. A simple model for the formation of the nanotubes is presented. The diameter of the wires and the wall thickness of the tubes can be controlled by the synthesis temperature. Photoluminescence measurements on individual wires show a strong polarization dependency. Moreover, the nanostructures exhibit a considerable blue shift with respect to bulk emission as a result of size-quantization. In addition, this blue shift indicates that the optical properties are not dominated by defect states.

INTRODUCTION

One-dimensional structures, such as semiconducting nanowires are attractive building blocks for bottom-up nanoelectronics. Nanowires can relatively easily be manipulated and contacted to form functional circuits.[1] The electronic structure of nanowires is determined by the chemical composition and the diameter and additionally these wires can be n-type or p-type doped.[2] Moreover, atomically abrupt III-V semiconductor heterojunctions have been synthesized in a nanowire.[3] Strong quantization effects in a nanowire were established by the incorporation of a quantum dot, having a different chemical composition than the rest of the wire.[4] However, the confinement for nanowires with a homogeneous composition, which still have macroscopic dimensions in one direction, is clearly less than for quantum dots. The confinement effect in one-dimensional structures can be enhanced by the formation of tubes instead of wires. Carbon nanotubes and metal chalcogenide nanotubes have a pseudographitic morphology and the electronic properties are determined by the diameter and the chirality of the tubes.[5,6] Therefore, it is difficult to predetermine the electronic properties of these nanotubes. In this paper the synthesis of crystalline InP nanowires and nanotubes is reported. The substrate temperature during growth establishes the shape of the resulting nanostructures and can be used to tune the electronic properties. The optical properties of the nanowires and nanotubes are discussed.

EXPERIMENTAL

One-dimensional nanostructures were grown via the VLS (vapor-liquid-solid) laser ablation method as schematically drawn in figure 1A.[7] The synthesis was carried out without the use of a template. A silicon sample with a native oxide layer covered with an equivalent of a 2 -

20 Å Au film was used as the substrate. The substrate was mounted onto an Al_2O_3 block at the downstream end of the tube oven. The quartz oven tube was evacuated to a pressure of 10^{-7} mbar and was subsequently flushed by argon (6N). The oven was heated to 730-850 °C. The substrate temperature was measured 0.5 mm below the substrate by the use of a thermocouple; the temperature was well stabilized before the ablation was started. Upon heating, the thin Au film broke up into small gold particles. The beam of an ArF laser (λ=193 nm, 100 mJ/pulse, 2.5-10 Hz) was focused on either a pressed InP target (density 65%) or on a piece of single crystalline InP. The In and P species were carried by the argon gas flow (210 sccm, 140 mbar) towards the substrate, where they dissolved into the Au particles. Once the Au dots were saturated by In and P, a one-dimensional, crystalline structure started to excrete. After the ablation was stopped (after 15-30 kshots) the sample was studied with electron microscopy. For TEM, a Cu-grid provided with a film of amorphous carbon was wiped over the substrate.

The nanostructures were dispersed in chlorobenzene by low power ultrasonification for 10 seconds. A droplet of the dispersion was deposited onto a silicon sample provided with 500 nm thermal oxide. Photoluminescence studies on individual nanostructures were performed by the use of a home-built fluorescence microscope. The excitation laser beam (λ=457 nm, 200 mW Melles Griot DPSS58) was reflected from a dichroic mirror and focused through the microscope objective onto the sample to obtain a power density of 10 kW/cm^2. The linearly polarized light from the laser could be rotated by a polarization rhomb. The photoemission was collected by the same objective and passed through the dichroic mirror and an emission filter. The light was either coupled into a glass fiber connected to a spectrophotometer or projected onto a CCD camera.

RESULTS AND DISCUSSION

Figure 1B shows a typical SEM image of the as-grown nanostructures. Over 95 % of the deposited material consisted of one-dimensional structures. When the substrate temperature was in the range 425-500°C and an undoped InP (6N) target was used, single-crystalline InP nanowires were formed. The growth direction and the crystal structure were determined from high-resolution TEM (TECNAI TF30ST, figure 1C). The long axis of most of the wires was perpendicular to the (111) lattice plane as has been reported by others[8], but also growth along the [211] direction was observed occasionally. Periodic arrays of twinning boundaries perpendicular to the long axis of the wire and with an irregular period were observed when the wires were grown in the [111] direction. Each wire was terminated by a particle containing Au and an amount (typically 40%, determined by EDX) of InP, indicating that the wires grow via the VLS mechanism. The diameter of the nanowires is initially determined by the thickness of the Au film. However, the substrate temperature during growth affects the resulting diameter as well. In figure 2A the resulting wire diameter, obtained from TEM measurements, is plotted versus the substrate temperature in the range 425-500 °C for a constant Au layer thickness of 5 Ångstroms. The wire diameter increases with temperature, which is expected considering the AuInP-phase diagram. An additional amount of InP will dissolve in the Au particle when the temperature is raised. This will result in an increase of the particle diameter leading to a thicker wire. The thinnest wires we have obtained starting with a 2 Å Au film had diameters of 4 nm.

When higher temperatures (>500°C) were applied InP nanotubes were formed. A closer examination by TEM, presented in figure 1D, showed that hollow tubes were formed. Judging

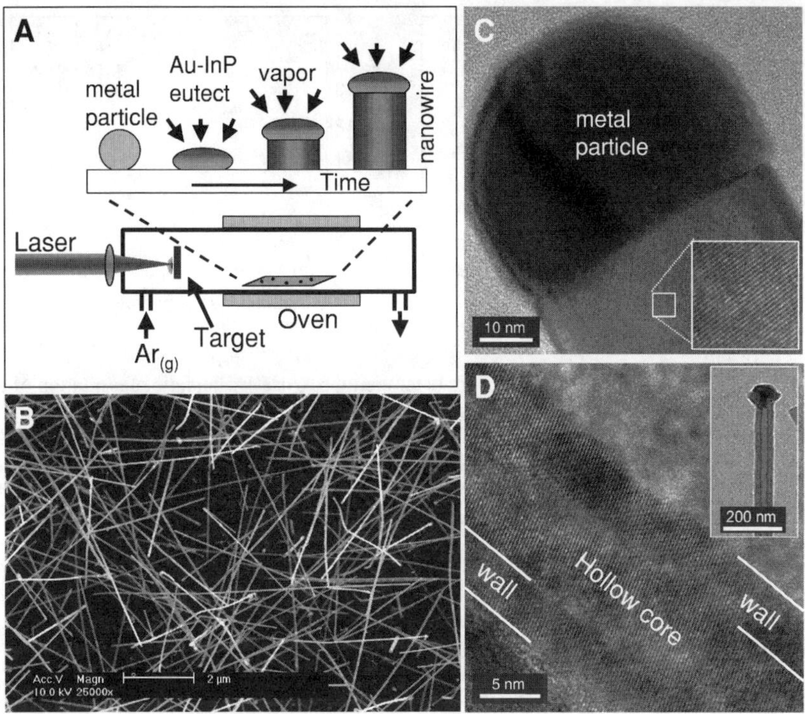

Figure 1 (A) Schematic image of the laser ablation set up for the VLS growth of nanowires. (B) SEM image of InP nanostructures grown on a silicon substrate provided with a 2 Å Au film. (C) HRTEM image of an InP nanowire. (D) HRTEM image from a InP nanotube grown at 520°C by the use of an InP target doped with 0.1 mol% Zn (inset) overview TEM image of an InP nanotube.

from the contrast from both bright field TEM and HAADF (high-angle annular dark field) imaging it was clear that there was no material present in the core of the tubes. The tubes were terminated by a particle, (inset figure 1D) which contained gold; this indicates that the tubes grow from the liquid InP-Au phase via the VLS mechanism. The diameter of the tubes was uniform along their length. Upon tilting the sample with respect to the electron beam, diffraction fringes were seen to move over the entire width of the tubes, implying a cylindrical shape of the crystals. The observation that the diffraction contrast is most pronounced in the walls again confirms the hollow nature of the tubes. The thickness of the wall of the nanotube shown in the HRTEM image in figure 1D was approximately 4 nm. From electron diffraction and X-ray diffraction measurements it was clear that the crystal lattice corresponded to the InP zinc blende lattice. The tubes did not oxidize upon exposure to ambient air for over a month.

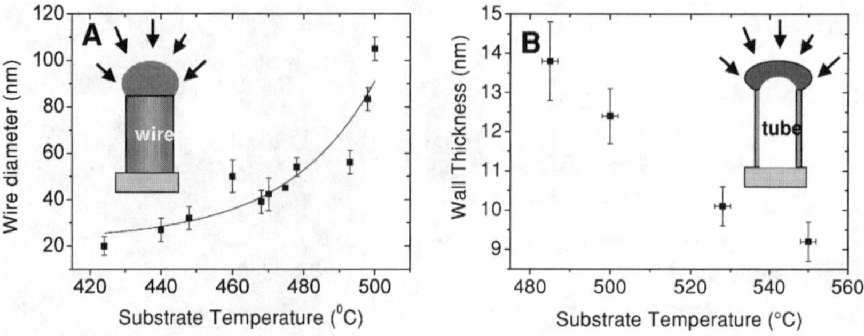

Figure 2 (A) Wire diameter versus substrate temperature (inset) schematic presentation of the formation of solid nanowires by a kinetically limited growth process at relatively low temperatures; the InP concentration in the Au particle is indicated by the gray scale. (B) Wall thickness of the InP nanotubes versus substrate temperature. An InP target with 0.1 mol% sulfur was used, leading to the formation of tubes in the applied temperature range (inset) the formation of hollow nanotubes by a diffusion limited growth process at higher temperatures.

A simple model for the growth dynamics is proposed. At relatively low temperature, the growth rate is controlled by the crystal growth rate at the liquid-solid interface (kinetically limited). Diffusion of atoms in the droplet is then relatively fast with respect to crystal growth, and the concentration is uniform throughout the droplet (inset figure2A). Growth takes place at the entire liquid-solid junction, giving rise to solid wires. At elevated temperatures the growth rate has increased with respect to the rate of diffusion; the concentration of atoms in the droplet has been depleted, and the growth becomes diffusion limited. At the liquid- solid interface, the nucleation front is ring-shaped, giving rise to the growth of tubes (inset figure 2B). The transition temperature is typically 500°C, but dopants added to the InP target influence this temperature.[9] An analogous growth mechanism was proposed for the growth of hollow and solid carbon nanofibers.[10] According to our proposed model the thickness of the wall is expected to depend on the temperature. At increasing temperatures, the growth rate increases with respect to the diffusion rate; the species have a decreasing amount of time to migrate through the droplet. In figure 2B the mantle thickness of the tubes is plotted versus the substrate temperature. For this experiment, an InP target with 0.1 mol% S was used, leading to the formation of tubes in the applied temperature range. The wall thickness was measured by using TEM. Obviously, the wall thickness decreased for higher temperatures.

Photoluminescence measurements were performed on individual nanostructures. Figure 3A shows a dark field image of a nanowire. In figures 3B and 3C photoluminescence images of the same wire are shown with the excitation polarization vector parallel and perpendicular to the orientation of the wire, respectively. Figure 3D gives the emission spectra of the wire for parallel and perpendicular excitation. In general, the measured excitation polarization ratio, $\rho=(I_{//}-I_{\perp})/(I_{//}+I_{\perp})$, with ($I_{//}$) the emission intensities with parallel and ($I_{\perp}$) perpendicular polarized

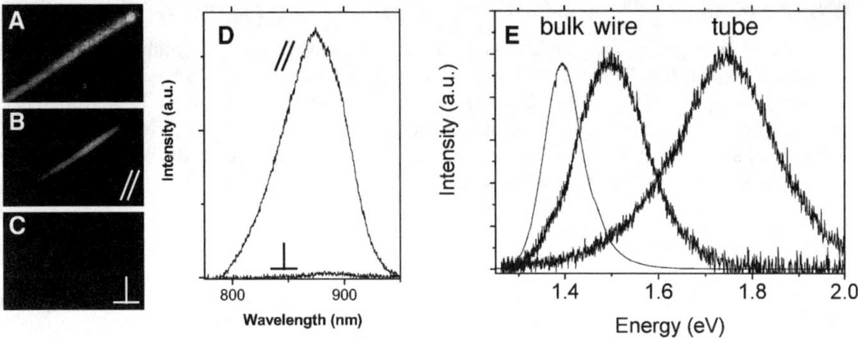

Figure 3 (A) dark field image of a single nanowire. (B) pholotoluminescence image of the same wire with parallel excitation and (C) perpendicular excitation. (D) The emission spectra for parallel and perpendicular excitation. (E) Emission spectra of a bulk InP crystal, an InP nanowire, and an InP nanotube demonstrating the enhanced size quantization effect for the tube compared to the wire.

excitation with respect to the wire axis, is 0.95±0.02 for thin and straight wires. A similar value has been reported by others.[11] Thicker wires and tubes showed a lower excitation polarization ratio of $\rho \approx 0.6$. The large polarization ratio is believed to originate from the large dielectric contrast between the nanowire and the air surrounding the wire. The maximum of the emission band of the nanowire (λ=880 nm) is blue-shifted with respect to the emission of bulk InP (λ=918 nm) due to size quantization. Effective-mass-approximation calculations show that the size quantization effect for a solid one-dimensional structure is determined by the diameter, whereas the size quantization for a hollow tube is determined by the wall thickness. In figure 3E the emission spectra of a bulk InP crystal, an InP nanowire (d~15 nm), and an InP nanotube (d~30nm, wall thickness~4nm) are plotted. The emission intensity is normalized for easy comparison. Other emission bands at lower energies were not observed, demonstrating that defect states, for instance at the surface of the structures, do not dominate the optical properties. The maximum of the measured emission of bulk InP is at a slightly higher energy than the bandgap energy of InP reported in the literature. Probably, this is caused by the limited sensitivity of the spectrophotometer at wavelengths > 900 nm. The emission maximum of the wire has blue-shifted by ~100 meV, whereas the tube emission has blue-shifted by ~400 meV. This enhanced blue-shift clearly demonstrates the strong size quantization effect for the tube. The width of the emission band has increased for the nanostructures with respect to that of the bulk InP crystal. This might be due to fluctuations in the diameter or the wall thickness, for the wire and the tube, respectively.

CONCLUSIONS

In summary, we have synthesised crystalline InP nanowires and nanotubes by using the VLS growth mechanism. The diameter of wires and the wall thickness of tubes can be tuned by the substrate temperature. The nanostructures show polarization dependent pholotoluminescence. Since nanowires of most of the III-V compound semiconductors have been grown[12], we believe that tubes consisting of these materials can also be synthesised. This gives the opportunity to tune the opto-electronic properties of these one-dimensional structures in an even wider range.

REFERENCES

1. Huang, Y.; Duan, X.; Cui, Y.; Lauhon, L.J.; Kim, K-H.; Lieber, C.M. *Science* **2001**, *294*, 1313
2. Haraguchi, K.; Katsuyama, K.; Hiruma, K.; Ogawa, K. *Appl. Phys. Lett.* **1992**, *60*, 745
3. Björk, M.T.; Ohlsson, B.J.; Sass, T.; Persson, A.I.; Thelander, C.; Magnusson, M.H.; Deppert, K.; Wallenberg, L.R.; Samuelson, L. *Nano Lett.* **2002**, *2*, 87
4. Björk, M.T.; Ohlsson, B.J.; Thelander, C.; Persson, A.I.; Deppert, K.; Wallenberg, L.R.; Samuelson, L. *Appl. Phys. Lett.* **2002**, *81*, 4458
5. Tenne, R.; Margulis, L.; Genut, M.; Hodes, G. *Nature* **1992**, *360*, 444
6. Seifert, G.; Terrones, H.; Terrones, M.; Jungnickel, G.; Frauenheim, T. *Phys. Rev. Lett.* **2000**, *85*, 146
7. Wagner, R.S.; Ellis, W.C. *Appl. Phys. Lett.* **1964**, *4*, 89
8. Gudiksen, M.S.; Wang, J.; Lieber, C.M. *J. Phys. Chem.* **2001**, *105*, 4062
9. Bakkers, E.P.A.M.; Verheijen, M.A. *J. Am. Chem. Soc.* **2003**, *125*, 3440
10. Snoeck, J.W.; Froment, G.F.; Fowles, J. *J. Catalysis* **1997**, *169*, 240
11. Wang, J.; Gudiksen, M.S.; Duan, X.; Cui, Y.; Lieber, C.M. *Science* **2001**, *293*, 1455
12. Duan, X.; Lieber, C.M. *Adv. Mat.* **2000**, *12*, 298

Mat. Res. Soc. Symp. Proc. Vol. 789 © 2004 Materials Research Society N12.7

HIGHLY EFFICIENT FORMATION OF $TiO_{2-X}N_X$–BASED PHOTOCATALYSTS - POTENTIAL APPLICATIONS FOR ACTIVE SITES IN MICROREACTORS, SENSORS, AND PHOTOVOLTAICS

James L. Gole[1], Clemens Burda[2], Andrei Fedorov[1], and S. M. Prokes[3]
1. Schools of Physics and Mechanical Engineering, Georgia Institute of Technology, Atlanta, GA 30332-0430
2. Department of Chemistry, Case Western Reserve University, Cleveland, Ohio 44106
3. Naval Research Laboratory, Washington, D. C.

Abstract

We have produced nitrogen doped, stable, and environmentally benign $TiO_{2-x}N_x$ photocatalysts whose optical response can be tuned across the entire visible region using a nanoscale exclusive synthesis route, performed in seconds at room temperature. This synthesis, which can be simultaneously accompanied by metal atom seeding, that can be accomplished through the direct nitration of anatase TiO_2 nanostructures with alkyl ammonium salts. Tunability throughout the visible spectrum depends on the degree of TiO_2 nano- particle agglomeration and the influence of metal seeding. The introduction of a small quantity of palladium, in the form of the chloride or nitrate, promotes further nitrogen uptake, appears to lead to a partial phase transformation, displays a counter ion effect with the acetate, and produces a material absorbing well into the near infrared. The nitridation process also induces the formation of oxygen hole centers. Silver introduced as the nitrate into a TiO_2 or $TiO_{2-x}N_x$ nanostructure framework, forms seeded Ag_xO - TiO_2 or $TiO_{2-x}N_x$ nanostructure mixtures which can be induced to self-assemble and to agglomerate into nano-needle and planar arrays using select metal probes. Surprisingly, no organics are incorporated into the final $TiO_{2-x}N_x$ products. These visible light absorbing photocatalysts readily photodegrade methylene blue and when used to create photocatalytic sites on a surface based microreactor induce ethylene oxidation with visible light. They can be transformed from liquids to gels and placed on the surfaces of sensor and microreactor based configurations to 1) facilitate a photocatalytically induced solar pumped sensor response, and 2) provide a possible means for the catalytically induced disinfection of airborne pathogens. In contrast to a process which is facile at the nanoscale, we find little or no direct nitridation of micrometer sized anatase or rutile TiO_2 powders at room temperature. Thus, we demonstrate an example of how a traversal to the nanoscale can vastly improve the efficiency for producing important submicron materials.

Introduction

An exciting aspect of research at the nanoscale results as nanostructures can display not only an enhanced activity but also an unexpected reactivity relative to that at the micron scale and bulk phase. Further, their formation and interaction may be accompanied by phase coexistence and transformations not commonly observed in bulk systems. These factors can lead to unusual surface oxidation states, routes for the highly efficient seeding of metals/metal ions into active submicron assemblies, sensitized interactions with external probes which can be used to induce self-assembly and the ready formation of metastable phases.

Highly Efficient Formation of Visible Light Absorbing $TiO_{2-x}N_x$ Photocatalysts

There has been a long, continued, interest in TiO_2 based photocatalysis[1] because of the relatively high reactivity and chemical stability of the oxide under uv light excitation ($\lambda < 387$ nm) where this energy exceeds the bandgap of anatase (3.2 eV) crystalline n-TiO_2. However, anatase TiO_2 is a poor absorber in the visible region. The cost and accessibility of uv photons makes it desirable to develop photocatalysts which are highly reactive under visible light excitation utilizing the solar spectrum or even interior room lighting.

Recently, Asahi et al.[2] prepared $TiO_{2-x}N_x$ films by (1) sputtering TiO_2 targets in an N_2 (40%)/Ar gas mixture and then annealing in N_2 gas at 550°C for 4 hours and (2) treating anatase TiO_2 powders in an NH_3 (67%)/Ar atmosphere at 600° for 3 hours. This study was important because it clearly demonstrated that nitrogen doping to form n-$TiO_{2-x}N_x$ shifts the optical absorption of TiO_2, and hence the ability to photodegrade methylene blue and gaseous acetaldehyde, to the visible region at wavelengths less than 500 nm. We have produced visible light absorbing $TiO_{2-x}N_x$ particles in a controlled size range (3-10 nm) *in seconds* at room temperature employing the direct nitridation of TiO_2 nanocolloids using alkyl ammonium compounds.[3-6] These titanium oxynitrides demonstrate a high quantum yield (> 0.5) for reductive photocatalysis with methylene blue and for ethylene oxidation on a surface where the quantum yield in the visible is easily comparable to the quantum yield under UV irradiation.

An advantage of the nanoscale doping procedure is that it produces slurries of the nitrided particles that form viscous solutions whose viscosity can be readily adjusted. This is of practical advantage as one places the agglomerates on a surface (for example a quartz reactor tube) to create a film of photocatalytically active sites. A further advantage of these colloidal solutions is the unique means by which they can incorporate and uniformly disperse transition metal compounds and the manner in which these transition metal compounds can be introduced in concert with the nitriding process. In m instances the introduction of the transition metal compounds produces a solution which, after some controlled period time, can be transformed to a gel. This process can be accelerated by minimal heating (e.g., ~ 50°C).

We have treated both the initial TiO_2 nanocolloid solution and partially agglomerated colloidal solutions with excess of triethyl amine at room temperature by stirring for a maximum of 2 minutes. We have also introduced a numbe transition metals during the nitriding process (e.g., transition metals from the early to late periods of the Periodic Ta including the Group VIII compounds). With subtle changes in the post synthesis treatment of the $TiO_{2-x}N_x$ slurries, we h produced powders ranging from yellow to orange in color with increasing agglomeration. Figure 1 compares the opti reflectance for Degussa P25 TiO_2 (reported at an average size of 30 nm), onsetting sharply at $\lambda \sim 380$ nm, the reflecta spectrum for the nitrided ($TiO_{2-x}N_x$) (>3, < 10) nm nanoparticles, rising sharply at 450 nm, and the corresponding spectr for the nitrided ($TiO_{2-x}N_x$) partially agglomerated nanoparticles rising sharply at 550 nm. This characterization reflectance spectroscopy emphasizes an ability to gain control of the absorption of light through nitriding and subsequ processing steps, producing a tuning through the visible region. Infrared spectra demonstrate that the colors and chan introduced upon doping are not associated with the incorporation of molecular alkyl groups. Further, we have introduced into a nitriding amine-TiO_2 mixture. The corresponding reflectance spectrum, obtained for palladium incorporation into $TiO_{2-x}N_x$, producing a structure with reduced Pd-based nanocrystallites[4] and the apparent transformation of some of $TiO_{2-x}N_x$ anatase structure,[4] displays an even broader response extending to considerably longer wavelength, albeit notably lower light absorption efficiency as a function of wavelength. The nitriding of TiO_2 to produce $TiO_{2-x}N_x$ also res in an enhanced argon ion laser induced photoluminescence which has been demonstrated to result from the formatior oxygen-hole centers (OHC's).[6]

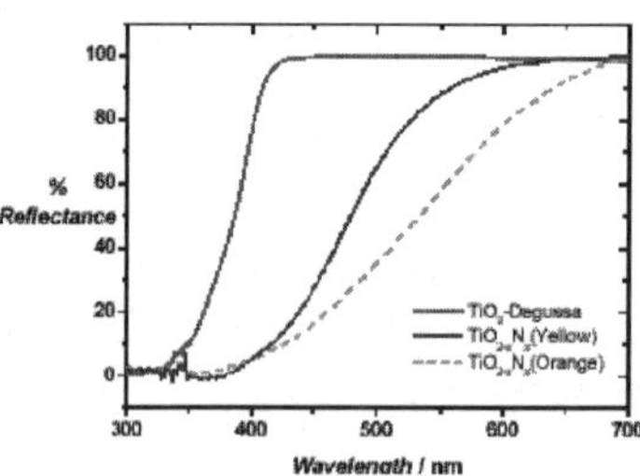

Figure 1. Reflectance measurements on different TiO_2 samples showing the red-shift effect of nitrogen-doping on the absorption of the nanocrystals

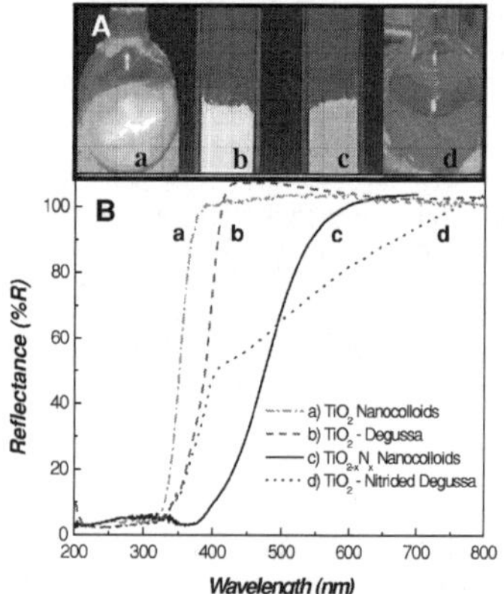

Figure 2. A) Visual comparison of a. TiO_2 nanopartic b. Degussa P25 TiO_2 powder; c. Nitrided TiO_{2-x} nano-powder; d. Degussa P25 TiO_2 powder nitrided w Triethyl-amine.
B) UV-visible reflectance spectra of (a) TiO_2 nanopartic ((white)-dash-dotted line), (b) Degussa P25 TiO_2 powder (dashed line), (c) nitrided TiO_{2-x} nanopowder ((yellow)-solid line) and (d) nitrided Degus P25 TiO_2 powder ((brown)-dotted line).

Figure 2 compares the reflectance spectra for TiO_2 and $TiO_{2-x}N_x$ nanocolloid and Degussa P25 TiO_2[5], the latte which is nitrided over a time period of days! The efficiency of the nitriding process for Degussa P25 is minimal compa to that of the colloid and the corresponding reflectance spectrum is notably more complex. A Debye-Scherrer analysis the XRD patterns of Degussa P25 and the nitrided and partially agglomerated TiO_2 nanocolloid (550 nm feature, Figure shows a broadened pattern for the nanocolloid that suggests a size close to 10 nm. In contrast to the outlined nanopart activity, no measurable reaction or heat release is observed as either distinct rutile or anatase TiO_2 micropowders are trea directly with an excess of triethylamine.

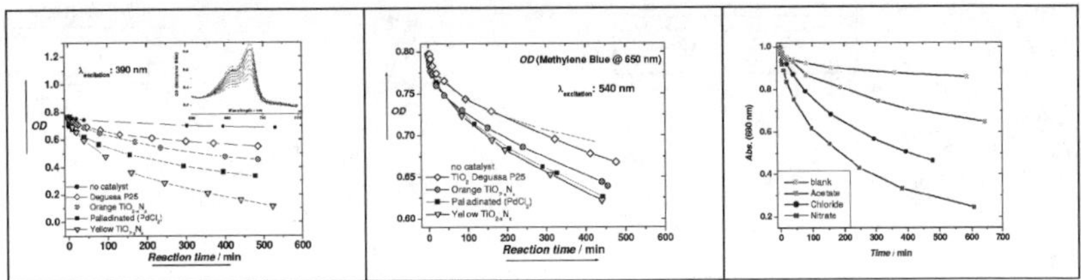

Figure 3. Comparison of the photocatalytic decomposition of Methylene Blue after a) 390 nm laser excitation, b) 540 nm monitored at 650 nm, catalyzed by undoped-TiO_2 (blank test, open diamonds) and Nitrogen-doped TiO_2 nanocrystals, and c) laser excitation at 390 nm as a function of the introduc- tion of palladium. The inset in 3a shows the photodegradation of methylene blue in water at neutral pH.

Methylene blue decolorization in water (either aerobic or anerobic) represents a probe reaction often cited to establish the photocatalytic activity of a novel photocatalyst. We have observed methylene blue decolorized under aerobic conditions for irradiation at 390 and 540 nm at a pH of 7 and for various $TiO_{2-x}N_x$ and Pd-seeded-Ti-O-N catalysts. Photocatalytic activity has been evaluated at 390 and 540 nm, respectively, using a laser producing a 1 KHz pulse train of 120 femtosecond pulses.[4-6] The laser output was used to pump either an optical parametric amplifier to obtain tunable wavelengths in the visible spectrum including 540 nm or a second harmonic generation crystal to produce 390 nm photons. Excitation powers were adjusted using a neutral density filter wheel. Figures 3 demonstrates the photodegradation observed at 390 and 540 nm for methylene blue in water at PH 7. The data for the nitrided samples as well as the palladium treated samples referred to above are consistent with a notably enhanced activity for the $TiO_{2-x}N_x$ constituencies at 390nm whereas undoped TiO_2 nanoparticles showed minimal activity under visible light irradiation. The significant decrease in the optical density for methylene blue in the presence of the synthesized $TiO_{2-x}N_x$ photocatalyst, upon visible light excitation, signals a conversion process that displays a notable activity relative to a blank and TiO_2 sample. Notable decreases in optical density are also observed at 540 nm, but differences in activity are muted in this less sensitive absorption region. The introduction of a small quantity of palladium in the form of the acetate, chloride, or nitrate not only produces a material absorbing well into the near infrared but also leads to further nitrogen uptake with introduction of the chloride or nitrate. This appears to lead to a partial phase transformation and displays a clear counterion effect[4] as photocatalytic activity (Figure 3(c)) increases from the acetate to chloride to nitrate. The behavior might be attributed to the surface/volume ratio of the resultant treated products[4]. However, the counterion effect and the relative efficiencies of the chloride and nitrate treated samples must be the subject of a further study. In contrast, at wavelengths below 350 nm, the activity of both the TiO_2 and nitrided samples is comparable. Thus nitrided $TiO_{2-x}N_x$ samples that can be generated in several seconds at room temperature are catalytically active at considerably longer wavelength than TiO_2

The results which we have obtained thus far demonstrate that by forming and adjusting an initial TiO_2 nanoparticle size distribution and mode of nanoparticle treatment, it is possible to tune and extend the absorption of a doped $TiO_{2-x}N_x$ sample well into the visible region. Further, the current study demonstrates that an important modification of a TiO_2 photocatalyst can be made considerably simpler and more efficient by extension to the nanometer regime. While these results are very encouraging for the dispersed colloids in solution, the gas phase oxidation of an olefin in contact with a photocatalytic surface is a more demanding test for the application of this science to the challenge of designing selective oxidation catalysts using low temperature, visible light activation.

Visible Light Active Photocatalytic Films

Films constructed from colloidal solutions of $TiO_{2-x}N_x$ and Pd-$TiO_{2-x}N_x$ are photocatalytically active for the total oxidation of ethylene to carbon dioxide under UV and incandescent light illumination at room temperature.[7] This is clearly demonstrated in Fig. 4, which shows the experimental results we have obtained for this process in a stop-flow microreactor. Specifically, CO_2 formation measured by a mass spectrometer after a one-hour long illumination of the reactor, filled with an oxygen/ethylene mixture, by a UV lamp (365 nm) and an incandescent lamp (primarily in the visible region) is compared for the reactor loaded with nitrided, Pd-impregnated, Pd-$TiO_{2-x}N_x$ nanostructures. The case of no illumination is also shown to provide a baseline CO_2 level. Unlike the untreated TiO_2 photocatalyst, films composed of nitrided TiO_2 colloid centers display a significant photocatalytic activity under illumination by the primarily visible light generated from the incandescent lamp at a much lower photon flux. The visible response is therefore greater than that for pure UV excitation.

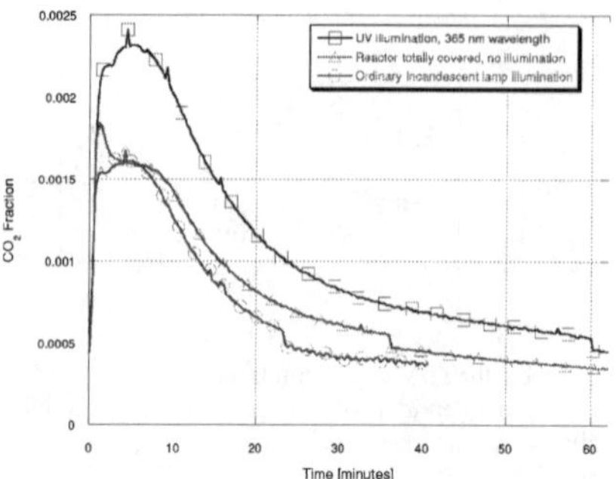

Figure 4a. CO_2 formation measuring the photocatalytic oxidation of ethylene in a stop-flow reactor loaded with TiO_2 nanostructures under UV (365 nm), incandescent lamp (primarily in the visible range), and no illumination at room temperature.

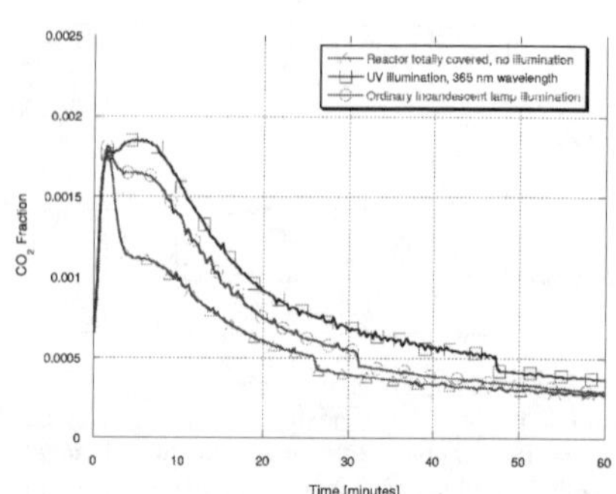

Figure 4b. CO_2 formation measuring the photocataly oxidation of ethylene in a stop-flow reactor loaded wi Pd-$TiO_{2-x}N_x$ nanostructures under UV lamp (365 nm incandescent lamp primarily in the visible range), and r illumination at room temperature (photon flux in visible ~ 1 flux in UV).

This demonstrates unambiguously that the nitriding procedures we have employed can produce surface based doped tita photocatalyst centers which are catalytically active within a broad range of optical excitation well into the visible. Th results provide strong evidence that novel materials can be synthesized at room temperature and used under visible li illumination to produce strongly oxidizing photocatalytic centers that could lead to a low temperature, selective, surf oxidation catalyst. The strategy of such an effort should be to generate uniquely optimized photocatalytic films modifying the nature of those centers which produce the photocatalytic activity at the nanoscale prior to surface film catalytic bed formation. These experiments are currently being pursued in our laboratory.

Further Applications of Doped and Metal/Seeded TiO_2

There are other potential applications of the TiO_2 based doping/seeding at the nanoscale. In view of the uniq ability of these nanocolloid solutions to incorporate transition metal compounds, we have considered the potent photovoltaic applications resulting from the incorporation of readily ionized metals such as silver into the TiO_2 and TiO_{2-x} lattice. Here, initial studies have revealed a very intriguing chemistry that might be generalized using a number of mix metal combinations. While silver crystallites can be incorporated and dispersed into TiO_2 or its nitrided analog, $TiO_{2-x}N$ Figure 5 demonstrates assembled silver oxide based arrays. They are formed as this crystalline material is interspers within colloidal TiO_2 as the entire assembly takes on a needle-like shape. These needle-like arrays are formed up introduction of a zinc metal probe into a silver nitrate impregnated lattice. We suggest that the self-assembly may induced through the formation of a zinc hydroxide hydrogel.[8] Experiments are currently underway to verify this suggesti As we are now beginning to investigate the potential photovoltaic applications of these materials, this striking self-assem may prove useful for future applications.

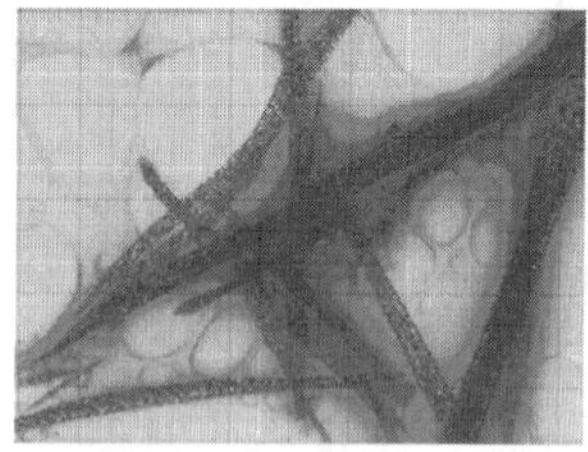

(a)

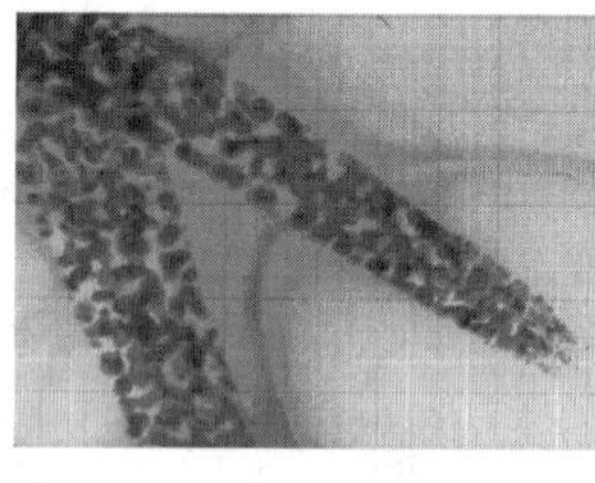

(b)

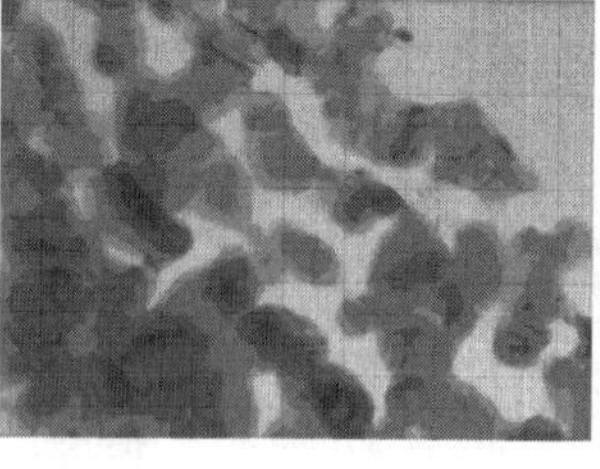

(c)

Figure 5. (a) TEM micrograph of the probe induced assembly of silver oxide crystallites on TiO_2. (b) Closer view of silv oxide crystallites assembled on TiO_2. (c) Closeup of silver oxide crystallites on which Moire Patterns are visible.

An important aspect of the seeding experiments which we have outlined is the ability to form a metal seeded slurry of the visible light absorbing $TiO_{2-x}N_x$. These slurries can be transferred from liquids to gels as they are placed on the surfaces of sensor and microreactor based configurations to 1) facilitate a photocatalytically induced solar pumped sensor response, and 2) provide a possible means for the catalytically induced disinfection of airborne pathogens. Once developed, the utilization of visible light absorbing photocatalytic materials can offer the potential for the significant improvement of sensor suites. Here solar and/or interior room lighting absorbed by a photocatalyst is used to produce electron-hole pairs, either of which might be used to enhance the sensing process by modifying the interfacial depletion layers of the chemically sensitive metal oxide layer of, for example, a CHEMFET. In our titanium-based studies, it has been possible to readily introduce Pt, Ru, Ir, Co, Ni, and Au into both the TiO_2 and $TiO_{2-x}N_x$ nanocolloid solutions. As the resulting materials are readily deposited to form films and gels on a surface, they are amenable to the creation of seed metal - $TiO_{2-x}N_x$ combinations whose interaction can also form the basis for a sensor.

The nitriding of TiO_2 to produce $TiO_{2-x}N_x$ as well as its subsequent metallization with gold and silver[6] leads to a vastly enhanced argon ion laser (488, 457.9 nm) induced photoluminescence from the oxynitride (vs. TiO_2). A strong emission between 550-560 nm, red shifts and drops in intensity with aging in the atmosphere. Electron Spin Resonance performed on these samples identifies a resonance at g = 2.0035, which increases significantly with the nitridation step. This resonance is attributed to an oxygen hole center created near the surface of the nanostructures which correlates well with the noted optical activity.[6,9] We suggest that the formation of the OHC's in the presence of metal oxides may also affect the interfacial depletion layers so as to enhance sensor response.

We have tried to exemplify the manner in which unique properties at the nanoscale can be applied to form visible light activated photocatalysts. We suggest that these materials will be useful for the development of microreactors, solar cells, and solar pumped sensors.

References:

1. See for example: a. A. Fujishima and K. Honda, Nature 238, 37 (1972), b. J. R. Bolton, Sol. Energy 57, 37 (1996), c. S. U. M. Khan and J. Akikusa, J. Phys. Chem. B 103, 7184 (1999). d. O. Khaselev and J. A. Turner, Science 280, 425 (1998). e. S. Licht et al., J. Phys. Chem. 104, 8920 (2000). f. M. R. Hoffman et al., Chem. Rev. 95, 69 (1995). g. D. S.Ollis, H. Al-Ekabi, eds., Photocatalytic Purification and Treatment of Water and Air, Elsevier, Amsterdam (1993).
2. R. Asahi, T. Morikawa, T. Ohwaki, K. Aoki, and Y. Taga, Science 293, 269 (2001).
3. C. Burda, Y. Lou, X. Chen, A. C. S. Samia, J. Stout, and J. L. Gole, "Enhanced Nitrogen Doping in TiO_2 Nanoparticles", Nano Lett. 3, 1049 (2003).
4. J. L. Gole, J. Stout, C. Burda, Y. Lou, and X. Chen, "Highly Efficient Formation of Visible Light Tunable $TiO_{2-x}N_x$ Photocatalysts and Their Transformation at the Nanoscale", J. Phys. Chem., in press.
5. X. Chen, Y. Lou, A. C. S. Samia, C. Burda, and J. L. Gole, "Synthesis and Characterization of Nitrogen-Doped TiO_2 Nanomaterial with High Photocatalytic Activity Under Visible Light Excitation: Comparison to a Commercial TiO_2 Photocatalyst Standard", submitted for publication.
6. S. M. Prokes, W. E. Carlos, J. L. Gole, C. She, and T. Lian, "Surface Modification and Optical Behavior of TiO_2 Nanostructure", MRS Proceedings "Spatially Resolved Characterization of Local Phenomena in Materials and Nanostructures" 738, 239 (2003).
7. S. Kumar, A. Fedorov, and J. L. Gole, submitted for publication.
8. L. A. Bottomley, private communication.
9. (a) S. M. Prokes, W. E. Carlos, and O. J. Glembocki, Phys. Rev. B 50, 17093 (1994), (b) S. M. Prokes and W. E. Carlos, J. Appl. Phys. 78, 2671 (1995), (c) S. M. Prokes, J. Mater. Res. 11, 305 (1996).

Electrical, Electronic Properties and Devices II

Mat. Res. Soc. Symp. Proc. Vol. 789 © 2004 Materials Research Society N13.2

Surface States in Passivated, Unpassivated and Core/Shell Nanocrystals: Electronic Structure and Optical Properties

Garnett W. Bryant[1] and W. Jaskolski[2]
[1]Atomic Physics Division, National Institute of Standards and Technology, Gaithersburg, MD 20899-8423
[2]Instytut Fizyki, UMK, Grudziadzka 5, 87-100 Torun, Poland

ABSTRACT

Surface effects significantly influence the functionality of semiconductor nanocrystals. A theoretical understanding of these surface effects requires models capable of describing surface details at an atomic scale, passivation with molecular ligands, and few-monolayer capping shells. We present an atomistic tight-binding theory of the electronic structure and optical properties of passivated, unpassivated and core/shell nanocrystals to study these surface effects.

INTRODUCTION

Quantum dots and nanocrystals have been studied intensely due to their enticing possibilities as artificial atoms with enhanced optical properties. Nanocrystals are being used now as building blocks for bio/nanohybrids for use in biological systems as biomarkers [1-3] and resonant energy-transfer sensors [4]. Construction of these bio/nanohybrids is a challenging, ongoing problem [3,4]. An understanding of how dots function in these structures is needed so that tailored nanooptics with these systems can be exploited. Surface effects significantly influence the functionality of semiconductor nanocrystals. Passivation with ligands or high band-gap semiconductor shells is necessary to reduce surface trap densities, enhance quantum yield and increase photostability [5-8]. Continuum effective mass theory [9,10] can provide an understanding of internal confined states in semiconductor nanocrystals, although atomistic theories are needed for a more complete understanding of these confined states [11-16]. However, a full theoretical understanding of surface effects requires atomistic models capable of describing surface faceting and relaxation; site-dependent, partial and random passivation; the molecular structure of passivants; and few-monolayer shells. We present an atomistic tight-binding theory of the electronic structure and optical properties of passivated, unpassivated and core/shell nanocrystals to study these surface effects.

THEORY

We calculate electron and hole states in nanocrystals by use of the empirical tight-binding (ETB) method [12-16]. The ETB approach is an atomistic approach well suited for calculating the electronic states of nanosystems with atomic-scale interfaces, surface structure and variations in composition and shape. In this paper, we consider zinc-blende CdS and CdS/ZnS nanocrystals. We assume that all atoms occupy a common, zinc-blende lattice with no relaxation in the bulk, at interfaces or surfaces. To date, lattice relaxation and surface reconstruction have been considered only for very small nanocrystals with a few hundred atoms [17-19]. Here, we consider a full range of nanocrystal sizes, including core/shell structures with up to 25,000 atoms. It is extremely time-consuming to include lattice relaxation and surface reconstruction for the full range of sizes that we consider. For that reason, we limit our attention to unrelaxed structures. We can model spherical, hemispherical, tetrahedral or pyramidal nanoparticles. Here, we consider spherical structures.

In our ETB model, each atom is described by its outer valence s orbital, the 3 outer p orbitals and a fictitious excited s* orbital, included to mimic the effects of higher lying states [20]. The empirical Hamiltonians are determined by adjusting the matrix elements to reproduce band gaps and effective masses of the bulk band structures. We include on-site and nearest-neighbor coupling between orbitals. Spin-orbit coupling is included in the theory but will not be considered for the results presented here. This simplification allows us to focus on how the spatial trapping at surfaces influences the nanocrystal optical response and how surface effects are suppressed in core/shell structures.

The electron and hole eigenstates close to the band edges are found by diagonalizing the Hamiltonians with an iterative eigenvalue solver. The nanosystems have on the order of 200-25,000 atoms. Typically, we find up to several hundred states closest to the band edge plus all states in the gap. Here we consider four models for the surface passivation: all surface dangling-bonds unpassivated, all surface-cation dangling bonds passivated, all surface-anion dangling-bonds passivated, and all surface dangling-bonds passivated. The passivation is modeled by shifting the energy of the dangling bonds above the conduction band so that the dangling bonds do not modify states near the band gap. Partial passivation can also be modeled by a partial shift of the dangling bond energy. We model CdS and CdS/ZnS nanocrystals with the CdS and ZnS ETB parameters taken from Lippens and Lannoo [12].

Optical spectra are calculated by evaluating the dipole matrix elements for electron and hole eigenstates found in the ETB calculations [13,14,16]. To calculate the spectra, we assume that all anion-derived surface states, due to unpassivated anion dangling bonds, are occupied and all cation-derived surface states are empty. In the ETB approach, dipole matrix elements are not needed to define the Hamiltonian. We find on-site dipole matrix elements by use of calculated atomic dipole matrix elements [21]. Dipole matrix elements between bonding orbitals on nearest neighbor sites are chosen by reasonable estimates. The qualitative structure in the calculated spectra is insensitive to variations in the dipole matrix elements for bond orbitals for choices of these matrix elements less than the bond length. This shows that allowed transitions are determined by symmetry considerations.

RESULTS

We consider spherical CdS and CdS/ZnS nanocrystals with unrelaxed, unfaceted surfaces that are passivated at all sites, only at cations, only at anions, or unpassivated at all sites. Different behavior is obtained for each case. Figure 1 shows the electronic states and spectra for a CdS nanocrystal with radius $4a$, where a is the bulk lattice constant. Fully passivated dots with all dangling bonds saturated have no surface states in the fundamental band gap and all near-band-edge states are quantum-confined internal states. The hole states have negligible probability to be at the surface, while electron states have a small but finite probability to be at the surface. The optical spectrum shows a few, discrete, optically allowed transitions with a 20 meV Stokes shift between the first strong transition and the lowest possible transition. Such a Stokes shift was recently seen experimentally [22].

When just surface anions are passivated, a broad band of mixed surface/internal states exists between the conduction band edge and the onset of internal states. 1S and 1P electron states are difficult to identify. However, hole states are weakly affected by the unpassivated Cd dangling bonds. The energy of the lowest strong optical transition is determined mostly by the energy of the lowest internal conduction state. However none of the transitions are assignable to the strong transitions for the fully passivated nanocrystal. Incomplete passivation of dangling bonds can push lowest internal conduction level well above the level for fully passivated dots or down to the conduction band edge. Because of this strong sensitivity to passivation, explicit models for surface effects are needed for a precise description of the lowest internal states.

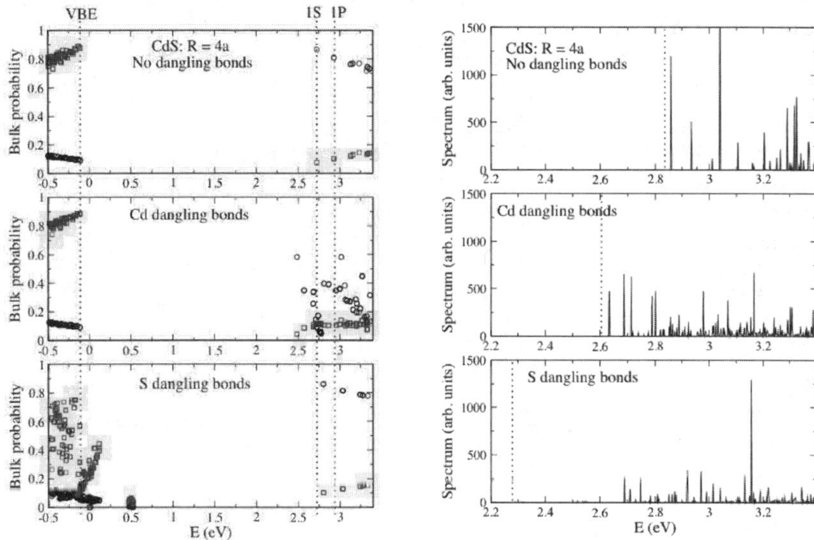

Figure 1. (left) Probability that states in a CdS nanocrystal (radius 4*a*, where *a* is the lattice constant) with energy *E* near the bulk band gap (bulk CdS valence (conduction) band edge at 0 eV (2.5 eV)) occupy the bulk cation (dark circles) and bulk anion (light squares) sites. Vertical lines indicate the valence band edge (VBE) and the 1S and 1P electron levels in the CdS nanocrystal with all dangling bonds passivated (top panel). The second (third) panel is for a CdS nanocrystal with unpassivated Cd (S) dangling bonds. (right) The corresponding optical spectra. The dashed lines indicate the energies of the lowest dark transitions.

When just surface cations are passivated, an anion-derived surface state band (near 0.5 eV) and a band of backbonded surface states exist above the valence band edge. Due to level repulsion with the gap states, conduction levels are blue-shifted from their energies in the fully passivated nanocrystal. Again, none of the transitions is assignable to transitions for the fully passivated nanocrystal. Instead, there are a series of weak low-energy transitions. At lower energy, there are additional transitions with negligible strength, barely discernible in the figure, involving the surface states. When both cations and anions are unpassivated, both anion-derived and cation-derived surface states occur. Internal confined states experience level repulsion from both types of surface states in this case.

Capping the CdS dot with a few monolayers of ZnS reduces the influence of the surface on the internal electronic states and optical properties, as shown for a CdS nanocrystal of radius $4a$ with a ZnS cap of thickness a and $2a$ in Figures 2 and 3 respectively. The states for the fully passivated nanocrystal red shift when the cap is added. The lowest hole state in the uncapped, fully passivated dot is a P_0 state, while the second hole state is a S_l level. As a consequence, the lowest transition in the uncapped dot is dark, with the 20 meV Stokes shift to the first allowed transition. However the S_l level is more sensitive to surface effects. When the dot is capped, the S_l level has a larger red shift and the Stokes shift is reduced. For thicker caps (see Figure 3) the ordering of P_0 and S_l levels is reversed and there is no Stokes shift. Otherwise, there are no significant changes to the optical response of the fully passivated dot when it is capped.

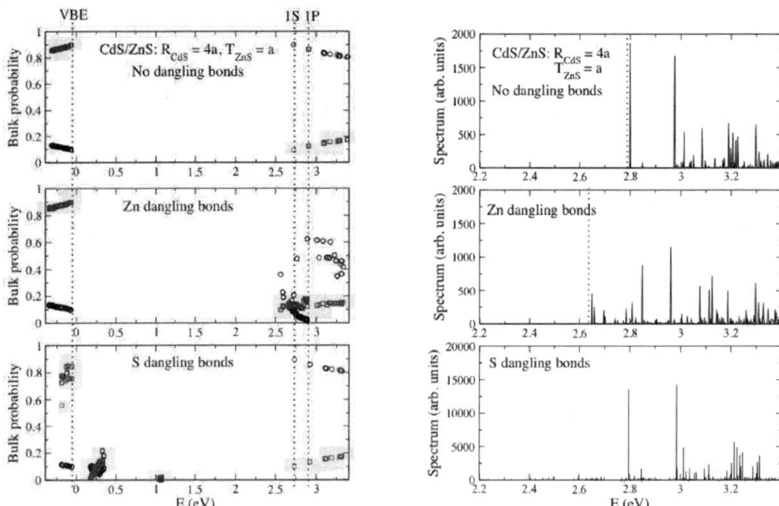

Figure 2. (left) Probability that states in a CdS/ZnS nanocrystal (CdS radius $4a$, ZnS thickness a) with energy E occupy the bulk cation (dark circles) and bulk anion (light squares) sites. Vertical lines indicate the valence band edge (VBE) and the 1S and 1P electron levels in the CdS/ZnS nanocrystal with all dangling bonds passivated (top panel). The second (third) panel is for a CdS nanocrystal with unpassivated Zn (S) dangling bonds. (right) The corresponding optical spectra. The dashed lines indicate the energies of the lowest dark transitions. For a nanocrystal with unpassivated S dangling bonds, dark transitions extend down to 1.5 eV.

Capping dots with partial passivation clearly reduces the surface effects both on the electronic states and on the optical response, as seen by comparing Figures 1, 2 and 3. The internal confined states have a higher probability to be in the bulk, making trapping at the surface less likely. The 1S and 1P electron states are easier to identify even when there are cation dangling bonds. For CdS/ZnS with the thickest cap, the blue shift of the 1S and 1P levels, due to level repulsion by the surface states, is negligible, also indicating that the surface has little effect on the internal states. Finally, the dominant transitions in the fully passivated dots are still clearly present in the partially passivated dots when the ZnS cap is included. However, a series of dark transitions involving surface states still exists below the dominant transitions. These surface states only weakly couple to the internal states so the dark transitions should be less important for the capped nanocrystals.

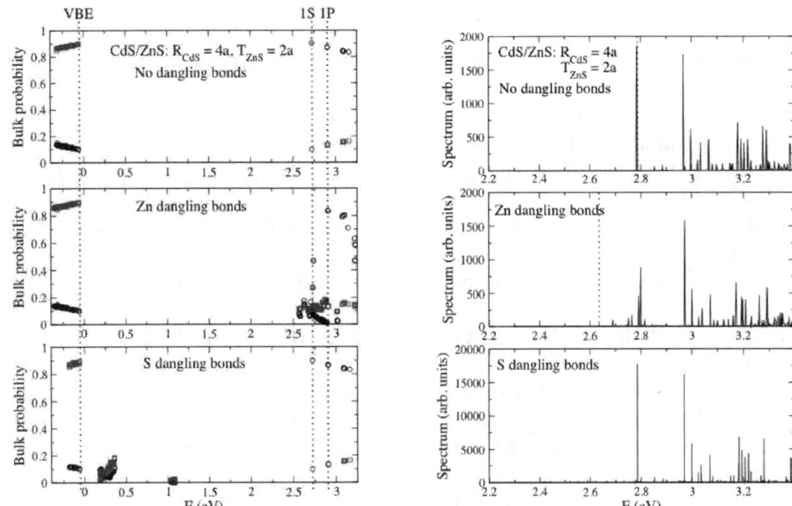

Figure 3. (left) Probability that states in a CdS/ZnS nanocrystal (CdS radius 4*a*, ZnS thickness 2*a*) with energy *E* occupy the bulk cation (dark circles) and bulk anion (light squares) sites. Vertical lines indicate the valence band edge (VBE) and the 1S and 1P electron levels in the CdS/ZnS nanocrystal with all dangling bonds passivated (top panel). The second (third) panel is for a CdS nanocrystal with unpassivated Zn (S) dangling bonds. (right) The corresponding optical spectra. The dashed lines indicate the energies of the lowest transitions. For a nanocrystal with unpassivated S dangling bonds, the dark transitions extend down to 1.5 eV. There is no Stokes shift for fully passivated CdS/ZnS with a 2*a* thick cap.

CONCLUSIONS

We consider spherical CdS nanocrystals with unrelaxed, unfaceted surfaces that are passivated at all sites, only at cations, only at anions, or unpassivated at all sites. Different behavior is obtained for each case. Fully passivated dots with all dangling bonds saturated have no surface states in the fundamental band gap and all near-band-edge states are quantum-confined internal states. When only surface cations are passivated, an anion-derived surface state band and a band of backbonded surface states exist near the valence band edge. When only surface anions are passivated, a broad band of mixed surface/internal states exists between the conduction band edge and the onset of internal states. The energy of the lowest strong optical transition is determined mostly by the energy of the lowest internal conduction state. Incomplete passivation of dangling bonds can push this level well above the energy for fully passivated dots or down to the conduction band edge. Because of this strong sensitivity to passivation, explicit models for surface effects are necessary for precise descriptions of the lowest internal states. Capping the CdS dot with more than two monolayers of ZnS significantly reduces the influence of the surface on the internal electronic states and optical properties.

REFERENCES

1. M. Bruchez Jr., M. Moronne, P. Gin, S. Weiss, and A.P. Alivisatos, Science **281**, 2013 (1998).
2. W.C.W. Chan and S. Nie, Science **281**, 2016 (1998).
3. B. Dubertret, P. Skourides, D.J. Norris, V. Noireaux, A.H. Brivanlou, and A. Libchaber, Science **298**, 1759 (2002).
4. I.L. Medintz, A.R. Clapp, H. Mattoussi, E.R. Goldman, B. Fisher, and J.M. Mauro, Nature Materials **2**, 630 (2003).
5. X. Peng, M.C. Schlamp, A.V. Kadavanich, and A.P. Alivisatos, J. Am. Chem. Soc. **119**, 7019 (1997).
6.B.O. Dabbousi, J. Rodriguez-Viejo, F.V. Mikulec, J.R. Heine, H. Mattoussi, R. Ober, K. F. Jensen, and M.G. Bawendi, J. Phys. Chem. B **101**, 9463 (1997).
7. Y.W. Cao and U. Banin, J. Am. Chem. Soc. **122**, 9692 (2000).
8. S.-Y. Lu, M.-L. Wu, and H.-L. Chen, J. Appl. Phys. **93**, 5789 (2003).
9. Al.L. Efros, M. Rosen, M. Kuno, M. Nirmal, D.J. Norris, and M.G. Bawendi, Phys. Rev. B **54**, 4843 (1996).
10. U. Banin, J.C. Lee, A.A. Guzelian, A.V. Kadavanich, A.P. Alivisatos, W. Jaskolski, G.W. Bryant, Al.L. Efros, and M. Rosen, J. Chem. Phys. **109**, 2306 (1998).
11. L.W. Wang and A. Zunger, Phys. Rev. B **53**, 9579 (1996).
12. P. E. Lippens and M. Lannoo, Phys. Rev. B **39**, 10935 (1989).
13. K. Leung and K. B. Whaley, Phys. Rev. B **56**, 7455 (1997).
14. K. Leung, S. Pokrant, and K. B. Whaley, Phys. Rev. B **57**, 12291 (1998).
15. R.B. Little, M.A. El-Sayed, G.W. Bryant, and S.E. Burke, J. Chem. Phys. **114**, 1813 (2001).
16. G. W. Bryant and W. Jaskolski, Phys. Rev. B **67**, 205320 (2003).
17. K. Leung and K.B. Whaley, J. Chem. Phys. **110**, 11012 (1999).
18. S. Pokrant and K.B. Whaley, Eur. Phys. J. D **6**, 255 (1999).
19. A. Puzder, A.J. Williamson, F.A. Reboredo, and G. Galli, Phys. Rev. Lett. **91**, 157405 (2003).
20. P. Vogl, H. P. Hjalmarson, and J. D. Dow, J. Phys. Chem. Solids **44**, 365 (1983).
21. S. Frage and J. Muszynska, *Atoms in External Fields*, (Elsevier, New York 1981).
22. Z. Yu, J. Li, D.B. O'Connor, L.-W. Wang, and P.F. Barbara, J. Phys. Chem. B **107**, 5670 (2003).

Mat. Res. Soc. Symp. Proc. Vol. 789 © 2004 Materials Research Society N13.5

Influence of the Network Geometry on Electron Transport in Nanoparticle Networks

K. D. Benkstein, N. Kopidakis, J. van de Lagemaat, and A. J. Frank
National Renewable Energy Laboratory
Golden, CO 80401, U.S.A.

ABSTRACT

Computer simulations are applied to understand the influence of network geometry on the electron transport dynamics in random nanoparticle networks, and the predicted results are compared with those measured in one class of random nanoparticle networks: dye-sensitized nanocrystalline TiO_2 solar cells. The model is applicable to all classes of random nanoparticle networks, such as highly disordered quantum dot arrays. The random nanoparticle networks are simulated by the step-wise condensation of a diffusion-limited aggregate. The fractal dimension of the nanoparticle films was estimated from the simulations to be 2.28, which is in quantitative agreement with gas-sorption measurements of TiO_2 nanoparticle films. Electron transport on the computer-generated networks is simulated by random walk. The experimental measurements and random-walk simulations are found to be in quantitative agreement. For both a power-law dependence of the electron diffusion coefficient D on the film porosity P is found as described by the relation: $D \propto | P-P_c |^{\mu}$. This power-law relation can also be derived from percolation theory, although only qualitatively. The critical porosity P_c (percolation threshold) and the conductivity exponent μ are found to be 0.76 ± 0.01 and 0.82 ± 0.05, respectively. It is estimated that during their respective transit through 50 and 75% porous 10-µm thick films as employed in the dye-cell, the average number of particles visited by electrons increases by 10-fold, from 10^6 to 10^7.

INTRODUCTION

Electron transport in mesoporous semiconductors has generally been modeled from the perspective that network topology has no influence on electron transport [1-5]. It is typically assumed that transport is only trap limited and that electrons diffuse in three dimensions, restricted only by the macroscopic dimensions of the film and electrostatic interaction with the electrolyte (ambipolar diffusion [6]). A recent study has shown, however, that there is a strong relation between the film porosity and the average coordination number of particles in a network [7]. An important implication of this study is that film morphology can affect the path length of electrons in the film. It has also been reported that the number of interconnects between particles [8] and the extent of overlap between neighboring particles [9] influence the electron transport rate. A comparison between real and simulated films suggest that real nanoparticle film can be accurately depicted as a random network with a random number of interconnections at each particle, which may be likened to a "hub" [7,10].

In this paper, we highlight a recent study [10] that describes the first clear evidence that the network geometry strongly influences the electron transport dynamics in mesoporous TiO_2 nanoparticle films and show that percolation theory accounts very well for the results. The electron transport dynamics was modeled using simulated mesoporous *random* nanoparticle films and the random-walk approach. Diffusion (random walk) in such networks can be described in terms of percolation theory, which predicts a power-law dependence (exponent μ)

of the diffusion coefficient on the difference between the film *porosity* (P) and some critical *porosity* (P_c).

$$D \propto |P_c - P|^{\mu} \quad (1)$$

EXPERIMENTAL DETAILS

TiO_2 nanoparticles were prepared and analyzed as described elsewhere [7,11]. The preparation method yields exclusively anatase TiO_2 with a narrow size distribution centered at a particle diameter of 19 nm. The porosity was varied by changing the wt % ratio of polyethylene glycol-to-TiO_2 in a paste. The resulting paste was spread on top of a conducting glass plate, which was used as the substrate for the deposited films. The TiO_2 covered glass was sintered in air to 450 °C for 30 minutes and then allowed to cool. The preparation method produced crack-free films with porosities ranging from 52-71% [10]. Subsequently, the TiO_2 electrodes were stained with $Ru[LL'(NCS)_2]$ (L = 2,2'-bypyridyl-4,4'-dicarboxylic acid, L' = 2,2'-bipyridyl-4,4'-ditetrabutylammoniumcarboxylate). Semitransparent counter electrodes were conducting glass plates coated with a thin Pt layer. The cells were filled with methoxypropionitrile containing either 0.8 M 1-hexyl-2,3-dimethylimidazolium iodide (C6DMII) with 50 mM iodine or 0.8 M tetrabutylammonium iodide (TBAI) with 50 mM iodine.

The surface area, porosity, and surface fractal dimensions of each film were determined with a high-speed gas sorption surface and pore-size analyzer. Film morphology was examined by field emission scanning electron microscopy (SEM). The thickness of the films was measured with a profilometer. The cells were probed with a weak laser pulse at 532 nm (10 ns pulse duration) superimposed on a relatively large, background (bias) illumination at 680 nm. From the results of the transient measurements, the diffusion coefficient of electrons D was determined using the equation [5]: $D = d^2/(2.35\tau)$, where τ is the time constant, corresponding to the exponential (long-time) decay of the transient photocurrent signal, d is the film thickness, and the factor 2.35 arises from the geometry of the diffusion problem. To a very good approximation, the ambipolar diffusion coefficient is the same as the diffusion coefficient of electrons in the mesoporous network [6]. From the measurement of τ and the short-circuit photocurrent density J_{SC} at different (bias) light intensities, the photocarrier density N was estimated with the relation: $N = J_{SC}\tau/(q_e d(1-P))$, where q_e is the unit of charge of an electron.

In simulating mesoporous films, the particles were placed at random locations inside a box after which their positions were optimized by performing a Brownian motion of all particles until each particle was in contact with at least one other particle, and there was minimal overlap between the particles. Following this step, the box was contracted stepwise to the desired porosity. At each step, the positions of the particles were optimized. Sintering of the films was simulated by requiring that the distance between two particles in contact be slightly less than twice their radii. At the boundaries of the box, the simulation was wrapped around so that larger films could be formed by repeating the particle coordinates. The resulting simulated films display the same pore size distribution as experimental films [7]. Full calculations for a 200 x 200 x 200 particle radii box take a few minutes to a few hours depending on the particle density. The interconnectedness of a nanoparticle network was evaluated by determining for each particle the number of its neighbors within 2 particle radii of it. Random-walk simulations for electron transport were performed using the simulated films as the matrix through which electrons travel. The boxes used for the simulations were 200 particle radii along each of their edges. For an average particle radius of 10 nm, this box corresponds to 2 μm along its edge. In the random walk simulations, electrons were placed on a random particle of the largest cluster in the film (typically, the infinite cluster) and were allowed to perform a random walk of a fixed number of steps (typically, 10^5 steps). The simulated film was repeated in all three dimensions to create an infinitely large simulation volume. The distance that the electron travels from its origin was then recorded and averaged over many electrons. From diffusion theory, the root-mean-square-average displacement R scales with the diffusion coefficient D and time t using the relation: $R^2 = 6\,D\,t$. Assuming that the average residence time of electrons t_r for the particles visited is independent of film porosity, the time t is proportional to the total number of interparticle steps. From the simulations, information about the number of particles through which each electron must travel from its point of origin in the box to its final destination can be assessed.

RESULTS AND DISCUSSION

Figure 1 shows the cluster radius dependence on the cluster size at the percolation limit for a simulated film. The radius of the clusters is the average distance between two particles on the cluster. The cluster size refers to the number of particles in a cluster. The fractal dimension of the clusters D_f was determined to be 2.28, which is the same value determined by gas-sorption

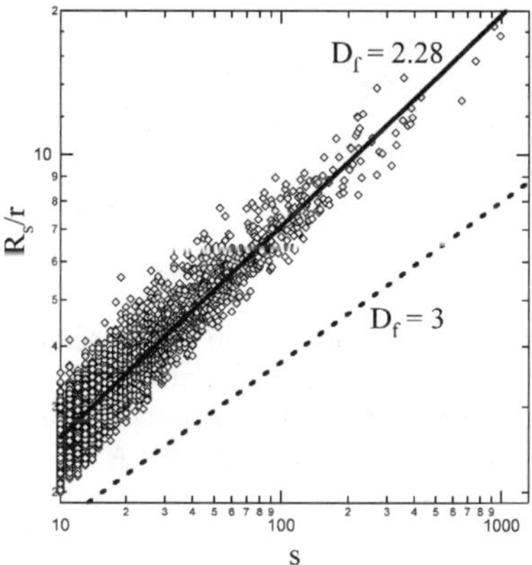

Figure 1. Dependence of cluster radius R_s on cluster size s for a simulated film of porosity P = 0.775 and l =200; r is the particle radius. The solid line is a fit to $R_S \propto s^{1/D_f}$ with $D_f = 2.28$. The dashed line corresponds to $D_f = 3$.

measurements, indicating that the simulated films have the same pore (shape) morphology and connectivity as actual nanoparticle films. Similarly, the simulated and real films display the same pore-sized distributions [7], providing further evidence that simulated films have the same basic structure as actual nanoparticle films.

Figure 2 shows the distribution of the coordination numbers of the particles for film porosities ranging from 0.5–0.8. It can be seen that the distribution of coordination numbers is broad and approximately Gaussian (dashed line) in accordance with a random fluctuation of the particle density. The films cannot, however, be described accurately as a purely random fluctuation of particle density. If the particles were distributed completely randomly, one would expect that the average coordination number *CN* dependence on the porosity for a homogeneous film would be given by the relation: $CN = 8(1 - P)$ [10]. The actual coordination number dependence on the porosity (inset of figure 2) is, however, slightly higher than that predicted for a homogeneous film and is better described by the power-law relation: $8.16(1 - P)^{0.839}$. This difference is attributed to the attractive force between the particles that causes them to "clump" together, resulting in a larger particle density around any particle than the mean value for a homogeneous film.

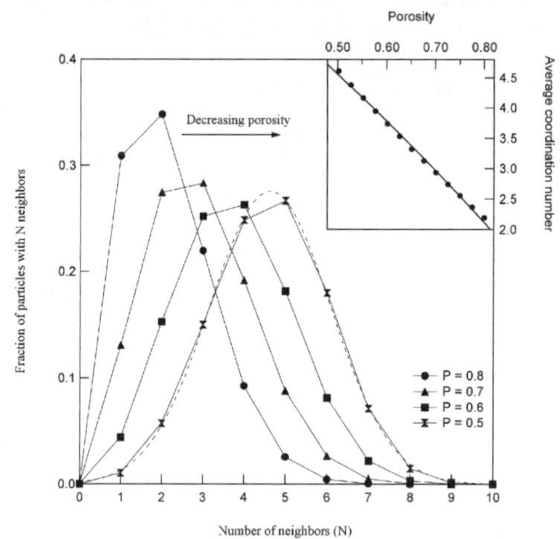

Figure 2. Dependence of the distribution of particle coordination numbers of simulated TiO_2 films on porosity. The dashed line represents a Gaussian fit of the data at P = 0.5. The inset shows the average coordination number as a function of film porosity. The solid line in the inset is a fit to a power-law variation of this relation: $CN = A\,(1\text{-}P)^{y}$, where the values of the fit parameters are A = 8.16 and y = 0.839.

Figure 3 compares the porosity dependence of the diffusion coefficient determined from random walk simulations and measurements for two redox electrolytes at a fixed photocharge density. It can be seen that the simulations describe the measurements accurately and that both measurements and simulations are described by the same power-law relation (equation 1). The diffusion coefficient decreases by a factor of 10 between 50% and 75% porosity. It is estimated that during their respective transit through 50 and 75% porous 10-μm thick films of 20 nm particles, the average number of particles visited by electrons increases by 10-fold, from 10^6 to 10^7, indicating that with higher porosity films the electron transport pathway becomes longer, even though there are less particles present per volume area. From the fits of equation 1 to the simulations, it follows that $P_c = 0.76 \pm 0.01$ and $\mu = 0.82 \pm 0.05$.

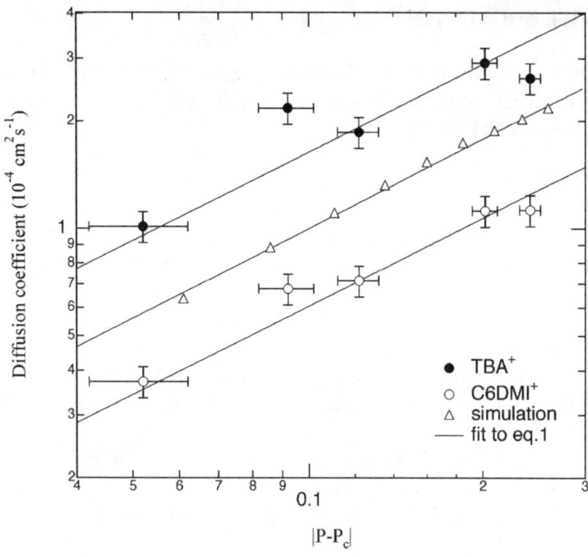

Figure 3. Comparison of the porosity dependence of the diffusion coefficient between the simulations (Δ) and measurements (●, O) for a photocharge density of 10^{17} cm^{-3} for two redox electrolytes. P_c is obtained from the fits of equation 1 to the simulation data. Solid lines are fits to equation 1 for fixed values of $P_c = 0.76$ and $\gamma = 0.81$. The simulations are scaled for comparison with the experimental data.

Figure 3 also shows that the cation used in conjunction with iodide in the electrolyte influences the magnitude of the electron diffusion coefficient. The rate of transport is larger in the presence of TBA$^+$-based electrolyte than in the case of the C6DMI$^+$-based electrolyte, suggesting that the two cations interact differently with the TiO_2 surface.

CONCLUSIONS

Photocurrent transient measurements combined with computer simulations provide the first clear evidence that the network geometry strongly influences electron transport dynamics in mesoporous TiO_2 films used in dye-sensitized solar cells. The film porosity was varied experimentally from 52 to 71%. Transport was modeled using simulated mesoporous TiO_2 films, consisting of a *random* nanoparticle network, and the random-walk approach. The electron diffusion pathway through the network distribution was correlated with the film porosity and the interparticle coordination numbers. Both the experimental measurements and simulations of electron transport are described very well by *percolation theory*. Increasing the film porosity is shown to shift the distribution of particle coordination numbers to lower values, which leads to a more tortuous and extended pathway for electrons during their passage through a film. The number of particles visited by electrons during their transit through a film increases with increased the film porosity.

ACKNOWLEDGMENTS

This work was supported by the Office of Science, Division of Chemical Sciences, and the Office of Utility Technologies, Division of Photovoltaics, U.S. Department of Energy, under contract DE-AC36-99GO10337.

REFERENCES

1. P. E. de Jongh and D. Vanmaekelbergh, *J. Phys. Chem. B* **101**, 2716 (1997).
2. L. Dloczik, O. Ileperuma, I. Lauermann, L. M. Peter, E. A. Ponomarev, G. Redmond, N. J. Shaw, and I. Uhlendorf, *J. Phys. Chem. B* **101**, 10281 (1997).
3. J. van de Lagemaat and A. J. Frank, *J. Phys. Chem. B* **104**, 4292 (2000).
4. Jenny Nelson, *Phys. Rev. B* **59**, 15374 (1999).
5. J. van de Lagemaat and A. J. Frank, *J. Phys. Chem. B* **105**, 11194 (2001).
6. N. Kopidakis, E. A. Schiff, N.-G. Park, J. van de Lagemaat, and A. J. Frank, *J. Phys. Chem. B* **104**, 3930 (2000).
7. J. van de Lagemaat, Kurt D. Benkstein, and A. J. Frank, *J. Phys. Chem. B* **105**, 12433 (2001).
8. N.-G. Park, J. van de Lagemaat, and A. J. Frank, *J. Phys. Chem. B* **104**, 8989 (2000).
9. M. J. Cass, F. L. Qiu, A. B. Walker, A. C. Fisher, and L. M. Peter, *J. Phys. Chem. B* **107**, 113 (2003).
10. K. D. Benkstein, N. Kopidakis, J. van de Lagemaat, and A. J. Frank, *J. Phys. Chem. B* **107**, 7759 (2003).
11. J. van de Lagemaat, N.-G. Park, and A. J. Frank, *J. Phys. Chem. B* **104**, 2044 (2000).

Poster Session IV

Mat. Res. Soc. Symp. Proc. Vol. 789 © 2004 Materials Research Society N15.2

Degree of Asymmetry of CdSe Quantum Dots Grown in Glass Probed by Four Wave Mixing

A. I. Filin[1], K. Babocsi[2], M. Schmitt[2], P. D. Persans[1], W. Kiefer[2], and V. D. Kulakovskii[3]
[1] Department of Physics, Rensselaer Polytechnic Institute, Troy, NY, USA.
[2] University of Wuerzburg, Wuerzburg, DE.
[3] Institute of Solid State Physics, RAS, Chernogolovka, Russia.

ABSTRACT

Information about quantum dot (QD) asymmetry is derived by analyzing the polarization properties of the time-integrated four wave mixing (FWM) signal. The lowering of QD symmetry results in the splitting of bright J= +/-1 exciton states. This causes the polarization oscillation of circularly excited excitons between these two split states. In a QD ensemble with a random distribution in the exciton level splitting, this results in the decay of the difference in FWM signals observed in scattering of σ^- and σ^+ polarized light on the population grating created by two σ^+ pulses, the decay time reflecting the degree of QD asymmetry. We have investigated the decay time of the difference in two polarized signals for quantum dots of equal size, grown in a glass matrix under different conditions. Increasing growth temperature and decreasing growth time lead to lowering of QD symmetry. We discuss this experimental result in terms of kinetics of nanoparticle growth in glass.

INTRODUCTION

During the last decade, semiconductor nanocrystals (NC's) or quantum dots (QD's) have attracted much attention due to the range of novel physical properties observed in the quantum confinement regime for carriers. Recently, spin-dependent effects in QD's have become a focus of particular interest due to possible applications utilizing the storage of spins in quantum computing [1]. It has been shown theoretically [2] and experimentally [3], that in the quantum confinement regime, spin relaxation can exhibit very long decay time in comparison with carrier depopulation.

Due to quantum confinement, the electron-hole interaction plays a very important role in all processes in QD's. In particular, in CdS and CdSe QD's, the lowest exciton state, formed by electron and hole with total spin J–1/2 and 3/2 respectively, splits by e-h exchange interaction into a set of states with J=0, 1, and 2 [4]. When spherical symmetry is broken, the lowest optically active state (spin doublet with J=1) splits into two states that are active in mutually orthogonal linear polarizations [5]. In the time-integrated four wave mixing (TI FWM) experiment, it causes in the appearance of a strong FWM-signal in the scattering of a σ^- circularly polarized beam on the grating created by two σ^+ polarized beams (such scattering is forbidden in a non-interacting oscillator model). In Ref. [6] it has been shown that the appearance of the forbidden signal is associated with a strong exciton-exciton coupling in the QD and with the inhomogeneous exciton spin dephasing due to a random exchange splitting of the J=1 exciton state, originating from the lowered QD symmetry. Thus, exciton spin dephasing can be used as a measure of QD asymmetry.

In this paper, we present the results of TI FWM experiment on investigation of exciton spin dephasing dynamics in CdSe QD's embedded in glass matrix. We have demonstrated strong dependence of the exciton spin dephasing time on QD growth conditions for the QD's of the equal size. We discuss the observed asymmetry in terms of the kinetics of QD's growth.

EXPERIMENTAL DETAILS

Samples were prepared from commercial RG695 filter glass from Schott Glass Technologies. As-received RG695 glass was melted at 1050^0C to dissolve particles and heat-treated after quench to grow QD's of different size (1.2 – 4 nm in radius). (For details about the preparation of similar samples see Ref. [7]). In the remainder of this letter we will discuss 1 pair of samples with an average particle size <R> = 1.3 nm. The first sample was heat treated for 280 hours at 590^0C (we will call it the "slow-growth" sample) and the second one was treated for 2 hours at 650^0C (we will call it the "fast-growth" sample).

In our TI FWM experiment, we use three laser pulses of 70 fs in duration with the wave vectors $\mathbf{k_1}$, $\mathbf{k_2}$ and $\mathbf{k_3}$. The first two pulses $\mathbf{k_1}$ and $\mathbf{k_2}$ interact simultaneously ($t_{21} = t_2 - t_1 = 0$) with the sample, preparing a transient population grating. The third pulse $\mathbf{k_3}$ is delayed in time $t_{31} = t_3 - t_1$ relative to the two time coincident pulses $\mathbf{k_1}$ and $\mathbf{k_2}$ in order to interrogate the dynamics of the transient population. The transient grating signal $\mathbf{k_s}$, scatters in the phase matched direction, $\mathbf{k_s} = -\mathbf{k_1} + \mathbf{k_2} + \mathbf{k_3}$. The laser system used to generate the femtosecond pulses is described in detail in Ref. [8]. Essential for the experiments presented here was the use of fully circular polarised laser pulses, which was achieved by placing quarter wave plates into the path of each of the three lasers $\mathbf{k_1}$, $\mathbf{k_2}$ and $\mathbf{k_3}$, and the FWM signal $\mathbf{k_s}$. The sample was mounted in an optical cryostat and kept at a temperature of 10 K. All three laser pulses had the same wavelength lying in the range of $1S_e$-$1S_h$ transition as shown in Fig.1 absorption spectra. Note, that the sizes of the QD's we are working with are defined by the laser wavelength rather than the peak of the absorption curve.

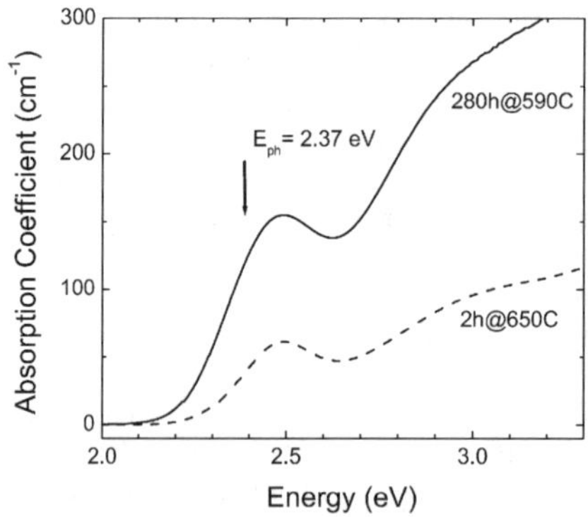

Fig.1. Optical absorption spectra of the 2 samples. Both samples have a maximum of the absorption peak near 2.45 eV; samples were heat treated 280 hrs at 590^0C ("slow-growth") and 2 hrs at 650^0C ("fast-growth"). The arrow indicates the energy position E_{ph} of laser excitation. For both samples E_{ph} = 2.37 eV and corresponds to QD size R = 1.3 nm.

As it was mentioned above, in CdSe QD, in the case of lowered QD symmetry, the lowest exciton state splits by e-h exchange interaction into two states optically active in mutually orthogonal linear polarizations [5]. Thus two σ^+ polarized beams $\mathbf{k_1}$ and $\mathbf{k_2}$ create a density grating in both states, and both σ^+ and σ^- $\mathbf{k_3}$ beams can be scattered on those gratings [6]. The scattered $\mathbf{k_s}$ beam contains both σ^+ and σ^- components. In our experiment, we always create the population grating with two σ^+ $\mathbf{k_1}$ and $\mathbf{k_2}$ polarized beams. The $\mathbf{k_3}$ beam was either σ^+ polarized (in this case we recorded the σ^+ polarized component of $\mathbf{k_s}$ signal), or σ^- polarized (in this case we recorded the σ^- polarized component of $\mathbf{k_s}$). We analyze dynamics of the sum ($\sigma^+ + \sigma^-$) and the difference ($\sigma^+ - \sigma^-$) of σ^+ and σ^- components of the $\mathbf{k_s}$ beam. The sum ($\sigma^+ + \sigma^-$) reflects the population dynamics. As it was shown in Ref. 6, the difference ($\sigma^+ - \sigma^-$) reflects the exciton spin dephasing processes: the splitting Δ of the J=1 exciton state causes spin precession with a rate $\Delta/2\pi\hbar$; in a QD ensemble with a random distribution in the exciton level splitting, this results in the decay of the difference in FWM signals observed in scattering of σ^+ and σ^- polarized light. Thus temporal behavior of difference ($\sigma^+ - \sigma^-$) characterizes the dynamics of exciton spin dephasing.

Fig.2 shows the population dynamics for the two samples. It is clear that the occupation decay rate is the same for both particle growth conditions. (We have observed that the population decay time is a strong function of particle size however.) At the same time, exciton spin dephasing dynamics drastically depends on growth conditions. Fig. 3 displays temporal behavior of the difference ($\sigma^+ - \sigma^-$). Occupation dynamics ($\sigma^+ + \sigma^-$) is shown for reference. Symbols correspond to experimental data, solid lines – deconvolution. The sample which was grown at higher temperature for shorter time ("fast-growth") demonstrates much shorter decay time of difference ($\sigma^+ - \sigma^-$), than the sample grown at lower temperature for longer time ("slow-growth").

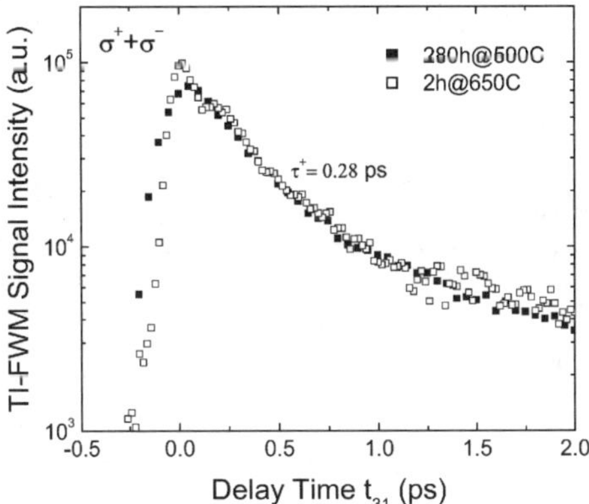

Figure 2. Decay curves of sum σ^+ and σ^- polarized FWM signals, scattered on the grating, created by 2 σ^+-polarized pulses. Empty squares correspond to "fast-growth" sample, dark squares - "fast-growth" sample.

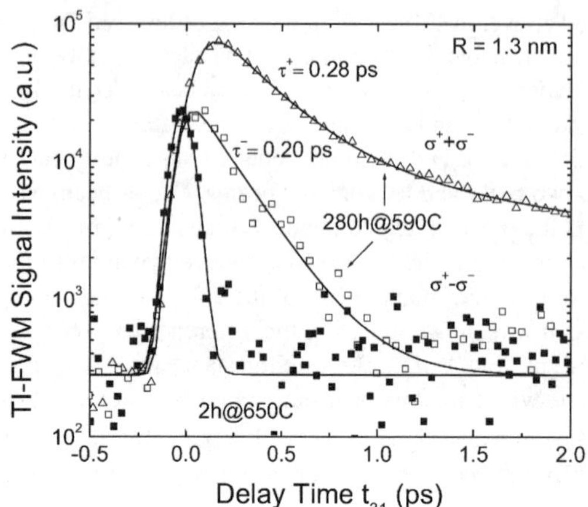

Figure 3. Decay curves of difference of σ^+ and σ^--polarized FWM signals, scattered on the grating, created by 2 σ^+ polarized pulses. Empty squares correspond to "slow-growth" sample, dark squares - "fast-growth" sample. Solid curves – deconvolution. Triangles correspond to sum $(\sigma^+ + \sigma^-)$ – for reference.

For the "slow-growth" sample (Fig.3), deconvolution gives the decay time of the difference $(\sigma^+ - \sigma^-)$ τ^-=0.20 ps. To estimate an exciton spin dephasing time τ^{deph}, we have to take into account a depopulation rate $1/\tau^+$ at that time interval: $\tau^{deph} = \tau^+\tau^- / (\tau^+ - \tau^-)$, where τ^+ corresponds to decay time of the sum $(\sigma^+ + \sigma^-)$, τ^+=0.28 ps. According to this procedure, τ^{deph}=0.70 ps. For the "fast-growth" sample τ^- is shorter than our time resolution, so, we can tell only, that τ^{deph}<0.07 ps.

DISCUSSION

Asymmetry in CdSe nanoparticles might arise from different sources, including: actual shape variation from sphericity, crystal defects such as defect planes, or point defects. All of these defects can arise from growth variations.

The crystalline quality of a nanocrystal is related to three physical effects: 1) entropy-induced disorder and 2) defects and roughness due to kinetic limitations, or 3) energy differences between crystallite facets. We believe that kinetics plays an important role in the formation of defects in CdSe in glass.

If the lowest energy state of a crystallite involves symmetric facets with smooth surfaces, then it is prevented from achieving that state if diffusion and relaxation are slow compared to the rate of arrival of atoms to the growth surface. If we assume that crystals grow by diffusion-limited growth from infinitesimal seeds, then we can estimate the growth rate for our samples at a particular size (results of estimation are summarized in Table 1). The rate of diffusion and relaxation for atoms within the crystallite is activated with energy 2.7 eV [9]. One can suppose, that the degree of crystal quality or structural symmetry is proportional to diffusion rate and proportional to the time to grow the last crystal layer. Under such supposition, we can introduce a parameter we have called "quality coefficient" Q which is proportional to diffusion rate and proportional to the growth time for the last layer of the crystallite. We set Q arbitrarily to 1 for our lower temperature "slow-growth" sample and scaled it for the other sample. The inverse of the quality coefficient (1/Q) reflects the degree of QD asymmetry.

Table 1. Growth parameters and experimental spin splitting. (See text).

R (nm)	Growth time (hrs)	Growth temperature (K)	Time to grow last layer (hrs)	Diffusion scaling factor	Quality coefficient Q	1/Q	Δ (meV)
1.3	280	890	116	$3.22*10^{-15}$	1	1	2
1.3	2	950	0.83	$2.65*10^{-14}$	0.059	17	>20

In Table 1 the data to calculate Q are summarized. Time to grow the last crystal layer was calculated taking into account, that in growth process the QD size increases as the square root of the growth time. The diffusion scaling factor was estimated using $D = D_0\exp(-2.7[eV]/kT)$. The splitting Δ of the J=1 exciton state was estimated from the exciton spin dephasing time τ^{deph}, $\Delta \approx \hbar/\tau^{deph}$.

The "fast-growth" sample has much larger 1/Q, than the "slow-growth" sample, consistent with increased splitting for the "fast-growth" sample. We can give only a lower bound of Δ for this sample (because the dephasing time is shorter than our time resolution). The Δ for the "slow-growth" sample is more than ten times smaller. This ratio corresponds very well to the ratio of 1/Q for these two samples.

CONCLUSION

We have investigated the exciton spin dephasing time in CdSe QD's grown in a glass matrix by studying of the polarization properties of TI-FWM signal. From the dephasing time, we deduce that asymmetry-induced splitting is 2 meV for 1.3 nm radius particles that were grown slowly and > 20 meV for particles that were grown rapidly. This splitting correlates with a structural quality coefficient Q, which is related to the diffusion speed and the growth rate of the nanoparticles. Future studies will address the effects of asymmetry for other particle sizes as well as for a wider range of growth conditions

ACKNOWLEDGMENTS

We gratefully acknowledge funding by the Department of Energy (DOE) Office of Basic Energy Sciences grant DE-FG0297ER455662

REFERENCES

1. D.D.Awschalom, *Physica* **E10**, 1 (2001)
2. A.V. Khaetskii, Yu.V. Nasarov, *Physica* **E6**, 470 (2000)
3. J. A. Gupta, D. D. Awschalom, X. Peng and A. P. Alivisatos, *Phys.Rev.* **B59**, R10421 (1999); P. Borri, W. Langbein, S. Schneider, U. Woggon, R. L. Sellin, D. Ouyang, and D. Bimberg,

Phys.Rev.Letters **87** 157401 (2001); D. Birkedal, K. Leosson, and J. M. Hvam, *Phys.Rev.Letters* **87** 227401 (2001)

4. M. Nirmal, D. J. Norris, M. Kuno, M. G. Bawendi, Al. L. Efros, and M. Rosen, *Phys.Rev.Letters* **75**, 3728 (1995)
5. V. D. Kulakovskii, G. Bacher, R. Weigand, T. Kümmell, A. Forchel, E. Borovitskaya, K. Leonardi, and D. Hommel, *Phys.Rev.Letters* **82**, 1780 (1999)
6. V. D. Kulakovskii, K. Babocsi, M. Schmitt, N. A. Gippius, and W. Kiefer, *Phys.Rev.* **B67**, 113303 (2003)
7. H. Yukselici, *PhD Thesis*, Rensselaer Polytechnic Institute, Troy, NY 1996.
8. T. Siebert, R. Maksimenka, A. Materny, V. Engel, W: Kiefer and M. Schmitt, *J. Raman. Spectrosc.* **33**, (2002) 844
9. H. Yükselici, P. D. Persans, T. M. Hayes, *Phys.Rev.* **B52** 11763 (1995).

Mat. Res. Soc. Symp. Proc. Vol. 789 © 2004 Materials Research Society N15.12

Diffusion-Limited Recombination in Dye-Sensitized TiO_2 Solar Cells

Nikos Kopidakis, Kurt D. Benkstein, Jao van de Lagemaat, and Arthur J. Frank
National Renewable Energy Laboratory,
Golden, CO 80401, U.S.A.

ABSTRACT

The effect of doping on the electron transport dynamics and recombination kinetics in dye-sensitized solar cells was investigated. A simple electrochemical method was developed to dope TiO_2 nanoparticle films with Li. Increasing the doping levels is found to slow electron diffusion. The electron diffusion time exhibits a light intensity dependence at all doping levels consistent with a multiple electron-trapping model involving native and doping-induced traps. Importantly, the diffusion time and recombination lifetime of photocarriers are observed to increase in unison with increased doping. This is the first observation that electron diffusion limits recombination with the redox electrolyte under normal working conditions of the dye cell. A model is presented that accounts for the observation. The implications of this mechanism on cell performance are also discussed.

INTRODUCTION

Since its discovery in 1991 [1], dye-sensitized nanocrystalline TiO_2 solar cells have received much attention because of their low cost and high efficiency [2,3]. A dye-sensitized solar cell typically consists of a highly porous film composed of sintered TiO_2 crystallites of 20 nm diameter sandwiched between a collecting F:SnO_2 electrode and a Pt counter electrode. The pores are filled with an electrolyte containing a redox couple, typically I_3^-/I^-. Dye sensitization at distributed TiO_2/redox electrolyte interfaces provides an efficient means of light harvesting and charge separation: the photoexcited dye molecule injects an electron into the conduction band of TiO_2 and a hole into the redox electrolyte solution. Owing to the screening of the macroscopic electric fields by the high ionic strength electrolyte, electron transport in TiO_2 occurs by diffusion [4]. Diffusion of photocarriers is ambipolar [5], involving the electrostatic coupling of the electron and ion motion. At light intensities up to 1 sun, the injected electron density in TiO_2 is orders of magnitude lower then the ion density in the electrolyte such that the ambipolar diffusion coefficient essentially equals the electron diffusion coefficient [5].

Recombination in the conventional dye-sensitized solar cell has been accounted for in terms of the following mechanism [6]:

$$I_3^- \overset{K_1}{\rightleftarrows} I_2 + I^- \qquad (1)$$

$$I_2 + e^- \xrightarrow{k_2} I_2^{\bullet -} \qquad (2)$$

$$2I_2^{\bullet -} \rightarrow I_3^- + I^- \qquad (3)$$

Triiodide ions in solution are assumed to be in equilibrium with molecular iodine on the TiO_2 surface (reaction 1) [7], which reacts with photoinjected electrons (reaction 2). The resulting iodine radical anion ($I_2^{\bullet -}$) undergoes dismutation (reaction 3) [6]. An alternative to reaction 3 has been proposed [8] but will not be discussed in this paper.

The main focus of the present work is on reaction 2. Taking the equilibrium constant of reaction 1 to be 10^{-10} mol/cm^3 [8], the density of I_2 is estimated to be 10^{13} cm^{-3} [9]. Assuming a TiO_2 nanoparticle density of $2x10^{17}$ cm^{-3} in the film [10], one can estimate that there is one potential I_2 recombination site for every $2x10^4$ nanoparticles. This means that an electron must traverse $2x10^4$ nanoparticles on average before encountering a potential recombination site. Considering the rarity of encountering a recombination site, it is plausible that reaction 2 is limited by the rate of electron transport through the nanoparticle film. The aim of this paper is to investigate this possibility. We will show that the recombination of photocarriers is indeed limited by electron diffusion. A causal link is found between transport and recombination when TiO_2 is doped electrochemically with Li.

EXPERIMENTAL DETAILS

TiO_2 nanoparticle films were deposited onto F:SnO_2 coated glass as detailed elsewhere [9]. TiO_2 films were doped electrochemically with Li [9]. The electrochemical cell consisted of a TiO_2 nanoparticle film as the working electrode, a Ag/AgCl wire as the reference electrode, and a Pt foil as the counter electrode. The electrolyte was 0.5 M $LiClO_4$ in ethanol. Lithium is known to intercalate into TiO_2 films with applied bias according to the reaction [11,12]:

$$TiO_2 + xe^- + xLi^+ \rightarrow Li_xTiO_2 \qquad (4)$$

Although reaction 4 is normally considered to be reversible [11], under some circumstances, a small residual amount of Li remains in the film after an intercalation/deintercalation cycle [12,13]. The density of residual Li increases with increasing number of cycles and with the maximum voltage of the scan [12]. A qualitative comparison of the amount of intercalated Li concentration as a function of the number of cycles was made. Doped and nondoped TiO_2 films were sensitized overnight in dye solution [9] and subsequently assembled in solar cells. The cells were filled with 0.5 M 1-hexyl-2,3-dimethylimidazolium iodide (C6DMII)/0.05 M I_2 in methoxypropionitrile.

Photocurrent transients of nonsensitized films were measured in the same three-electrode cell arrangement used for Li intercalation. The photocarriers were excited by pulses from a N_2 laser (337 nm) incident on the outermost surface of the TiO_2 film. The sensitized cells were probed with a weak laser pulse at 532 nm superimposed on a relatively large, background (bias) illumination at 680 nm at short circuit and open circuit. From the short-circuit photocurrent transient measurements, information about the characteristic time τ_C, which represents the collection (or diffusion) time of electrons [14], was extracted from the exponential decay signal at long time following the probe pulse [14]. Similarly, the characteristic photovoltage transient decay at open circuit yielded information about the recombination lifetime of photocarriers τ_R [9].

EXPERIMENTAL RESULTS

Figure 1 shows UV photocurrent transients of a nonsensitized TiO_2 film that was subject to several consecutive Li intercalation-deintercalation cycles. The peak in the transients at short times represents carriers generated near the substrate by the portion of the probe light reflected at the $F:SnO_2$ surface. The peaks at later times correspond to the arrival of the main packet of electrons generated at the outermost region of the film. Figure 1 shows that with increased Li insertion, the arrival times of electrons increase dramatically.

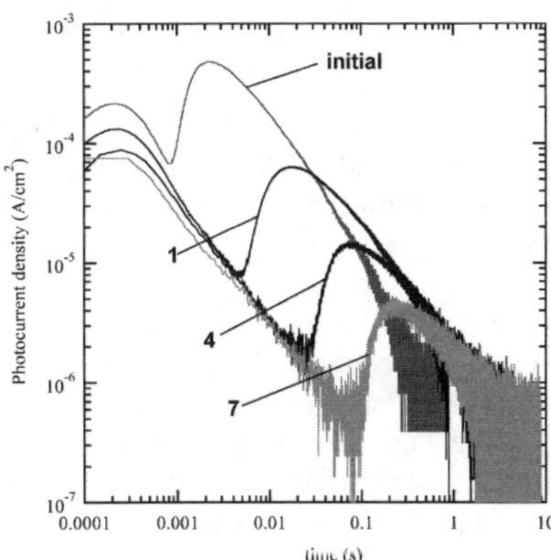

Figure 1. Photocurrent transients of a nonsensitized TiO_2 film under 337nm laser pulses, immediately after cell assembly (initial), and after 1, 4 and 7 intercalation-deintercalation cycles.

The left panel of Figure 2 shows the dependence of the electron diffusion time on bias light intensity, represented by the short circuit photocurrent J_{SC}, for sensitized cells containing nondoped and doped films. Increased doping is seen to increase the electron diffusion time, especially at low light intensities. A power-law dependence between τ_C and J_{SC} is observed at all doping levels although the logarithmic slope ranges from –0.55 for the nondoped film to –0.82 for the most heavily doped film.

The corresponding recombination times versus short circuit photocurrent density for the same cells are shown in the right panel of Figure 2. The increase in collection time is followed by a comparable increase in recombination time.

DISCUSSION

Optical absorption measurements have confirmed that Li intercalation-deintercalation cycles introduce a density of residual Li in the TiO_2 film [9]. Therefore, the photocurrent transients in

Figure 1 suggest that increasing the residual Li density in the TiO_2 film causes an increase in the collection time of electrons. The power law dependence of the diffusion time on short circuit

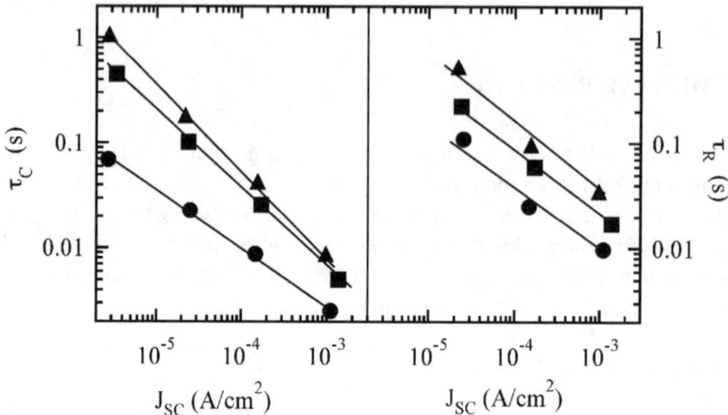

Figure 2. Characteristic times for collection (τ_C – left panel) and recombination (τ_R – right panel) for dye sensitized solar cells with nondoped (circles) and doped (squares and triangles) TiO_2 films. The triangles represent the most heavily doped film. All films had the same thickness. The lines are power-law fits to the data.

photocurrent density, commonly observed in these cells [5], is usually attributed to the presence of an exponential distribution of traps adjacent to the conduction band [5,14]. In this model, electrons undergo many trapping and detrapping events in transit to the collecting substrate [14]. With increased bias light intensity, trap levels fill and the quasi-Fermi level moves closer to the conduction band. Correspondingly, the detrapping times for traps in the vicinity of the quasi-Fermi level become shorter, resulting in faster transport. This model predicts that $\tau_C \propto J_{SC}^{\alpha-1}$, where $\alpha = \frac{kT}{m_c}$, m_c is the width of the trap state distribution [5,9,14], T is the temperature, and k is the Boltzmann's constant. Within the framework of this model, a change in slope (α-1) from –0.55 (nondoped sample) to –0.82 (highly doped sample) implies that m_c increases from 55 meV to 138 meV. An exponential trap density is usually ascribed to disorder [15]. Therefore, the widening of the trap state distribution is attributed to increased disorder caused by randomly placed Li atoms in the material [9].

Assuming that Li-induced traps only alter the diffusion kinetics and do not affect the interfacial electron transfer to the electrolyte, the causal link between the collection (diffusion) of electrons and recombination, as depicted by the characteristic times displayed in Figure 2, is attributed to transport limited recombination. Trap-limited recombination has been observed in cells without a redox couple, where the electron can only recombine with an oxidized adsorbed dye molecule [16]. The present study establishes for the first time that transport limits recombination in fully functional dye-sensitized solar cells.

A transport-limited recombination mechanism explains a number of observations. From reactions 1 and 2, the rate of recombination R is given by

$$R = K_1 k_2 \frac{[I_3^-]}{[I^-]} N \qquad (5)$$

where $[I_3^-]$ and $[I^-]$ are the concentrations of the redox couple and N is the electron density in the film. Following the arguments presented in the Introduction, we propose that reaction 2 is limited by the ability of the electron to find an I_2 site. For this situation, k_2 is proportional to the electron diffusion coefficient D, which is inversely proportional to the diffusion time τ_C. The recombination lifetime τ_R, defined by the pseudo-first order kinetics equation $R=N/\tau_R$, can be written with the aid of equation 5 as

$$\tau_R = \frac{N}{R} \propto 1/k_2 \propto 1/D \propto \tau_C \qquad (6)$$

Equation 6 describes the connection between the recombination lifetime and the diffusion time of electrons in TiO_2 films. It can be shown that the diffusion-limited recombination model preserves the nonlinear dependence of the recombination rate on electron density N [7] via the dependence of the diffusion time on N [5,14].

Because Li is commonly used in the electrolyte of dye-sensitized TiO_2 solar cells, we explored whether Li insertion occurs under normal light intensity (AM 1.5) in a working cell at open circuit. It was found that, in the presence of a Li containing electrolyte, photocharge-induced Li intercalation occurs in dye-sensitized TiO_2 films. A comparable increase in diffusion and recombination times was observed as a function of photocharge-induced Li doping levels resulting from consecutive increasing illumination intervals [9].

The photovoltaic properties of the cell are largely unaffected by Li doping because doping leaves the ratio τ_C/τ_R, and therefore the charge collection efficiency [17], almost unchanged (equation 6). The latter determines the short circuit photocurrent. The rate of recombination (equation 5) at 1 sun light intensity is expected to change very little because the diffusion time, and accordingly, the rate constant k_2, display only minor changes with doping (figure 2) at high light intensities. Therefore, the open circuit voltage of the cell [6,7] at 1 sun light intensity is predicted to remain almost constant with doping. These predictions have been verified by experiments [9].

CONCLUSIONS

Electrochemical doping of TiO_2 nanoparticle films with Li increases the collection time of photoinduced electrons in both nonsensitized and dye sensitized films. Evidence is presented for transport-limited recombination in working dye-sensitized solar cells, and a mechanism is proposed. Because the transport and recombination times change to the same extent with Li doping, they do not have a significant effect on the photovoltaic properties of the cell.

ACKNOWLEDGEMENTS

This work was supported by the Office of Science, Division of Chemical Sciences, and the Office of Utility Technologies, Division of Photovoltaics, U.S. Department of Energy, under contract DE-AC36-99GO10337.

REFERENCES

1. B. O'Regan and M. Grätzel, *Nature* **353**, 737 (1991).
2. M. K. Nazeeruddin, P. Péchy, T. Renouard, S. M. Zakeeruddin, R. Humphry-Baker, P. Comte, P. Liska, L. Cevey, E. Costa, V. Shklover, L. Spiccia, G. B. Deacon, C. A. Bignozzi, and M. Grätzel, *J. Am. Chem. Soc.* **123**, 1613 (2001).
3. P. Wang, S. M. Zakeeruddin, J. E. Moser, M. K. Nazeeruddin, T. Sekiguchi, and M. Grätzel, *Nature Materials* **2**, 402 (2003).
4. Arie Zaban, Andreas Meier, and Brian Gregg, *J. Phys. Chem. B* **101**, 7985 (1997).
5. N. Kopidakis, E. A. Schiff, N.-G. Park, J. van de Lagemaat, and A. J. Frank, *J. Phys. Chem. B* **104**, 3930 (2000).
6. S. Y. Huang, G. Schlichthörl, A. J. Nozik, M. Grätzel, and A. J. Frank, *J. Phys. Chem. B* **101**, 2576 (1997).
7. G. Schlichthörl, S. Y. Huang, J. Sprague, and A. J. Frank, *J. Phys. Chem. B* **101**, 8141 (1997).
8. A. C. Fisher, L. M. Peter, E. A. Ponomarev, A. B. Walker, and K. G. U. Wijayantha, *J. Phys. Chem. B* **104**, 949 (2000).
9. N. Kopidakis, K. D. Benkstein, J. van de Lagemaat, and A. J. Frank, *J. Phys. Chem. B* **107**, 11307 (2003).
10. K. D. Benkstein, N. Kopidakis, J. van de Lagemaat, and A. J. Frank, *J. Phys. Chem. B* **107**, 7759 (2003).
11. R. van de Krol, Albert Goossens, and Eric A. Meulenkamp, *J. Electrochem.Soc.* **146**, 3150 (1999).
12. R. van de Krol, Albert Goossens, and Eric A. Meulenkamp, *J. Appl. Phys.* **90**, 2235 (2001).
13. L. Kavan, M. Grätzel, J. Rathousky, and A. Zukal, *J. Electrochem. Soc.* **143**, 394 (1996).
14. J. van de Lagemaat and A. J. Frank, *J. Phys. Chem. B* **105**, 11194 (2001).
15. M. Silver, L. Pautmeier, and H. Bässler, *Sol. State Communications* **72**, 177 (1989).
16. Jenny Nelson, Saif A. Haque, David R. Klug, and James R. Durrant, *Phys. Rev. B* **63**, 205321 (2001).
17. G. Schlichthörl, N.-G. Park, and A. J. Frank, *J. Phys. Chem. B* **103**, 782 (1999).

Mat. Res. Soc. Symp. Proc. Vol. 789 © 2004 Materials Research Society N15.23

The Magnetic Behavior of Triangular Shaped Permalloy Nanomagnet Arrays

J. Y. Shiu[1], M.F. Tai[2], Y. D. Yao[3], C. W. Kuo[1] and P. Chen*[1]
[1] Institute of Applied Science and Engineering Research, Academia Sinica, Taipei 115, Taiwan
[2]Department of Physics, National Chung-Cheng University, Chia-Yi 621, Taiwan
[3]Institute of Physics, Academia Sinica, Taipei 115, Taiwan

Abstract

During the past few decades, the density of magnetic storage has been improved considerably. To increase the storage capacity, it is necessary to reduce the size of magnetic grains. However, as the domain size decreases, their thermal stability will also decrease, which results in the loss of magnetization. To overcome the limit imposed by such superparamagnetic behavior, lots of recent research attentions have been focused on the patterned magnetic media. To maximize the storage density, it is preferable to create periodical magnetic patterns, in which single-domain magnetic dots are well separated from each other. In this experiment, we have utilized nanosphere lithography to create large-area well-ordered two dimension arrays of permalloy ($Ni_{80}Fe_{20}$) nanoparticles. Nanosphere lithography is an inexpensive, simple, parallel, and high throughput fabrication technique. We have employed monodisperse polystyrene beads with diameter of 650, 560, 440, 350, 280 nm to fabricate triangle-shaped permalloy ($Ni_{80}Fe_{20}$) nano-arrays with lateral dimension in the region of 170~90 nm, and thickness in the region of 10~50 nm. The magnetic behavior of these triangle-shaped nanomagnet arrays have been investigated by longitudinal magnetic optic Kerr effect (LMOKE) and magnetic force microscopy (MFM). It was found that the coercivity of the permalloy nanoparticle arrays increases with decreasing the thickness of the nanoparticle. This can be attributed to the interface effect between the arrays and the substrate.

Introduction

Patterning materials with nanometer resolution has attracted a lot of research attention, because these patterned materials have exhibited various novel optical[1], magnetic[2], catalytic[3] and electric transport [4] properties. In order to utilize these noble properties, it requires systematic investigation of the size dependent properties of nano-patterned materials. Therefore, it is very important to develop inexpensive, large-scale nanofabrication techniques for the study of size dependent behavior. To produce nanostructures with satisfactory size control in the sub-100 nm region, e-beam lithography is the most popular method. However, it is not practical to employ e-beam lithography for large-scale fabrication. Self-organization process, on the other hand, provides an alternative approach to fabricate large-area nanostructures with sufficient size control [5]. It has been demonstrated that both two-dimensional [6-11] and three-dimensional [12-14] crystalline structures can be obtained on the substrate surfaces using monodispersed colloid solutions. These crystalline structures can then be used as the templates to produce patterned materials [15-16].

Among various noble properties of nano-patterned materials, patterned magnetic structures at nanometer scale has drawn lots of research attention recently, because of their potential applications in high-density data storage. As the lateral dimension of magnetic films reduces to some internal characteristic length of magnets, the size and shape of magnets may have strong influence on their magnetic properties because geometric restriction controls the balance between exchange field and demagnetization field. Here we report the fabrication of size-controlled nanomagnet arrays by a simple approach: nanosphere lithography. Nanosphere lithography, a simple, parallel process, utilizes monodispersed polystyrene beads to form large scale close packed structures on the substrate surface. These close packed patterns are then served as deposition masks. Various materials can be deposited into the interstice spacing among polystyrene beads via evaporation or sputtering. After removing the polystyrene beads, well-ordered arrays of triangle shaped islands are formed on the substrate surface. The size control of the nanostructure can be achieved by selecting the size of polystyrene beads. The area covered with nanomagnet arrays can be as large as 1 cm^2, which makes it easy to utilize conventional magnetic optical Kerr effect (MOKE) instruments to investigate their magnetic behavior.

In this study, we have employed monodisperse polystyrene beads with diameter of 650, 560, 440, 350, 280 nm to fabricate triangle-shape permalloy ($Ni_{80}Fe_{20}$) nanomagnet arrays with lateral dimension in the region of 170~90 nm, and thickness in the region of 10~50 nm. Depending on their concentration, polystyrene beads can form single-layer or double-layer close packed patterns. The nanomagnets fabricated using single layer template exhibit triangle shape in hexagonal arrangement while double layer template produces smaller triangle nanomagnets with periodicity equal to the diameter of polystyrene beads.

EXPERIMENT

A typical fabrication procedure for nanomagnet arrays is to fill the interstice of the close packed polystyrene templates produced by nanosphere lithography. The detail procedure of nanosphere lithography can be found in the literatures [5-7]. In short, N-doped silicon (100) wafers (Gredmann) were cut into several ~1 cm^2 pieces. These silicon substrates were first cleaned in piranha solution (3:1 concentrated H_2SO_4:30% H_2O_2) for 30 minutes and then rinsed repeatedly with ultrapure water (18.2 MOhms, Millipore Simplicity). These substrates were further cleaned with acetone, and methanol before used. Monodisperse polystyrene beads of various diameters purchased from Bangs Laboratories, Inc. (Fishers, IN) were diluted in a solution of surfactant Triton X-100 (Aldrich) and methanol (1:400 by volume). Polystyrene solution was then spin-cast on to silicon substrates to form hexagonally close-packed two-dimensional colloidal crystals. By adjusting the speed of spin-coater (800-3600 RPM) and the surfactant concentration, both single layer (SL) and double layer (DL) large-area (up to 1 cm^2) close packed structures have been obtained similar to those reported previously [5-7]. These two dimensional periodic nanosphere arrays were then used as the deposition templates.

Thin films of permalloy ($Ni_{80}Fe_{20}$) were deposited in a modified Consolidated Vacuum Corporation vapor deposition system with a base pressure of 10^{-7} Torr. After deposition of permalloy, the polystyrene nanospheres were removed from substrate by CH_2Cl_2 solution and sonicating for 1-4 min.

To investigate the magnetic properties of nanomagnet arrays. The samples prepared by nanosphere lithography were first examined by magnetic force microscope (MFM). From the MFM image we can see the magnetic behavior of individual nanomagnet. The overall magnetic behavior of these triangle-shaped nanomagnet arrays has also been studied by, longitudinal magnetic optical Kerr effect (LMOKE).

RESULTS

Fig. 1 shows the SEM images of the permalloy nanomagnet arrays formed by both single-layer and double-layer nanosphere lithography. It can be clearly seen that both templates produce triangular shaped nanomagnets. To investigate the magnetic behavior of individual nanomagnet, MFM was used. Figure 2 show the AFM and MFM images of nanomagnet arrays using a single-layer 650 nm in diameter polystyrene template. The lateral dimension of nanomagnets using 650 nm polystyrene beads was measured to be 170 nm. From the MFM image it is clear that the magnetic dipole of triangular shaped nanomagnet is pointing toward the normal of the substrate. To investigate the size-dependent magnetic behavior of nanomagnet arrays, nanomagnet arrays with different lateral dimension (170, 150, 130, 110 and 90 nm) and thickness (10, 20, 30, 40 and 50 nm) were prepared and probed by LMOKE. To remove the effect of thickness, array size and array filling factor, the LMOKE signals were normalized in height. Figure 3 shows the result of LMOKE measurement. One thing worth of mentioning was that for the nanomagnets fabricated using single layer 350 nm polystyrene beads, the measured H_c varied from 10 to 120 Oe, as the thickness decreased from 40 nm to 20 nm. For samples with the same thickness, it was observed that the coercivity was unchanged even the amplitude of the LMOKE signal was varied due to ration. Two general trends were observed from LMOKE measurements: first, as the size of nanomagnets decreases, the coercivity increases. Second, the coercivity of the triangle shaped nanomagnets decreases as thickness of permalloy increases.

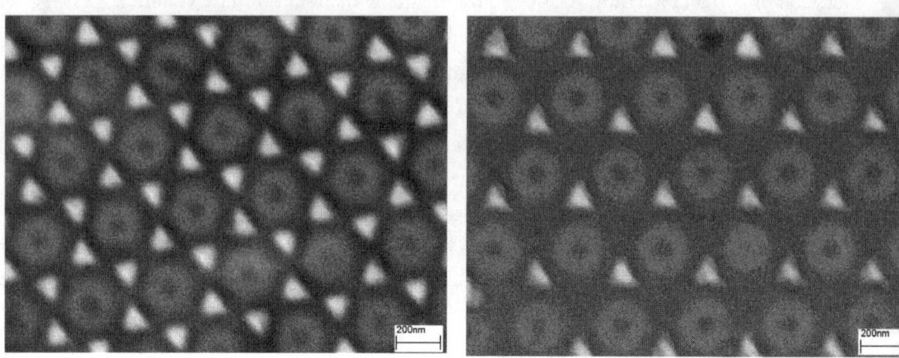

Figure 1: SEM images of nanomagnet arrays formed by single-layer (left) and double-layer (right) nanosphere lithography using polystyrene with diameter of 280 nm.

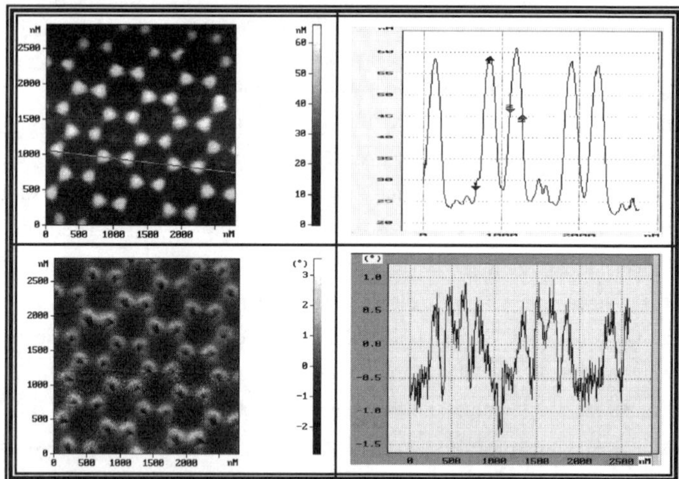

Figure 2: AFM and MFM images of nanomagnet arrays formed by single-layer nanosphere lithography using polystyrene with diameter of 650 nm. The thickness of $Ni_{80}Fe_{20}$ permalloy is 50 nm.

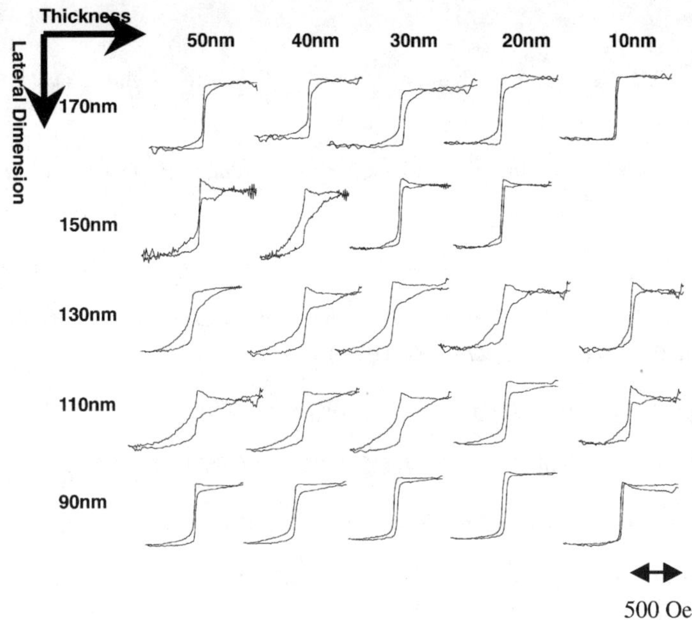

Figure 3: The hystersis measured by LMOKE.

CONCLUSIONS

In summary, we have successfully fabricated large-area triangle-shaped $Ni_{80}Fe_{20}$ arrays with lateral dimension from 90 to 170 nm by nanosphere lithography. We have measured the MFM images of nanomagnet array and investigated the magnetic behavior of nanomagnet array by LMOKE. It was found that the coercivity increases with decreasing in the size and the thickness of the nanomagnets.

ACKNOWLEDGMENTS

This research was supported in part by National Science Council under contract 92-2113-M-001-036.

REFERENCES

1. M.J. Feldstein, C.D. Keating, Y.-H. Liau, M.J. Natan, N.F. Scherer, *J. Am. Chem. Soc.* **119**, 6338 (1997).
2. M. Hehn, K. Ounadjela, J.-P. Bucher, F. Rousseauz, D. Decanni, B. Bartenlian, *Science*, **272**, 1782 (1996).
3. U. Heiz, F. Vanolli, A.Sanchez, W.-D. Schneider, *J. Am. Chem. Soc.* **120**, 9668 (1998).
4. R.P. Andres, J.D. Bielefeld, J.I. Henderson, D.B. Janes, V.R. Kolagunta, C.P. Kubiak, W.J. Mahoney, R.G. Osifchin, *Science*, **273**, 1690 (1996).
5. C.L. Haynes, R.P. Van Duyne, *J. Phys. Chem. B*, **105**, 5599 (2001).
6. U.C. Fischer, H.P. Zingsheim, *J. Vac. Sci. Technol.*, **19**, 881 (1981).
7. H.W. Deckman, J.H. Dunsmuir, *J. Vac. Sci. Technol. B*, **1**,1109 (1983).
8. J.C. Hulteen, R.P. Van Duyne, *J. Vac. Sci. Technol. A*, **13**,1553 (1995).
9. R. Micheletto, H. Fukuda, M. Ohtsu, *Langmuir*, **11**, 3333 (1995).
10. F. Lenzman, K. Li, A.H. Kitai, H.D.H. Stover, *Chem. Mater.*, **6**, 156 (1994).
11. J. Boneberg, F. Burmeister, C. Schafle, R. Leiderer, D. Reim, A. Frey, S. Herminghaus, *Langmuir*, **13**, 7080 (1997).
12. Y.A. Vlasov, X.Z. Bo, J.C. Sturm, D.J. Norris, *Nature*, **414**, 289 (2001).
13. P.V. Braun, P. Wiltzius, *Nature*, **402**, 603 (1999).
14. A. Blanco, E. Chomski, S. Grabtchak, M. Ibisate, S. John; S.W. Leonard, C. Lopez, F. Meseguer, H. Miguez, J.P. Mondia, G.A. Ozin, O. Toader, H.M. Van Driel, *Nature*, **405**, 437 (2000).
15. P. Jiang, J.F. Bertone, V.L. Colvin, V.L., *Science*, **291**, 453 (2001).
16. S. Han, X. Shi, F. Zhou, *Nano Lett.* **2**, 97 (2002).

Mat. Res. Soc. Symp. Proc. Vol. 789 © 2004 Materials Research Society N15.27

Development of Nanoparticles with Tunable UV Absorption Characteristics

Daniel Morel, Imad Ahmed and Henry Haefke
Micro and Nanomaterials Section,
CSEM Swiss Center for Electronics and Microtechnology, Inc.
CH-2007 Neuchâtel, Switzerland

ABSTRACT

Engineering of nanoparticles to detect ultraviolet light within a specified range and its feasibility to make a device has been demonstrated. It is shown that the absorption edge of a material can be shifted to significantly lower wavelengths in the UV range by using nanoparticles and that this feature can be incorporated within a device. All experimental work was focused on ZnO. Both commercially obtained ZnO nanoparticles as well as in-house synthesized ZnO nanoparticles were examined. For the in-house developed particles it was shown that varying the diameter of the ZnO nanoparticles could vary the absorption wavelength from 315 to 365 nm. Commercially available nanoparticles did not show this shift due to their relatively larger sizes (diameter ≅ 20 nm) as well as their broad size distributions.

A photocurrent effect of UV light on thin films prepared with nanoparticles has been demonstrated. Not only the optical band-gap value depends on the size of the nanoparticles but also the mobility gap of the material and, as a consequence, the onset of photocurrent.

INTRODUCTION

Ultraviolet filters for wavelength selection and selective UV detectors are in demand devices. A well-known and commercially important use of UV detectors is for flame detection and monitoring in heating installations. Here detection is performed with UV sensitive tubes or solid-state sensors based on silicon technology. The reason for this is that the most specific feature in a burner flame is the emission band at 310 nm attributed to OH radicals, whose monitoring allows optimum flame detection and monitoring.

Reports in literature have already demonstrated that materials when reduced to dimensions of a few nanometers show modified or new properties with regard to bulk materials [1-3]. So, the optical and opto-electrical characteristics of an absorbing material show a shift in the absorption wavelength towards lower wavelengths depending of the size of the particles. The idea to prepare and use nanoparticles with tailored sizes of a semiconductor materials in such a way to get controlled absorption within a given wavelength range in the ultraviolet region was not systematically studied up to now.

Screening among semiconductor materials with a wide band-gap has allowed selecting a good promising candidate material, zinc oxide, which is already used in products involving UV applications as filtration, absorption and skin protection cream [4].

EXPERIMENTALS

Sample preparation

At the beginning of this study, commercially available nanoparticles were obtained in the hope of using these for this project. Unfortunately, it was not possible to locate commercial nanoparticles with diameters less than 10 nm. Colloids 1 % in ethanol were prepared with all the acquired nanopowders to measure their optical absorption in UV and visible domain.

ZnO nanoparticles smaller than 10 nm were synthesized adapting a described process [5,6]. $Zn(CH_3COO)_2{\cdot}2H_2O$ is dissolved in boiling pure ethanol at a concentration of 0.1 M. A solution 0.14 M $LiOH{\cdot}H_2O$ in pure ethanol is also prepared. Two equal volumes of both solutions are mixed at room temperature. ZnO nanoparticles immediately form and growth of nanoparticles starts. The colloid remains translucent. Growth is followed by UV absorption measurement.

To separate nanoparticles from the reactive medium, a small quantity of water is added to the suspension. Strong aggregation and precipitation occur. The solid is separated by centrifugation and then washed twice with ethanol.

Structural characterization

Characterization of the structure of the nanoparticles has been performed by Transmission Electron Microscopy (TEM) and by Electron Diffraction Patterning at high resolution.

Optical Absorption

Optical absorption measurements have been performed with a two channel Perkin-Elmer Lambda 14 instrument in 1 cm length cells. This method was used to measure stable colloids prepared with the commercial nanoparticles and to follow the growth of the nanoparticles smaller than 10 nm.

Photocurrent measurement

Clean aggregates of ZnO nanoparticles are put in suspension in ethanol by strong ultrasonic agitation and films of these nanoparticles on quartz plates are prepared by dip-coating. Films are dried at 80 °C. To measure photocurrent effect, films are enclosed in a dark cabinet on a X-Y positionning table and two lamps at wavelengths of respectively 366 nm and 254 nm provide UV light. Hard metal electrodes directly contact films with a 500 μm separation in between. I/V data are acquired through a HP 4145 semiconductor parameter analyzer.

RESULTS AND DISCUSSION

Optical absorption

All commercial nanoparticles showed similar behavior with maximum absorption peaks ranging between 364 and 368 nm (figure 1). A particularly negative feature of these particles was a broad threshold of absorption, which was attributed to their large size distributions as seen in TEM images of 37 nm mean size nanoparticles in figure 2.

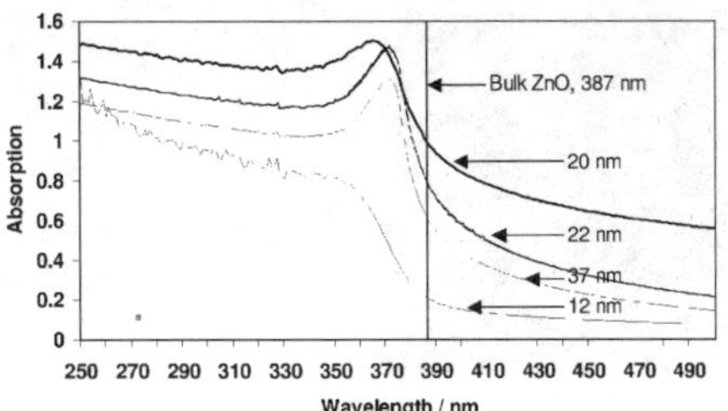

Figure 1. Reduced blue-shift of UV absorption band-gap of different sizes commercial ZnO nanoparticles compared with the band-gap of bulk ZnO.

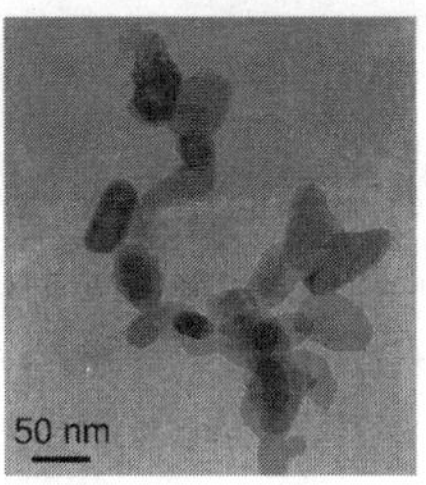

Figure 2. TEM image of a commercial 37 nm mean size ZnO nanoparticles.

Another lot with a mean size of 12 nm was synthesized by flame spray pyrolysis (Institute of Process Engineering in ETHZ, Switzerland). The absorption maximum of that powder is at 350 nm (figure 1), which was lower than all the other commercially obtained nanoparticles. The shift of the absorption threshold of these powders was found to be insufficient for the foreseen target application and the variation between the powders of different diameters was so small that it is not possible to speak of tunable absorption in the strict sense of the word.

Growth of ZnO nanoparticles smaller than 10 nm is followed by UV absorption measurement and is presented in figure 3. A red-shift of the absorption threshold is observed as a function of growth time. Particles size distribution has also been measured by light scattering. After 4.5 min of reaction time, the absorption threshold is at 315 nm, what corresponds to a particle diameter of 3.0 nm [7,8]. About 10 hours later the threshold shifts to 365 nm, which corresponds to a nanoparticle diameter of 5.0 nm. TEM observation and electron diffraction patterning (figure 4) confirm evidence of crystallized ZnO nanoparticles.

The absorption spectrum of nanoparticles, which are larger than 12 nm shows a well-developed maximum near the onset of the absorption, which is ascribed to the first exitonic transition also described for other nanoparticulate materials like CdS [9]. That transition no longer appears in the absorption spectrum of smaller ZnO nanoparticles, but other transtions are observed on growing particles in the wavelength range of 320 - 330 nm (Figure 3). Clear attribution of these absorption maxima is not possible. At this size range, shape of the crystallites, surface energy, and correlation effect will influence the relative spacing of the energy levels [10].

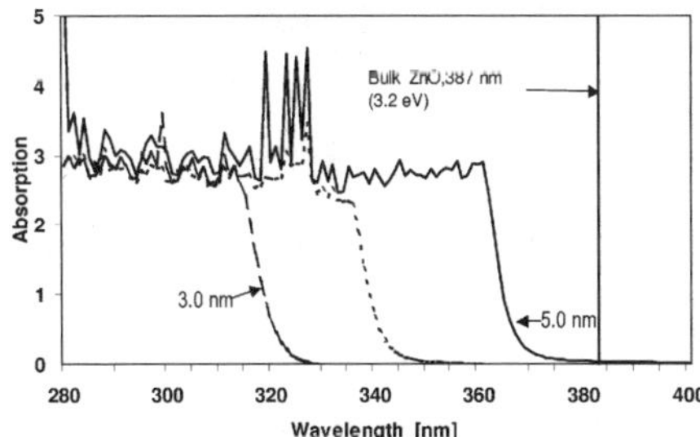

Figure 3. UV absorption measurements carried out during growth of nanoparticles. Red-shift in the absorption edge of about 50 nm is observed between particles of 3.0 nm particles, 4.5 minutes after the beginning of the reaction and particles of 5.0 nm obtained after 10 hours.

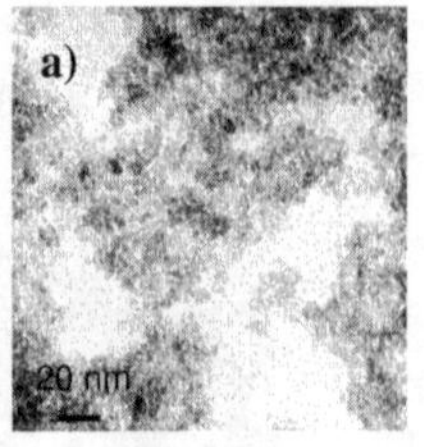

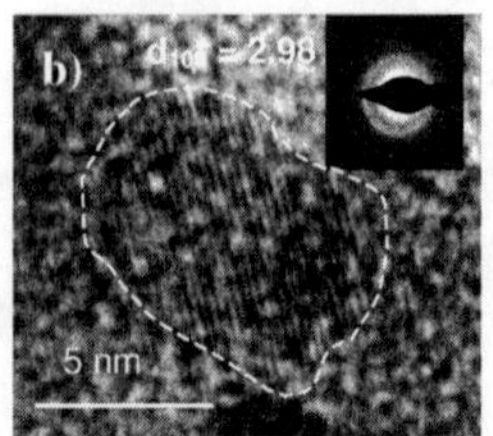

Figure 4
a) TEM pictures of ZnO nanoparticles with a mean size of 5.0 nm
b) High resolution electron micrograph with electron diffraction pattern (EDP) of individual ZnO nanoparticles. Evidence for crytalline structure is given by the presence of lattice fringes in the circled zone corresponding to (100) planes and the EDP shown in the inset.

Photocurrent measurement

Figure 5 a) shows the variation of the current through a film prepared with commercial nanoparticles (diameter = 22 nm). The main increase of the photocurrent appears when light at a wavelength of 366 nm strikes the sample, which corresponds to the maximum absorption of the colloid preparation. At lower wavelengths, activation of some sites with higher energies still takes place. However, the total photocurrent effect remains weak in intensity. One explanation is that the percolation level of the particles is weak. The film prepared by dip coating is formed of nanoparticle aggregates and the contacts between the aggregates occur only at a very few points. The photocurrent effect on a film prepared from CSEM nanoparticles (diameter = 5 nm) is different depending on the wavelength of the radiation. Figure 5 b) shows that there is no photocurrent generated by radiation of 366 nm (E_g = 3.39 eV), which is just at the upper limit of the optical absorption presented in figure 3. But at a higher energy (λ = 254 nm, E_g = 4.88 eV), a photoelectronic effect takes place. The built-in potential observed on 5 nm particles is actually not clearly understood, and needs further investigation.

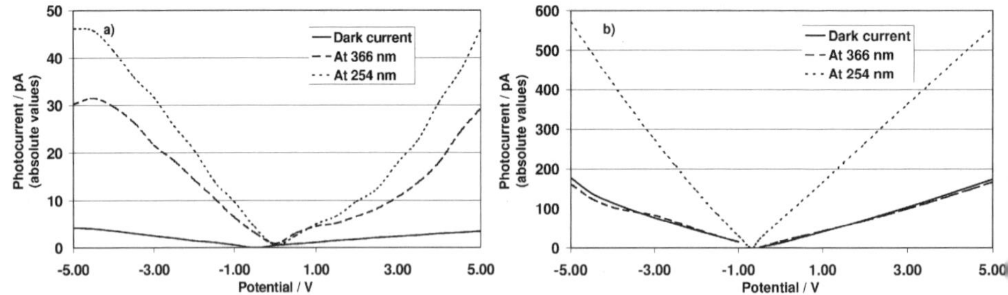

Figure 5.
a) Photocurrent generated by UV light incident on a film prepared from acquired ZnO nanoparticles with a mean size of 22 nm (Nanophase Inc.)
b) Photocurrent generated by UV light incident on a film prepared with CSEM ZnO nanoparticles with a mean size of 5 nm.

This measurement leads to the conclusion that not only the optical band-gap value depends on the size of the nanoparticles but also the mobility gap. This measured photocurrent is a global effect of the nanoparticulate ZnO system and it cannot be exclusively attributed to the interband absorption of the nanoparticles. Energy levels linked to surface and grain boundaries play an important role and can also be activated by light.

CONCLUSION

Reduction of ZnO nanoparticle size achieved by various methods results in a clear blue shift of the absorption edge. Thus, tunability of the absorption edge is feasible in a range of about 60 nm. Besides the optical band-gap it is also demonstrated that the mobility gap depends on the size of the nanoparticles, and that semiconductor nanoparticles can be used for tunable photoelectronic sensing.However, to achieve tunability within a large UV range, nanoparticles with smaller diameters need to be synthesized and isolated. Remaining key challenges are the agglomeration of the nanoparticles, and purification processes during synthesis.

ACKNOWLEDGEMENTS

The authors would like to thank Prof. S. E. Pratsinis and Dr. L. Mädler, ETH Zürich, for providing ZnO nanopowder, and Prof. P. Schurtenberger, University of Fribourg (Switzerland), for light scattering measurements. This work was partly funded by the National Research Program TOP NANO 21, project No. 5707.

REFERENCES

1. A. Henglein, *Topics Current Chem.*, **143**, 113 (1988).
2. L. E. Brus, *Appl. Phys.* **A53**, 465 (1991).
3. S. J. Oldenburg, R. D. Averitt, S. L. Westcott, and N. J. Halas, *Chem. Phys. Lett.* **288**, 243 (1998).
4. T. Lyon and T. Turney, *Australiasian Science*, **Nov./Dec. 1999,**
5. L. Spanhel and M.A. Anderson, *J. Am Chem. Soc.*, **113,** 2826 (1991).
6. E. A. Meulenkamp, *J. Phys Chem. B,* **102,** 5566 (1998).
7. L. E. Brus, *J. Phys. Chem.,* **90**, 2555 (1986).
8. E. M. Wong, P. G. Hoertz, C.J. Liang, B.-M. Shi, G.J. Meyer, and P.C. Searson, *Langmuir,* **17**, 8362 (2001).
9. T. Vossmeyer, L. Katsikas, M. Giersig, I.G. Popovic, K. Diesner, A. Chemseddine, A. Eychmüller, and H. Weller, *J. Phys. Chem.,* **98**, 7665 (1994).
10. A.P. Alivisatos, *J. Phys. Chem.,* **100**, 13226 (1996).

Mat. Res. Soc. Symp. Proc. Vol. 789 © 2004 Materials Research Society N15.30

Optical Properties of Polymer-Embedded Silicon Nanoparticles

William D. Kirkey[1], Alexander N. Cartwright[1], Xuegeng Li[2], Yuanqing He[2], Mark T. Swihart[2], Yudhisthira Sahoo[3], and Paras N. Prasad[3]
Departments of [1]Electrical Engineering, [2]Chemical Engineering, and [3]Chemistry and Institute for Lasers, Photonics, and Biophotonics
State University of New York at Buffalo, U.S.A.
[1]332 Bonner Hall, Buffalo, NY 14260, U.S.A.

ABSTRACT

We seek to use electrically conducting polymers, such as those commonly utilized in polymeric LEDs, as hosts for silicon nanoparticles. The proper design of multilayered devices based on these materials will yield efficient light-emitters in which charge carriers localize and recombine within the nanoparticles. Furthermore, these may combine the flexibility and processability of polymeric LEDs with the reliability of inorganic materials. We have synthesized luminescent silicon nanoparticles and have characterized their photoluminescence (PL) using continuous-wave and time-resolved spectroscopy. These particles have been incorporated into a variety of transparent solid hosts. The photoluminescence obtained from particle-containing poly(methyl methacrylate) (PMMA) matrices is very similar to that of the particles in solution, both in spectral content and PL decay characteristics. However, when incorporated into a variety of conducting polymers, such as poly(N-vinylcarbazole) (PVK), the nanoparticles do not retain their photoluminescence properties. A variety of chemical species have been reported as effective PL quenchers for porous silicon. We believe that these polymers quench the luminescence through similar mechanisms. Protective passivation of the nanoparticle surface is suggested as a strategy for overcoming this quenching.

INTRODUCTION

Nanocrystalline silicon has received a great deal of attention in recent years as a material that might be easily integrated into silicon wafer processing and utilized for biological and chemical sensing or for light-emitting devices. The well-established effects of the surface termination and the surrounding environment on the silicon photoluminescence (PL) are advantageous for sensing applications, but present a problem for optoelectronic applications requiring stable light emission. One solution is to produce silicon nanoparticles within SiO_2, for example by implanting Si ions into a glass substrate and thermally annealing the composite. These materials show promise for use as optical devices; optical gain has been reported in nanoparticles produced by this method[1]. An alternate approach is to produce colloidal nanoparticles that can then be incorporated into transparent hosts such as polymers or xerogels.

Various researchers have incorporated luminescent nanoparticles within electrically conducting polymers to yield hybrid electroluminescent devices[2-5]. However, to our knowledge, no study of luminescence from silicon nanoparticles embedded in such materials has been reported. The work presented in this paper is focused on the application of surface modification techniques to silicon nanoparticles produced by high-yield aerosol synthesis as a route toward obtaining such hybrid materials and devices.

EXPERIMENTAL DETAILS

The nanoparticle synthesis procedure has been described in detail elsewhere[6-8]. Briefly, the nanoparticles are synthesized in an aerosol reactor by dissociating silane gas using a CO_2 laser. Helium and hydrogen gas flows confine the reactant mixture, yielding silicon nanoparticles with a mean diameter that can be varied from 5 to 20 nm by controlling the growth conditions. Following the reaction, the nanoparticles are collected on cellulose nitrate membrane filters and dispersed in methanol. The nanoparticles are then etched in solutions containing HF and HNO_3, which reduces the nanoparticle size and passivates the surface.

Continuous wave (CW) and time-resolved spectroscopic techniques were used to study the basic emission properties of the nanoparticles. Specifically, the CW PL data reported in this work was excited using either a Perkin Elmer LS 50 fluorescence spectrometer (355 nm excitation) or the 351 nm and 364 nm lines of a Coherent Innova 300 Ar^+ laser. The source for the time resolved spectroscopy was 400 nm, ~300 fs, pulses, with repetition rates varying between 9 and 250 kHz, that were generated by frequency doubling the 800 nm, 200 fs pulses output from a Coherent RegA 9000 regenerative amplifier. The time resolved photoluminescence (TRPL) data was collected and recorded using a spectrometer and a C4334 Hamamatsu streak camera. All measurements reported in this work were conducted at room temperature.

RESULTS AND DISCUSSION

The production method described above generates a high yield of silicon nanoparticles, on the order of 100 mg/hr. The nanoparticle luminescence is similar to that commonly observed from nanocrystalline silicon (including porous silicon), with a luminescence peak that can be varied from about 550 nm to the near infrared by varying the etch conditions, as illustrated in Figure 1a. TRPL measurements of the nanoparticles show emission with multi- or stretched-exponential behavior, and with lifetimes on the order of tens of microseconds, which vary substantially across the emission spectrum. For low energy emission, the decay can be reasonably approximated by a single exponential on long time scales. However, for high energy emission, the decay is more complicated and requires stretched-exponential fitting. The emission decay of a freshly prepared sample is shown in Figure 1b.

When dispersed in most common organic solvents, the PL of the nanoparticles degrades and spectrally shifts with time. To prevent such degradation, the nanoparticles can be embedded in poly(methyl methacrylate) (PMMA). Figure 2 shows the PL spectrum and lifetime of a particular nanoparticle sample both in chloroform solution and when embedded in a PMMA film. The film was prepared by drop-casting a polymer-nanoparticle mixture on a glass slide. The characteristics of the nanoparticle luminescence are not greatly affected by the presence of the polymer. The polymer shields the nanoparticles and prevents degradation of the PL. A second advantage is that these materials facilitate a variety of optical studies that are commonly performed on solid materials but are difficult to perform on solutions.

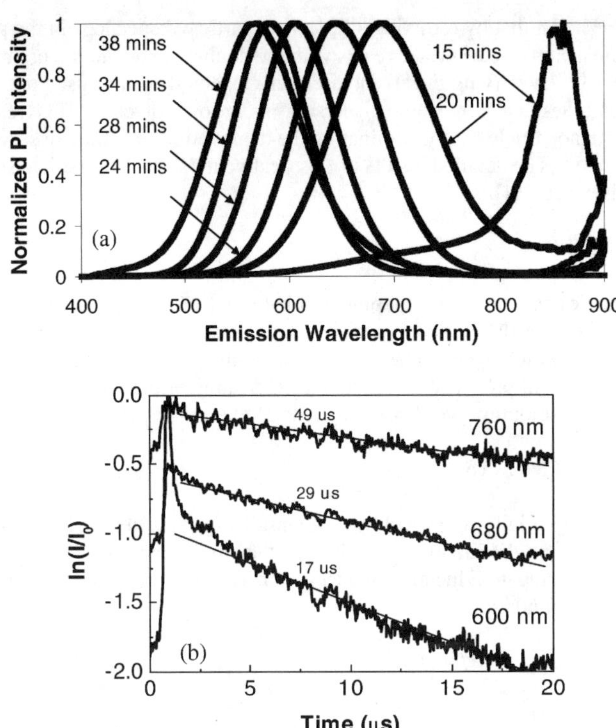

Figure 1. (a) Normalized photoluminescence of nanoparticles after etching for various amounts of time in a 3% HF / 26% HNO_3 / 69% methanol etching solution. (b) PL decay of a typical sample at various emission wavelengths. The plots have been offset for clarity.

PMMA is not an electrically conducting polymer, and therefore it is not suitable for use in electroluminescent devices. Alternative polymer hosts include materials such as poly(N-vinylcarbazole) (PVK), which has been used as a hole-transporting host for a luminescent dye in an LED structure[9], and poly(phenylene-vinylene) (PPV), which is commonly used as the emissive material in polymer LEDs or as a host or hole-transport layer in nanoparticle LEDs. However, the as-produced nanoparticles do not disperse well in most common organic solvents, which is a necessary step in the processing of these polymer films. Further, the PL of unprotected nanocrystalline silicon has been shown by many authors to be severely quenched by the presence of various chemical species, including many amines[10-12]. We have observed such quenching due to the presence of the PVK polymer. Thus, it is necessary to modify the surface of the nanoparticles in order to improve their solubility, and, if possible, to provide some shielding of the surface to prevent PL quenching while not drastically impeding the ability of holes and electrons to be injected into the nanoparticles.

It has been shown that by refluxing silicon nanoparticle dispersions in the presence of octadecene, the nanoparticle surface reacts with the double bond, and the organic becomes bound to the surface[13]. By varying the alkene used, this technique can be used to create organic-capped nanoparticles that form stable dispersions in various solvents. This treatment also serves to stablize the nanoparticle PL by eliminating the degradation over time observed for the as-produced particles. The detailed results of this treatment on the nanoparticle surface and PL will be reported elsewhere[14].

CONCLUSIONS

Many variables exist in the development of hybrid electroluminescent materials. To utilize silicon nanoparticles as the emitting species in such devices, it will be necessary to establish a device structure in which the nanoparticles are accessible for carrier injection, while retaining good photoluminescent properties. Effective modification of the semiconductor surface is the first step toward developing useful combinations of these materials.

ACKNOWLEDGEMENTS

This work was partially supported by a Defense University Research Initiative on Nanotechnology grant #F496200110358 through the Air Force Office of Scientific Research, and by an Integrative Graduate Education and Research Traineeship #DGE0114330 through the National Science Foundation.

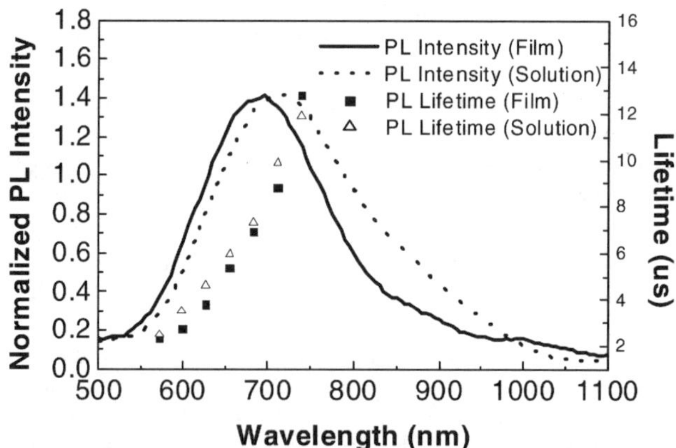

Figure 2. CW Emission and photoluminescence lifetime (400 nm excitation) of a silicon nanoparticle sample when embedded in PMMA and when dispersed in chloroform solution. The lifetimes were obtained using a single-exponential fit.

REFERENCES

1. L. Pavesi, L. Dal Negro, C. Mazzoleni, G. Franzo and F. Priolo, *Nature* **408**, 440-444 (2000).
2. B. O. Dabbousi, C. B. Murray, M. F. Rubner and M. G. Bawendi, *Chemistry of Materials* **6**, 216-219 (1994).
3. S. Coe, W. K. Woo, M. Bawendi and V. Bulovic, *Nature* **420**, 800-803 (2002).
4. L. Bakueva, S. Musikhin, M. A. Hines, T. W. F. Chang, M. Tzolov, G. D. Scholes and E. H. Sargent, *Applied Physics Letters* **82**, 2895-2897 (2003).
5. N. Tessler, V. Medvedev, M. Kazes, S. H. Kan and U. Banin, *Science* **295**, 1506-1508 (2002).
6. X. Li, Y. He and M. T. Swihart, *Proceedings of the Electrochemical Society* **2003-08**, 1161-1167 (2003).
7. X. Li, Y. He, S. S. Talukdar and M. T. Swihart, *Langmuir* **19**, 8490-8496 (2003).
8. M. T. Swihart, X. Li, Y. He, Z. Li, E. Ruckenstein, W. Kirkey, A. Cartwright, Y. Sahoo and P. Prasad, *Proceedings of SPIE* **5222B-27**, (2003).
9. M. Pan, Patra, A., Friend, C., Tzu-Chau, L., Cartwright, A., and Prasad, P., *Mat. Res. Soc. Symp. Proc.* **734**, (2003).
10. J. K. M. Chun, A. B. Bocarsly, T. R. Cottrell, J. B. Benziger and J. C. Yee, *Journal of the American Chemical Society* **115**, 3024-3025 (1993).
11. B. Sweryda-Krawiec, R. R. Chandler-Henderson, J. L. Coffer, Y. G. Rho and R. F. Pinizzotto, *Journal of Physical Chemistry* **100**, 13776-13780 (1996).
12. J. M. Rehm, G. L. McLendon and P. M. Fauchet, *Journal of the American Chemical Society* **118**, 4490-4491 (1996).
13. L. H. Lie, M. Duerdin, E. M. Tuite, A. Houlton and B. R. Horrocks, *Journal of Electroanalytical Chemistry* **538**, 183-190 (2002).
14. X. Li, Y. He and M. T. Swihart, submitted to *Langmuir* (2003).

Mat. Res. Soc. Symp. Proc. Vol. 789 © 2004 Materials Research Society N15.35

Growth of InGaN Nanorods as the Blue Light Source for White Light Emitting Devices

Hwa-Mok Kim[1,*], Tae Won Kang[1] and Kwan Soo Chung[2]
[1]Quantum-functional Semiconductor Research Center,
Dongguk University,
Seoul 100-715, Korea
[2]School of Electronics and Information, College of Electronics and Information
Kyunghee University
Yongin 449-701, Korea

ABSTRACT

This work demonstrates the properties of InGaN nanorod arrays with various In mole fractions by modified HVPE using In metal at the low growth temperature. The nanorods grown on (0001) sapphire substrates are preferentially oriented in the c-axis direction. We found that the In mole fractions in the nanorods were linearly increased at $x \leq 0.1$. However, In mole fractions were slightly increased at $x > 0.1$ and then were gradually saturated at $x = 0.2$. CL spectra show strong blue emissions from 428 nm ($x = 0.1$, 2.89 eV) to 470 nm ($x = 0.2$, 2.64 eV) at room temperature. From this result, we found the fact that increasing In mole fraction in the InGaN nanorod shift the peak position of CL spectrum emitted from InGaN nanorod to the low energy region.

INTRODUCTION

Recent results have challenged the long-standing band gap value for indium nitride (InN) (~2 eV) and asserted 0.9 eV as the true band gap for high quality InN films.[1] This alone makes indium gallium nitride (InGaN) a promising material for blue/violet emission (and even a candidate for deep ultraviolet (UV) and infrared (IR) emission). If InGaN's indium (In) mole fraction can be successfully turned from 0 to 1, InGaN represents a ternary alloy capable of emitting photons from 0.9 eV to 3.4 eV – that is, from GaN's UV emission (~365 nm) to InN's emission (~1380 nm). The most common method to achieve white light emission is to combine a phosphor wavelength down-converter with a blue InGaN/GaN light emitting diode (LED). A blue LED is typically placed in a parabolic mirror and subsequently coated with a phosphor-containing epoxy.[2] The LED emits blue light which is absorbed by the phosphor and re-emitted as longer-wavelength phosphorescence. This occurs via a process known as down-conversion. These two wavelength (generally blue and yellow) combine form white right.[3] Unfortunately, many threading dislocations (TDs) are produced in InGaN/GaN layer, used for blue light source, due to lattice mismatch with the substrate and difference of thermal expansion coefficient with substrate, and thus they affect significantly the device performance as non-radiative recombination centers.[4] On the other hand, the growth mechanism for nanorods is completely different, and threading dislocations can be all but non-existent in nanorods. Therefore, the nanorods have the potential for negligible non-radiative recombination loss, and thus the efficiency of down-conversion is much higher than bulk InGaN/GaN. However, no one has successfully grown InGaN and InGaN/GaN nanorods to data. Here, we demonstrate the optical properties of InGaN nanorod arrays on (0001) sapphire substrate by modified hydride vapor phase epitaxy (HVPE) with various In mole fractions. Ga and In metals were used as Ga and In precursors (group III), respectively. NH_3 was used as N precursor (group V). InGaN nanorods in this study were characterized by scanning electron microscope (SEM), energy dispersive x-ray spectrometer (EDS), wavelength dispersive x-ray spectrometer (WDS), transmission electron microscope (TEM) and cathodoluminescence (CL).

EXPERIMENTAL DETAILS

The InGaN nanorods were grown by modified HVPE similar to our previous work.[5] Sapphire (0001) wafer was used as a substrate. Substrate was cleaned by an HCl solution and rinsed with deionized water before use. The Ga and In precursors were synthesized via a reaction of HCl gas (5N) (in an N_2 diluent gas) with Ga (7N) and In (6N) metal at 600 °C. These precursors were then transported to the substrate area where it is mixed with NH_3 (6N4) to form InGaN ternary nanorods at 510 °C. In this process, HCl flow rate for Ga precursor was in the range of 100-300 sccm and for In precursor was in the range of 5-100 sccm. NH_3 flow rate was 4000-5000 sccm and growth time was 1 hours. InGaN nanorods were grown directly on (0001) sapphire substrate. After furnace was cooled to room temperature, a light yellow layer was found on the surface of the substrate. The morphologies of the resulting materials were characterized using field emission (FE) SEM (FEI, SIRION). Structural and elemental analyses of a single nanorod were performed using TEM (JEOL, JEM_2010 at 200 kV), selective area electron diffraction pattern (SAEP), EDS (EDAX, NEW XL-30) and WDS (EDAX, LambdaSpec). For the optical characterization of InGaN nanorods with various In mole fractions, CL spectroscopy was carried out using a FESEM (FEI, XL 30 SFEG) with a CL system (Gatan, MONOCL2). The emission from the samples was dispersed by a monochromator with a 1200 lines/mm grating blazed at 500 nm and detected by a Peltier-cooled photomultiplier. The maximum spectral resolution is 0.2 nm. The experimental conditions in CL were carefully established to minimize the undesired influence of electron-beam bombardment on the conditions of the specimens.

RESULTS AND DISCUSSION

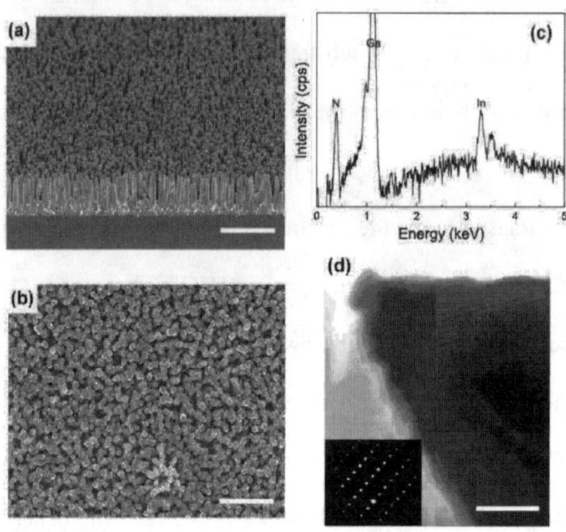

Figure 1. SEM, EDS and TEM characterizations of straight InGaN nanorods on (111) silicon substrate grown by HVPE at 510 °C. (a) 30 ° tilted view, scale bar, 2 μm, (b) plan view SEM image, scale bar, 1 μm. (c) EDS spectrum of a single crystalline InGaN nanorod and (d) TEM image of the InGaN nanorod and corresponding electron diffraction pattern (inset). Scale bar is 8 nm.

Figure 1 (a) and (b) show the InGaN nanorod arrays grown on (0001) sapphire substrate by using modified HVPE system. The growth temperature for growing InGaN nanorods was about 510 °C compared to the growth temperature of GaN nanorods. The average diameter and length were 70 nm and 2 μm, respectively. The control of the InGaN nanorod's diameter and length were achieved by adjusting the growth temperature and growth time, respectively. Figure 1 (c) shows EDS spectrum of a single InGaN nanorod. From this spectrum, we could observe InL line in a single nanorod. Figure 1 (d) shows TEM image of a single InGaN nanorod. In this image, the [0001] direction was parallel to the long axis of rod, indicating that the [0001] direction is a common growth direction in an InGaN nanorod. SAEP of the same nanorod from [2110] direction confirmed that it was a single crystal, and that the electron diffraction pattern was indexed to the reflection of hexagonal InGaN crystal.

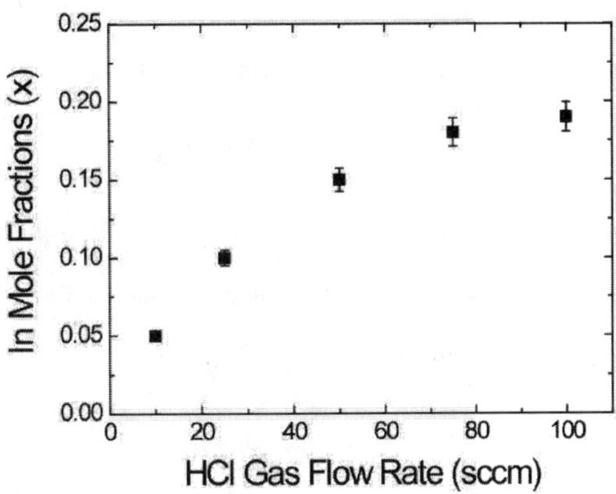

Figure 2. Variation of In mole fractions (x) in the InGaN nanorods with various HCl reactant gas flow rates by WDS measurement.

For investigating In mole fractions in the InGaN nanorods, we measured WDS spectrum of our InGaN nanorods samples as shown in Fig. 2. Increasing HCl gas flow rate for reacting In metal, that is In precursor, from 10 sccm to 100 sccm, In mole fractions (x) were increased from 0.04 to 0.20 in the nanorods. In this case, NH_3, main carrier N_2 and HCl carrier N_2 gas flow rate were kept 4000 sccm, 5000 sccm and 2000 sccm, respectively. Growth temperature was 510 °C.

When HCl reactant gas flow rate was small (≤ 25 sccm), In mole fractions were linearly increased with HCl gas flow rate. As HCl reactant gas flow rate was increased, however, In mole fractions in the InGaN nanorods were slightly increased and then these were saturated. One of the reasons for saturating In mole fraction is thought to be the consumption of $InCl_3$ in the reactor by hydrogen reduction, because the hydrogen partial pressure in the reactor increases with the decomposition of HCl with increasing HCl gas flow rate. With increasing of hydrogen partial pressure, the following reaction of $InCl_3$ consumption proceeds to the right-hand side.

$$2In + 6HCl \rightarrow 2(InCl_3) + 3H_2$$

$$InCl_3 + H_2 \rightarrow InCl + 2HCl$$

Since the produced InCl does not react effectively for the deposition of InN,[6] it is thought that the consumption of $InCl_3$ by the above reaction leads to the saturation of the In mole fraction in the InGaN nanorods as seen in the In-HCl-NH_3 system.

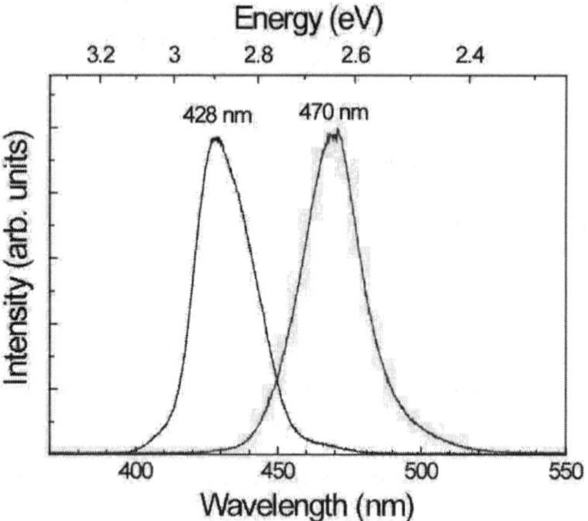

Figure 3. Optical properties of two individual InGaN nanorods with In mole fractions of x = 0.1, and 0.2 by CL measurement at 300 K, respectively.

To investigate the optical properties of two individual InGaN nanorods with In mole fractions of x = 0.1 and 0.2, respectively, we carried out the CL experiments at 300 K. Figure 3 shows the CL spectra of two individual InGaN nanorods with In mole fractions of x = 0.1 and 0.2. The CL peak positions of 428 nm (2.89 eV) and 470 nm (2.64 eV) were observed for the InGaN nanorods with In mole fraction (x) of 0.1 and 0.2, respectively. These values are very close to that observed from the optical properties of InGaN alloys.[7]

CONCLUSIONS

This work demonstrates the optical properties of InGaN nanorod arrays with various In mole fractions by modified HVPE at the low growth temperature. The nanorods grown on (0001) sapphire substrates are preferentially oriented in the c-axis direction. We found that the In mole fractions in the nanorods were linearly increased at $x \leq 0.1$. However, In mole fractions were slightly increased at $x > 0.1$ and then were gradually saturated at x = 0.2. CL spectra show strong blue emissions peaking at around 428 nm (2.89 eV) and 470 nm (2.64 eV) at room temperature, respectively. We believe that present approach is a simple one for practical application to nanoscale white light emitting devices.

ACKNOWLEDGMENTS

This work was supported by KOSEF through the QSRC at Dongguk University.

REFERENCES

1. S. H. Wei, X. Nie, I. G. Batyrev, S. B. Zhang, *Phys. Rev. B* **67**, 165209 (2003).
2. S. Nakamura, D. Pearton, G. Fasol, in *The Blue Laser Diode: the complete story*, 2nd edition, Springer, Berlin (2000) Ch. 10.
3. S. Nakamura, G. Fasol, in *The Blue Laser Diode*, Springer, Berlin 216 (1996).
4. D. Cherns, S. J. Henley, F. A. Ponce, *Appl. Phys. Lett.* **78**, 2691 (2001).
5. H. –M. Kim, W. C. Lee, T. W. Kang, K. S. Chung, C. S. Yoon, C. K. Kim, *Chem. Phys. Lett.* **380**, 181 (2003).
6. N. Takahashi, J. Ogasawara, A. Koukitu, *J. Cryst. Growth* **172**, 298 (1997).
7. (a) K. P. O'Donnell, *Mat. Res. Soc. Symp.* **595**, W11.26.1 (2000). (b) M. –H. Kim, J. –K. Cho, I. –H. Lee, S. –J. Park, *phys. stat. sol. (a)* **176**, 269 (1999). (c) W. Shan, B. D. Little, J. J. Song, Z. C. Feng, M. Schurman, R. A. Stall, *Appl. Phys. Lett.* **69**, 3315 (1996).

Mat. Res. Soc. Symp. Proc. Vol. 789 © 2004 Materials Research Society N15.37

Trap Effects in PbS Quantum Dots

Peter D. Persans[1], Aleksey Filin[1], Feiran Huang[1], Andrew Vitek[2], Pratima G.N. Rao[3], Robert H. Doremus[3]

[1]Department of Physics, Applied Physics and Astronomy, Rensselaer Polytechnic Institute, Troy, New York
[2]Department of Physics, Bethel College, Minnesota
[3]Department of Materials Science and Engineering, Rensselaer Polytechnic Institute, Troy, New York

ABSTRACT

We have found that traps can dominate important relaxation processes in PbS nanoparticles prepared by precipitation in borosilicate glass. The primary photoluminescence peak energy falls about 50 meV below the lowest absorption peak energy and has a long decay time of ~4 μs. Photoinduced bleaching of the lowest absorption peak also has a long decay lifetime of ~2 μs. Photoinduced bleaching also exhibits a surprisingly long rise time of hundreds of nanoseconds. Such long-lifetime effects must be attributed to multiple traps.

INTRODUCTION

In PbS quantum dots with diameter of ~3 nm, the lowest excited state falls at 1550 nm. Such particles are therefore potentially important as saturable absorbers, optical switching elements, or emitters for telecommunications [1-9]. The speed and magnitude of nonlinear optical and luminescence responses will determine the utility of this system. We have therefore studied these properties using a slow-repetition excitation source that assures complete relaxation of trapped carriers between pulses. Specifically, in this paper we report time dependent and cw luminescence and pump-probe spectra of 3 nm PbS nanoparticles grown in glass.

EXPERIMENTAL DETAILS

Growth of PbS nanoparticles in glass for quantum size effect studies has been previously reported and details on growth appear elsewhere [10]. In our preparation process, PbS nanoparticles are grown from quenched PbS-doped borosilicate samples by heat-treating at the temperature range from 540 to 580°C for 12 to 36 hours. In quenched glass, particles are smaller than 1 nm. We have prepared a set of samples in which the average particle radius is varied from ~1 nm to >3 nm. The optical absorption spectra of a sample with average radius of 3 nm is shown in Fig. 1. Particle size was deduced by comparison of optical properties to theory [11, 12]. Particle size was also directly measured by TEM imaging on similar samples [10].

The pump-probe system is described in detail elsewhere [13]. A dye laser pumped by a nitrogen laser is used as the pump source for both luminescence and pump-probe time-dependence. Laser pulses have a wavelength of 640 nm (1.94 eV), duration of 10 ns, and repetition rate of ~10 Hz. Maximum excitation intensity of pump applied to the sample is 420 μJ/cm^2.

The probe for time-resolved pump-probe studies is a 1550 nm cw semiconductor laser. The response time for detectors and electronics is less than 20 ns.

Time-resolved luminescence was excited with the same dye laser and detected with either an InAs diode or a Ge diode detector. CW luminescence was excited with a chopped HeNe (633 nm) laser, dispersed with a 0.75 m f8 monochromator, and detected with an uncooled Ge photodiode. The spectral response of the measurement systems was calibrated using a tungsten halide lamp.

RESULTS

In Fig. 1 we show the room temperature absorption and luminescence spectra for a typical sample with lowest absorption (HOMO-LUMO) peak centered at 0.8 eV. The HOMO-LUMO absorption peak is well-defined. A second excited state is observed at about 1.25 eV.

The luminescence was excited with photon energy well above the gap and therefore excites all particle sizes in the distribution. The integrated room temperature luminescence efficiency is approximately 2% and the spectrum is clearly shifted about 50 meV below the absorption peak. Excitation with 1550 nm light yields luminescence that is peaked ~50 meV below the excitation energy.

The intensity of the integrated luminescence decays exponentially after excitation with a lifetime of ~4 μs.

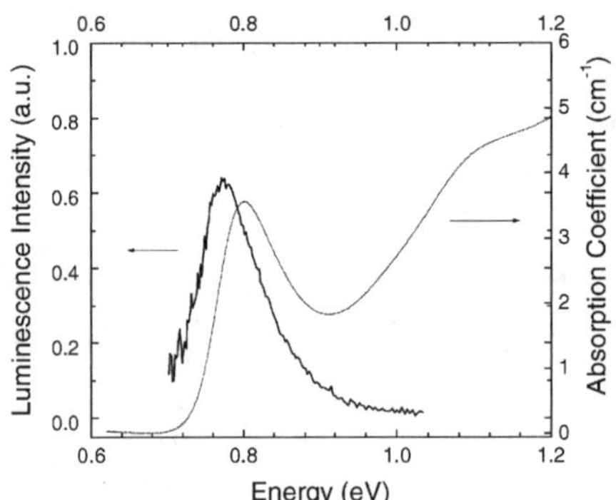

Figure 1 - Optical absorption and luminescence spectra of quenched and heat treated glass samples with PbS nanocrystals. Average particle size is 3 nm. The left axis corresponds to the luminescence spectrum. The right axis corresponds to the absorption spectrum.

In Fig. 2 we show the spectrum of the CW photoinduced bleaching (circles), along with a fit to the deconvolved HOMO-LUMO absorption peak (line). The two sharp features between 0.87 and 0.9 eV are due to thermomodulation of O-H absorption bands in the glass. The peak in absorption modulation is in close correspondence to the absorption peak, within 10 meV of the peak in absorption. The magnitude of the cw bleaching peak is about one part in 10,000.

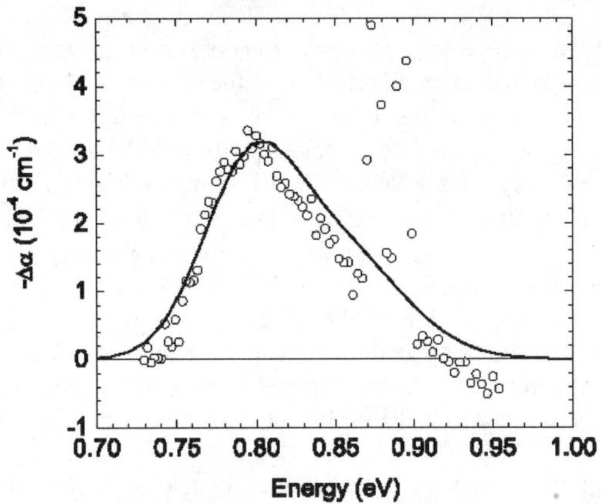

Figure 2 - CW induced bleaching spectrum for the same sample as in Fig. 1. The solid line is a model fit to the absorption coefficient in the lowest peak. The open symbols are the change in absorption due to excitation with a chopped cw Ar^+ laser (514.5 nm) at room temperature.

Time-resolved transmission pump-probe measurements on similar samples have been reported previously [14]. Example time dependence spectra are shown in Fig. 3. We observe photoinduced transmission at 0.8 eV for all measurement conditions. Rise and decay times of the bleaching are nearly independent of pump fluence, as seen in the figure. The 10%-90% rise time of ~ 200 ns and the decay lifetime of ~2 μs are much longer than either the pump pulse or the response time of the electronics. The absorption bleaching saturates at $\Delta\alpha/\alpha$~0.04 for pump energy density of ~200 $\mu J/cm^2$. When the magnitude of the absorption change is corrected for sample thickness and area overlap between pump and probe, the saturation magnitude is close to 25%.

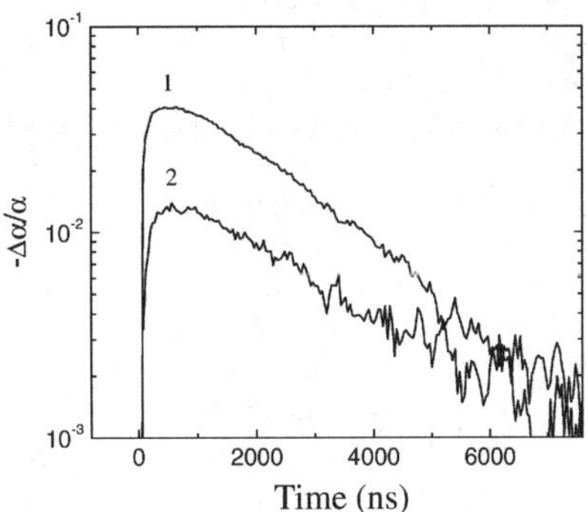

Fig. 3 - Kinetics of bleaching at different pump pulse intensities: (1) 420 $\mu J/cm^2$, (2) 10 $\mu J/cm^2$. The pump energy is 1.98 eV and the probe energy is 0.8 eV.

DISCUSSION

We have several interesting features of the observations that must be explained. i) The photoluminescence spectrum is red-shifted from the HOMO-LUMO absorption peak by ~50 meV. ii) The luminescence lifetime is long (~4 μs) compared with expected HOMO-LUMO lifetimes. iii) Bleaching is spectrally coincident with HOMO-LUMO absorption. iv) The bleaching rise time is surprisingly long (~200 ns) compared with intraband carrier decay rates. v) The bleaching decay time is long (~2 μs) compared with expected HOMO-LUMO lifetimes. vi) The magnitude of the saturation level of bleaching is large, of order $\Delta\alpha/\alpha$~20% after correction for absorption volume.

The results are discussed using the schematic energy level diagram in Fig. 4. The quantum states derived from conduction and valence bands are labeled LUMO and HOMO respectively. E and G represent the high energy states through which excitation occurs. T_h is a hole trap which lies between G and HOMO and T_e is an electron trap which lies below the LUMO state.

The long photoluminescence decay time can also be explained by the presence of a trap within the gap, or by the presence of a dark exciton state within the gap [15]. The lifetime observed here is significantly longer than that observed for PbSe [16], so we propose the presence of a trap. The choice of hole or electron trap is somewhat arbitrary at this point because we do not have sufficient information to make a decision. We place the electron trap T_e 50 meV below the LUMO state to be consistent with the luminescence spectrum. The large magnitude of the bleaching effect suggests that this trap is present in a large fraction of the quantum dots.

The shape of the bleaching spectrum implies change in oscillator strength with little or no shift [17, 18]. When it is corrected for excited volume in the sample, the magnitude of the change in absorption is relatively large ($\Delta\alpha/\alpha$~20%). There are several possible explanations for this observed behavior, including i) filling of both HOMO and LUMO states with paired electron-holes, ii) filling of only one of the two quantum states (HOMO or LUMO) with the second carrier in a trap, or iii) occupation of some other state which exerts a large effect on the HOMO-LUMO transition. Case i) would lead to a fast decay of induced bleaching, which is not observed. Case ii) is consistent with both magnitude and decay time observations. Case iii) is possible if there exists a state which exerts large effect on absorption in the HOMO-LUMO state and which has a long lifetime. It is unlikely that a state with both carriers trapped could affect the HOMO-LUMO transition to any significant extent. We cannot exclude the possibility of a dark-exciton like state, but the mechanism for bleaching in this situation has yet to be worked out in detail.

While the long decay time for photomodulation and photoluminescence can be explained easily by the presence of a trap for one carrier, the rise time is problematic. It is expected that carriers excited into states G and E will decay to states H and L within one picosecond. This implies that the bleaching rise time should be equally short. Fast rise and decay time has indeed been observed by us and by others using sub-picosecond degenerate pump-probe spectroscopy [7]. Our experiment differs from earlier experiments in three important regards: 1) We use relatively low pump flux. 2) The repetition rate for the current experiment is very low, allowing the particle to fully relax between pulses. 3). The excitation photon energy is relatively high, forcing the excited carriers to relax through intermediate states before arriving at the HOMO or LUMO state.

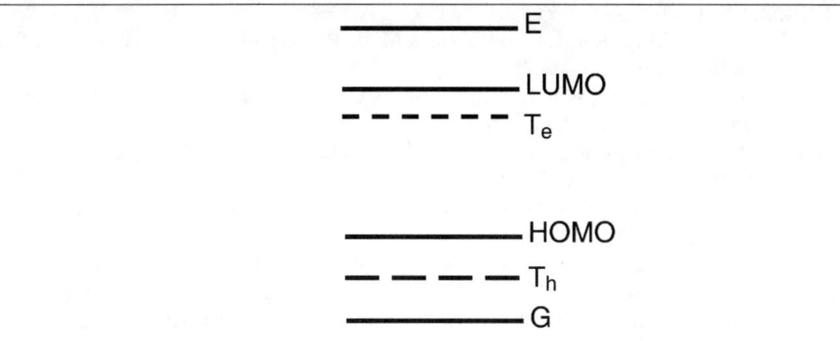

Figure 4 - Schematic of the important energy states in a 3 nm PbS quantum dot. Solid lines represent quantum states of the particle and dashed lines represent localized traps or resonant states.

This simplest explanation for the 200 ns rise time is the presence of a long-lived intermediate state. If this state were a trap, it would have to be for the opposite carrier from the trap that slows the decay. We have represented it as T_h in the schematic. We note that the bleaching rise dynamics would be complex in this situation, and the observed slow rise time requires that the electron be trapped very rapidly into T_e. Another possibility for the long-lived intermediate state is the presence of a long-lived exciton-like state associated with a higher quantum transition, much like the "dark exciton" in CdSe is associated with the HOMO-LUMO transition [15].

SUMMARY AND CONCLUSIONS

Time-resolved two-color pump-probe spectroscopy reveals the existence of at least two surprisingly long-lived states in PbS quantum dots with HOMO-LUMO gap of 0.8 eV. We have proposed that these states are traps. The first state must lie within the HOMO-LUMO gap and may be the state through which defect luminescence occurs. If so, the defect state is shallow and has a lifetime of 2 μs. The probability for capture into this defect is high, implying that the capture time is of the order of picoseconds. The second trap state must fall outside of the HOMO-LUMO gap because it affects the rise time for one carrier to fall into its lowest excited state. The probability for capture into this state is high, implying that capture is fast but escape to the HOMO or LUMO state takes 200 ns. At present, we cannot exclude the possibility of new "dark exciton" states.

The nature of these states is as yet unclear. Further work using excitation spectroscopy, size dependence, electric and magnetic field, and temperature dependence should help to address open questions.

ACKNOWLEDGEMENTS

This work is supported in part by the Department of Energy, Office of Basic Energy Sciences grant DE-FG0297ER455662. AV was supported by the NSF REU program at Rensselaer.

REFERENCES

1. C.H.B. Cruz, C.L. Cesar, L.C. Barbosa, A.M. de Paula, and S. Tsuda, Appl. Surf. Sci., 1997. **109**(110): p. 30-5.
2. P.T. Guerreiro, S. Ten, N.F. Borrelli, J. Butty, G.E. Jabbour, and N. Peyghambanan, Appl. Phys. Lett., 1997. **71**(12): p. 1595.
3. V. Gulbinas, A. Dementjev, L. Valkunas, G. Tamulaitis, L. Motchalov, and H. Raaben, Proc. SPIE, 2001. **4318**: p. 26-30.
4. A.M. Malyarevich, I.A. Denisov, V.G. Savitsky, K.V. Yumashev, and A.A. Lipovskii, Appl. Opt., 2000. **39**(24): p. 4345.
5. A.M. Malyarevich, V.G. Savitsky, I.A. Denisov, P.V. Prokoshin, K.V. Yumashev, E. Raaben, A.A. Zhilin, and A.A. Lipovskii, Phys. Stat. Sol. B, 2001. **224**(1): p. 253-6.
6. A.M. Malyarevich, V.G. Savitski, P.V. Prokoshin, N.N. Posnov, K.V. Yumashev, E. Raaben, and A.A. Zhilin, J. Opt. Soc. Am. B, 2002. **19**(1): p. 28-32.
7. G. Tamulaitis, V. Gulbinas, G. Kodis, A. Dementjev, L. Valkunas, I. Motchalov, and H. Raaben, J. of Appl. Phys., 2000. **88**(1): p. 178.
8. K. Wundke, J.M. Auxier, A. Schulzgen, N. Peyghambarian, and N.F. Borrelli, Technical Digest. CLEO '99, 1999: p. 318.
9. K. Wundke, S. Potting, J. Auxier, A. Schulzgen, N. Peyghambarian, and N.F. Borrelli, Appl. Phys. Lett., 2000. **76**(1): p. 10-12.
10. P.N. Rao, *Cadmium sulfide and lead sulfide quantum dots in glass: processing, growth, and optical absorption*. 2001, Rensselaer Polytechnic Institute: Troy.
11. I. Kang and F.W. Wise, J. Opt. Soc. Am. B, 1997. **14**(7): p. 1632-46.
12. F.W. Wise, Acc. Chem. Res., 2000. **33**: p. 773-780.
13. F. Huang, *Nonlinear Optical Studies of the Properties of Nanoparticles*, in *Physics*. 2003, Rensselaer Polytechnic Institute: Troy. p. 122.
14. F. Huang, A. Filin, R. Doremus, P.N. Rao, and P.D. Persans, Mat. Res. Soc. Symp. Proc., 2003. **737**: p. 163.
15. M. Nirmal, D.J. Norris, M. Kuno, M.G. Bawendi, A.L. Efros, and M. Rosen, Phys. Rev. Lett., 1995. **75**: p. 3728.
16. H. Do, C. Chen, R. Krishnan, T.D. Krauss, J.M. Harbold, F.W. Wise, M.G. Thomas, and J. Silcox, Nano Lett., 2002. **2**: p. 1321.
17. K.L. Stokes, *Modulation spectroscopy of semiconductor nanocrystals*. 1995, Rensselaer Polytechnic Institute: Troy, NY.
18. K.L. Stokes and P.D. Persans, Phys. Rev. B, 1996. **54**: p. 1892.

Mat. Res. Soc. Symp. Proc. Vol. 789 © 2004 Materials Research Society N15.38

Low Temperature Gas Phase Synthesis of Germanium Nanowires

Sanjay Mathur*, Hao Shen and Ulf Werner
Institute of New Materials
D-66041 Saarbruecken, Germany

ABSTRACT

Single crystal Ge nanowires (NWs) were obtained in high yield by gas phase decomposition of germanium di-cyclopentadienylide ([Ge(C_5H_5)$_2$]), at 325 °C on iron substrates. High-resolution electron microscopy (SEM/TEM) showed Ge NWs to be uniform in terms of diameter (20 nm) and length (> 25 μm). The wire growth is selective and appears to be governed by a Ge-Fe alloy epilayer formed by the reaction between Ge clusters and iron substrate, during the initial stages of the CVD process. The supersaturation of Ge-Fe solid-solution with respect to Ge content induces the spontaneous formation of single crystal germanium nuclei that act as templates for the nanowire growth. X-ray and electron diffraction revealed the NWs to be single crystals of cubic germanium with a preferred growth direction [11-2]. The proposed base-growth model on Fe substrate is supported by TEM, EDX and XPS studies.

INTRODUCTION

Non-carbon one-dimensional (1D) nanostructures such as nanotubes (NTs) and nanowires (NWs) are receiving increasing attention due to the finite size effects and their possible applications in designing nanodevices [1-4]. Semiconductor materials such as Ge and Si display interesting electronic, optical and mechanical properties due to the quantization of electronic states in confined (nano)geometries [5]. Further, Germanium is an important semiconductor with high carrier mobility and a narrow band gap and owing to the larger excitonic Bohr radius in Ge (24.3 nm), the quantum size effects are expected to be more prominent when compared to silicon (4.9 nm). Although much effects have been made to obtain Ge nanowires, the successful growth of 1D Ge nanostructures is limited to a few reports [6-12]. Lieber and Morales have employed laser ablation (ca. 800 °C)for synthesizing Ge nanowires [8]. Yang et al. have used vapor transport reaction at rather high temperatures (900-1100 °C) to grow random and aligned Ge nanowires by *in situ* synthesis and disproportionation of GeI_2 on Au islands [10]. Hanrath et al. have used a solvothermal method to synthesize Ge nanowires by decomposing mixed Au-Ge precursors under high temperature and pressure conditions (300-400 °C, 100 atm) [11]. The chemical vapour deposition of Ge nanowires has recently been achieved by the reduction of tetrahydrogermane (GeH_4) through H_2 addition at 275 °C [12]. The control of GeH_4 decomposition through H_2 is inevitable in order to obtain the high quality Ge nanowires.

One-dimensional structures can be generated from the gas phase synthesis by well-known Vapor-Liquid-Solid (VLS) [13] growth process where a molten catalyst is used to promote axial instead of radial growth. Chemical vapour deposition (CVD) is an effective technique to produce single or core-shell type nanowires, because it allows an easy regulation of the precursor feedstock, in the vapor phase. The VLS type metal nanoparticle-catalyzed growth demands a homogeneous nucleation of growth-templates on the substrate. However, the nanowires obtained are 'contaminated' due to the presence of a eutectic tip. Therefore, we have used a *base-growth* method to grow Ge nanowires on an adherent bed that can be reused, and allows a high throughput for the synthesis of freestanding nanowires. Further, the precursor used was a single molecular species, $[Ge(Cp)_2]$ ($Cp = -C_5H_5$), that enables the growth of single crystalline Ge nanowires at low temperatures [14].

EXPERIMENTAL DETAILS

(i) Chemicals

The precursor, bis-cyclopentadienylide germanium ($Ge(C_5H_5)_2$), was synthesized by a salt elimination reaction between $GeCl_2{*}$Dioxan and two equivalents of $Na(C_5H_5)$, in a tetrahydrofuran and n-hexane mixture (eq. 1). The experimental manipulations were performed in a modified Schlenk type vacuum assembly, taking stringent precautions against atmospheric moisture. The reaction product was sublimed in vacuum (50 °C, 10^{-2} Torr) to obtain pure $Ge(C_5H_5)_2$.

$$GeCl_2{*}\text{Dioxan} \quad + \quad 2Na(C_5H_5) \quad \longrightarrow \quad Ge(C_5H_5)_2 \quad + \quad 2NaCl \qquad (1)$$

(ii) Nanowire Synthesis and Characterization

The CVD experiments were performed in a cold-wall tubular quartz reactor [14]. An inductive coupling of radio frequency was used to heat the substrate. $[Ge(C_5H_5)_2]$ was introduced in the reactor and heated to the desired temperature. Germanium nanowires were grown on polished Fe substrates at 325 °C at 10^{-2} Torr, producing a one-dimensional growth rate of ca. 100 nm/min. The substrate-bound nanowires were mechanically scrapped and sonicated in ethanol for TEM measurements performed on a JEOL 200CX transmission electron microscope. The nanowire morphology and elemental distribution (Energy Dispersive X-ray, EDX) across the length were recorded on the scanning electron microscope JSM-6400F (JEOL). The characterization of the CVD deposits and texture analysis was done by X-ray diffraction measurements performed on as-deposited films using a Siemens 5000 diffractometer. The micro-Raman spectra were recorded on Bruker RFS100 (1.06 μm Nd:YAG). The XPS analysis was performed on a Surface Science Instrument (M-Probe) operating with an Al Kα radiation and an instrumental resolution of ca. 0.8 eV.

DISCUSSION

The precursor $Ge(C_5H_5)_2$ is suited for NW synthesis due to the lability of the metal-ligand bond. Moreover, a single chemical source provides an easy monitoring of Ge feedstock in the gas phase. Recently, Ge nanowires have been synthesized by the reduction of tetrahydrogermane (GeH_4) through H_2 addition [12]. Although chemical vapor deposition is widely used in the carbon nanotube research [1,2], its application in the synthesis of non-carbon 1D-systems is less explored majorly due to the unavailability of precursors with suitable vapor pressure. A controlled gas phase chemistry of the precursor is a pre-requisite for the chemical synthesis of nanomaterials [15].

The decomposition pathway outlined in eq. 2 is based on an in-situ mass spectral analysis that revealed metal ion and $C_nH_m^+$ fragments [14].

$$[Ge(C_5H_5)_2] \longrightarrow Ge(0) + C_nH_m^+ + H_2 \qquad (2)$$

The nanowire synthesis through CVD relies on metal nanoparticles that act as growth templates. To circumvent substrate preparation and generation of catalyst nanoparticles, we investigated Ge NWs formation on steel substrates because Ge-Fe alloys exist in the binary phase diagram although at higher temperatures (400-1170 °C) [16] than that used for the CVD process. The nanowire growth occurs on Fe at a substrate temperature of 325 °C as shown

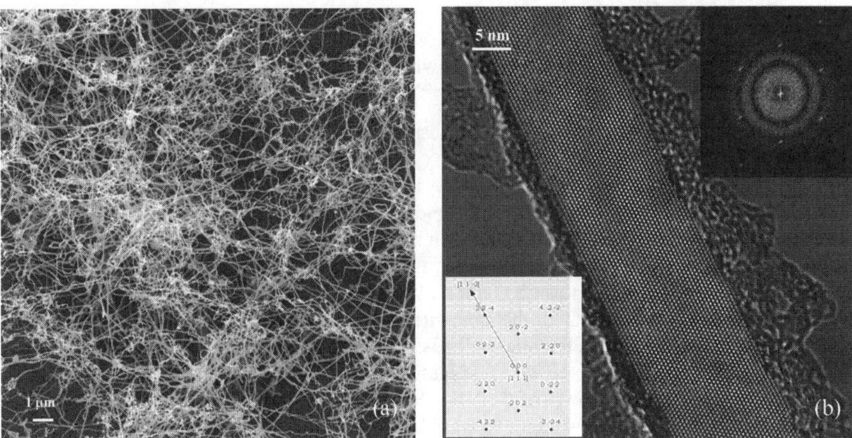

Figure 1. (a) Typical SEM images of Ge NWs on steel substrate and (b) HR-TEM image of a Ge nanowire. Top inset shows the SAED pattern recorded along <111> zone axis.

in figure 1a. The average NW diameter was found to be ca. 20 nm while lengths ranged in tens of micrometers, which illustrates that growth along the wire axis is several magnitudes faster than that in the radial direction. The formation of Ge-Fe eutectic at 325 °C in our case is possibly an effect of low-pressure conditions or melting point lowering in nanostructured systems [17]. The transmission electron microscopy of the nanowires does not show defects such as stacking faults and micro-twins that are commonly observed along the growth axis. The selective area electron diffraction of individual nanowires exhibits the preferred growth direction of nanowires to be [11-2] (Inset, fig. 1b).

Upon increasing the precursor flux, formation of amorphous overlayers (2 – 20 nm) on crystalline cores (16 nm) is observed that is a promising possibility to grow core-shell nanowire structures [18]. The TEM analysis (fig. 2a) shows single crystalline Ge core conformally covered by a thick amorphous Ge layer. The appearance of lower energy surface, (111), parallel to the growth axis will reduce the surface energy in a 1D system. This is supported by the X-ray diffraction analysis (fig. 2b) of CVD deposits that exhibit diffraction peaks due to cubic diamond structure (fig. 2). The peak broading observed is caused by the presence of amorphous Ge.

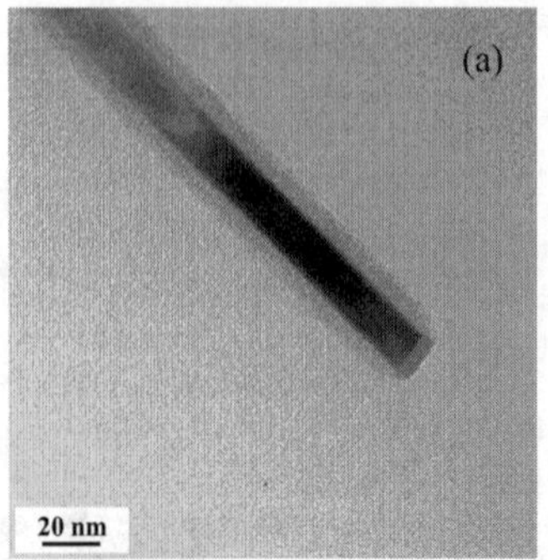

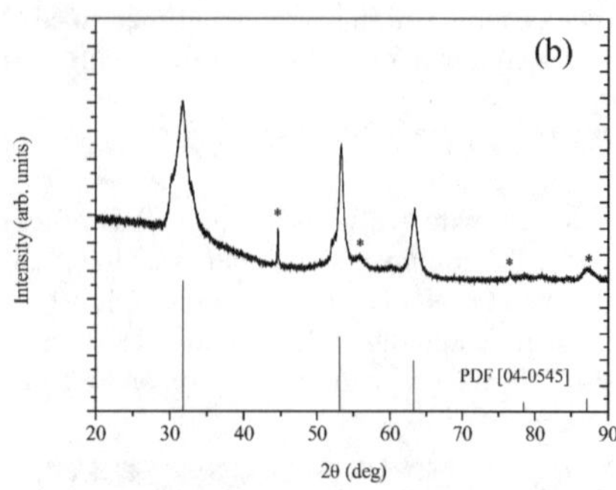

Figure 2. (a) TEM image of a c-Ge/a-Ge nanowire with core-shell configuration and (b) the corresponding XRD pattern.

The energy dispersive X-ray analysis performed on several samples revealed the single-crystalline germanium deposits are devoid of oxygen. Since the nanowires show no catalyst particle at their tip in both SEM and TEM analysis of the CVD deposits, a base growth mechanism seems to be responsible for the one-dimensional growth. Based on the experimental observation, it can be envisaged that interaction between the Fe surface and the incoming Ge vapor forms a Ge-Fe liquid layer thereby creating solid-liquid and liquid-vapor interfaces. Once the Ge-Fe epilayer is saturated toward Ge, it will undergo phase separation to nucleate Ge nanocrystals that act as catalysts for the nanowire growth. During the further growth, the in-coming germanium clusters preferentially deposit on the protruding nuclei embedded in the Ge-Fe solid-solution and are saturated with respect to Ge intake.

When compared to the growth of Ge NWs on Si achieved by vapour-liquid-solid (VLS) mechanism, the base growth is straight forward because it needs no substrate preparation and the size of the Fe-Ge nuclei in the epilayer can be easily tuned by the substrate temperature. Moreover, the base-growth method allows a higher throughput of nanowires without a

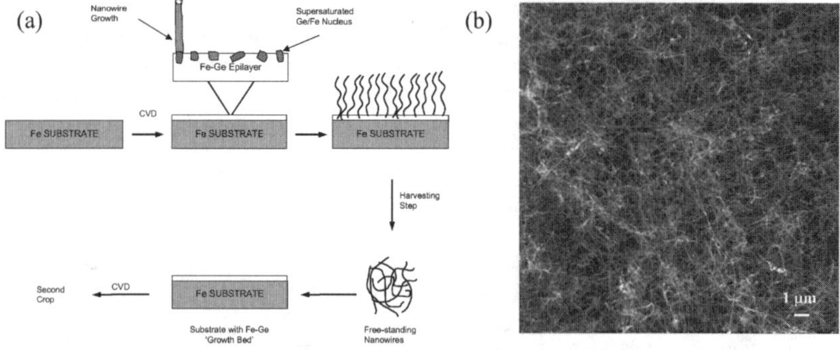

Figure 3. (a) Proposed *base-growth* mechanism for the synthesis of germanium nanowires and (b) SEM image of a second crop of Ge NWs on the steel substrate.

contaminating 'head' that is typical for the VLS growth. An added advantage of this method is that the Fe-Ge bed can be re-used for harvesting second and more crops of nanowires on the same substrate (fig. 3a). Fig. 3b shows the SEM image of a second crop of nanowires grown on the same iron substrate.

A number of nanowire samples were investigated to confirm the presence or absence of the alloy droplet at the tip of the nanowires, however, the metal particle attached nanowire typical for Au-Ge system could not be identified in high resolution TEM analyses. The X-ray photoelectron spectra (XPS) (fig. 4a) of the Ge NWs were recorded to determine the surface composition of the CVD deposits. No evidence for the presence of Fe could be detected in NWs grown on iron substrates despite long acquisition times (fig. 4a).

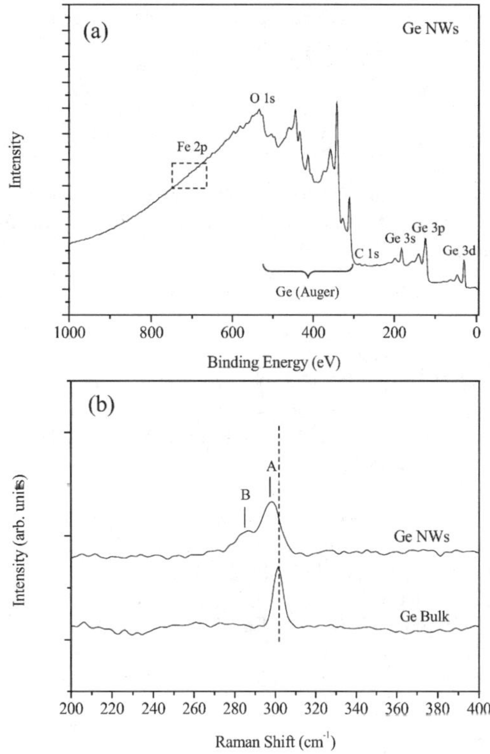

Figure 4. (a) XPS overview spectra of (a) as-deposited Ge whiskers and Ge NWs, and (b) Ge NWs after Ar^+ sputtering and (b) Raman spectra of Ge NWs and Ge standard.

Figure 4b shows the Raman spectra of bulk and NWs specimens of germanium. When compared to the characteristic peak of bulk Ge (300.5 cm^{-1}) with a full width at half-maximum (FWHM) of 2.7 cm^{-1}, the Raman spectrum of Ge NWs shows two spectral features (A and B), which are shifted to lower frequency. The strong and sharp peak (A) can be interpreted as the contribution of the crystalline Ge core whereas the broad shoulder (B), in view of the published data [19], can be assigned to a-Ge or small Ge nanocrystals.

In summary, single crystalline Ge NWs are synthesized by low pressure chemical vapour deposition of a molecular precursor, $[Ge(C_5H_5)_2]$. The low decomposition temperature can be attributed to the weak metal-ligand interaction in the C_5H_5-Ge-C_5H_5 unit. The clean stripping of organic ligands generates Ge(0) atoms and clusters in the gas phase that are responsible for the growth of Ge NWs. The growth occurs through a base-growth model based on a Ge-Fe epilayer formed by the interaction of energetic Ge species with Fe substrate under reduced pressure and elevated surface temperature. The formation of a quasi-liquid Fe-Ge epilayer and the landing of Ge ions and clusters on the metallic surface (due to higher charge transfer rate) suggest the overall growth process to be a combination of VLS and charged cluster growth models. Our efforts to investigate the electrical and optical properties of individual nanowires are currently underway.

ACKNOWLEDGEMENT

Authors are thankul to the German Science Foundation (DFG) for the financial assistance. S.M. is thankful to Prof. M. Veith and Prof. H. Schmidt for providing the necessary infrastructural facilities.

REFERENCES

1. S. Iijima, *Nature* **354**, 56 (1991).
2. P. M. Ajayan, *Chem. Rev.* **99**, 1787 (1999).
3. J. T. Hu, T. W. Odom, and C. M. Lieber, *Acc. Chem. Res.* **32**, 435 (1999).
4. Y. N. Xia, P. D. Yang, Y. G. Sun, Y. Y. Wu, B. Mayers, B. Gates, Y. D. Yin, F. Kim, and H. Q. Yan, *Adv. Mater.* **15**, 353 (2003).
5. M. T. Björk, B. J. Ohlsson, T. Sass, A. I. Persson, C. Thelander, M. H. Magnusson, K. Deppert, L. R. Wallenberg, and L. Samuelson, *Nano Lett.* **2**, 87 (2003).
6. J. R. Heath, and F. K. Legoues, *Chem. Phys. Lett.* **208**, 263 (1993).
7. H. Omi, and T. Ogino, *Appl. Phys. Lett.*, **71**, 2163 (1997).
8. A. Morales, and C. M. Lieber, *Science* **279**, 208 (1998).
9. Y. F. Zhang, Y. H. Tang, N. Wang, C. S. Lee, I. Bello, and S. T. Lee, *Phys. Rev. B* **61**, 4518 (2000).
10. Y. Wu, and P. D. Yang, *Chem. Mater.* **12**, 605, (2000).
11. T. Hanrath, and B. A. Korgel, *J. Am. Chem. Soc.*, **124**, 1424 (2002).
12. D. W. Wang, and H. J. Dai, *Angew. Chem. Int. Ed.* **41**, 4783 (2002).
13. R. S. Wagner and W. C. Ellis, *Appl. Phys. Lett.* **4**, 89 (1964).
14. S. Mathur, H. Shen, V. Sivakov, and U. Werner, *Chem. Mater.* (submitted).
15. S. Mathur, M. Veith, H. Shen, and S. Hüfner, *Mater. Sci. Forum* **386-388**, 341 (2002).
16. O. Bosholm, H. Oppermann, and S. Däbritz, *Zeitschrift für Naturforschung* 329 (2001).
17. P. Buffat, and J. P. Borel, *Phys. Rev. A* **13**, 2287 (1976).
18. L. J. Lauhon, M. S. Gudiksen, D. Wang, and C. M. Lieber, *Nature* **420**, 57 (2002).
19. V. A. Volodin, E. B. Gorokhov, M. D. Efremov, D. V. Marin, and D. A. Orekhov, *JETP Lett.* **77**, 411 (2003).

Mat. Res. Soc. Symp. Proc. Vol. 789 © 2004 Materials Research Society N15.54

Vapor-Phase Synthesis and Surface Functionalization of ZnSe Nanoparticles in a Counterflow Jet Reactor

Christos Sarigiannidis[1], Athos Petrou[2] and T. J. Mountziaris*[1, 3]
Departments of [1]Chemical and Biological Engineering, and [2]Physics, University at Buffalo
The State University of New York, Buffalo, N.Y 14260, U.S.A.
[3]The National Science Foundation, 4201 Wilson Boulevard, Arlington, VA 22230, U.S.A.

ABSTRACT

Compound semiconductor nanocrystals (quantum dots) exhibit unique size-dependent optoelectronic properties making them attractive for a variety of applications, including ultrasensitive biological detection, high-density information storage, solar energy conversion, and photocatalysis. There is presently a great need for developing scalable techniques that allow efficient synthesis, size control, and functionalization of quantum dots, without a loss of the desirable optical properties. We report experimental results on the properties and surface modification of ZnSe nanoparticles grown by a continuous vapor-phase technique utilizing an axisymmetric counterflow jet reactor. Luminescent ZnSe nanocrystals were obtained at room temperature by reacting vapors of dimethylzinc:triethylamine adduct with hydrogen selenide, diluted in a hydrogen carrier gas. The two reactants were supplied from opposite inlets of the counterflow jet configuration and initiated particle nucleation in a region near the stagnation point of the laminar flow field. Surface modification of nanoparticles by adsorption of 1-pentanethiol was used to control the rate of particle coalescence. The counterflow jet technique can be scaled up for commercial production and is compatible with other vapor-phase processing techniques used in the microelectronics industry.

INTRODUCTION

II-VI semiconductor nanocrystals, e.g., CdSe, CdS, ZnSe and ZnS, are exciting materials exhibiting size-dependent luminescence when their size becomes smaller than the mean separation of an optically excited electron-hole pair (Bohr radius). These materials, also called "quantum dots", exhibit quantum confinement effects that shift the onset of absorption and emission maxima towards higher energy with decreasing particle size. The bright luminescence and small size of quantum dots makes them ideal luminescent tags for studying biomolecular interactions and for developing multiplexed optical biosensors [1, 2]. New functional materials are also being envisioned, having nanocrystals, instead of atoms, as the building blocks of a lattice. Nanocrystals can be also used for developing novel inorganic/organic nanocomposites.

There is currently a great need for developing flexible, scalable, and cost-effective techniques for controlled synthesis and surface functionalization of semiconductor nanocrystals. The desirable properties of the synthesized nanocrystals, that any preparation technique must address, include a defect-free crystalline structure, narrow size distribution, chemical purity, and a surface structure that allows functionalization by chemisorption of organic molecules, without significant loss in the luminescence quantum yield [3].

*Corresponding author. E-mail: tjm@eng.buffalo.edu or tmountzi@nsf.gov

Over the past decade, a considerable effort has been devoted to the synthesis of compound semiconductor nanocrystals using a variety of techniques. Several solution-based chemical methods have been reported and can be classified into three main categories [4]:

(a) Molecular precursor methods in which the precursors are decomposed in a coordinating solvent at relatively high temperatures (~300°C).

(b) Controlled precipitation in solution.

(c) Synthesis in a structured medium using various templates, such as: zeolites, micelles, microemulsions, gels, and glasses.

These techniques have been used for producing luminescent nanocrystals and core-shell structures of various II-VI semiconductors, with the first approach being the most successful.

On the other hand, most of the reported vapor-phase techniques have either failed to produce optically-active nanoparticles or do not provide the necessary control over particle size, with a few exceptions [5]. However, these techniques are compatible with the current infrastructure of the microelectronics industry, provide better control on product purity and are easier to scale up.

This paper discusses the growth of ZnSe nanocrystals. ZnSe has a "wide" band gap of 2.7 eV at room temperature that makes it useful for optoelectronic applications in the UV-blue-green part of the spectrum. High-quality ZnSe quantum dots have been synthesized from organometallic precursors in a hot alkylamine coordination solvent [6]. Other reported ZnSe preparation methods include arrested precipitation [7] and precipitation from sol-gel solutions [8].

The objective of our work is the development of a vapor-phase technique for controlled synthesis of semiconductor nanocrystals using a counterflow jet reactor [9]. To grow ZnSe nanocrystals in such a reactor, vapors of dimethylzinc:triethylamine adduct and hydrogen selenide gas, both diluted in hydrogen, are introduced from opposite sides of a counterflow jet configuration under laminar flow conditions (Figure 1). Particle nucleation takes place when the reactants mix in a region near the stagnation point. The initial ZnSe nuclei grow by surface reactions with unreacted precursors and by cluster-cluster coalescence. The clusters continue to grow by coalescence and they form primary nanocrystals through an annealing mechanism. The energy release during coalescence is responsible for the annealing. The primary nanocrystals subsequently grow by coalescence into larger polycrystalline nanoparticles [9]. Even larger nanoparticle aggregates eventually form as the characteristic time for sintering becomes greater than the characteristic time for particle-particle collisions [10].

The ultimate objective of our investigation is the elucidation of the links between processing conditions and particle properties, such as size distribution, crystallinity, and morphology. We are also interested in controlling the particle growth rate by introducing coalescence inhibitors that passivate and functionalize their surface. Surface passivation and functionalization (without significant loss of luminescence) are key factors for successful utilization of the nanocrystals in practical applications.

EXPERIMENTAL DETAILS

A schematic of the counterflow jet reactor used in this study is shown in Figure 1. The reactor is an 8-inch O.D. stainless steel vacuum chamber, equipped with two vertical 3/4-inch O.D. 316 seamless stainless steel tubes that are aligned with the axis of the cylindrical vacuum chamber. A 1/4-inch O.D. vertical tube has been installed inside the lower 3/4-inch tube. Two horizontal 7.5-inch diameter aluminum plates that are separated by a distance of 2/3-inch form the upper and lower walls of the counterflow region. The reactor is designed to produce an

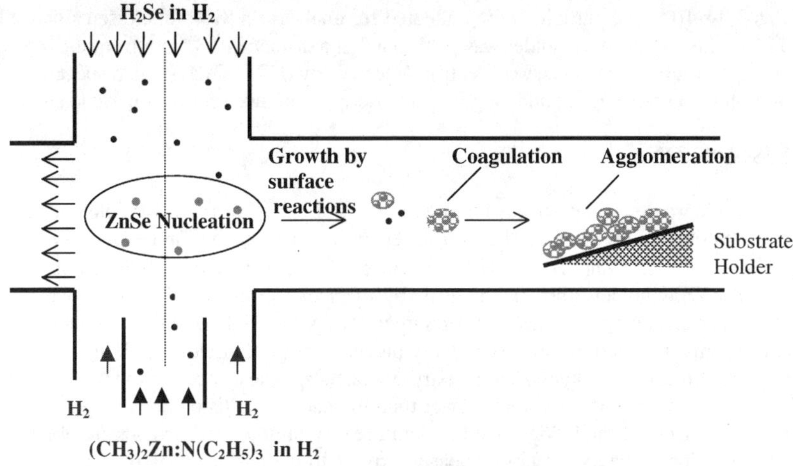

Figure 1. Counterflow jet reactor and mechanism of ZnSe nanoparticle synthesis.

axisymmetric flow pattern. The downstream ends of the inlet tubes are fitted with a fine stainless steel mesh to generate plug flow conditions at the two reactor inlets. The gases from the two inlets form a laminar stagnation-flow pattern at the reactor center (verified by flow visualization using TiO_2 smoke particle tracing) and subsequently flow radially towards six exhaust ports that are evenly distributed around the perimeter of the cylindrical vessel. A gas-handling manifold typical for Metalorganic Vapor Phase Epitaxy (MOVPE) reactors was used to deliver the Zn and Se precursors, both diluted in hydrogen [11]. The entire system was placed in a vented enclosure because the precursors are toxic materials. A hydride detector and a hydrogen detector were used for safety reasons.

Hydrogen selenide, H_2Se, 5% mixture in H_2 was introduced from the upper tube. The lower jet contained an inner tube (1/4-inch O.D.) and an outer tube (3/4- inch O.D.). The inner tube was used to deliver vapors of dimethylzinc:triethylamine (DMZ-TEA) adduct, i.e. $(CH_3)_2Zn$: $N(C_2H_5)_3$, both diluted in hydrogen. The annular region was used to deliver additional hydrogen for dilution of the Zn precursor. It was also used for the delivery of vapors of 1-pentanethiol for controlling the rate of particle coalescence. The specific Zn and Se precursors were chosen because of their reactivity, even at room temperature. The experimental observations indicate that there are multiple steps involved in the formation of nanocrystals, polycrystalline nanoparticles, and eventually nanoparticle aggregates (shown schematically in Figure 1).

The reactor was operated at pressures between 74 and 360 Torr and room temperature. The flow rates of each precursor were controlled using electronic mass flow controllers (Edwards). The typical flow rate for the H_2Se/H_2 mixture entering the reactor through the upper jet was 50 sccm. The corresponding flow rate of H_2 through the Zn precursor bubbler was 20 sccm and was delivered to the reactor via the inner tube of the lower jet. The flow rate of hydrogen through the annulus of the lower jet was 30 sccm. The same flow rate was used when this stream was passed through a bubbler containing 1-pentanethiol before flowing through the annulus. Under typical conditions, the VI/II ratio was 3.2 to 1 and the mole fraction of the Zn precursor in the lower jet was 3.9 x 10^{-2}. The particles were collected for analysis on Si wafers, fused silica substrates, and TEM grids. A substrate holder was positioned at a distance of 3.5 cm from the reactor centerline. High-Resolution Transmission Electron Microscopy (HR-TEM), Raman spectroscopy, photoluminescence (PL), and Auger spectroscopy were used for nanoparticle characterization.

DISCUSSION

The particle formation was initiated in the gas-phase by an irreversible nucleation reaction between the two precursors. The overall reaction is spontaneous and exothermic:

$$Zn(CH_3)_2 + H_2Se \rightarrow ZnSe_{(s)} + 2\,CH_4\ ; \quad \Delta H = -395\ kJ/mol$$

The ZnSe nuclei subsequently grow by surface reactions involving unreacted precursors and by cluster-cluster coalescence to form larger nanoparticles. Due to the fast nucleation rate, surface growth reactions are a secondary mechanism for particle growth in this case. The most important mechanism appears to be particle-particle coalescence. As the particle size increases, the rate of coalescence becomes slower than the particle collision rate and particle aggregates are produced. Despite the low growth temperature, crystalline nanoparticles are obtained, most probably due to the local energy release during particle-particle coalescence [12]. The reactions between the two precursors (and the surface reactions between the precursors and the particles) are also exothermic, thus supplying a secondary energy source to the growing particles. The local energy release is apparently high enough to heat the particles and increase the mobility of the Zn and Se atoms, thus producing crystalline structures through an annealing mechanism.

TEM images of nanoparticles collected at three different operating pressures are shown in Figure 2. At 74 Torr (Figure 2a), dendritic features are observed consisting of amorphous and crystalline nanoparticles with sizes smaller than 100nm. The amorphous particles obtained at low pressures are probably the result of a slower nucleation rate, due infrequent collisions between unreacted precursors. The slow nucleation rate results in higher availability of precursors for surface growth reactions. This enables the surface growth mechanism to become significant at lower pressures and limits the coalescence growth mechanism. The latter can yield mostly crystalline particles, due to the significant energy release associated with it. The smaller energy release during the reactions of precursors on the surface of the particles is usually not sufficient for annealing them. As a result, more amorphous particles form at low pressures. At two higher pressures of 120 and 360 Torr (Fig. 2b and 2c) dendritic structures are not obtained. Spherical aggregates are observed at 120 Torr and a finer deposit containing nanocrystals at 360 Torr.

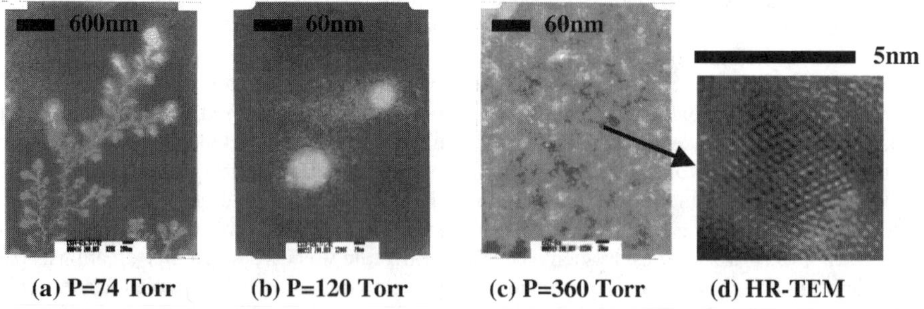

(a) P=74 Torr **(b) P=120 Torr** **(c) P=360 Torr** **(d) HR-TEM**

Figure 2. TEM micrographs of ZnSe nanoparticles synthesized at different operating pressures.

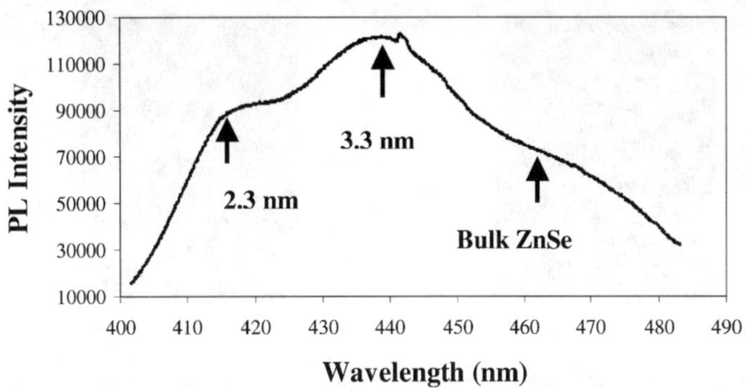

Figure 3. Photoluminescence spectrum of ZnSe nanoparticles (reactor pressure: 120 Torr).

As the reactor pressure increases the nucleation rate increases too, resulting in higher particle number density and smaller average particle size. The critical size for the onset of confinement effects in ZnSe is 9nm. At 120 Torr (Figure 2b) a typical deposit collected on the TEM grids consists of primary nanocrystals, polycrystalline nanoparticles formed by nanocrystal coagulation, and larger spherical aggregates of these nanoparticles. At 360 Torr (Figure 2c) the deposit collected is denser and consists of primary nanocrystals with sizes smaller than 10 nm and larger nanoparticles. The crystalline structure of a typical primary nanocrystal, embedded in the deposit collected at 360 Torr, is shown in the HR-TEM micrograph of Figure 2d.

The room-temperature photoluminescence spectrum of a deposit collected on fused silica at 120 Torr (Figure 3) exhibits a bimodal nanocrystal size distribution, consistent with a coalescence growth mechanism that progressively produces larger nanocrystals. The average size of the two nanocrystal populations was estimated from the PL data to be 2.3nm and 3.3nm using the procedure described in [13]. The nanocrystals exhibit size-dependent luminescence with the smaller particles emitting at shorter wavelengths. Nanocrystals with sizes larger than the critical one emit at the wavelength of bulk ZnSe, which is 460nm at room temperature. Auger electron spectroscopy was performed to determine the elemental composition profile of the deposits. The Zn to Se ratio was found to be 1:1 and remained approximately constant throughout the profile.

To control the rate of particle coalescence, vapors of 1-pentanethiol (diluted in hydrogen) were introduced from the annular region of the lower jet (Figure 1). The molecules of 1-pentanethiol can adsorb on the surface of ZnSe to form a self-assembled monolayer, thus producing a barrier during particle coalescence. The effects of using the coagulation inhibitor on the morphology of the deposits obtained at 120 Torr are shown in Figure 4. The HR-TEM micrographs indicate a significant increase in the concentration of smaller particles during growth in the presence of the coagulation inhibitor. Experiments are underway to assess the effects of using 1-pentanethiol and other thiols on the primary nanocrystal size and luminescence characteristics. The use of bifunctional thiols as surface functionalizing units is also planned, in an effort to simultaneously control the particle size distribution and functionalize the surface of the particles, thus enabling subsequent dispersion of nanocrystals in water or organic solvents. This is a necessary step for linking the nanocrystals with biomolecules or macromolecules.

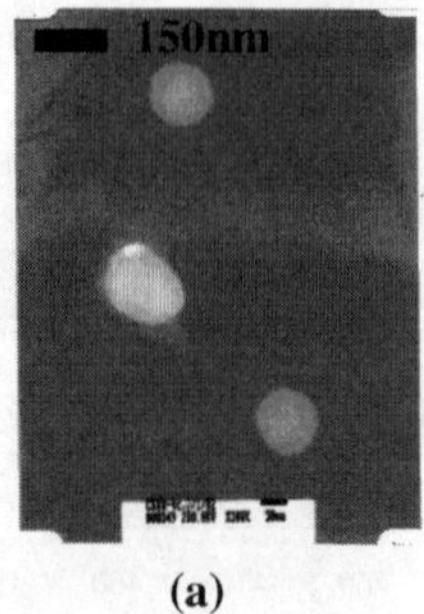

(a)

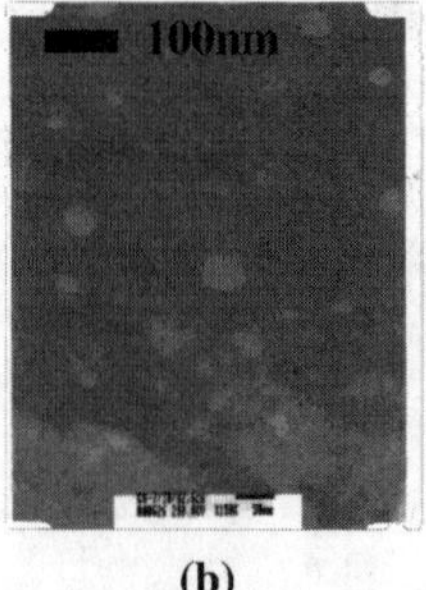

(b)

Figure 4. TEM pictures of ZnSe nanoparticle deposits obtained at 120 Torr: (a) Without using a coagulation inhibitor. (b) In the presence of a coagulation inhibitor (1-pentanethiol).

CONCLUSIONS

In conclusion, a vapor-phase technique for synthesis of luminescent II-VI nanocrystals with sizes below the confinement threshold has been developed. The technique utilizes a counterflow jet reactor operating at room temperature [9]. ZnSe nanocrystal synthesis has been performed under various operating conditions to identify the most attractive ones for maximizing the yield of crystalline particles that exhibit size-dependent luminescence. The use of 1-pentanethiol as a coagulation inhibitor resulted in the formation of smaller particle aggregates. Experiments are underway to reveal the fundamental links between processing conditions and particle properties.

ACKNOWLEDGMENTS

The authors wish to acknowledge contributions from M. Koutsona, G. Kioseoglou, G. Itskos, G. Karanikolos, J. Wang, B. Kostova, T. C. Lee, P. Bush, and L. Guo, and financial support from the National Science Foundation and from SUNY-Buffalo (IRCAF Program).

REFERENCES

1. A. P. Alivisatos, *Science*, **271**, 933 (1996).
2. C. B. Murray, C. R. Kagan, and M. G. Bawendi, *Annu. Rev. Mater. Sci.*, **30**, 545 (2000).
3. A. Eychmuller, *J. Phys. Chem. B*, **104**, 6514 (2000).
4. T. Trindade, P. O'Brien, and N. L. Pickett, *Chem. Mater.*, **13** (11), 3843 (2001).
5. M. T. Swihart, *Curr. Opin. Colloid In.*, **8**, 127 (2003).
6. M. A. Hines and P. Guyot-Sionnest, *J. Phys. Chem. B*, **102** (19), 3655 (1998).
7. N. Chestnoy, R. Hull, L. E. Brus, *J. Phys. Chem.*, **85** (4), 2237 (1986).
8. G. Li and M. Nogami, *J. Appl. Phys.*, **75** (8), 4276 (1994).
9. D. Sarigiannis, J. D. Peck, G. Kioseoglou, A. Petrou, and T. J. Mountziaris, *App. Phys. Lett.*, **80** (21), 4024 (2002).
10. M. R. Zachariah and M. J. Carrier, *J. Aerosol Sci.*, **30** (9), 1139 (1999).
11. J. D. Peck, T. J. Mountziaris, S. Stoltz, A. Petrou, and P. Mattocks, *J. Cryst. Growth*, **170** (1-4), 523 (1997).
12. K. E. J. Lehtinen and M. R. Zachariah, *J. Aerosol Sci.*, **33**, 357 (2002).
13. J. Warnock and D. D. Awschalom, *Phys. Rev. B*, **32**, 5529 (1985).

Mat. Res. Soc. Symp. Proc. Vol. 789 © 2004 Materials Research Society N15.55

Synthesis and Size Control of Luminescent II-VI Semiconductor Nanocrystals by a Novel Microemulsion-Gas Contacting Technique

Georgios N. Karanikolos[1], Paschalis Alexandridis[1], Athos Petrou[2] and T.J. Mountziaris*[1]
Departments of [1]Chemical and Biological Engineering, and [2]Physics, University at Buffalo, The State University of New York, Buffalo, New York 14260, U.S.A.

ABSTRACT

A scalable, room-temperature technique for controlled synthesis of luminescent II-VI nanocrystals has been developed by using the dispersed phase of stable, well-characterized microemulsions as templates for nanoparticle synthesis. The microemulsions were formed by self-assembly of poly (ethylene oxide)-poly (propylene oxide)-poly (ethylene oxide) (PEO-PPO-PEO) amphiphilic block copolymer and heptane in formamide. By adjusting the surfactant to dispersed phase ratio, stable microemulsions were obtained with droplet diameter of ~40nm. These microemulsions avoid problems or rapid droplet-droplet coalescence that hamper reverse micelles and lead to polydisperse particle populations. Luminescent ZnSe quantum dots were synthesized by reacting diethylzinc (dissolved in the heptane dispersed phase) with hydrogen selenide gas (diluted in hydrogen). The gas was bubbled through the microemulsion, dissolved in the formamide, and diffused to the nanodroplet interfaces to react with diethylzinc. The experiments indicate that a single nanocrystal is formed in each nanodroplet by coalescence of clusters (nuclei) and smaller crystals. The energy released during coalescence is sufficient to anneal the clusters into high-quality crystals. The process allows precise control of nanocrystal size by adjusting the initial concentration of diethylzinc in heptane. The "as grown" nanocrystals exhibit size-dependent luminescence, narrow and symmetric emission, good monodispersity (confirmed by TEM analysis), and excellent photochemical stability. The technique is currently being extended to the synthesis of CdSe nanocrystals with promising preliminary results.

INTRODUCTION

Compound semiconductor nanocrystals (quantum dots) with sizes between 2-10 nm, or having ~10-50 atoms along the nanocrystal diameter, are exciting materials exhibiting size-dependent properties due to quantum confinement effects [1-3]. Nanocrystals of CdSe, CdS, ZnS or ZnSe have been shown to exhibit size-dependent luminescence and absorption, broad excitation by almost all wavelengths smaller than the emission wavelength, high brightness, narrow and symmetric emission, photochemical stability, negligible photobleaching, and long fluorescence lifetime. In addition to playing an important role in fundamental studies on solid-state physics, materials science, and thermodynamics [4], quantum dots can be used in a wide range of applications, e.g. in photovoltaic devices [5], and as fluorescent biological labels [6]. ZnSe is a wide band gap material that can be used for developing devices operating in the UV-blue-green part of the spectrum. Highly luminescent quantum dots of ZnSe have been synthesized from organometallic precursors in a hot alkylamine coordination solvent [7]. Diluted magnetic semiconductor nanocrystals, e.g. (Zn,Mn)Se, have been identified as materials for "spintronic" devices, i.e. devices that manipulate the spin of individual electrons [8].

* Corresponding author. E-mail: tjm@eng.buffalo.edu

The most commonly used synthesis routes for compound semiconductor nanocrystals are: (1) Organometallic synthesis in a trioctylphosphine (TOP)/trioctylphosphine oxide (TOPO) and/or hexadecylamine (HDA) coordination solvent at ~ 300 °C, and (2) Synthesis in reverse micelles, which are dispersions of an aqueous phase in a non-polar solvent, stabilized by amphiphilic molecules. To obtain monodisperse nanocrystal populations by the first technique, instantaneous injection of the reactants, uniform nucleation over the entire mass of the coordination solvent, and perfect mixing are required. These conditions are difficult to achieve in practice, but selective precipitation techniques can narrow down the particle size distribution [2]. Reverse micelles on the other hand could provide a template for precise control of particle size, under ideal conditions. In practice, the fast dynamics of droplet-droplet coalescence in water-in-oil microemulsions lead to the formation of clusters and polydisperse particle populations [9].

The technique presented here utilizes stable, well-characterized heptane-in-formamide microemulsions that avoid the rapid droplet-droplet coalescence problem hampering the reverse micelles. A group II alkyl is dissolved in the heptane of the dispersed phase and a group VI hydride gas is bubbled through the microemulsion to yield II-VI particles inside the nanodroplets. Coalescence of the clusters formed inside the droplets yields a single nanocrystal per nanodroplet, whose crystalline structure can be attributed to the energy released during coalescence. Templating of the nanocrystal synthesis by the microemulsions allows precise control of particle size by adjusting the initial concentration of group II alkyl in heptane.

EXPERIMENTAL DETAILS

Diethylzinc ($(C_2H_5)_2Zn$, 1M solution in n-heptane), formamide (CH_3NO, 99.5+ %), and n-heptane (n-C_7H_{16}, 99%) were purchased from Aldrich. Electronic-grade hydrogen selenide gas (H_2Se, 5% mixture with H_2) was purchased from Solkatronic Chemicals. Pluronic P105, poly (ethylene oxide)-poly (propylene oxide)-poly (ethylene oxide) or PEO-PPO-PEO block copolymer ($EO_{37}PO_{58}EO_{37}$, with MW of 6,500 and 50% PEO content), was provided by BASF Corporation. All chemicals were used "as received". Care was taken to avoid exposure of the hygroscopic formamide and PEO-PPO-PEO to atmospheric moisture. Standard airless techniques were used to avoid exposure of diethylzinc to oxygen and moisture.

The diethylzinc-containing microemulsions were formed as follows: (a) 3.33 gr PEO-PPO-PEO was added to 20 ml formamide and the mixture was stirred for 1.5 hrs. (b) 0.5 ml of diethylzinc-heptane solution was added to PEO-PPO-PEO/formamide solution under nitrogen. (c) The final mixture was sonicated for 1.5 hrs. The resulting liquid was transparent and homogeneous, an indication that a microemulsion was formed [10]. It was subsequently transferred to the reactor (located in a vented enclosure) under nitrogen. A flow of 20 sccm 5% hydrogen selenide in hydrogen was bubbled through the microemulsion for 15 min, sufficient time for converting all the diethylzinc to ZnSe. The reactor was subsequently purged with nitrogen for 1 hr to remove all traces of hydrogen selenide. The gases exiting the reactor were passed through a cracking furnace and a bed of adsorbents before released into a fume hood. A hydride detector was used to ensure personnel safety.

Photoluminescence (PL) emission spectra were obtained by loading samples from the reactor into quartz cuvettes and analyzing them using a 0.5 m single-stage spectrometer (CVI Laser Corp.), equipped with a thermoelectrically-cooled multi-channel CCD detector (Camera AD-205 working in the wavelength range of 200 – 1100 nm). A 325 nm line of a 20 mW He-Cd UV laser (Melles Giot) was used to excite the nanocrystals. Transmission spectra from samples loaded in quartz cuvettes were obtained by using a 150 W Xenon lamp (UV) and a 0.35 m scanning monochromator (McPherson) with PMT phase sensitive detection. The samples used for Transmission Electron Microscopy (TEM) were prepared by placing a drop of the processed microemulsion on a 400-mesh carbon-coated copper grid (Ernest F. Fullam, Inc.) and leaving it under vacuum for 48 hours to evaporate all solvents. The instrument used was a JEOL JEM 2010 high-resolution TEM, operated at 200 kV, with a point-to-point resolution of 0.193 nm. Dynamic light scattering was performed to estimate the microemulsion droplet size, prior to each synthesis experiment, using a model 95 argon ion laser (Lexel Corp.) with a BI-200SM goniometer detector (Brookhaven Instruments Corp.) operating at an angle of 90°. X-ray diffraction was performed using a Siemens D500 XR diffractometer.

DISCUSSION

The mechanism of ZnSe nanocrystal growth by using the microemulsion-gas contacting technique is shown schematically in Fig. 1. The microemulsions were formed by self-assembly using a solution of diethylzinc in n-heptane as the dispersed oil phase, formamide as the polar continuous phase, and an amphiphilic block copolymer, PEO-PPO-PEO, as the surfactant. PEO is the formamide-soluble block and PPO the heptane-soluble block. PEO-PPO block copolymers represent an exciting class of amphiphilic molecules with high versatility in terms of self-assembly and corresponding practical applications [11]. Imhof et al. [12] tested a variety of different combinations of amphiphiles with non-aqueous polar solvents, and reported that PEO-PPO-PEO block copolymers form very stable emulsions in formamide. Formamide was used instead of water, because it is sufficiently polar to be immiscible in heptane and does not react with diethylzinc. Hydrogen selenide gas diluted in hydrogen was bubbled through the microemulsion at room temperature and atmospheric pressure, diffused through the interface of the nanodroplets, and reacted with diethylzinc to yield ZnSe and ethane gas. A single ZnSe nanocrystal was formed in each droplet, thus allowing precise particle size control by manipulation of the initial diethylzinc concentration in heptane.

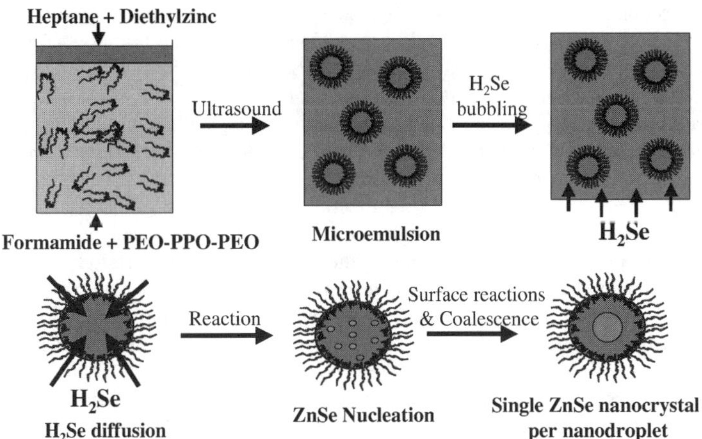

Figure 1. Growth of ZnSe nanocrystals by microemulsion-gas contacting.

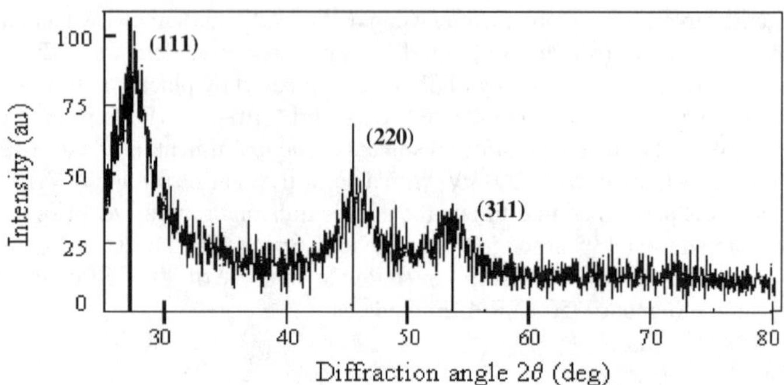

Figure 2. X-ray diffraction pattern from ZnSe particle aggregates obtained by reacting diethylzinc (diluted in heptane) with hydrogen selenide gas (diluted in hydrogen) at room temperature. The vertical lines at 27.2°, 45.2°, and 53.6° correspond to the expected diffraction angles by the (111), (220), and (311) planes of cubic ZnSe.

The overall reaction is: $H_2Se_{(g)} + Zn(C_2H_5)_{2(l)} \rightarrow ZnSe_{(s)} + 2C_2H_{6(g)}$. It is spontaneous and exothermic with heat of reaction –380 kJ/mole [13,14]. The reaction can yield crystalline cubic (zinc blende) ZnSe at room temperature in the vapor phase [15] and liquid phase. The latter was demonstrated by bubbling H_2Se gas as a 5% mixture with H_2 through a 0.1M solution of diethylzinc in heptane at room temperature. The X-ray diffraction pattern of particle aggregates taken by this "proof of principle" experiment is shown in Fig. 2. The three main diffraction peaks exactly match the standard peaks for cubic ZnSe.

A mixture of 12.6 wt% PEO-PPO-PEO and 1.3 wt% n-heptane in formamide was found to form very stable microemulsions with a droplet diameter of ~40 nm (as confirmed by light scattering analysis) and was used for the ZnSe nanocrystal growth experiments. Particle nucleation probably occurs simultaneously at different locations inside each nanodroplet and the nuclei subsequently grow by addition of precursors to their surface (i.e. by surface growth reactions) and by particle-particle coalescence. Coalescence of all primary particles formed inside each droplet eventually leads to a single ZnSe particle. The energy released by the reaction between the precursors and by particle-particle coalescence is sufficient to increase the temperature locally [16], thus enabling the formation of crystalline particles through annealing. The macroscopic temperature of the mixture remains constant (room temperature). The apparent melting point of nanoparticles is suppressed as their size decreases [15] and, as a result, annealing requires a smaller increase in temperature than that for bulk crystals.

Based on the hypothesis that a single nanocrystal is formed in each nanodroplet, we calculated the amount of diethylzinc in heptane required to obtain certain nanocrystal sizes under ideal conditions. For initial diethylzinc concentrations in heptane of 0.3, 0.03, and 0.003 M, and for a fixed microemulsion droplet diameter of 40 nm, the target particle size was estimated to be 8 nm, 3.7 nm and 1.7 nm, respectively. The ZnSe nanocrystals obtained by the series of experiments exhibit size dependent luminescence as shown in Fig. 3a. There is a systematic blue shift of the emission peak from sample 1 to sample 3, corresponding to 0.3, 0.03, and 0.003 M diethylzinc concentration in heptane, which clearly indicates that the average size of the nanocrystals decreases. The expected emission wavelength from bulk ZnSe is 460 nm at room temperature. The peaks are single, symmetric, and narrow, indicating good control on particle

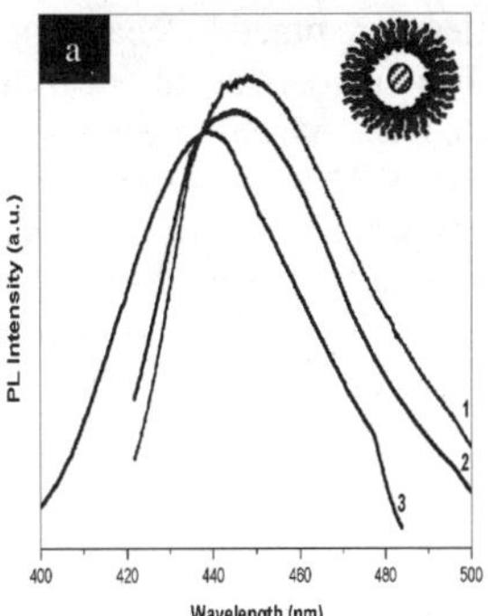

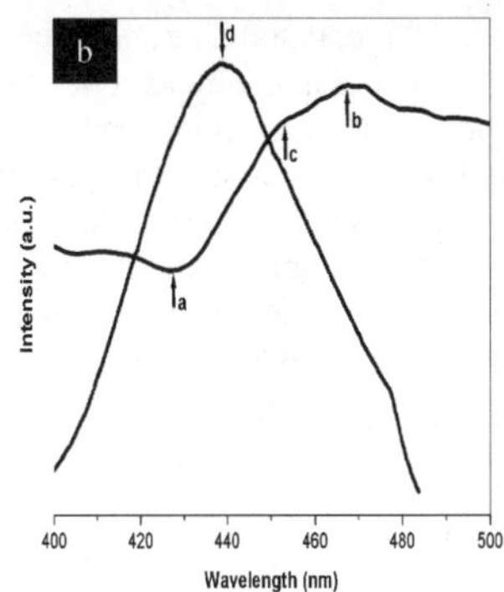

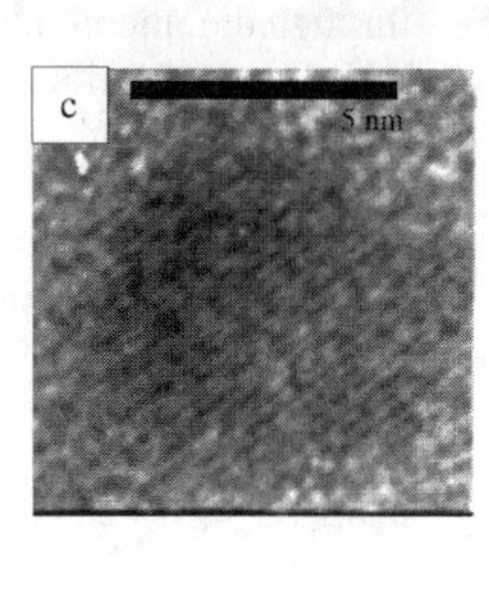

Figure 3. (a) Photoluminescence spectra of ZnSe quantum dots obtained by microemulsion-gas contacting. The emission wavelength is blue-shifted compared to bulk ZnSe (460nm) as the particle size decreases. The three curves, (1), (2), and (3) correspond to 0.3, 0.03, and 0.003 M diethylzinc concentration in heptane. The inset is a schematic of a ZnSe quantum dot encapsulated in a heptane nanodroplet. (b) Superposition of photoluminescence and transmission spectra for curve 3 in (a). (c) High-resolution TEM image of ZnSe nanocrystal obtained by processing a 0.3 M diethylzinc solution in a 40 nm heptane nanodroplets (curve 1 in (a)).

size and crystal structure without any post-processing step. A comparison between the PL and corresponding transmission spectra for sample 3 is shown in Fig. 3b. The nanoparticles of this sample absorb at wavelengths between 428 nm (a) and 453 nm (c). Feature (b) (at 467 nm) is attributed to absorption from larger ZnSe particles, because its wavelength coincides with the room-temperature absorption from bulk ZnSe. The observed emission peak (d) lies within the absorption wavelength interval (a-c). Since the emission spectrum reflects the distribution of particle sizes, the fact that its peak (i.e. the most probable size) is within the absorption feature indicates that the emission and absorption spectra coincide [17]. TEM images from sample 1 showed almost spherical nanocrystals with diameters of 6 ± 0.7 nm. A high resolution TEM image of such a nanocrystal is shown in Fig. 3c. This indicates that the initial hypothesis for single nanocrystal formation per nanodroplet is reasonable. The technique can also offer flexibility in functionalization. For example, thiol-conjugated molecules can be dissolved in the heptane nanodroplets for *in situ* functionalization of the nanocrystals upon synthesis.

CONCLUSIONS

In conclusion, a novel microemulsion-gas contacting technique has been developed for controlled synthesis of compound semiconductor nanocrystals that utilizes the dispersed phase of a heptane/PEO-PPO-PEO/formamide microemulsion to form numerous identical nanoreactors, thus enabling precise control of particle size. The technique employs reactions between group-II alkyls and group-VI hydrides, similar to those used by the microelectronics industry for Metalorganic Vapor Phase Epitaxy of high quality thin films [18,19], and can be scaled up for industrial production. The growth of luminescent ZnSe quantum dots was demonstrated by reacting diethylzinc, dissolved in heptane nanodroplets, with hydrogen selenide gas bubbled through the microemulsion. Nanocrystal size and optical properties were tuned by changing the initial concentration of diethylzinc in the heptane nanodroplets. The microemulsion-encapsulated nanocrystals exhibit single, symmetric, and narrow photoluminescence, indicating good control on particle size and crystal structure without any post-processing step. They are also very stable with photoluminescence and absorption spectra that remain unchanged over a period of several months. The technique is currently being extended to CdSe nanocrystal synthesis. X-ray diffraction patterns obtained from CdSe particles synthesized by bubbling hydrogen selenide diluted in hydrogen through a dimethylcadmium solution in heptane at room temperature confirm the formation of cubic CdSe. Experiments are underway to synthesize CdSe nanocrystals by the microemulsion-gas contacting technique.

ACKNOWLEDGEMENTS

We thank C. Sarigiannidis, J. Wang, M. Koutsona, L. Guo, D. Borden, G. Itskos, K.-T. Yong, and K.-K. Chain for assistance with the experiments. We also thank BASF Corporation for supplying the Pluronic™ block copolymer. This work was partially supported by SUNY-Buffalo (IRCAF Program) and by the National Science Foundation.

REFERENCES

1. A.P. Alivisatos, *Science* 271, 933 (1996).
2. C.B. Murray, C.R. Kagan, and M.G. Bawendi, *Annu. Rev. Mater. Sci.* 30, 545 (2000).
3. M. T. Swihart, *Curr. Opin. Colloid In.* 8, 127 (2003).
4. S.A. Empedocles et al., *Adv. Mater.* 11, 1243 (1999).
5. W.U. Huynh, X. Peng, and A.P. Alivisatos, *Adv. Mater.* 11, 923 (1999).
6. X. Michalet et al., *Single Mol.* 2, 261 (2001).
7. M. A. Hines and P. Guyot-Sionnest, *J. Phys. Chem. B*, **102** (19), 3655 (1998).
8. D.J. Norris, N. Yao, F.T. Charnock, and T.A. Kennedy, *Nano Lett.* 1, 3 (2001).
9. H. Zhao, E. P. Douglas, B. S. Harrison, and K. S. Schanze, *Langmuir* 17, 8428 (2001).
10. I. Danielsson and B. Lindman, *Colloids Surf.* 3, 391 (1981).
11. P. Alexandridis, U. Olsson, P. Linse, and B. Lindman, in *Amphiphilic Block Copolymers: Self-Assembly and Applications*, edited by P. Alexandridis and B. Lindman (Elsevier Science B.V., Amsterdam, 2000), p. 169.
12. A. Imhof and D.J. Pine, *J. Colloid Interface Sci.* 192, 368 (1997).
13. *National Institute of Standards and Technology Chemistry WebBook.* (NIST Standard Reference Database Number 69, 2001).
14. *CRC Handbook of Chemistry and Physics*, 3rd Electronic Ed. (CRC Press, 2000).
15. D. Sarigiannis, J.D. Peck, G. Kioseoglou, A. Petrou, and T.J. Mountziaris, *Appl. Phys. Lett.* 80, 4024 (2002).
16. K.E.J. Lehtinen and M.R. Zachariah, *Phys. Rev. B* 63, 205402 (2001).
17. G.N. Karanikolos, P. Alexandridis, G. Itskos, A. Petrou, and T.J. Mountziaris, *Langmuir*, 20(5), in press (2004).
18. J. Peck, T.J. Mountziaris, S. Stoltz, A. Petrou, and P.G. Mattocks, *J. Crystal Growth* 170, 523 (1997).
19. A. C. Jones, *J. Crystal Growth* 129, 728 (1993).

Carbon Nanotubes (CNT) and Related Properties

Mat. Res. Soc. Symp. Proc. Vol. 789 © 2004 Materials Research Society N16.2

Photo- and Thermal Annealing-Induced Processes in Carbon Nanotube Transistors

Moonsub Shim, Giles P. Siddons, Jae Kyeong Jeong, and David Merchin
Department of Materials Science and Engineering, University of Illinois at Urbana-Champaign, 1304 W. Green St. Urbana, IL 61801

ABSTRACT

Photoinduced conductivity changes and effects of thermal annealing in carbon nanotube transistors have been examined. Low-intensity ultraviolet light significantly reduces the p-channel conductance while simultaneously increasing the n-channel conductance. A combination of optical absorption and electron transport measurements reveals that these changes occur without variations in dopant concentrations. Measurements with different metals reveal that UV induces oxygen desorption from the electrodes rather than from nanotubes. In Ti-nanotube contact where the Schottky barrier plays an important role, photodesorption of oxygen mainly occurs from the native oxide of Ti electrodes. Decrease in the p-channel conductance arises from the metal work function change which causes larger hole Schottky barrier. Non-Schottky Pd-contacted nanotube transistors do not show photodesorption effects with low intensity UV. Thermal annealing of nanotube transistors with Ti/Au electrodes also leads to the disappearance of the photodesorption effects. However, a noticeable p-doping is observed to upon air exposure after thermal annealing.

INTRODUCTION

Single-walled carbon nanotubes (SWNTs)[1] with diameter and chirality dependent electronic structure have provided a unique system to study 1-D electron transport properties.[2] A variety of potential applications ranging from computer logic circuits[3] to chemical sensors[4] has also been demonstrated. With an increasing interest in their optical properties,[5] a combination of their unique electrical conductivity with optical effects should open up new opportunities. However, relatively little is known about how light affects the transport properties of SWNTs.

A striking decrease in the electrical conductance of SWNT field-effect transistors (FETs) has recently been observed and has been attributed to oxygen photodesorption.[6,7] The extreme sensitivity of electrical conductivity of nanotubes to oxygen is an important and currently debated issue.[8,9,10] One of the limiting consequences of oxygen adsorption has been the suppression of n-channel conduction. Initially, p-doping has been suggested to occur due to oxygen adsorption on nanotubes.[8,10] More recently, Schottky barriers at the metal-nanotube contacts have been shown to determine many transport properties of SWNT FETs[11,12] and the apparent p-type-only observation has been attributed to work function changes brought on by oxygen adsorption on metal electrodes resulting in asymmetric Schottky barriers rather than doping effects.[9] Understanding how photoinduced desorption of oxygen alters nanotube transistor characteristics will provide new insights into the role of oxygen on the observed properties of carbon nanotubes and will be required for any potential optoelectronic applications. Here, we show that photodesorption of oxygen occurs from the metal electrodes and does not cause doping level changes. Thermal annealing followed by air exposure, on the other hand, induces p-doping.

EXPERIMENTAL DETAILS

Individual SWNT FETs were fabricated by patterned chemical vapor deposition (CVD) as described elsewhere.[13] Metal electrodes (Au with Ti wetting layer or direct Pd contact) were deposited on top of the nanotubes after lithography. The device geometry is shown on the left side of fig. 1. Films of nanotubes were made by spraying ethanol suspension on glass slides for arc-discharge tubes (Carbolex) or grown directly on sapphire substrates for CVD tubes. Samples were equilibrated in air prior to measurements. Unless otherwise noted, monochromatic light source from a SPEX Fluoromax-3 spectrofluorometer with a 150 W Xe arc lamp collected with a quartz fiber optic was used. In all cases, light intensities are < 2 mW/cm^2.

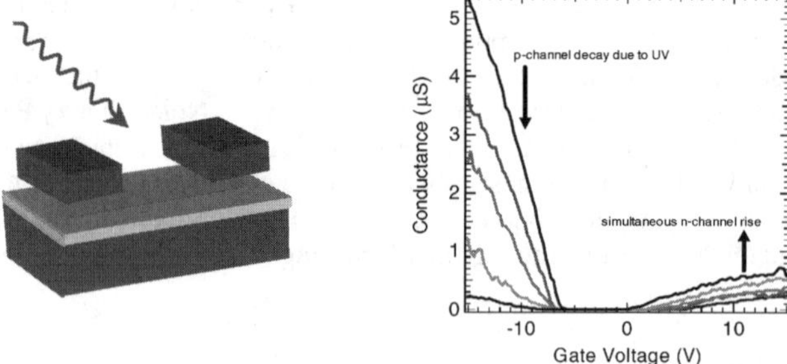

Figure 1. Nanotube transistor geometry (left) and changes brought on by low intensity UV exposure to the gate dependence of conductance (right).

DISCUSSION

Figure 1 (right) shows the gate dependence of the conductance of a nanotube transistor at different times of UV exposure (photon energy = 4.14 eV, 0.21 eV full-width at half maximum, and intensity = 1.8 mW/cm^2) under Ar. There are no significant shifts in the threshold voltage in both forward and reverse gate voltage sweeps upon UV irradiation (only negative to positive sweep is shown for clarity). The prominent changes are the nearly complete disappearance of the p-channel at negative gate voltages and the simultaneous increase in the n-channel. If the conductivity change arises from doping level variations, a shift in the threshold voltage without an increase or a decrease in both conduction channels is expected. Decrease in the work function of the metal contact, on the other hand, should result in larger hole Schottky barrier and smaller electron barrier[9] leading to gate dependence change similar to that shown in fig.1. While these results suggest that photodesorption of oxygen occurs without doping level changes, it should be noted that the determination of dopant concentrations based solely on gate-dependence can be misleading especially due to large hysteresis.[14,15,16] The threshold voltage can vary as much as 10 V depending on the gate voltage sweep direction, rate and range.[17] Therefore, we have kept these measurement parameters constant. For more conclusive evidence, optical measurements are compared.

Figure 2a shows the near-IR absorption spectra of a film of arc-discharge tubes before (1) and after (2) UV irradiation (intensity ~ 2 mW/cm^2 from a Hg arc lamp with photon energy maximum at 4.9 eV). The main peak at ~0.67 eV corresponds to the transition between the first two van Hove singularities of semiconducting nanotubes (often referred to as S_{11} transition). The second peak at ~1.2 eV is the transition between the second pair of van Hove singularities (S_{22} transition). When there is an increase in the charge carrier density caused by doping (e.g. by acid treatment, alkali metal or halogen doping, and electrochemical doping), a bleach in the S_{11} transition is expected. Conversely, a decrease in the number of charge carriers should result in an increased absorption at this transition. If we consider O_2 as a p-dopant, then the latter case applies. Although more than 30% decrease in the conductance is observed in fig. 2(b) inset, no change in the S_{11} transition is seen in the difference spectrum in fig. 2(b). Films of CVD tubes show similar response to UV exposure. These results combined with measurements on nanotube transistors show that low intensity UV-induced desorption of oxygen does not lead to doping level changes.

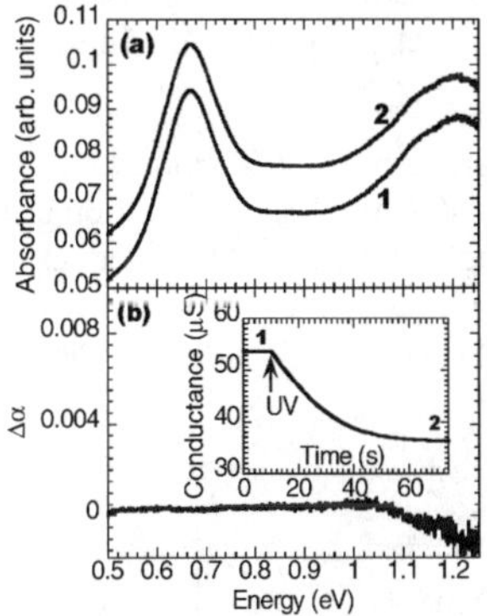

Figure 2. (a) Near-IR absorption spectra of a film of arc-discharge nanotubes before (1) and after (2) UV irradiation. Spectrum 2 is offset for clarity. UV exposure is carried out in an N_2 purged FTIR spectrometer. The difference spectrum is shown **(b)**. The inset is the corresponding conductance change indicating that there is over a 30% decrease in the conductance without any dopant concentration change.

Since oxygen desorption with low intensity UV does not cause doping changes, we explore the role of Schottky barriers directly by comparing different metal electrodes. Figure 3 compares the UV response of Ti- contacted nanotube transistor (a) with Pd-contacted transistor (b). With Ti-contact, nearly complete disappearance of p-channel conductance is observed in ~10 s due to exposure to UV (photon energy = 4.14 eV, ~0.21 eV FWHM, and intensity < 2 mW/cm^2). No significant changes are seen in Pd-contacted nanotube transistors under same UV exposure conditions. The overlaid curves in fig. 3(b) are measurements every 10 s during UV exposure. The extremely high on-state conductance (resistance at -15 V gate ~ 30kΩ at room temperature with ~ 4 μm channel length) in these transistors can be attributed to non-Schottky contacts. This non-Schottky contact along with 3 to 4 orders of magnitude smaller photodesorption cross sections for Pd compared to TiO_2[18] leads to the disappearance of the UV-induced conductance changes (in our low intensity limit) and shows that photodesorption of oxygen occurs at the metal electrodes rather than from nanotubes.

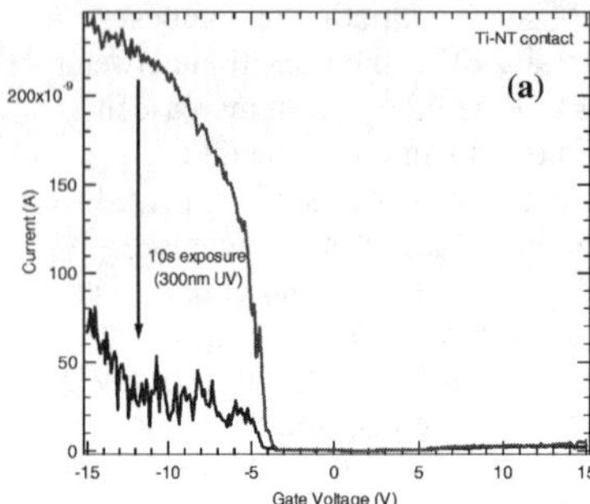

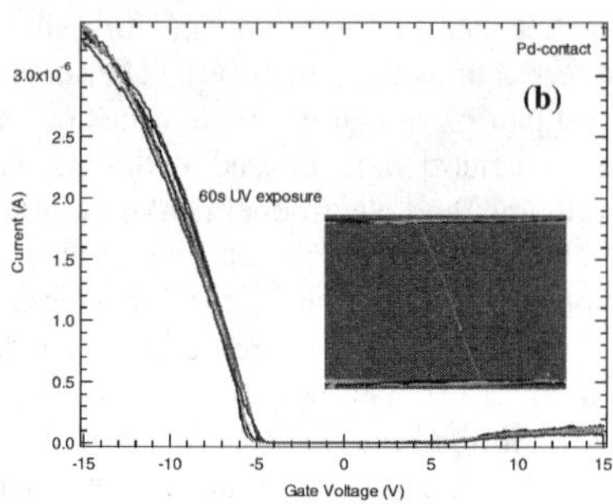

Figure 3. UV induced effects in Ti-contacted nanotube transistor (**a**). No significant effect is observed for non-Schottky Pd-contacts (**b**). The inset in (**b**) is an AFM image of the device.

Similar disappearance of photodesorption effects are seen when Ti-contacted nanotube transistors (with Au on top of Ti) are annealed under Ar. This observation can be attributed to thermally induced diffusion of Au resulting in Au-nanotube contact as suggested by Yaish et al.[19] We expect Au to have much smaller photodesorption cross sections than TiO_2 and non-Schottky contacts leading to the same explanation for the

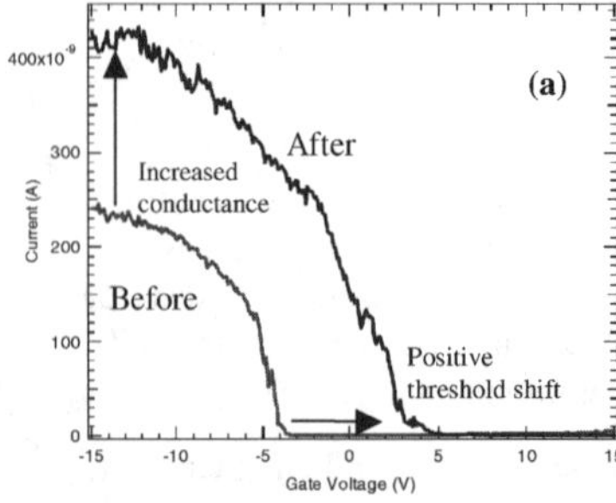

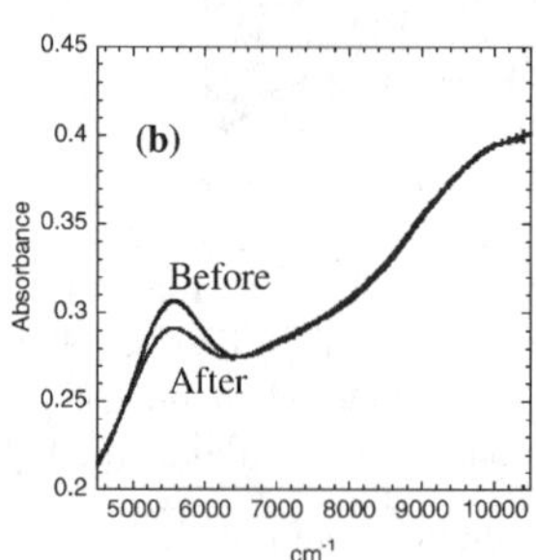

Figure 4. Effect of annealing (under Ar at 500°C for 20 min) on the nanotube transistor characteristics (**a**). Annealing and equilibrating in air also results in bleach of the band edge transitions in nanotube films (**b**).

disappearance of UV effect as in Pd-contacted transistors. We note that thicker Ti layer requires longer and higher temperature annealing to remove the UV-induced desorption effects further confirming that the photodesorption of oxygen occurs at the metal electrode. However, we also note that thermal annealing leads to increased p-doping as indicated by the positive shift in the threshold voltage in fig. 4a. Bleach in the optical absorption also confirms this effect as shown in fig. 4b. While the threshold shifts in single tube transistors occur instantaneously (i.e. within the time it takes to remove the sample from the annealing furnace to the measurement setup), the bleach in the band edge transition of nanotube films occurs over several days.

CONCLUSION

We have shown that low intensity UV irradiation leads to photodesorption of oxygen from metal electrodes rather than from nanotubes. In Ti-contacted nanotube transistors, photodesorption occurs from the native oxide layer of Ti electrodes. Decay in p-channel conductance arises from increased Schottky barrier at Ti-nanotube contacts. With low intensity UV (< 2 mW/cm^2) utilized here, no significant effects are seen in Au- and Pd-contacted nanotube transistors due to orders of magnitude smaller photodesorption cross sections and non-Schottky contacts. We have also observed that thermal annealing followed by air exposure leads to noticeable positive shift in the threshold voltage as well as bleach in the band edge absorption suggesting that possible p-doping effects of oxygen need to be more carefully examined.

ACKNOWLEDGMENT

This work was funded by UIUC. Characterization of the samples was carried out in the Center for Microanalysis of Materials, University of Illinois, which is partially supported by the U.S. Department of Energy under grant DEFG02-91-ER45439.

REFERENCES

1. *Carbon Nanotubes: Synthesis, Structure, Properties and Applications*, edited by M. Dresselhaus, G. Dresselhaus, and Ph. Avouris (Springer, Berlin, 2001).
2. M. Bockrath, D. H. Cobden, J. Lu, A. G. Rinzler, R. E. Smalley, L. Balents, P. L. McEuen, *Nature* **397**, 598 (1999).
3. A. Bochtold, P. Hadley, T. Nakanishi, and C. Dekker, *Science* **294**, 1317 (2001).
4. J. Kong, N. R. Franklin, C. Zhou, M. G. Chapline, S. Peng, K. Cho, and H. Dai, *Science* **287**, 622 (2000).
5. M. J. O'Connell, S. M. Bachilo, C. B. Huffman, V. C. Moore, M. S. Strano, E. H. Haroz, K. L. Rialon, P. J. Boul, W. H. Noon, C. Kittrell, J. Ma, R. H. Hauge, R. B. Weisman, and R. E. Smalley, *Science* **297**, 593 (2002).

6. R. J. Chen, N. R. Franklin, J. Kong, J. Cao, T. W. Tombler, Y. Zhang, and H. Dai., *Appl. Phys. Lett.* **79**, 2258 (2001).
7. M. Shim and G. P. Siddons, *Appl. Phys. Lett.* **83**, 3564 (2003).
8. P. G. Collins K. Bradley, M. Ishigami, and A. Zettl, *Science* **287**, 1801 (2001).
9. S. Heinze J. Tersoff, R. Martel, V. Derycke, J. Appenzeller, and Ph. Avouris, *Phys. Rev. Lett.* **89**, 6801 (2002).
10. G. U. Sumanasekera, C. K. W. Adu, S. Fang, and P. C. Eklund, *Phys. Rev. Lett.* **85**, 1096 (2000).
11. J. Park and P. L. McEuen, *Appl. Phys. Lett.* **79**, 1363 (2001).
12. V. Derycke, R. Martel, J. Appenzeller, and Ph. Avouris, *Appl. Phys. Lett.* **80**, 2773 (2002).
13. H. T. Soh, C. F. Quate, A. F. Morpurgo, C. M. Marcus, J. Kong, and H. Dai, *Appl. Phys. Lett. 75*, 627 (1999).
14. M. S. Fuhrer, B. M. Kim, T. Durkop, T. Brintlinger, *Nano Lett.* **2**, 755 (2002).
15. M. Radosevljevi, M. Freitag, K. V. Thadani, A. T. Johnson, *Nano Lett*. **2**, 761 (2002).
16. J. B. Cui, R. Sordan, M. Burghard, and K. Kern, *Appl. Phys. Lett.* **81**, 3260 (2002).
17. W. Kim, A. Javey, O. Vermesh, Q. Wang, Y. Li, H. Dai, *Nano Lett.* **3**, 193 (2003).
18. L. Hanley, X. Guo, and J. T. Yates, *J. Chem. Phys.* **91**, 7220 (1989). C. N. Rusu and J. T. Yates, *Langmuir* **13**, 4311 (1997).
19. Y. Yaish, J. –Y. Park, S. Rosenblatt, V. Sazonova, M. Brink, P. L. McEuen, preprint.

Mat. Res. Soc. Symp. Proc. Vol. 789 © 2004 Materials Research Society N16.3

Integration of fullerenes and carbon nanotubes with aggressively scaled CMOS gate stacks

Udayan Ganguly[1], *Chungho Lee*[2] *and Edwin C. Kan*[2]
[1]Department of Materials Science and Engineering, [2]School of Electrical and Computer Engineering, Cornell University

ABSTRACT

Here we report the first study towards the integration of fullerenes and carbon nanotubes (CNT) in the gate stack of CMOS technology, which is a promising hybrid approach of top-down and bottom-up fabrication process. Prospective processes for C_{60} and CNT deposition over an aggressively scaled 2 nm gate oxide in the MOS capacitor structure have been monitored. CV measurements show minimal silicon contamination and interface states. Step charging at a specific voltage that corresponds to a fixed number density of C_{60} is used to establish the structural integrity and size-mono-dispersion of C_{60}. The CV method can be further used to probe the charge injection into C_{60} and its anions to establish fundamental understanding of their molecular orbital (MO) structure.

INTRODUCTION

Integration of carbon nanotubes and fullerenes in silicon electronics is a new promising direction for device miniaturization towards the nanoscale. Conventional complimentary metal oxide semiconductor (CMOS) technology with the deep sub-50 nm drawn gate length [1] is pushing the limit of lithographic dimensions that can be mass produced. At this size scale and smaller, nature manufactures structures by self assembly with great precision and reliability. However, combination with top-down design patterns will collectively determine the functional density, because self-assembled homogeneous structures do not contain asymmetry to represent designer-defined information. Hence, a hybrid of bottom-up and top-down approach for device fabrication can provide critical links to engineering in the nanoscale. Carbon nanotubes (CNT) and fullerenes have also stimulated great scientific interest. Carbon has rich chemistry and it self-assembles into these structures of low dimensions. The C_{60} molecules can be used to replace nanocrystals in non-volatile memory devices. The mono-disperse nature and small size of C_{60} in comparison with self-assembled semiconductor and metal nanocrystals [2-4], which have non-negligible size variation, will lead to large and accurate step charging into molecular orbitals (MO) and hence can potentially provide reliable multiple-level storage in a single device. On the other hand, CNT can be used as floating gates in CvMOS for nanometer scale sensor applications [5] and scaling studies beyond lithographic dimensions. However, there are imminent challenges towards the goal of integration. MOS processing may involve high energy plasma and high temperature processes that can adversely affect the single or few shells of atoms that constitute the C_{60} and CNT. Conversely, the new materials in the CMOS gate stack can potentially contaminate and degrade the MOS structure during processing. Hence a mutually compatible process development is necessary for any integration plan. In this paper, C_{60} is integrated into the dielectric of a MOS capacitor structure as shown in Fig. 1a. The current study also lends itself to experimentally characterizing the electronic structure of C_{60} in a dielectric by measuring C_{60} energy levels as a function of successive charge injection [6].

EXPERIMENTAL SETUP

The three immediate challenges for process integration are a) deposition of fullerene and nanotubes on thin gate oxide without damage, b) the deposition of control oxide on the CNT and C_{60} while ensuring their molecular integrity, and c) to monitor defects and contamination level at the Si/SiO_2 interface and in SiO_2. Firstly, there are several possible approaches to dispense the molecules onto the substrate. C_{60} and CNT can be dissolved in suitable solvents like 1,2-dichlorobenzene or ortho-dichlorobenzene (o-DCB), isopropyl alcohol (IPA), and acetone in a spin-on process [7, 8]. Fullerenes can be thermally evaporated. Nanotubes can be grown in situ using metal catalyst [9]. The *in situ* growth is not included in our initial investigation reported here. Furthermore, deposition of SiO_2 on carbon molecules was investigated with plasma enhanced chemical vapor deposition (PECVD), RF sputter deposition and e-beam evaporation of silica. Carbon nanotube field effect transistors (CNFET) were obtained for explicit process tests [10]. Oxide was deposited through different methods and drain current versus gate voltage sweep was monitored at a 10 mV drain voltage. CNT under evaporated oxide shows almost unaffected IV characteristics while other processes caused large degradations in IV characteristics. Hence, evaporated oxide was selected for control gate oxide growth. Finally, to monitor defect formation, contamination and process compatibility, the MOS capacitors with the different process variations were subjected to charge injection and CV analysis.

Square MOS capacitor structures as shown in Fig. 1a were fabricated with lateral dimension varying from 10 μm to 250 μm as follows. Standard LOCOS isolation structure was fabricated on a boron implanted p-type wafer with a $2\times10^{17}cm^{-3}$ target surface doping [3]. After RCA clean, trichloroethane (TCA) dry oxidation is carried out for 2 nm SiO_2 growth at 750°C for 10 minutes [4]. After tunnel oxide growth, the wafers were subjected to different processes for studying C_{60} deposition. Different solvents and a solution: (a) IPA (b) acetone and IPA (c) o-DCB and (d) a solution of C_{60} in o-DCB, were spun on 4 different wafers. In another two wafers, C_{60} was thermally evaporated to an effective thickness of (e) 4 Å and (f) 6 Å measured on quartz crystal microbalance using graphite material parameters. A control wafer (g) with no such additional processing was also present. Evaporated oxide was then deposited on all the wafers to a thickness of 20 nm by e-beam evaporation of silica. Chromium gates were deposited by e-beam evaporation and patterned by wet etching. The backside nitride and oxide was etched by reactive ion etching (RIE) and finally the substrate contact was made by sputtering aluminum. All devices underwent a rapid thermal anneal (RTA) for 10 minutes in H_2/Ar gas at 450°C.

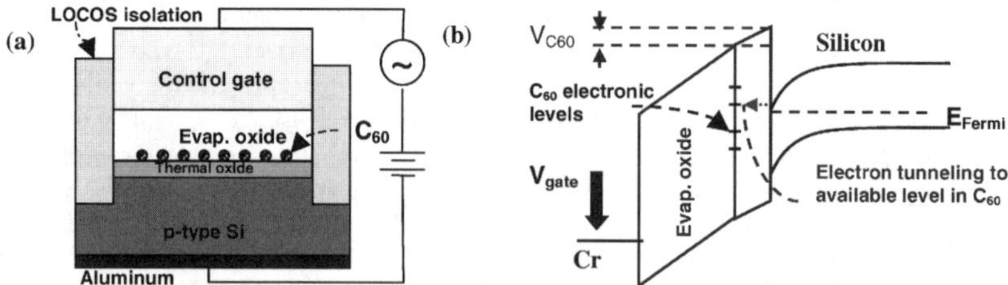

Fig. 1. (a) MOS capacitor structure and CV measurement setup for sample (e) with C_{60} evaporat on the 2 nm tunnel oxide **(b)** Energy band structure (not to scale) of MOS capacitor (e)

ROOM TEMPERATURE MEASUREMENTS

High frequency capacitance voltage (HFCV) analyses at 1 MHz ac test voltage were performed at room temperature for all 7 process wafers. Ultra-violet (UV) erasure for 20 minutes is performed to attempt erasing all trapped charges from the oxide and resetting the capacitors as practiced in the electrically erasable programmable read-only memory (EEPROM) devices [2-4]. A charging voltage (V_{charge}) is applied for 10 s followed by a CV measurement sweep. CV measurements show steep accumulation to inversion transition, close to the theoretical CV curve, which translates to acceptably low interface states. Deep depletion is easily visible for all devices at 0.5 V/s sweep rate, which indicated low concentration of generation sites and contamination free silicon. However the control oxide quality has a strong dependence on the C_{60} dispension process. All the devices show injection of carriers into the oxide through the control gate except for the wafers treated with (a) IPA and (b) acetone and IPA. Since the gate and channel have charge carriers of opposite polarity during CV sweeps, gate injection shifts CV curves in the opposite direction compared to when injection occurs through the channel side. Control gate injection makes any charge injection into the C_{60} molecules difficult to characterize. Fig. 2(a) shows the shift in CV curves due to gate injection and subsequent motion in the reverse direction due to channel injection. Fig. 2(b) shows CV measurements of the trap-free oxide obtained from IPA spun on the tunnel oxide before control oxide evaporation. Charge injection and conduction in a thick insulator is dominated by trap based charge hopping mechanism that results in Poole-Frenkel currents, which is a strong function of temperature and the electric field. To deal with this non-ideal situation in our preliminary setup, low temperature measurements were used to eliminate conduction in the evaporated oxide.

LOW TEMPERATURE MEAUREMENTS (T=10 K)

Low temperature measurements were done on an ultra high vacuum (UHV) helium cooled probe station at 10 K. Field ionization of dopants can still give normal CV measurements. The CV measurements do not show charge injection for wafers (a-d) and (g). Only wafers (e) and (f) with evaporated C_{60} show charge injection (see Fig. 3a (i) and 3b (i)), which confirms the elimination of hopping transport. The capacitors with evaporated C_{60} were charged at a V_{charge} for 10s and then the CV sweep is performed to extract the flat band voltage (V_{fb}). Microscope light was used to generate minority carriers for electron injection in inversion. V_{charge} was stepped to study the charge injection into C_{60}. Fig. 3a (i) shows CV curves of the square capacitors with 250 μm

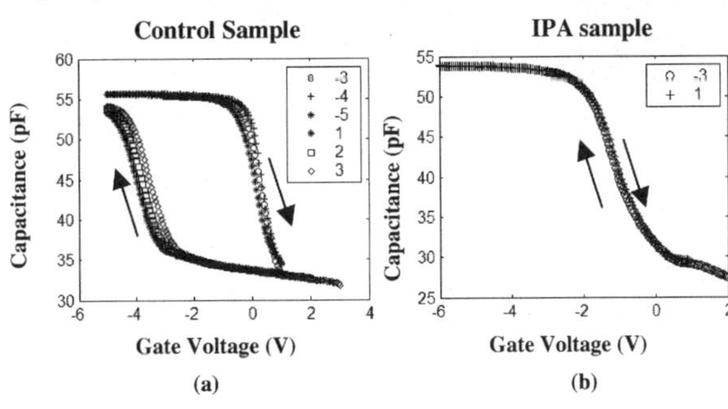

Fig. 2. Room temperature CV measurements: **(a)** control sample (g) showing charge injection from the gate into the oxide. **(b)** samples (a) and (b) treated with IPA show ideal CV curves. Legend shows different charging voltage for injection. Arrows show sweeping direction.

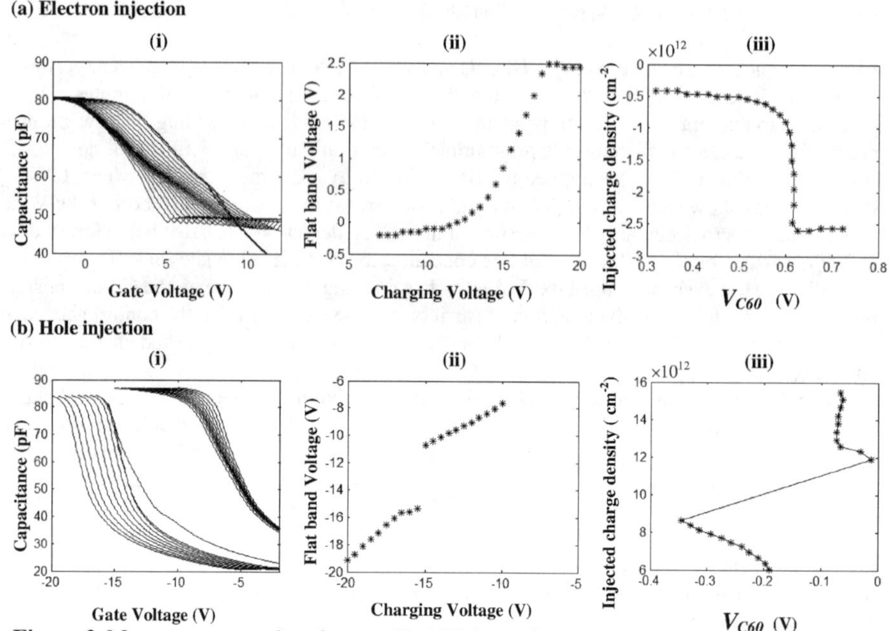

Figure 3. Measurements and analyses at T=10K for wafer (e) (i) family of CV sweeps (ii) V_{fb} vs V_{charge} (iii) Q_f vs V_{C60}. The figures in set (a) are for electron injection, while the set (b) for hole injection.

sides and V_{charge} was stepped from 5-20V in steps of 0.5V. The CV curve moves to the right of the graph as electrons are injected from the Si channel into the C_{60}. Clustering of CV curves can be observed, i.e., for some range even though the V_{charge} is increased, there is no further charge injection. Similar experiments for hole injection have been shown in Fig. 3b (i). V_{fb} is a measure of the weighted sum of charge between the control gate and the channel and can be extracted from the CV curves [11] for both electron and hole injections and plotted versus the V_{charge} as shown in Figs. 3a (ii) and 3b (ii). To extract fundamental MO positions of the C_{60} with respect to the silicon energy bands during charging, we consult the energy band diagram of the capacitor as shown in Fig. 1b. The effect of stepping the control gate voltage during charge injection from Si channel into the C_{60} is equivalent to scanning the energy levels of fullerene in energy with respect to the energy band of silicon. Hence, upon an alignment of an unfilled MO level to the silicon band, under favorable electric field there will be a high probability of electron tunneling into the MO of the C_{60} by direct tunneling through thin tunnel oxide. The voltage on C_{60} (V_{C60}) with respect to the conduction energy band of Si/SiO_2 interface can be calculated by considering a simple parallel plate capacitor model. Here, the gate charge Q_{total} is screened by the charge in the oxide Q_f (i.e. charge injected into either C_{60} or oxide traps near evaporated oxide thermal oxide interface if present as estimated from the extracted V_{fb}) to produce a constant electric field in the tunnel oxide with capacitance C_{tunnel}:

$$V_{C60} = \frac{Q_{total} - Q_f}{C_{tunnel}} \quad (1)$$

Using the above formulation, we extract Q_f versus V_{C60} as shown in Fig. 3a (iii) for electron injection. For different devices, we observe the same sharp step charging at V_{C60}= 0.6V and the step size is about $2\text{x}10^{12}\,\text{cm}^{-2}$. This step charging observed as an average over a large 250 μm square area is clearly due to charge confinement in the same energy level in a mono-disperse quantum dot or a MO. The number density of $2\times10^{12}\,\text{cm}^{-2}$ is also consistent during step charging in all devices. Fig. 3b (iii) shows Q_f versus V_{C60} in hole injection. Here the slope of Q_f versus V_{C60} indicates continuously distributed trap states. Comparatively such traps were not observed during electron injection. The hole injection into these traps is about 5 times higher than electron injection. However, there is a clear step charging at -0.36V of about $2\times10^{12}\,\text{cm}^{-2}$, which screens the gate charge sufficiently that the field in the tunneling oxide drastically decreases, and hence decreases the magnitude of V_{C60}.

DISCUSSIONS

The sharp step charging during electron injection at V_{C60} of 0.6V and number density of 2×10^{12} cm^{-2} consistently in different devices is a clear proof of the existence of C_{60} in molecular form in our device. Theoretically, MO alignment with silicon bands can be constructed as in [6]. A correction for the dielectric stabilization ΔE as given by the Eq. (2) should be subtracted from the MO calculation in vacuum to obtain the MO energy of C_{60}

$$\Delta E = \frac{qe\left(1-\frac{1}{\kappa}\right)}{4\pi\varepsilon_0 r} = 2.5q \quad (in\ eV) \quad (2)$$

where κ (~4 for SiO_2) is the oxide dielectric constant and q is the number of electrons in the final C_{60} anion [6]. Since the theoretical calculations do not reliably calculate absolute energies but give rather accurate relative energies of injection, the corroboration as shown in table 1 is remarkable. In future work, the experimental determination of MO alignment of C_{60} with respect to silicon energy bands can provide an alternative probe into the electronic property of C_{60} inside a dielectric [13, 14]. Finally, control oxide quality can be improved by using hybrid deposition process of evaporated and PECVD oxide for room temperature operations.

Table 1. Comparison of theoretical and experimental results on C_{60} MO energies

Initial charge state of C60	MO energy for injection in vacuum (eV)	Energy Correction in dielectric (eV)	Dielectric stabilization corrected MO energy level for injection (eV)	Theoretical VC60 (eV) (***shifted by 0.23eV to zero for 0.6eV***)	Experimental observation VC60 (eV) (details)
0	-1.77	2.5	-4.27	-0.22 (***-0.45***)	-0.45 to -0.35 (hole from CB)
-1	1.29	5	-3.71	0.34 (***0.11***)	0.1 to 0.2 (electron from CB)
-2	4.28	7.5	-3.22	0.83 (***0.6***)	0.6 (electron from CB)
-3	7.31	10	-2.69	1.36	-
-4	10.3	12.5	-2.2	1.85	-
-5	13.24	15	-1.76	2.29	-

CONCLUSIONS

In conclusion, we report our preliminary effort at integrating CNT and C_{60} in the gate stack of CMOS devices. We have developed a MOS-compatible process that does not affect the integrity of the C_{60} during processing as shown by step charging at a specific potential for the fullerene MO levels with respect to the silicon band energies. Interface traps and silicon contamination are minimal. Further investigations of the MO of C_{60} and it anions are promising from both scientific and engineering perspectives. The applications in nonvolatile memory, MOSFET scaling studies and CvMOS are other interesting directions.

ACKNOWLEDGEMENTS

This project is supported by NSF NIRT ECS-0304483. The authors would like to thank Prof. Paul McEuen, his assistants, Sami Rosenblatt, and Vera Sazanova, and Dr. Ken Bosnick for fruitful discussions and supply of test nanotube transistors. The technical staff of Cornell Nanoscale Facilities is also acknowledged for their helpful discussions on fabrication.

REFERENCES

[1] D. Hisamoto, W-C. Lee, J. Kedzierski, H. Takeuchi, K. Asano, C. Kuo, E. Anderson, T-J. King, J. Boko and C. Hu, IEEE Trans. Electron Devices **47** (12) 2320 (2000)

[2] Z. Liu, C. Lee, G. Pei, V. Narayanan and E. C. Kan, Mat. Res. Soc. Symp. Proc. **686**, A5.3 (2001)

[3] Z. Liu, C. Lee, V. Narayanan, G. Pei and E. C. Kan, IEEE Trans. Electron Devices **49** (9) 1606, (2002)

[4] C. Lee, Z. Liu and E. C. Kan, Mat. Res. Soc. Symp. Proc. **737**, F8.18 (2002)

[5] Y. N. Shen, Z. Liu, B. A. Minch, and E. C. Kan, IEEE Trans. Electron Devices **50** (10), 2171 (2003)

[6] W. H. Green Jr., M. G. G. Fitzgerald, P. W. Fowler, A. Ceulemans and B. C. Titeca, J. Phys. Chem. **100**, 14892 (1996)

[7] R.S. Ruoff, D.S. Tse, R. Malhotra, D.C. Lorents., J. Phys. Chem. **97**, 3379 (1993)

[8] J. L. Bahr, E. T. Mickelson, M. J. Bronikowski, R. E. Smalley, J M. Tour, Chemical Communications, p.193, (2001)

[9] N. R. Franklin, Q. Wang, T. W. Tombler, A. Javey, M. Shim,and H. Dai, Appl. Phys. Lett. **81** (5), 913 (2002)

[10] M. Fuhrer, H. Park, and P. L. McEuen, IEEE Trans. on Nanotech. **1**, 78 (2002).

[11] D.K. Schroeder, *Semiconductor Material and Device Characterization,* 3rd Ed. (John Wiley & Sons), p. 365 (1998)

[12] H. Park, J. Park, A. K.L. Lim, E. H. Anderson, A. P. Alivisatos, and P. L. McEuen, Nature **407**, 57 (2000).

[13] D. Porath and O. Millo, J. Appl. Phys. **81**, (5), 2241 (1997)

[14] M. A. Greaney and S. M. Gorun, J. Phys. Chem. **95**, 7142 (1991)

Mat. Res. Soc. Symp. Proc. Vol. 789 © 2004 Materials Research Society N16.9

Probing the Long Range Distance Dependence of Noble Metal Nanoparticles

Amanda J. Haes and Richard P. Van Duyne
Northwestern University, Department of Chemistry, 2145 Sheridan Road
Evanston, Illinois 60208-3113, U.S.A.

ABSTRACT

The localized surface plasmon resonance (LSPR) of noble metal nanoparticles has recently been the subject of extensive studies. Previously, it has been demonstrated that Ag nanotriangles that have been synthesized using nanosphere lithography (NSL) behave as extremely sensitive and selective chemical and biological sensors. The present work reveals information regarding the long range distance dependence of the localized surface plasmon resonance (LSPR) of silver and gold nanoparticles. Multilayer adsorbates based on the interaction of $HOOC(CH_2)_{10}SH$ and Cu^{2+} were assembled onto surface-confined nanoparticles. Measurement of the LSPR extinction peak shift versus number of layers and adsorbate thickness is non-linear and has a sensing range that is dependent on the composition, shape, in-plane width, and out-of-plane height of the nanoparticles. Theoretical modeling confirms and offers a mathematical interpretation of these results. These experiments indicate that the LSPR sensing capabilities of noble metal nanoparticles can be tuned to match the size of biological and chemical analytes by adjusting the aforementioned properties. The optimization of the LSPR nanosensor for a specific analyte will improve an already sensitive nanoparticle-based sensor.

INTRODUCTION

For ~20 years, surface plasmon resonance (SPR) sensors, that is, copper, gold, or silver planar films have been used as refractive index based sensing devices to detect analyte binding at or near a metal surface.[1] This sensor exhibits an extremely large refractive index sensitivity ($\sim 2\times10^6$ nm/RIU) and modest decay length (200-300 nm),[2] and this sensitivity is proportional to the square of the electric field that extends from the metal film. Recently, it was realized that this refractive index sensitivity also exists for noble metal nanoparticles. Although it has been demonstrated to operate successfully for nanoparticles, details regarding the aforementioned properties of planar SPR sensors still need a more complete explanation.

Advancements in technology due to nanoscale phenomena of materials will be made and/or optimized when the chemical and physical properties of materials are more thoroughly understood. Prior to the realization of this technology, methods to synthesize isolated monodisperse nanoparticles in a controlled environment must be developed and their properties must be thoroughly characterized.

The development of nanoparticle-based optical sensors is an extremely active area of nanoscience research. One such nanoparticle based optical sensor is known as the localized surface plasmon resonance (LSPR) nanosensor. The LSPR of noble metal nanoparticles arises when electromagnetic radiation induces a collective oscillation of the conduction electrons of the individual nanoparticles and has two primary consequences: (1) selective photon absorption which allows the optical properties of these nanoparticles to be monitored with UV-visible spectroscopy and (2) the enhancement of the electromagnetic fields surrounding the nanoparticles which is responsible for all surface-enhanced spectroscopies.

Recently, we found that a result of this phenomenon is the extreme sensitivity of nanoparticles to changes in its surrounding dielectric environment. For this reason, we have developed a sensing scheme based on this property. In order to understand and optimize this sensing platform, the goal of this work is to achieve a better comprehension of the electromagnetic fields surrounding the nanoparticle. Specifically, the elucidation of the long range distance dependence of the LSPR of surface confined noble metal nanoparticles will assist in the optimization of the LSPR nanosensor.

EXPERIMENTAL DETAILS

Materials

11-Mercaptoundecanoic acid (11-MUA) and $Cu(ClO_4)_2$ were purchased from Aldrich (Milwaukee, WI). Absolute ethanol was purchased from Pharmco (Brookfield, CT). Hexanes and methanol were purchased from Fisher Scientific (Pittsburgh, PA). Ag wire (99.99%, 0.5 mm diameter) and Au wire (99.9%, 0.025 mm diameter) was obtained from D. F. Goldsmith (Evanston, IL). Borosilicate glass substrates, No. 2 Fisherbrand 18 mm circle coverslips were purchased from Fisher Scientific (Pittsburgh, PA). Tungsten vapor deposition boats were acquired from R. D. Mathis (Long Beach, CA). Polystyrene nanospheres with diameters of 280±4 nm, 310±9 nm, 400±8 nm, 450±5 nm, and 510±11 nm were received as a suspension in water (Interfacial Dynamics Corporation, Portland, OR) and were used without further treatment. Millipore cartridges (Marlborough, MA) were used to purify water to a resistivity of 18 $M\Omega cm^{-1}$. All materials were used without further purification.

Substrate Preparation

Glass substrates were cleaned in a piranha solution (1:3 30 % H_2O_2:H_2SO_4) at 80°C for 30 minutes. Once cooled, the glass substrates were rinsed with copious amounts of water and then sonicated for 60 minutes in 5:1:1 H_2O:NH_4OH:30% H_2O_2. Next, the glass was rinsed repeatedly with water and was stored in water until used.

Nanoparticle Preparation

NSL was used to fabricate monodisperse, surface-confined Ag nanoparticles (Figure 1).[3, 4] For these experiments, single layer colloidal crystal nanosphere masks were prepared by drop coating ~2 μL of nanosphere solution onto glass substrates. Once the nanosphere masks were dry, the substrates were mounted into a Consolidated Vacuum Corporation vapor deposition system. A Leybold Inficon XTM/2 quartz crystal microbalance (East Syracuse, NY) was used to measure the thickness of the Ag or Au film deposited over the nanosphere mask, d_m. Following metal deposition, the nanosphere mask was removed by sonicating the sample in ethanol for 3 minutes. The perpendicular bisector of the nanoparticles, a, was varied by changing the diameter, D, of the nanospheres used. The samples were either thermally annealed for 1 hour at ~600°C in the aforementioned chamber or solvent annealed. Unless otherwise noted, all samples were stabilized by solvent annealing. While the solvent slightly restructures the nanoparticles, that is, causes slight rounding of the nanotriangle tips, the overall triangular footprint is retained, and in this process, also produces stable extinction maxima readings in a

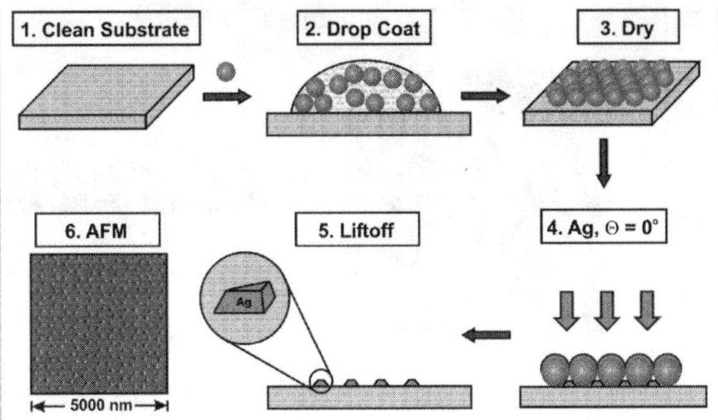

Figure 1. Ag nanoparticles were fabricated using NSL. Six steps are required for the synthesis of the nanoparticles: (1) glass or mica substrates are cleaned, (2) monodisperse polystyrene nanospheres are drop-coated onto the substrate, (3) a single layer of hexagonally close packed nanospheres dries creating a nanosphere mask, (4) Ag metal is vapor deposited onto the sample, (5) the nanosphere mask is removed via sonication in ethanol, and (6) the Ag nanoparticle sample is prepared for sensing experiments.

given environment. Thermal annealing is performed to convert the nanotriangles into nanohemispheres. A comparison between the two annealing methods enables the study of LSPR sensing as a function of nanoparticle shape.

Ultraviolet-visible Extinction Spectroscopy

Macroscale UV-vis extinction measurements were collected using an Ocean Optics (Dunedin, FL) SD2000 fiber optically coupled spectrometer with a CCD detector. All spectra collected are macroscopic measurements performed in standard transmission geometry with unpolarized light. The probe beam diameter was approximately 2 mm.

Nanoparticle Annealing

A home built flow cell[5] was used to control the external environment of the Ag nanoparticle substrates. Prior to modification, the Ag nanoparticles were solvent annealed.[5] Dry N_2 gas and solvent were cycled through the flow cell until the λ_{max} of the sample stabilized.

Nanoparticle Functionalization

After solvent annealing, the nanoparticle samples were functionalized with 1 mM 11-MUA (ethanol) solutions. Within 10 minutes, the extinction maximum of the sample stabilized and the sample was rinsed with copious amounts of ethanol. At this point, the sulfhydryl group binds to the Ag surface, thus placing the carboxylic acid group away from the surface. Next, a 1 mM $Cu(ClO_4)_2$ (ethanol) was exposed to the sample until the extinction maximum peak had stabilized (less than 3 minutes). The sample cell was flushed with ethanol, and the extinction spectrum of the sample in N_2 was collected. The Cu^{2+} ions interact with the surface-confined carboxylic acid

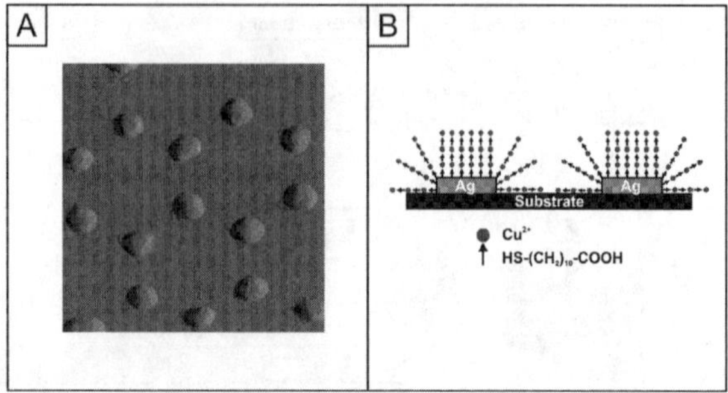

Figure 2. (A) AFM image of thermally annealed Ag nanoparticles. Ag nanoparticles were first synthesized using NSL. Following that process, the samples were then placed in a high vacuum chamber and heated to induced nanoparticle annealing. The resulting nanoparticle have in-plane widths of 110 nm and out-of-plane heights of 61 nm. The image is 1 μm x 1 μm. (B) Binding chemistry of the multilayer SAMs. One layer consists of 11-mercaptoundecanoic acid and Cu^{2+}.

groups. This represents one layer. A new layer is initiated by the introduction of 11-MUA. The sulfhydryl groups interact with the surface-confined Cu^{2+} ions, etc. These functionalization steps were repeated until the extinction maximum no longer shifted.

Atomic Force Microscopy (AFM)

AFM images were collected using a Digital Instruments Nanoscope IV microscope and Nanoscope IIIa controller operating in tapping mode. Etched Si nanoprobe tips (TESP, Digital Instruments, Santa Barbara, CA) were used. These tips had resonance frequencies between 280 and 320 kHz and are conical in shape with a cone angle of 20° and an effective radius of curvature at the tip of 10 nm. The image shown in Figure 2A that represents a thermally annealed sample is unfiltered data that were collected in ambient conditions. The height of the nanoparticles, b, was determined using AFM.

RESULTS AND DISCUSSION

The metal ion, carboxylated alkanethiol self-assembled monolayer (SAM) has many desirable characteristics (Figure 2B):[6, 7] (1) the first alkanethiol layer forms with the thiol group on the metal surface with no significant mixing of molecule orientation, (2) the second (and additional) alkanethiol layer forms with the same orientation as the first, (3) the first two alkanethiol layers have approximately equal number of molecules, (4) 20+ uniform layers can be formed, and (5) the refractive index of the layer is constant and assumed to be ~1.5 to 1.6 depending on layer ordering.[6 8] The exact binding of the alkanethiol to the metal ions is unclear. X-ray photoelectron spectroscopy studies indicate that 1 to 8 metal ions coordinate to every 2 carboxyl groups. Despite this unknown interaction, this multilayer structure offers a simple, controlled way to probe the long range distance dependence, and correspondingly, the electromagnetic field strength 10+ nm away from the nanostructured surfaces.[9]

Shape Influences on LSPR Long Range Dependence

The UV-vis extinction spectra for thermally annealed Ag nanoparticles (a = 110 nm, b = 61 nm) with 0 - 20 SAM layers is found in Figure 3A. Several distinctive features can be discerned from this data. As the adsorbate layer thickness increases, the extinction maximum red shifts, the full width half maximum (FWHM) remains approximately constant (100±10 nm), and the peak intensity increases. At layer 16, the extinction maximum stops shifting, the FWHM begins increasing, and the peak intensity begins to decrease. These three diagnostic features signal that of the sensing volume of the nanoparticle had been saturated.

An example of the LSPR extinction maximum shift from bare nanotriangles and nanohemispheres versus multilayer thickness (as determined by AFM) is displayed in Figures 3B. In this study, nanotriangles were synthesized on a glass substrate using nanospheres (D = 400 nm) and 50.0 nm of Ag metal. After nanosphere removal (in the case of Figure 3B), the samples were placed in a high vacuum chamber where they were thermally annealed for 1 hour at ~600°C. The nanotriangles (a = 114 nm, b = 54 nm) were converted into nanohemispheres (a = 110 nm, b = 61 nm) during this process. In both cases, the samples were stabilized in solvent and exposed to the appropriate molecular solution. In all cases, incubation in 11-MUA and $Cu(ClO_4)_2$ constitutes one layer. It is evident that even at low multilayer thicknesses, the nanotriangles give larger LSPR responses than the nanohemispheres. This result is consistent with studies done on identically prepared samples where the LSPR shift induced from hexadecanethiol was larger for nanotriangles than for nanohemispheres.[10] Additionally, theoretical studies support that as nanoparticles become more spherical, the electromagnetic field strength at their surface decreases.[11, 12] This indicates that the electromagnetic fields at the nanoparticle surface are more intense for nonhemispherical nanoparticles. Finally, because

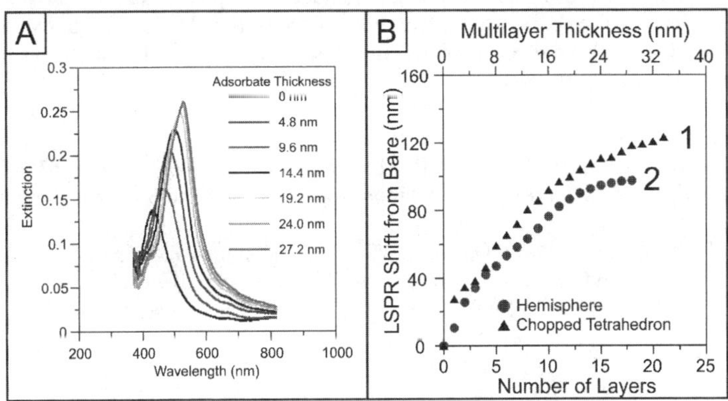

Figure 3. (A) Long range LSPR spectroscopy of Ag nanoparticles (a = 110 nm, b = 61.0 nm) for 0 - 17 layers of Cu^{2+}/HS-$CH_2)_{10}$COOH. All extinction measurements were collected in a N_2 environment. (B) Shape dependence on the long range LSPR distance dependence for Ag nanoparticles (D = 400 nm). (B-1) LSPR shift vs. Number of SAM layers/layer thickness for solvent annealed Ag nanoparticles. (B-2) LSPR shift vs. Number of SAM layers/layer thickness for thermally annealed (600°C for 1 hour) Ag nanoparticles.

nanotriangles exhibit larger total LSPR shifts and can detect molecules larger distances away from the nanoparticle surface, their electromagnetic fields strength and sensing volumes are both larger.

CONCLUSION

It was suspected that the linear distance dependence found in our previous alkanethiol self-assembled monolayer (SAM) formation studies was the thin shell limit of a longer range, nonlinear dependence. To verify this, a multilayer SAM shell approach based on the interaction of $HOOC(CH_2)_{10}SH$ and Cu^{2+} was used and were allowed to assemble onto surface-confined noble metal nanoparticles, and monitored their properties using UV-visible spectroscopy. Measurement of the LSPR extinction peak shift versus number of layers and adsorbate thickness is non-linear and has a sensing range that is dependent on the composition,[13] shape,[13] in-plane width,[13] and out-of-plane height[13] of the nanoparticles. Specifically, the following long range aspects of the LSPR nanosensor was demonstrated: (1) the LSPR shift versus number of adsorbate layers and adsorbate thickness is non-linear and (2) nanotriangles have larger sensing volumes and are more sensitive to multilayer adsorbates than nanohemispheres of equal volume. This optimization when coupled with additional size and composition studies and theoretical studies[13] is allowing us to significantly improve an already sensitive nanoparticle-based sensor.

ACKNOWLEDGEMENTS

The authors gratefully acknowledge support from the Nanoscale Science and Engineering Initiative of the National Science Foundation under NSF Award Number EEC-0118025. Any opinions, findings and conclusions or recommendations expressed in this material are those of the authors and do not necessarily reflect those of the National Science Foundation. A. Haes also wishes to acknowledge the American Chemical Society Division of Analytical Chemistry and Dupont for fellowship support.

REFERENCES

1. J. M. Brockman, B. P. Nelson and R. M. Corn, Annu. Rev. Phys. Chem. **51**, 41 (2000).
2. L. S. Jung, C. T. Campbell, T. M. Chinowsky, M. N. Mar and S. S. Yee, Langmuir **14**, 5636 (1998).
3. J. C. Hulteen and R. P. Van Duyne, J. Vac. Sci. Technol. A **13**, 1553 (1995).
4. C. L. Haynes and R. P. Van Duyne, J. Phys. Chem. B **105**, 5599 (2001).
5. M. D. Malinsky, K. L. Kelly, G. C. Schatz and R. P. Van Duyne, J. Am. Chem. Soc. **123**, 1471 (2001).
6. S. D. Evans, T. M. Flynn and A. Ulman, Langmuir **11**, 3811 (1995).
7. T. L. Freeman, S. D. Evans and A. Ulman, Langmuir **11**, 4411 (1995).
8. S. D. Evans, Personal Communication (2003).
9. A. Hatzor and P. S. Weiss, Science **291**, 1019 (2001).
10. A. J. Haes and R. P. Van Duyne, J. Am. Chem. Soc. **124**, 10596 (2002).
11. E. J. Zeman and G. C. Schatz, J. Phys. Chem. **91**, 634 (1987).
12. T. R. Jensen, K. L. Kelly, A. Lazarides and G. C. Schatz, J. Cluster Sci. **10**, 295 (1999).
13. A. J. Haes, S. Zou, G. C. Schatz and R. P. Van Duyne, J. Phys. Chem. B In Press (2003).

Poster Session V:
Carbon Nanotubes

Mat. Res. Soc. Symp. Proc. Vol. 789 © 2004 Materials Research Society N17.7

Quantum Dots from Carbon Nanotube Junctions

Fabrizio Cleri[1], Pawel Keblinski[2], Inkook Jang[3] and Susan B. Sinnott[3]

[1] Ente Nuove Tecnologie, Energia e Ambiente (ENEA), Unità Materiali e Nuove Tecnologie, Centro Ricerche Casaccia, 00100 Roma A. D. (Italy)
[2] Department of Materials Science and Engineering, Rensselaer Polytechnic Institute, Troy 12180-3590 (USA)
[3] Department of Materials Science and Engineering, University of Florida, Gainesville 32611-6400 (USA)

ABSTRACT

A tight-binding hamiltonian is used to study the electronic properties of covalently-bonded, crossed (5,5) metallic nanotubes with increasing degree of disorder in the junction region. At one extreme, ideal junctions between coplanar nanotubes with a minimal number of topological defects show a good ohmic behavior. Upon increasing disorder, ohmic conduction is suppressed in favor of hopping conductivity. At the opposite extreme, strongly disordered junctions as could be obtained after electron-beam irradiation of overlayed nanotubes, display weak localization and energy quantization, indicating the formation of a quantum dot contacted to metallic nanowires by tunnel barriers.

INTRODUCTION

Carbon nanotubes (CNTs) have been proposed as a main component of future nanoscale electronics, either as fully active circuit elements in the form of linear, Y-shaped or T-shaped heterojunctions [1-3], or as interconnects in molecular-level devices [4]. Such extended capabilities are made possible by the fascinating electronic properties of CNTs, which can behave either as metals or semiconductors as a function of their molecular symmetry.

Junctions between CNTs are a key factor in the design of such nanoscale integrated devices. In particular, the properties of crossed nanotube junctions bonded by van der Waals interactions have been already investigated in some detail by both experimental and theoretical methods [5-7]. Conduction in such weakly-coupled junctions occurs purely by tunneling, as determined by two- and four-terminal measurements aimed at factoring out the effect of the large contact resistance between the CNT and the metallic electrodes [8-9]. A different, and much less understood situation is encountered in the strongly-coupled, covalently-bonded junctions obtained by, e.g., electron-beam irradiation of crossed CNTs [10-11]. In this case a whole range of electrical conduction mechanisms, from purely ohmic, to phonon-assisted hopping, to quantum tunneling, is expected depending on both the electrical properties of pristine nanotubes and the nature of the covalent bonds formed upon irradiation.

In the present work we investigate the electronic properties of covalently-bonded ("e-beam welded") crossed-CNT junctions by means of tight-binding molecular dynamics (TBMD) simulations, based on a well-established orthogonal TB model for carbon [12]. We calculate and analyse the local density-of-states (LDOS) and localization coefficient ("participation ratio") for a number of different atomic structures. We find that a strongly disordered junction obtained by a simulated electron-beam irradiation welding may give rise to a quantum dot (QD). Such all-carbon QD structures appears very promising for several applications.

SIMULATION RESULTS

In all our studies we used an orthogonal TB model [12] including four valence orbitals $\{s, p_x, p_y, p_z\}$ for each C atom. The electronic structure is obtained by analyzing the eigenvector spectrum $c^{(n)}$ for the eigenvalues $\varepsilon^{(n)}$, $n \in 4N$, with N the number of C atoms and $4N$ the number of electrons in the supercell.

The "participation ratio", proportional to the second moment (mean) of the probability density, is defined as:

$$P^{(n)} = 4N \sum_{i,\alpha} \left| c_{\alpha}^{(n)}(r_i) \right|^4 , \tag{2}$$

the sum running over all sites r_i and orbitals α. Its deviation from the ideal value $(4N)^{-1}$ gives an overall measure of the localization of the n-th eigenvalue.

In the following we will discuss the simulation results for three representative covalent junction configurations, respectively displaying purely ohmic conduction, hopping-dominated charge transport and QD behavior with tunnel barriers. We start by describing the arrangement of our three model systems:

(a) an "ideal" junction between two coplanar (5,5) CNTs, with a diameter of 0.67 nm each. The system is periodic in the x and y directions and practically infinite along z. Two interpenetrating (5,5) CNT segments of 32 units each (corresponding to a length of about 3.9 nm between two periodic images of the junction along both x and y) are placed in the z=0 plane, with the cylinder axis parallel to the x and y direction, respectively. When two atoms are closer than 1.2 Å, one is removed by construction. The final atomic configuration, containing 568 C atoms, was relaxed by TBMD at T=0 K. The junction is asymmetryc, i.e. all the connections between the four branches of the (5,5) CNTs are different from each other. Overall, the junction region contains 2 twofold-coordinated C atoms, 6 octagonal rings, 4 heptagonal rings and 2 pentagonal rings.

(b) a "covalent" junction obtained by crossing two overlayed (5,5) CNTs of length 2.95 nm each (24 units), respectively placed at z=0 and parallel to the x-axis, and at z=+7.2 Å and parallel to the y-axis. The initial intertube distance is, therefore, about 0.5 Å. Three atoms were removed from the junction since too close to each other. After the structural relaxation of both the atomic positions and the supercell shape, the junction contains 3 heptagonal rings and 14 fourfold coordinated C atoms, all with bond angles largely distorted with respect to an ideally tetrahedral sp^3 bonding.

(c) a more "realistic" junction, as could be obtained after electron-beam irradiation of two crossed CNTs. As in the previous junction configuration, two segments of (5,5) CNTs containing 340 atoms each were overlayed at a wall-to-wall distance of about 3 Å along the z axis. In this case, however, a simulated electron-beam irradiation procedure was firstly carried out by combined microcanonical and Langevin molecular dynamics with a Tersoff-Brenner empirical

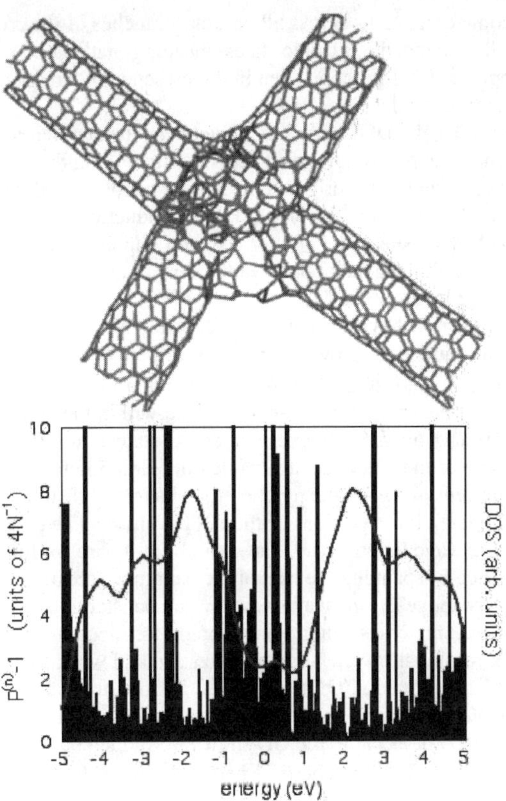

Figure 1. Top panel: Relaxed atomic structure of the junction obtained by simulating the electron-beam irradiation of two overlayed (5,5) carbon nanotubes. Bottom: DOS (full curve) and eigenvalue participation ratio (histogram) around the Fermi level (shifted at E=0).

potential [14]. The resulting atomic configuration was subsequently relaxed by TBMD. The final junction appears highly disordered (see Fig. 1-a), with a sort of amorphous-like particle embedded within four CNT arms. Such a junction includes a large number of miscoordinated atoms, and several four-, five-, seven- and eightfold rings.

Let us now discuss the electronic structure of the three model systems. For the junction (a), the DOS remains quite flat around the Fermi level. No sign of localization is apparent for an extent of about 0.5 eV above E_F, while a moderate localization is seen for a few "shallow" states lying immediately below E_F. This implies that the structural defects induce only a moderate perturbation on the conduction-electron wavefunctions. Moreover, if we cut away from the junction those sites at which the shallow states are mostly localized, the atomic structure still preserves several connecting paths across all the four branches of the cross. These paths, by contrast, support fully-extended electronic states, meaning that a good ohmic conduction is preserved in the coplanar CNT junction even in the presence of a moderate concentration of topological and coordination defects.

For the junction (b), the DOS shows a distinct peak around the Fermi energy, corresponding to whole set of strongly localized states including the eigenvalue at $E = 0$. Practically all such localized states originate from the fourfold coordinated sites connecting the two CNTs. In this kind of junction, therefore, ohmic conduction is suppressed and the only viable transport mechanism should be phonon-assisted hopping across localized states. Such a behavior is consistent with the general observation that an increasing concentration of defects brings about increasing localization.

The most interesting behavior is definitely shown by the junction (c). The electronic structure of this kind of junction, shown in Fig. 1-b (full curve) together with the participation ratio $P^{(n)}$ (histogram), is quite surprising. A whole band around the Fermi level is localized at the junction region. All the energy levels are quantized within the junction, although being finely spaced by $\Delta E \approx 10-50$ meV. Such levels are defect states, but are weakly localized (i.e., spread) over the whole of the junction sites, while not belonging to the rest of the system.

This makes the central part of the junction to behave as a distinct object, in some way disconnected from the adjoining portions of the CNTs. This can be clearly seen by studying the square modulus of the wave function (i.e., the sum of the squares of the coefficients) as a function of the distance x_i, spanning the axis of the nanotube (see also [15]). Starting from the first few energy levels above E_F, the wavefunctions are substantially different from zero only in the junction region, in some cases being more strongly localized just at a few atomic sites within the junction. Such a combination of energy quantization and weak spatial localization is the typical signature of a quantum dot (QD). Moreover, by looking at the local-DOS (obtained by projecting the total DOS on selected subsets of atomic sites) it can be seen that the first few carbon rings immediately adjacent to the QD show a forbidden band of states immediately above E_F, i.e., these rings behave as tunnel barriers between the junction and the adjoining perfect-CNT branches.

The above results suggest that by combining nanomanipulation techniques and electron irradiation of CNTs, arrays of all-carbon-based QDs as small as 1 nm, spaced by 5-10 nm, could be obtained. These should be experimentally realizable by manipulating CNTs with the tip of an atomic-force microscope over properly arranged metallic contacts, and by subsequently irradiating the junction region as demonstrated in [10-11] to create a single QD, pairs of coupled QDs and, eventually, arrays of regularly spaced QDs by means of tailored CNT-growth techniques [13]. Such QDs would be connected to ballistic conductors (the CNT arms) by construction, separated by a tunnel-barrier gate possibly allowing quantized charge injection into the QD. The junction region would be made chemically inert by saturation with hydrogen, which should not however alter its basic QD behavior. The application capabilities of such all-carbon nanoelectronic devices as building blocks of more complex molecular-scale electronic circuits should be immediately evident. For example, reactive substitution of hydrogen with organic radicals should be feasible, to explore possible connections with the domain of electronically-active molecules contacted to QD arrays.

ACKNOWLEDGEMENTS

FC thanks for RPI hospitality and support of his visit to RPI. Discussions with prof. G. Ramanath (RPI) and R. Raimondi (University RomaTre) are much appreciated. SBS acknowledges the support of the NSF Grant No. CHE-0200838. Work of PK was supported by the NSF Grant No. DMR 134725.

REFERENCES

1. S. J. Tans, A. R. M. Verschueren, C. Dekker, *Nature* **393**, 49 (1998).
2. Z. Yao, H. W. Ch. Postma, L. Balents, C. Dekker, *Nature* **402**, 273 (1999)
3. Y.-G. Yoon, M. S. C. Mazzoni, H. J. Choi, J. Ihm, S. G. Louie, *Phys. Rev. Lett.* **86**, 688 (2001).
4. P. W. Chiu, G. S. Duesberg, U. Dettlaff-Weglikowska, S. Roth, *Appl. Phys. Lett.* **80**, 3811 (2002).
5. M. S. Fuhrer, J. Nygård, L. Shih, M. Forero, Y.-G. Yoon, M. S. C. Mazzoni, H. J. Choi, J. Ihm, S. G. Louie, A. Zettl, P. L. McEuen, *Science* **288**, 494 (2000).
6. T. Rueckes, K. Kim, E. Joselevich, G. Y. Tseng, C.-L. Cheung, C. M. Lieber, *Science* **289**, 94 (2000).
7. H. W. Ch. Postma, M. de Jonge, Z. Yao, C. Dekker, *Phys. Rev.* B **62**, 10653 (2000).
8. A. Buldum, J. P. Lu, *Phys. Rev.* B **63**, 161403 (2001).
9. P. J. de Pablo, C. Gómez-Navarro, J. Colchero, P. A. Serena, J. Gómez-Herrero, A. M. Baró, *Phys. Rev. Lett.* **88**, 036084 (2002)
10. F. Banhart, *Nano Lett.* **1**, 329 (2001).
11. M. Terrones, F. Banhart, N. Grobert, J.-C. Charlier, H. Terrones, P. M. Ajayan, *Phys. Rev. Lett.* **89**, 075505 (2002).
12. C. H. Xu, C.Z. Wang, C.T. Chan and K.M. Ho, *J. Phys. Cond. Matter* **4** (1992) 6047
13. B. Q. Wei, R. Vajtai, Y. Jung, J. Ward, R. Zhang, G. Ramanath, and P. M. Ajayan, *Nature* **416**, 495 (2002).
14. I. Jang, S. B. Sinnot, P. Keblinski, in preparation
15. F. Cleri, P. Keblinski, I. Jang, S. B. Sinnot, *Nano Lett.* (submitted).

AUTHOR INDEX

SUBJECT INDEX